Two Dimensional Materials and Heterojunctions

二维材料和异质结构

田春华 王金刚 全军 著

内 容 简 介

本书详细介绍了二维材料及异质结构（水平和垂直异质结）的合成、结构、能带以及物理和化学特性。详细介绍了它们在光学、电学、热学、光电、热电以及磁学中的具体应用和最新科研进展。本书对材料的物理、结构、性能以及各种应用的解释深入浅出，简明易懂，紧跟二维材料及异质结构科学前沿的科研进展和应用。是相关科研领域的基础指导用书。

本书可作为科研工作者、研究生和高年级本科生在二维材料及异质结构领域学习的入门教材，也是探寻二维材料和异质结构科研前沿的重要参考资料。

本书所有图均附有彩图，可扫 46 页二维码观看。

版权所有，侵权必究。举报：010-62782989，beiqinquan@tup.tsinghua.edu.cn。

图书在版编目（CIP）数据

二维材料和异质结构＝Two Dimensional Materials and Heterojunctions：英文/田春华，王金刚，全军著．—北京：清华大学出版社，2021.12
（纳米光子学丛书）
ISBN 978-7-302-57874-1

Ⅰ. ①二… Ⅱ. ①田… ②王… ③全… Ⅲ. ①纳米材料－异质结－英文 Ⅳ. ①TB383

中国版本图书馆 CIP 数据核字（2021）第 056485 号

责任编辑：鲁永芳
封面设计：常雪影
责任校对：王淑云
责任印制：杨　艳

出版发行：清华大学出版社
网　　址：http://www.tup.com.cn, http://www.wqbook.com
地　　址：北京清华大学学研大厦 A 座　邮　　编：100084
社 总 机：010-62770175　邮　　购：010-62786544
投稿与读者服务：010-62776969，c-service@tup.tsinghua.edu.cn
质量反馈：010-62772015，zhiliang@tup.tsinghua.edu.cn

印 装 者：三河市东方印刷有限公司
经　　销：全国新华书店
开　　本：170mm×240mm　印　张：39.75　字　数：754 千字
版　　次：2021 年 12 月第 1 版　印　次：2021 年 12 月第 1 次印刷
定　　价：238.00 元

产品编号：089108-01

纳米光子学丛书
编委会

主编：孙萌涛　北京科技大学

编委：梁文杰　中国科学院物理研究所
　　　陈佳宁　中国科学院物理研究所
　　　杨志林　厦门大学
　　　肖湘衡　武汉大学
　　　徐　平　哈尔滨工业大学
　　　刘立伟　中国科学院苏州纳米技术研究所
　　　石　英　吉林大学
　　　孙树清　清华大学深圳研究生院
　　　方蔚瑞　大连理工大学
　　　黄映洲　重庆大学
　　　张正龙　陕西师范大学
　　　董　军　西安邮电大学
　　　李　敬　中国科学院理化技术研究所

CONTENTS

Chapter 1 Graphene, Hexagonal Boron Nitride, and Heterostructure: Properties and Applications ⋯⋯ 1

1.1 Introduction to the 2D materials ⋯⋯ 1
    1.1.1 Introduction to graphene ⋯⋯ 1
    1.1.2 Introduction to graphene-like 2D crystals
         —hexagonal boron nitride ⋯⋯ 4
    1.1.3 Introduction of graphene/h-BN, a 2D composite ⋯⋯ 5
1.2 Graphene ⋯⋯ 7
    1.2.1 Structure of graphene ⋯⋯ 7
    1.2.2 Preparation of graphene ⋯⋯ 7
    1.2.3 Physical properties of graphene ⋯⋯ 9
    1.2.4 Raman spectrum of graphene ⋯⋯ 12
1.3 2D h-BN ⋯⋯ 14
    1.3.1 Structure of h-BN ⋯⋯ 14
    1.3.2 Preparation of h-BN ⋯⋯ 15
    1.3.3 Physical properties of h-BN ⋯⋯ 15
    1.3.4 Raman spectroscopy of h-BN ⋯⋯ 19
1.4 Composite structure of graphene/h-BN ⋯⋯ 19
    1.4.1 Research rise and process of graphene/h-BN
         heterojunction structure ⋯⋯ 19
    1.4.2 Composite mode structure of graphene/h-BN ⋯⋯ 21
    1.4.3 Preparations of graphene/h-BN heterostructures ⋯⋯ 27
    1.4.4 The properties of graphene/h-BN heterostructures ⋯⋯ 33
    1.4.5 Potential applications of graphene/h-BN
         heterostructure ⋯⋯ 38

1.5 Summary and outlook 45
References 46

Chapter 2 Electrical Properties and Recent Electrical Applications of Graphene, h-BN, Graphene/h-BN Heterostructures 60

2.1 Graphene 60
    2.1.1 The structure of 2D graphene 61
    2.1.2 The electronic structure of graphene 61
    2.1.3 The electronic property of graphene 63
    2.1.4 The recent application of graphene in electronic property 67
2.2 h-BN 72
    2.2.1 The structure of 2D h-BN 73
    2.2.2 The electronic structure of h-BN 74
    2.2.3 The electronic property of h-BN 75
    2.2.4 The recent electrical application of h-BN 81
2.3 Graphene/h-BN heterostructures 85
    2.3.1 The structure of graphene/h-BN heterostructures 86
    2.3.2 The electronic structure of graphene/h-BN heterostructures 89
    2.3.3 The electronic properties of graphene/h-BN 92
    2.3.4 The recent application progress of graphene/h-BN heterostructure in electronics 102
2.4 Summary and outlook 113
References 114

Chapter 3 Optical, Photonic and Optoelectronic Properties of Graphene, h-NB and Their Hybrid Materials 130

3.1 Introduction to graphene 130
    3.1.1 Graphene's structure, electronic band 131
    3.1.2 Electronic properties of graphene, which impact the optical properties 132
    3.1.3 Optical properties of graphene 133
    3.1.4 The application of photonics and optoelectronics 139

3.2 Introduce of h-BN 144
   3.2.1 The electronic band structure of 2D h-BN 145
   3.2.2 The optical properties of h-BN 146
   3.2.3 Potential applications of h-BN 152
3.3 The introduce of graphene/h-BN van der Waals heterostructure 157
   3.3.1 The structure of graphene/h-BN van der Waals heterostructure 157
   3.3.2 The energy bandgap structure of graphene/h-BN van der Waals heterostructure 159
   3.3.3 The optical and photoelectric properties of graphene/h-BN van der Waals heterostructure 160
   3.3.4 Potential applications of graphene/h-BN heterostructures in optical property 169
3.4 Summary and prospect 178
References 179

Chapter 4 Optoelectronic Properties and Applications of Graphene-Based Hybrid Nanomaterials and van der Waals Heterostructure 193

4.1 Introduction 193
4.2 The optoelectronic properties of graphene 195
   4.2.1 The intrinsic optoelectronic properties of graphene nanomaterials 195
   4.2.2 The optoelectronic properties of hybrid graphene or heterostructure 203
4.3 Recent optoelectronic applications of graphene nanomaterials 214
   4.3.1 Optoelectronic modulator (OM) 214
   4.3.2 Photodetector 218
   4.3.3 Graphene-based light-emitting diodes(LEDs) and solar cells 224
   4.3.4 Graphene-based solar cell 226
   4.3.5 Graphene-based ultrafast lasers 227
   4.3.6 Graphene-based broadband image sensor array 228
4.4 Summary and outlook 231
References 232

Chapter 5 Magnetics and Spintronics of 2D Graphene/h-BN Composite Materials ·········· 242

5.1 Graphene ·········· 242
    5.1.1 Lattice structure and electronic structure ·········· 242
    5.1.2 The properties of graphene in magnetics and spintronics ·········· 244
    5.1.3 The application of graphene in magnetic properties and spin electronics ·········· 252

5.2 Hexagonal boron nitride ·········· 255
    5.2.1 Lattice structure and electronic structure ·········· 256
    5.2.2 Magnetic properties and spintronic of h-BN ·········· 257
    5.2.3 Application of h-BN in magnetics and spintronics ·········· 264

5.3 Graphene/h-BN heterostructure ·········· 268
    5.3.1 Lattice structure and electronic structure ·········· 268
    5.3.2 Magnetism and spintrons of graphene/h-BN van der Waals heterostructure ·········· 274
    5.3.3 The recent application of graphene/h-BN van der Waals heterostructure in magnetic device and spintronics ·········· 296

5.4 Summary and outlook ·········· 304

References ·········· 305

Chapter 6 The Thermal and Thermoelectric Properties of In-Plane C-BN Hybrid Structures and Graphene/h-BN van der Waals Heterostructures ·········· 318

6.1 2D nanomaterials: graphene and h-BN ·········· 319
    6.1.1 Structure and thermal properties of graphene ·········· 319
    6.1.2 Structure and thermal properties of h-BN ·········· 320

6.2 In-plane C-BN hybrid structure ·········· 321
    6.2.1 The structure of monolayer C-BN hybrids ·········· 322
    6.2.2 The thermal properties of in-plane C-BN hybrid structures ·········· 325

6.3 Graphene/h-BN van der Waals heterostructures ·········· 347
    6.3.1 Structures of van der Waals heterostructures ·········· 347

6.3.2 Thermal properties of graphene/h-BN van der Waals heterostructures ········· 350
6.3.3 Recent applications of thermal and thermoelectric in vertically stacked graphene/h-BN heterostructures ······ 365
6.4 Summary and outlook ········· 369
References ········· 370

Chapter 7 The Thermal, Electrical, and Thermoelectric Properties of Graphene Nanomaterials ········· 380

7.1 Introduction of graphene ········· 380
7.2 The crystal structure and electronic structure of graphene ······ 381
7.3 Graphene's novel electronic properties ········· 383
 7.3.1 Current vortices, electron viscosity, and negative nonlocal resistance ········· 383
 7.3.2 Transition between electrons and photos ············ 385
 7.3.3 Electron transport properties in nitrogen-doped graphene ········· 388
 7.3.4 Strong current tolerance ········· 390
 7.3.5 Novel electrical properties of graphene/graphene van der Waals heterostructure ········· 392
 7.3.6 The interaction between plasmons and electrons in graphene ········· 394
7.4 The thermal and thermoelectric properties of graphene ········ 397
 7.4.1 The TC's measurement of graphene ········· 397
 7.4.2 Length-depended and temperature-depended TC of graphene ········· 399
 7.4.3 Influence of boundary or configuration on thermal property and thermal rectification effect ············ 401
 7.4.4 The effect of atomic edge variation and size change on TC ········· 402
 7.4.5 The thermoelectric properties of graphene ············ 403
7.5 The recent applications in electronic and thermal properties of graphene ········· 409
 7.5.1 High-efficient TC composite film and flexible lateral

heat spreaders 409
7.5.2 Thermal conductance modulator 410
7.5.3 Graphene microheaters based on slow-light-enhanced
energy efficiency 412
7.5.4 Hybrid graphene tunneling photoconductor 413
7.5.5 Graphene electrodes 415
7.5.6 Dirac-source field effect transistors (DS-FET) 415
7.6 Conclusion and prospect 419
References 419

Chapter 8 Properties and Applications of New Superlattices: Twisted Bilayer Graphene 430

8.1 Twisted bilayer graphene (TwBLG) 430
8.1.1 Graphene and BLG 430
8.1.2 The lattice structure of TwBLG 432
8.1.3 The band structure of TwBLG 432
8.1.4 Superlattices with different symmetric structures 435
8.2 The properties of TwBLG 437
8.2.1 Electronic properties of TwBLG 437
8.2.2 Optical properties of TwBLG 448
8.2.3 Magnetic properties of TwBLG 450
8.2.4 Thermal properties of TwBLG 450
8.3 TwBLG preparation methods 453
8.3.1 SiC-based epitaxial growth 453
8.3.2 Chemical vapor deposition 454
8.3.3 Folding SLG 455
8.3.4 Vertically stacking SLG 456
8.3.5 Cutting-rotation-stacking (CRS) 458
8.4 TwBLG's latest research results 459
8.4.1 Optoelectronic device of TwBLG 459
8.4.2 Photonic crystals for nano-light 461
8.4.3 Tuning superconductivity of TwBLG 462
8.5 Summary and prospect 464
References 465

Chapter 9 Two Dimensional Black Phosphorus: Physical Properties and Applications 475

9.1 Introduction 475
    9.1.1 2D crystal structure of BP 476
    9.1.2 Electronic structure of BP 478
    9.1.3 Electronic structure of BP-based heterostructures with TMDCs 479
    9.1.4 Electronic structure of BP and blue phosphorus heterostructures 480

9.2 Preparation for BP 483
    9.2.1 Mechanical exfoliation 483
    9.2.2 Liquid phase exfoliation (LPE) 484

9.3 Anisotropy of BP's properties and application 486
    9.3.1 Anisotropic characteristics of band structures 487
    9.3.2 Anisotropic mechanical properties 487
    9.3.3 Anisotropic electrical properties 489
    9.3.4 Anisotropic thermal and thermoelectric properties 492
    9.3.5 Anisotropic optical properties 493
    9.3.6 Optoelectronic properties 507
    9.3.7 Magnetic properties 509

9.4 Summary and outlook 513
References 514

Chapter 10 Graphitic Carbon Nitride Nanostructures 521

10.1 Introduction 521
10.2 Materials and synthesis methods 523
    10.2.1 Materials 523
    10.2.2 Synthesis methods 526
    10.2.3 Characterization methods 527
10.3 Applications 529
    10.3.1 Based on $g\text{-}C_3N_4$ nanostructures nanocatalysts driven highly the ORR 529
    10.3.2 Based on $g\text{-}C_3N_4$ nanostructures driven for HER 542

10.3.3 g-C_3N_4 measurement of the gas sensing properties 556
10.3.4 g-C_3N_4 nanostructure used to wastewater treatments 581
10.4 Summary and outlook 595
References 599

Acknowledgements 622

Chapter 1

Graphene, Hexagonal Boron Nitride, and Heterostructure: Properties and Applications

1.1 Introduction to the 2D materials

1.1.1 Introduction to graphene

Graphene is a single layer of graphite sheet, constituting the basic unit of graphite, carbon nanotubes, fullerenes and other carbon materials (Figure 1-1(a))[1-2]. Before the experimental discovery of graphene, because of the effects of thermal expansion, theoretical and experimental circles believe that strict two dimensional(2D) crystals cannot be stable at finite temperatures. In 2004, Geim and Novoselov[3], produced a single layer of carbon in atoms-level thickness by using micromechanical exfoliation (microexfoliation), studied electric field effect (Figure 1-1(b)), and regarding the graphene, carried out a series of studies which broke the previous hypothesis[4-7].

Graphene is honeycomb-dimensional crystals closely arranged by sp^2 hybridized carbon atoms, and its hexagonal geometry makes it's structure very stable[8]. Each interlayer carbon atom bonds with the surrounding carbon atoms by sp^2 hybridized, and contributes a non-bonding electron to form a large π bond, making electrons move freely between the layers.

Graphene is the thinnest and hardest nano-materials[9], its tensile strength is 125 GPa, and its elastic modulus is 1.1 TPa. The 2D ultimate plane strength is 42 N/m^2. Carrier mobility is 2×10^5 cm^2/(V·s)[10-11], and it only affected by impurities and defects. Graphene's thermal conductivity is up to 5.5×10^3 W/(m·K)[12-13]. Theoretically, graphene specific surface area is up to 2,630 m^2/g. These unique physical properties make it widely apply to many

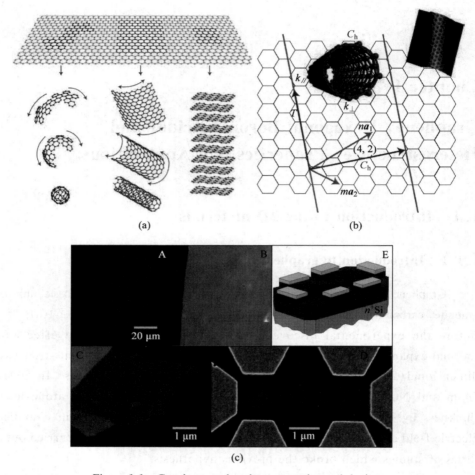

Figure 1-1 Graphene and carbon nanotube, and devices

(a) The basic unit of the other C material——graphene[1]; (b) The relational graph between graphene and carbon nanotubes[7]; (c) Graphene film and devices(Flow chart of graphene prepared by mechanical stripping method from A to E)[3]

areas of nanoelectronic devices, spin electronics, energy storage, and thermal conductivity materials.

Carbon nanotubes (deformation of graphene): properties, compared with graphene.

Carbon nanotubes (CNTs) are one-dimensional(1D) nanomaterials with a special structure[7,14]. It can be regarded as scale hollow tubular structure, which is made of a single or multi-layer graphene sheet (Figure 1-1(b)).

Chapter 1 Graphene, Hexagonal Boron Nitride, and Heterostructure: Properties and Applications

According its number of layers, CNTs can be divided into SWNTs[15-17] and MWNTs[18]. According to the arrangement of carbon atoms in the cross section, the single-wall CNTs can be categorized into the armchair and the zigzag[19-20]. And again according to the electronic structures, SWNT can be categorized into the metallic and the semi-conductive, both being non-integral[21-22].

The C = C covalent bond of carbon nanotubes makes their axial Young's modulus reaches 5TPa[20,23-24]; the length-diameter ratio is as high as 10^4, the comparison area is greater than 1,500 m^2/g[25-26], and the current carrying capacity is up to 10^9 A/cm^2 [27-28]. One of the optical properties of carbon nanotubes is wide-band absorption[29-30]. As a good thermal conductivity, the axial thermal conductivity of carbon nanotubes is up to 6,600 $W/(m \cdot K)$, an excellent field emission characteristic, and its emission current mainly comes from the occupied states slightly lower than the Fermi level.

As representatives of 1D and 2D nanomaterials, while graphene is composed only by a single carbon atomic layer, that is the true sense of the 2D crystal structure. Compared with graphene, carbon nanotubes increase the total amount of carbon atoms, making themselves have lower energy of edge dangling bonds than graphene, which can stabilize the molecules in the air without the reaction with the air. From the performance standpoint, graphene has similar or more excellent characteristics than carbon nanotubes in conductivity, carrier mobility, thermal conductivity, free-electron moving space, strength and stiffness.

Graphene and carbon-nanotubes have different applications for many reasons, but ultimately can be attributed to the difference between 1D and 2D materials. For example, a single carbon nanotube can be regarded as a single crystal with high length-diameter ratio; however, the current synthesis and assembly technology cannot prepare the carbon nanotube crystals on macroscopic scale, which limits its applications. While, the advantage of graphene is its 2D crystal structure, and its strength, conductivity and thermal conductivity are the best among other 2D crystal materials, and it has broad application prospects because of the ability of a large area of continuous growth.

1.1.2 Introduction to graphene-like 2D crystals—hexagonal boron nitride

Hexagonal boron nitride (h-BN)[31], is white block or powder. Its layered structure is similar to graphene lattice constant and characteristics, so-called "white graphene"[32-34]. The h-BN is a lattice alternately arranged by B atoms and N atoms in the 2D plane by a hexagonal lattice formation laws, showing a honeycomb structure (Figure 1-2). N atomic nucleus and B atom are combined with sp^2 orbital to form a strong σ bond[35-39], the interlayer is combined by weak van der Waals forces, and slide easily between the layers with soft lubricating properties.

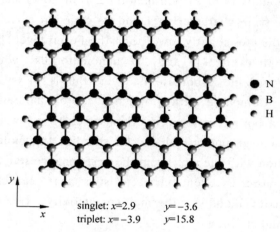

singlet: x=2.9   y=-3.6
triplet: x=-3.9 y=15.8

Figure 1-2 The flat structure diagram of h-BN. Blue on behalf of nitrogen atoms and pink on behalf of boron atoms, respectively[34]

In 1995, Nagashima et al. used an epitaxial growth on a variety of metal surfaces to obtain h-BN crystals[40]. The team in the University of Manchester adopting a micro-mechanical peeling method in 2005, successfully prepared a 2D h-BN[41]. The band gap of h-BN is about 5.9 eV, the Mohs' scale of hardness is about 2, the bulk modulus is about 36.5 GP, the heat conductivity layer up 600~1,000 W/(m·K), the coefficient of thermal expansion layer is about -2.7×10^{-6}/℃ (interlayer about 30×10^{-6}/℃), the refractive index is about 1.8, and having a neutron absorption ability[42-45]. Its anti-oxidation temperature is 900℃, the thermos ability is up to 2,000℃, an inert environment remaining stable at 2,700℃. H-BN has good process abilities, such as thermal shock resistance to electrical vibration, high resistance to breakdown electric field strength, non-toxic and environmentally friendly, no wettability to various metals, chemical corrosion and other excellent physical and chemical properties[46-53].

1.1.3 Introduction of graphene/h-BN, a 2D composite

Graphene has very good electronic properties. However, it will be a challenging to make graphene into nano-electronic devices. Moreover, mechanical stability is required when using a scanning probe technology to detect micro graphene[54]. A substrate is introduced to solve this problem, while examining the substrate can open graphene band gap, thereby improving the switching performance of graphene electronics.

There have been already a lot studies of graphene and the substrate materials, such as Co[55], Ni[56-59], Ru[60-61], Pt[62-63], as well as semiconductors SiO_2[64-66], SiC[67-69], each of them can be used as a substrate graphene. However, experiments show that the graphene is on the top of these substrates is very uneven, and have a lot of wrinkles, restraining the properties of graphene. For example, SiO_2 is the most common graphene substrate, but on its surface there are impurities, which cause the scattering of charge and a charge trap. Therefore, the growth and charge density distribution of graphene on the SiO_2 substrate are very uneven (Figure 1-3)[70], which result in significant suppression on the carrier mobility of graphene. Recently, a lot of researches have proved that h-BN is an ideal substrate, which suits for graphene to maintain geometrical and electrical properties.

Monolayer graphene and h-BN have similar lattice structure, and lattice mismatch in-between is only about 1.5%[72]. As a base, h-BN has smooth surface without charge trap, and also has a low dielectric constant, high-temperature stability, high thermal conductivity and other properties. Again, h-BN with dielectric constant ε of 3~4, breakdown electric field strength of about 0.7 V/nm, is great gate insulating layer for graphene. The surface optical phonon energy of h-BN is two orders of magnitude greater than that of SiO_2, which indicates the using of h-BN as the substrate is likely to improve the performance of graphene device in conditions like higher temperature and higher electric field. In the single cell of h-BN, the difference of grid energy between nitrogen and boron atoms leads to a broad band gap of about 5.9 eV[73], which is conducive to open the band gap of graphene. All above shows that h-BN is an ideal substrate material for graphene.

Dean et al. for the first time using h-BN as a supporting substrate of graphene, fabricated graphene transistor devices with high mobility[73], on which clear graphene quantum Hall effect was observed. Ponomarenko et al. (Figure 1-3)[71] used the technology of physical transfer to combine thin sheets of graphene and h-BN crystals produced hetero junction device with two single layers of graphene crystals. At the same time, researchers succeeded in achieving a variety of graphene heterojunctions and super lattice structures with more complex structures[74-76]; the United Kingdom and the United States researchers observed novel Hofstadter Butterfly phenomenon on graphene/h-BN heterojunction devices respectively[77-78]. Since then, researchers have used a variety of methods to fabricate graphene/h-BN heterostructure

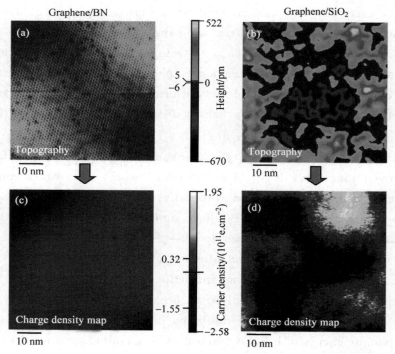

Figure 1-3 Comparing topography and charge density for graphene/BN vs. graphene/SiO$_2$ [71]

functional devices with their original purpose of improving the quality and the speed in the preparation of graphene instead of studying the performance of graphene/h-BN heterostructures.

1.2 Graphene

1.2.1 Structure of graphene

Ideal graphene is a single layer of 2D atomic crystal with orthohexagonal lattice structure. The length of C—C bond is around 0.142 nm, and the thickness of the layer is 0.35 nm. In each state, single carbon atom forms strong σ bond with three nearest neighbours respectively by sp^2 orbital hybridization, causing occupied and unoccupied states keep away from each other. Each unit cell of graphene has two types of sub-lattices, type A and B (Figure 1-4(a)), and there are chiral characteristics in the spin of between electrons of A and B[79].

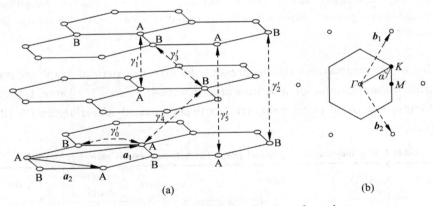

Figure 1-4 Structure and Brillouin zone of graphene
(a) The relationship between distributions of A and B atoms in the unit cell of graphene;
(b) Corresponding Brillouin zone. The Dirac cones are located at the K and M points[79]

1.2.2 Preparation of graphene

Preparation of graphene can be classified into physical and chemical methods, including mechanical peeling method[1] and epitaxial growth method[80], method of chemical cleavage[81], chemical vapor deposition

(CVD)[11], low-thermal expansion method[82], nanotubes cutting method[83], metal catalysis and so on[84-85], as shown in Figure 1-5.

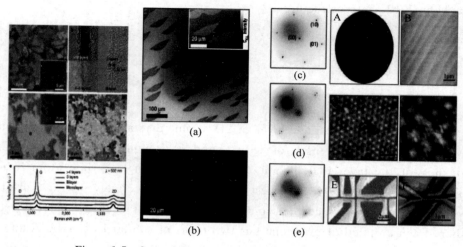

Figure 1-5 Several kinds of preparing methods of graphene

Spectral analysis of graphene prepared by CVD method is shown from a to e in the left group figures. Sample image of graphene by epitaxial growth is shown from a to e in the middle of group figures. Epitaxial growth of graphene samples and their characteristics is shown from A to F in the right group figures[11,84-85]

Yu et al. summarized the progress of the direct growth of graphene on the substrate of semiconductor and insulator in recent years[86]. Table 1-1 shows the comparison of physical properties between several materials and h-BN as substrate.

Table 1-1 Comparison of physical properties between several materials and h-BN

	Al_2O_3	SiO$_2$		Graphite	Graphite diamond	h-BN
Lattice constant	Face-centered: 1.27	Orthogonal	$a = 1.383$ $b = 1.741$ $c = 0.504$	$a = 0.246$ nm $c = 0.667$ nm	0.3567	$a = 0.2504$ nm $c = 0.6661$ nm
		Face-centered	1.936			
Thermal conductivity/ (W/(m·K))	40	1.4		25~470	22	25.1
Dielectric constant	6.8	3.9		8.7	5.7	4

1.2.3 Physical properties of graphene

1. Mechanical and thermal characteristics of graphene

2D honeycomb-shaped crystal structure endows graphene excellent in-plane mechanical properties. Lee et al. found that the Young's modulus of graphene can reach (130 ± 10) GPa under the assumption of the thickness of the graphene layer 0.35 nm[9], as shown in Figure 1-6 and the 2D ultimate plane strength is 42 N/m^2. Gómez-Navarro et al.[87] and Poot et al.[88] obtained intensity values of different graphene layers by various methods. As shown in above studies, graphene as novel nano-materials has excellent mechanical properties.

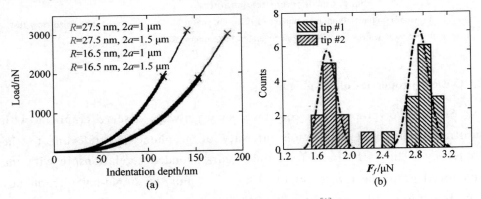

Figure 1-6 Fracture test results[9]
(a) The test line corresponding to different film diameters and tip radiis; (b) Histogram and Gauss distribution of two kinds of fracture forces[9]

Because of high modulus of elasticity and long mean free path of electrons, the thermal conductivity of graphene can reach up to 3,000 ~ 6,000 W/(m·K). Hong et al.[11] showed that graphene's thermal conductivity is 5,300 W/(m·K) and would decrease while temperature increasing[12,89]. Seol et al. have studied graphene's thermal conductivity on the substrate of SiO_2[90], and the results showed that the value of thermal conductivity was still as high as 600 W/(m·K) even the phenomenon of phonon scattering happens due to the transfer of heat across the interface between the two materials. In addition, defects and the unordered arrangement of edges would reduce the thermal conductivity of graphene (Figure 1-7)[91].

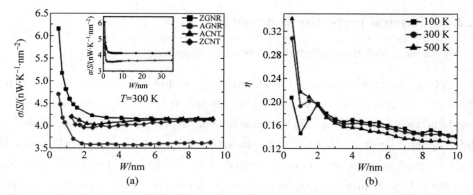

Figure 1-7 Different widths of graphene nanosheets's thermal conductivity changes at different temperatures

(a) The thermal conductivity change images of different materials at 300 K; (b) The anisotropy test images of different widths of nanoribbons at different temperatures[91]

2. Optical properties of graphene

Single-layer graphene is colorless, so we always observe graphene with substrate. The light absorption intensity of graphene is irrelevant to the frequency of light, but appears a linear correspondence relationship with the amount of graphene layers (Figure 1-8)[92-93]. For the single-layer graphene, the absorption and transmittance of visible light are 2.3% and 97.7%, respectively[94-95]. Graphene has characteristics of wide-band absorption and

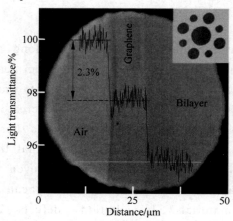

Figure 1-8 Looking through one-atom-thick crystals[93]

zero-energy gap, so in the irradiation of near-infrared powerful light, the absorption of light by graphene gets saturated and exhibits nonlinear optical behavior. The unique zero-band-gap structure of graphene makes it absorbing light with no selection. In addition, the defects, shape, and quality of graphene all have effects on optical performances.

3. Electrical characteristics of graphene

Graphene is a zero-band-gap semiconductor with six high-symmetrical K points in the Brillouin zone, and its conduction and valence bands intersect at Dirac point (Figure 1-9)[79]. Near the Dirac point, the energy and momentum present a linear dispersion relation $E = \hbar v_F K$[95], making the effective mass of electrons in graphene equal zero, that is, the electron in graphene is type of massless Dirac Fermion[7].

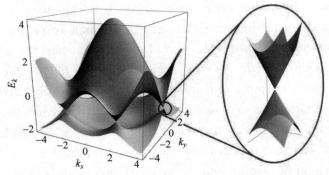

Figure 1-9 Electronic dispersion in the honeycomb lattice[96]

The unique band configuration of Dirac point gives graphene excellent conductivity. Electrons in graphene pass through the barrier with the probability of 100 %[96]. Due to a strong force between carbon atoms, their collision has little influence on the motion of electron. Chen et al. in 2008 found that the carrier mobility of graphene based on SiO_2 substrate could reach $4 \times 10^4 \text{cm}^2/(\text{V} \cdot \text{s})$ by limiting external conditions[64], which is comparable to the value of the best field-effect transistor, and at room temperature the carrier mobility of suspended graphene is possible to reach up to $2 \times 10^5 \text{ cm}^2/(\text{V} \cdot \text{s})$, which is as 140 times as the value of silicon. Biel et al. demonstrated that in graphene the mobility of electrons and holes was equal[82]. The conductivity of graphene can reach 10^6 S/m, the best conductivity among known materials at room temperature.

Meanwhile, graphene has bipolar field-effect[1], the transmission characteristics of the ballistic and so on[82]. Through graphene atomic doping[97-99], the control of edge structure of graphene nano-strips[100], and the imposing of external electric field on the graphene and related materials[101], people can regulate the size of energy gap by adjusting the interaction between graphene and substrate materials[102-105].

4. The magnetic properties of graphene

Graphene does not have d or f electrons, but this does not mean that graphene is not magnetic. On the contrary, under certain conditions, graphene can display paramagnetism and even ferromagnetism. In the case of a low temperature and zero magnetic field, the electrical conductivity of graphene exists a minimum value, close to $4e^2/h$. Graphene is able to exhibit quantum Hall effect at room temperature, or anomalous quantum Hall effect at low temperatures (below 4 K)[106-107]. Graphene has the characteristics called *pseudo-spin* (*pseudospin*) and *pseudo-magnetic field*, which appear the properties of electron spin and magnetic field. Electron spin-current injection and detection of Graphene were recently confirmed at room temperature[108-109].

Wang et al. applied various annealing temperatures to oxidize graphene reappeared its ferromagnetism based on room temperature[110]. Geim et al. obtained diamagnetic nano-graphene at room temperature and paramagnetic nano-graphene at low temperature (Figure 1-10)[111]. Two recent articles have pointed out that the doped graphene under various conditions has paramagnetism[112].

1.2.4 Raman spectrum of graphene

In 2006, Ferrari et al. first proposed the Raman spectroscopy identification method of single-layer graphene (Figure 1-11)[113]. The major Raman spectrum characteristic peak for graphene is G peak caused by in-layer transverse vibrations of sp^2 hybridized carbon atoms, and it appears in the vicinity of 1,580 cm^{-1}. The peak can effectively reflect the number of layers of graphene sheets, and is vulnerable to stress. With the increase of the number of layers N, the position of G peak will move to lower frequency, and the displacement of the movement is related to $1/N$[114]. G peak is susceptible

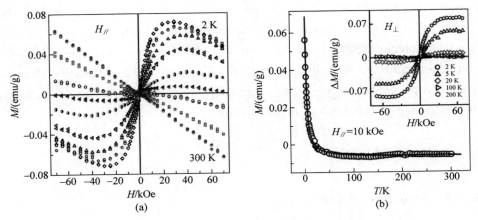

Figure 1-10 Magnetic response of graphene
(a) Magnetic moment as a function of magnetic field intensity at different temperatures from 2 K to 300 K; (b) Magnetic moment images with the changing temperature for the same sample[111]

to doping, and the peak frequency/width can be used to detect doping level[115].

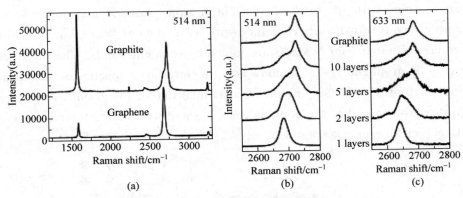

Figure 1-11 The Raman spectra of ideal graphene, different layers under the different wavelengths[113]

D peak usually refers to disordered vibration peak of graphene. The peak appears at 1,270~1,450 cm^{-1}, and its specific location relates to the excitation wavelength[116]. The peak is caused by the lattice vibration of graphene keeping away from the center of Brillouin zone, and is used to characterize defects or edge of the graphene sample. 2D peak is two-phonon resonance second-order Raman peak, and the double resonance process is connected with phonon wave vector, and the electronic band makes the frequency of 2D peak susceptible to excitation wavelength[117].

Characteristic peaks can be used to determine the different layers of graphene, and as an optical means of almost zero damage, Raman spectroscope, widely recognized and used by researchers, has greatly enhanced the efficiency of this kind of identification.

1.3 2D h-BN

1.3.1 Structure of h-BN

Analogy of graphene, B atoms and N atoms of a 2D h-BN are alternately arranged to form a honeycomb structure following the law of hexagonal lattice (Figure 1-12)[118-119]. The B-N bond length is 1.45 Å, which forms through sp^2 hybridization. Three sp^2 orbits of each B atom combine with sp^2 orbit of adjacent N atoms to form strong σ bond, likewise, three sp^2 orbits of each N atom combine with sp^2 orbit of adjacent B atoms to form a strong σ bond. Adjacent layers of h-BN are combined with weak van der Waals forces, and in each layer B atoms and N atoms are joined by covalent bonds. The interlayer spacing of graphene is 0.335 nm, and the interlayer spacing of hexagonal boron trichloride is 0.333 nm, slightly less than that of the graphite. In the c-axis direction of h-BN, the bonding force is small while the interlayer spacing is large, making interlayer slide easily[120-134].

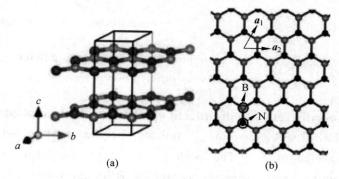

Figure 1-12 The structure of h-BN
(a) Stereogram; (b) The 2D floor plan[118]

1.3.2 Preparation of h-BN

Based on experimental or theoretical discussion, people prepared h-BN crystals in many ways[40,133-143]. Some basic methods include: mechanical separation[38], chemical vapor deposition[133-134], aqueous solvent thermal synthesis methods[135-137], solvent stripping methods and so on (Figure 1-13)[129,138].

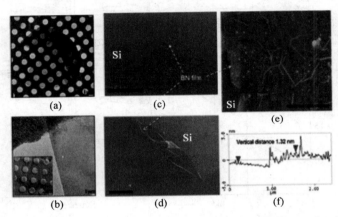

Figure 1-13 The manufacture methods of h-BN with the mechanical stripping method and CVD method

The left group figures: (a) Low magnification and large defocus of the h-BN TEM images. (b) The folding images for h-BN. The right group figures: (c) Photograph of a large h-BN film on a silicon substrate. Scale bar is 1 cm. (d) SEM image shows a h-BN film (scale bar 10 μm). (e),(f) AFM image and line-scan profile indicate that h-BN film has uniform thickness of a 1 nm. Scale bar is 2 μm[38,133]

1.3.3 Physical properties of h-BN

1. Electrical characteristics of h-BN

Surface of a 2D h-BN is smooth; its lattice structure is very similar to graphene. 2D h-BN has a large optical phonon mode, and a wide band gap, no dangling bonds or electron traps on surface, (LDA calculated band gap of about 4.5 eV, GW calculated band gap of 6.0 eV)[118,139], which belongs to the wide-band-gap insulators. Figure 1-14 is a band structure of 2D h-BN[118]. As Figure 1-14 shown, h-BN is a direct band gap, and both the top and rewinding bottom of valence band are in high-symmetry point—K. HOMO and LUMO levels of the system are determined by π and π* located on N atom nuclei B, respectively[118].

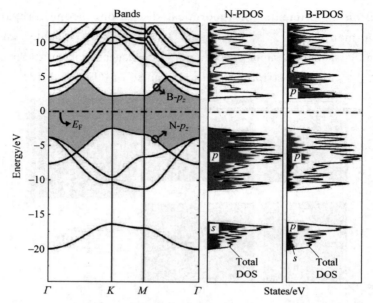

Figure 1-14 Orbital property and electronic structure of h-BN[118]

2. Force thermal characteristics of h-BN

Mechanics of 2D h-BN is prominent[140-141]. Song et al. measured the mechanical properties of h-BN film by using a diamond tip to emboss h-BN film center[133]. Shown in Figure 1-15, E2D is post-elastic coefficient, and 2D mσ is pretensioned stress. A large number of experimental results and theoretical calculations have shown that, Young's modulus of the 2D h-BN is large, approximately 270 N/m.

At room temperature, thermal conductivity of h-BN is up to 400 W/(m·K), higher than majority of metals and ceramic materials. h-BN has a typical anisotropy: having a high-thermal conductivity in the direction perpendicular to the axis c, 300 W/(m·K); low-thermal expansion coefficient of $0 \sim 2.6 \times 10^{-4}$/K; and relatively high-tensile strength (41 MPa). While parallel to the c axis direction, h-BN has a lower-thermal conductivity, $20 \sim 30$ W/(m·K); high-compressive strength and so on[138,142-145].

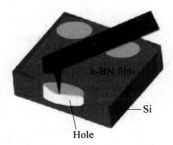

Figure 1-15 Schematic of nanoindentation on suspended h-BN membrane[133]

3. Optical properties of h-BN

In both the experimental and theoretical calculations, 2D h-BN has no absorption in the visible range, has absorption spectroscopy in the ultraviolet region, and has a good photoluminescence property. Geim et al. changed the silicon-rays to detect the polishing process, using an optical microscope to a single-atomic-layer thickness of h-BN[1]. The electronic state of the materials can be reflected by the optical absorption properties[35], which are widely used to calculate the band gap of semiconductors. We can know the range from 3.6 eV to 7.1 eV of h-BN band gap energy, which related to different h-BN experiments with different structures by analyzing the available literature data[131].

Gao et al. measured the UV-vis optical absorption spectrum of h-BN by investigating the as-prepared h-BN nanosheets. From the Figure 1-16, we can see that the absorption peaks around 251 nm, 307 nm, 365 nm correspond to a band gap energy of 4.94 eV, 4.04 eV, 3.40 eV, respectively. These two absorption peaks of 307nm and 365nm come owing to optical transitions and the redistribution of h-BN's electron-hole density between van Hove singularities of excited state[131]. As a wide-band-gap material[35,146], h-BN is transparent in the infrared and visible light and in ultraviolet light, has a strong optical absorption at 251 nm in the strong exciton peak[131].

4. Magnetic properties of h-BN

2D h-BN is non-magnetic semiconductors with wide band gap[146]. Studies have found that the introduction of doping or defects to 2D h-BN system can lead to spontaneous magnetization of h-BN. Wu et al. calculated the condition

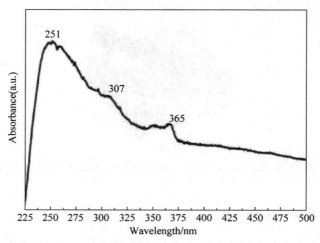

Figure 1-16 UV-vis optical absorption spectra of BN[131]

that B atoms and N atoms are replaced by carbon atoms[147-148], resulting in the generation of spontaneous magnetism in 2D h-BN (Figure 1-17). Si et al. found that the vacancy defects of B atoms or N atoms in h-BN can also lead to spontaneous spin magnetization[149]. These studies suggest that boron nitride (BN) nanoribbons can present excellent half-metallic magnetism in a variety of states[150].

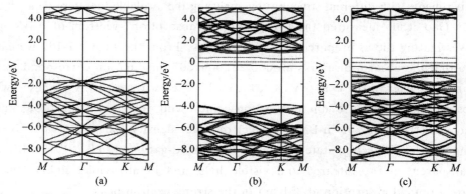

Figure 1-17 The calculated band structures of the pristine graphitic-BN with different atoms
(a) With a boron atom; (b) With a nitrogen atom; (c) Substituted by a carbon atom[148]

By researching no passivation and the passivation with zigzag BN nanoribbons, Lai et al. found that such systems present semi-metallic[150], while the nitrogen atom edge passivation is not magnetic. Barone et al. found spontaneous spin polarization in no passivated serrated BN nanoribbons[146], and under the effect of external electric field, the material can realize metal-semiconductor-semimetal transition. Zheng et al. found that only when B atoms are edge passivated[151], electronic polarization rate at the Fermi level is 100%, showing a semi-metallic properties with band gap of 0.38 eV, and the conductivity of the system is decided by metallic spin electrons, and when the N atoms are edge passivated, antiferromagnetic structure is more stable than the ferromagnetic structure, as the energy of the former is approximately 33 meV / (edge atom) more than the latter.

1.3.4 Raman spectroscopy of h-BN

Raman characteristic peak of h-BN crystals is about 1,366 cm^{-1}, and a single-layer peak will blue shift 4 cm^{-1}, which is due to the BN bond of a single-layer h-BN slightly shorter phonon modes causing E_{2g} hardening (Figure 1-18)[152]. Red shift depends on the random strain introduced by the stripping process, which dominates in the double h-BN. Therefore, h-BN Raman characteristic peak red shift 1~2 cm^{-1}.

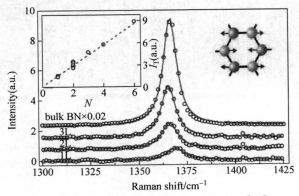

Figure 1-18 Raman spectra of 2D h-BN[152]

1.4 Composite structure of graphene/h-BN

1.4.1 Research rise and process of graphene/h-BN heterojunction structure

Linear dispersion relation of graphene near the Dirac point makes the high carrier mobility characteristics, which draws wide attention (Figure 1-19)[70], and it is considered as an important electronic device producing carbon Keener materials[106,153-154]. However, its zero band gap hindered its application in the construction of nano electronic devices, because these devices require a valid strip gap. Thus, whether graphene can open an effective band gap becomes critical. People have used a variety of methods to achieve this objective. For example, hydrogenated graphene[155], isoelectronic co-doped of B and N atoms[156], adsorption of metal atoms[157], double extra electric field components[2,158], a graphene-substrate interaction[159], the level of the tensile stress and so on[71]. All the methods were double nano-structures exceptionally compelling, because the interaction between the layers and stacking mode provide an additional degree of freedom to regulate the electronic structure. C materials and h-BN have matched lattice constant. By forming graphene/h-BN heterostructure to regulate graphene electronic structure and energy band, the goal that the graphene is applied to optoelectronic devices is achieved.

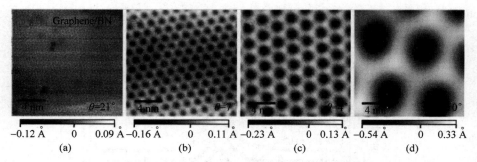

Figure 1-19　STM pictures of graphene/h-BN heterostructure with different rotational orientations[70]

In 2000, Oshima et al.[160] have prepared two-layer structure graphene/h-BN in the Ni(111) direction, and studied the influence of monolayer graphene on Ni(111) substrate grown h-BN single-layer electronic structure, ignoring the effects of h-BN on the electronic structure of graphene. Stewart et al. calculated the most stable configuration graphene/h-BN electronic properties through the tight-binding and density functional theory[161]. Their research shows that a level perpendicular to the applied electric field can regulate the electronic structure of the whole system, including offset, anti-cross, and other forms of energy deformation, and control the size of the band gap according to different needs. In 2010, Bjelkevig et al. used chemical vapor deposition method on Ru(0001) direction to prepare the double heterojunction structure graphene/h-BN[162]. The study found that there was a strong charge transferring between the BN and the graphene, which occupied part of the π^* band, and made occupied σ^* close to the Fermi level. This job makes that by forming double heterojunction structure of graphene/h-BN to regulate the electronic structure of graphene is possible. In 2012, Tang et al. made single-crystal graphene grow on h-BN[163], although the growth size is not large, the use of Raman technology has been demonstrated that nucleation growth hexagonal graphene was single-crystal structure. This finding enriches the means of BN hetero-junction preparing, and makes rapid preparation of high-quality graphene/h-BN is possible. Especially recently, Tang et al. used CVD method to make graphene of graphene/h-BN efficient large area growth[164], so that the preparation of graphene has reached a new level. Meanwhile, applications of graphene/h-BN heterostructure in nano-device obtain technical support.

1.4.2 Composite mode structure of graphene/h-BN

1. Stacking of graphene/h-BN heterostructure

Graphene/h-BN heterostructure is formed by the 1×1 lattice matched stacking of monolayer graphene and h-BN. The major three stacking patterns in studies are: ① AA stacking, which means C atoms are positioned immediately above B and N atoms; ②AB stacking, which means one C atom is positioned above the N atom, and the other is positioned above the h-BN ring; ③AB stacking, which means one C atom is positioned above the B atom, and the other is positioned above the h-BN ring.

Zhong et al. verified that the AB stacking of graphene/h-BN double-layer heterogeneous nanostructure is the most stable by first-principles calculation, and they found that as stress increases, the band gap increases linearly[165]. Fan et al. applied LDA and PBE + VDW methods to study the structural energy variations of graphene/h-BN double-layer heterogeneous nanostructures of four stacking patterns with various interlayer distances[166], and the results showed that in either method, the AB stacking, which means N atoms are positioned towards the center of graphene hexagon, and B atoms are positioned towards C atoms is the most stable at various interlayer distances, a stable model and calculation results are shown in Figure 1-20. They have studied 4 kinds of stacking configurations (Ⅰ～Ⅳ). Ⅰ has hexagonal configuration (AA), which means C atoms are positioned above the B atoms or N atoms. Ⅱ and Ⅲ are Bernal configuration (AB), which means C atoms of one sub-lattice of graphene are positioned above N atoms (configuration Ⅱ) or B atoms (configuration Ⅲ), while C atoms of the other sub-lattice of graphene are positioned above the center of the BN hexagon. Configuration Ⅳ can be obtained from the translation of h-BN layer in configuration Ⅰ along the direction of C-C bonds by a distance of 1/6 graphene lattice constant[166]. Kan et al. also found that the band gap of graphene/h-BN double-layer heterogeneous AB stacking stable structure can be modulated through the interlayer distances by first principles[167].

According to VASP package local density approximation (LDA) the changes of binding energy (BE) for each original cell with the variations of the interlayer distance were calculated[168], and the results showed that various stacking structures have different equilibrium positions, specifically, structure balancing interlayer spacing of 3.5 Å, 3.2 Å, 3.4 Å and 3.4 Å for configuration Ⅰ-Ⅳ respectively, and the results showed that the various stacking structures have different equilibrium positions, specifically, structure balancing interlayer spacing of 3.5 Å, 3.2 Å, 3.4 Å and 3.4 Å for configuration Ⅰ-Ⅳ respectively, which is consistent with previous calculations of the deposition of graphene on BN[71]. For this double-layer heterostructure, the order of stability under the same interlayer spacing is configuration Ⅰ＜configuration Ⅲ＜configuration Ⅳ＜configuration Ⅱ. This phenomenon is related to the attractive interaction between the electrons in graphene and the cations in BN and to the repulsive interaction between the electrons in graphene and the anions in BN. N anions tend to be directly above the center of the C hexagon, because the electron density of this position is low, and B cations tend to be directly above C atoms to enhance the effect of attracting (configuration Ⅱ).

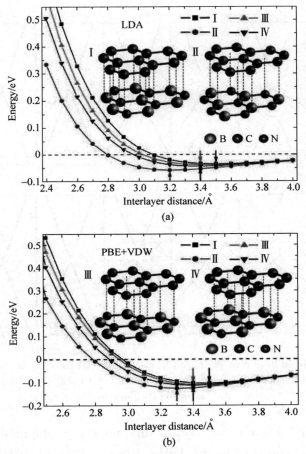

Figure 1-20 BE per unit cell of graphene/h-BN with different stacking manners[166]

2. Relations between graphene/h-BN heterostructure configurations, spacing and electronic structure properties

The above results show that configuration I ~ III have similar energy band structure and Figure 1-21 (a) presents under various interlayer distances the energy band structure of configuration II near Dirac point. The energy band structure of configuration IV is shown in Figure 1-21 (b)[166].

When the interlayer distance is large enough, the energy band structure near the Fermi level is consistent with that of monolayer graphene, because the interlayer interaction under such distance is very weak. With the decrease

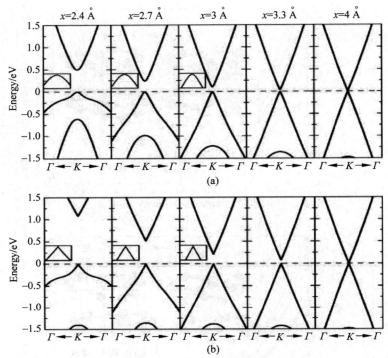

Figure 1-21   Band lines near the Fermi level of C/BN HBLs with interlayer distances $x = 5$ Å, 3.3 Å, 3 Å, 2.7 Å, and 2.4 Å for (a) the pattern II and (b) IV The Fermi level is set to zero[166]

of the distance between the layers, the interlayer interaction increases, resulting in a reconstruction of HBLs charge. Due to different chemical environments, the equivalence of two different sub-lattices of graphene is destroyed, and at the Dirac point a small band gap is opened between the valence and conduction bands. Obviously, the smaller the interlayer distance, the bigger the opened band gap, which provides a possible approach for the band-gap regulation of graphene.

For HBLs of configuration I ~ III, at the Dirac point there is a parabolic dispersion relation (nearly free electrons) between the valence and conduction bands near the Fermi level, which means the effective mass of electron is non-zero. However, for HBLs of configuration IV, a band gap is opened near the Dirac point, and the energy-band structure maintains the linear dispersion relation of graphene, which means the effective mass of electron and hole is 0.

The high mobility of carriers in this heterogeneous double layer, of which the band gap is not 0, is comparable to that in graphene monolayer, which has significance to the preparation of graphene devices that require both the band gap and high carrier mobility.

The variation of HBLs band gap is studied by adjusting the interlayer distance, as shown in Figure 1-22[167]. The band gap increases progressively with the decrease of interlayer distance, but the extent of change is different for the 4 configurations. Under the same interlayer distance, configuration I and II have the maximum and minimum band gap respectively, and the other two have the medium band gap.

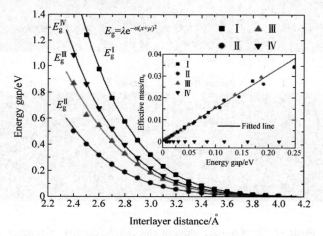

Figure 1-22 Variation in the energy gap of HBLs as a function of interlayer spacing. The inset shows the variation of the effective masses vs. band gap[166]

From Figure 1-23, we can see that the charge density near the two sub-lattices of graphene is no longer equivalent, and as interlayer distance decreases, this asymmetry is getting more pronounced. This shows that when the interlayer distance is compressed, the difference between the potential energy of the two sub-lattices of graphene becomes larger. Furthermore, for the 4 configurations, the degree of sensitivity to the change of interlayer distance is related to the way of atomic arrangement of HBLs. Carbon atom of configuration I is exactly located above the B atom or N atom of the BN monolayer, causing the highest asymmetry of the two graphene sub-lattices, and therefore configuration I is the most sensitive to the interlayer distance.

C atoms of one sub-lattice of configuration Ⅱ are located directly above the B atoms, however, all C atoms symmetrically distribute around the N atoms. In this arrangement the asymmetry of the two carbon atoms, the sub-lattice is the lowest, so configuration Ⅱ is the least sensitive to the interlayer distance. This is also consistent with both the trends of charge density and the variation of interlayer distance for the 4 configurations shown in Figure 1-23 (b).

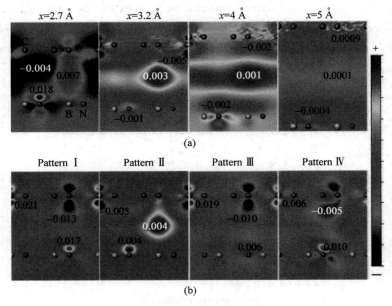

Figure 1-23 Contour plots of the charge density difference
(a) For stacking pattern I at the interlayer distances $x = 2.7$ Å, 3.2 Å, 4 Å, and 5 Å, respectively;
(b) For four stacking patterns (Ⅰ-Ⅳ) at the interlayer distances $x = 3$ Å, respectively[198]

DFT calculations show that for graphene/h-BN HBLs, the band gap and effective mass of carriers can be regulated by changing interlayer distance and stacking patterns. With the decrease of interlayer distance, the band gap gradually increases. With the increase of the band gap, for AA and AB stacking patterns (configuration Ⅰ-Ⅲ), the effective mass of carriers increases linearly; for configuration Ⅳ, the linear dispersion relation in monolayer graphene reserves in graphene/h-BN HBLs. The quality of controllable band gap and small effective mass (even zero for configuration Ⅳ) of C/BN HBLs makes it have potential important applications in the construction of high-performance graphene nano-devices.

1.4.3 Preparations of graphene/h-BN heterostructures

From 2010, the methods of preparing Graphene/h-BN heterostructures become a hot focus of the research, more and more reliable methods have been appeared.

1. Wet-transfer method

Dean et al. first prepared graphene transistor device with high mobility by transferring mechanically stripped graphene to the h-BN substrate and using h-BN as the support substrate for the graphene (Figure 1-24)[73]. Its electron mobility is measured one order of magnitude higher than in the same conditions as in the case of silica, and a clear graphene quantum Hall effect is observed. This method involves graphene in contact with an aqueous solution. Therefore, it is referred to as *wet transfer*.

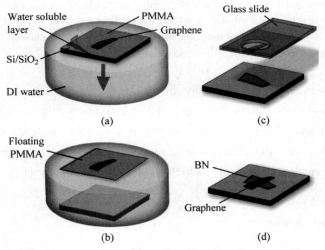

Figure 1-24 Schematic illustration of the transfer process to fabricate graphene-on-BN devices[72]

2. Liquid phase exfoliating method

Gao et al.[169] used liquid phase exfoliating to achieve that van der Waals structures of different layers of h-BN and graphene, and observed that the energy level of the hybrid structure was affected by the number of layers and stacking mode of the two layers (Figure 1-25).

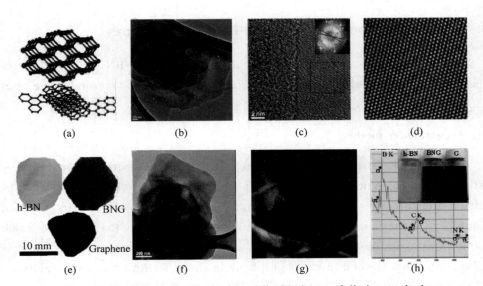

Figure 1-25 Schematic illustration of liquid phase exfoliating method
(a) The illustration of the stacked structure of graphene and h-BN nanosheets; (b)～(d) The TEM images of stack structure with different resolutions; (e) Graphene, h-BN and BNG hybrid free standing films; (f) The h-BN/GTEM image for mapping; (g) Boron mapping of image (f); (h) EELS spectrum of h-BN/graphene hybrid, K-shell excitations of B, C, and N[169]

These two methods inevitably have impurities appear on the top and bottom of graphene, low success rate of the lattice corresponding paste, poor precision, and easy to fold. In the preparation of this method, the larger the BN, the higher the preparation success rate; the larger the graphene, the easier for device preparation, which is tough in the mechanical stripping process.

3. Dry transfer method

In order to eliminate the graphene problems encountered in the transfer of folds, in 2011 Zomer et al.[170] designed a method of the preparation of graphene/h-BN heterostructure (Figure 1-26). Similarly, Leon et al. used this method to prepare heterostructures in 2014, and performed the corresponding tests[171]. From the method and the test results, the only drawback of this method is very difficult to find the graphene without the silicon oxide substrate, so that the rate of graphene preparation is very low, and the quality and efficiency are difficult to coexist.

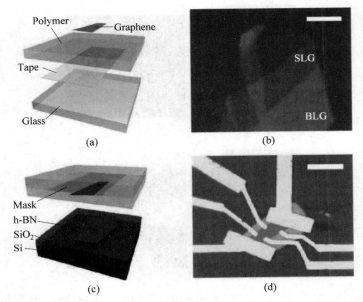

Figure 1-26 Dry preparation graphene/h-BN schematic drawing and sample picture (a) Schematic of the transfer mask; (b) SLG and BLG areas on the mask, the scale bar equals 5 μm; (c) Graphene alignment and transfer to a h-BN crystal; (d) Graphene flake in a Hall bar geometry on h-BN[170]

4. Chemical vapor deposition method

The researchers used chemical vapor deposition, through the adjustment of the order of preparation to generate artificial hybrid structure of graphite and h-BN[172-175].

Gannett et al. prepared the graphene/h-BN heterojunction by using CVD method and run appropriate tests in 2011. After that, many researchers gradually improved the preparation and made the heterostructure grow faster and faster, and the quality is getting better and better (Figure 1-27)[176].

Liu et al. realized the superposed growth of graphene and h-BN by a two-step gas-phase process[177]. Subsequently, Yan et al. used PMMA as a solid carbon source to achieve the growth of bilayer graphene on a chemically vapor-grown h-BN substrate[178]. Tang et al.[179-180] used methane as gaseous carbon source on h-BN substrate to achieve graphene lamellar nucleation growth with superlattice structure[164]. Recently, Kim et al. synthesized

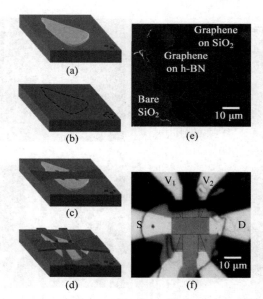

Figure 1-27 (a)~(d) Sample fabrication steps shown schematically; (e) SEM image of graphene on a candidate h-BN flake; (f) Optical image of finished graphene-on-BN device, with electrodes labelled and the graphene outlined[176]

graphene/h-BN heterostructures with different hybridization modes by modifying the vapor-phase growth process[174]. In 2016, Song et al. used PMMA particles to control the nucleation density and grain size, which provides the technical direction for high-quality mass production of graphene/h-BN[181].

Transition metal catalysis: In 2010, Cameron Bjelkeving made an atomic layer h-BN and first deposited on the Ru (0001) surface by an atomic deposition method (Figure 1-28)[162], and then use CVD to grow a layer graphene on the h-BN surface to form a GNR/h-BN/Ru (0001) heterostructures, but whether the graphene is directly grown on the h-BN or due to the role of Ru underneath the h-BN, the growth mechanism is not figured out. Oshima et al. have achieved the growth of graphene/h-BN/Ni heterostructures in Ni (111) transition metal catalyzed in 2000[160].

Physical transfer method: In 2011, Ponomarenko et al. combined the graphene crystals and h-BN thin layers by physical transfer technique to fabricate heterojunction devices with two monolayer graphene crystals[71]. Afterwards, through the improvement of physical transfer technique, researchers

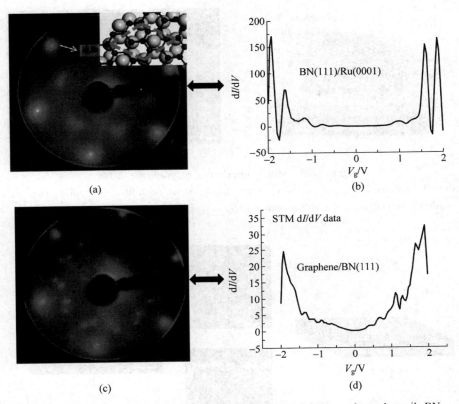

Figure 1-28　LEED and STM dI/dV data for BN monolayer and graphene/h-BN heterojunction grown on Ru(0001)[162]

successfully fabricated more complex heterostructures and superlattice structures of graphene/h-BN[75,182-183]. In particular, the two research groups successfully observed the novel Hofstadter butterfly phenomenon in graphene/h-BN heterostructures (Figure 1-29)[184-186].

The physical transfer method can realize the transfer of graphene on the surface of h-BN. However, this technique also brings the issues of structural inhomogeneity and interfacial contamination. Now, researchers are exploring further optimized methods.

Gas-phase epitaxy technology method: In 2013, Yang et al. implemented a controlled van der Waals heterostructures epitaxial growth of graphene on h-BN inactive substrate by using methane as a gas source and a remote plasma-enhanced vapor-phase epitaxy technique (Figure 1-30)[187].

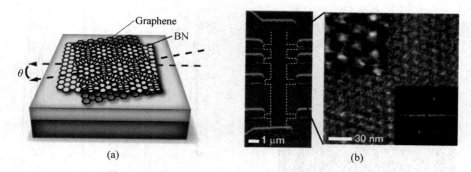

Figure 1-29 Moiré superlattice of graphene/h-BN
(a) The schematic diagram of moiré pattern for graphene on h-BN; (b) AFM image of a multiterminal Hall bar and a high-resolution image of a magnified region[185]

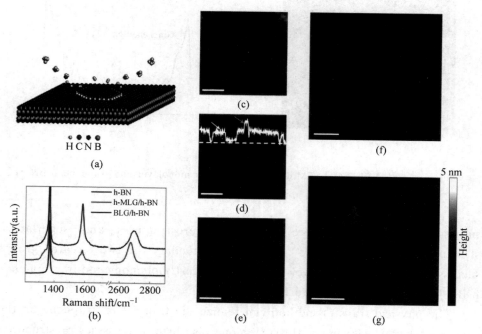

Figure 1-30 The method of epitaxial graphene growth and AFM image[187]

Co-segregation method: In 2015, Zhang et al. introduced a method, named co-segregation method, reaching directly large growth on the h-BN to produce the heterostructures (Figure 1-31)[188].

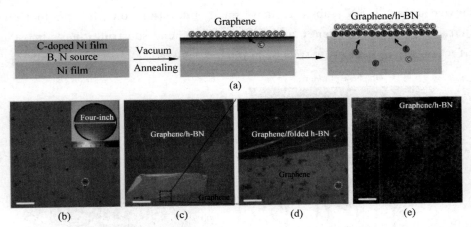

Figure 1-31 Growth of graphene/h-BN stacked heterostructures, schematic illustration, optical image of the as-grown sample on the growth substrates[188]

1.4.4 The properties of graphene/h-BN heterostructures

The graphene/h-BN heterostructures have attracted more attentions because of their novel electrical, morphological, optoelectronics, mechanical and thermal properties.

1. Optical characteristics of graphene/h-BN heterostructures

A 2D crystal has a large specific surface area, and the 2D graphene and h-BN have honeycomb structures. However, optical and electrical properties of the two materials are very special and completely different. Therefore, when graphene/h-BN heterostructures appeared, people quickly focus on the electro-optical properties of the structure. The h-BN is typical material, which is only excited luminescent in ultraviolet region[35,131,189-190], and graphene is excited in the visible region[92-93]. Therefore, the composition of graphene/h-BN heterostructure has little influence on each other from optical properties standpoint[191].

Yang et al.[192] experimentally studied the imaging properties the phonon spectrum of far and near-fields of graphene/h-BN heterostructures photon polarization and polarization (Figure 1-32), they also compared graphene excitation characteristics based on different substrates. Experimental results showed that, h-BN substrate has small inhibition on the photoelectric characteristics of graphene, and h-BN as graphene substrate, has almost no effect on the optical properties of graphene.

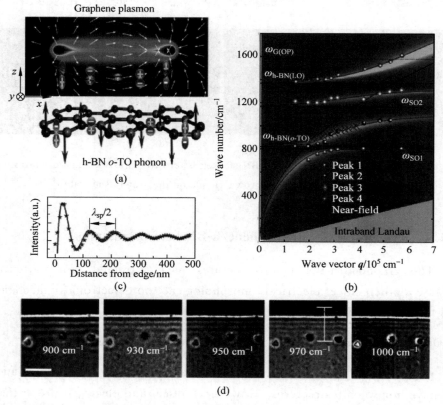

Figure 1-32　The coupling mechanism of graphene plasmons and the h-BN optical phonons

(a) The schematic diagram of graphene plasmon and h-BN o-TO phonon; (b) Plasmon frequencies as a function of wave vector for different hybridized mode peaks; (c) Line profile of s-SNOM optical signals at the excitation wavelength of 970 cm^{-1}; (d) s-SNOM images from 2D scan of the tip position near the graphene edge with varied excitation wavelengths[192]

2. Electrical characteristics of graphene/h-BN heterostructures

The h-BN with single crystal has a similar lattice structure to graphite. As insulating materials with the band gap of 5.97 eV, h-BN has atomically flat surface, very low roughness, no dangling bonds on its surface, combination with graphene showing weak van der Waals force, minimal impact on graphite dilute carrier transport properties, mismatch of 1.84% with graphene lattice[72], and no doping effect to graphene. Table 1-2 presents the difference in their lattice[193]. Mobility in graphene on BN substrate is up to 10^5 cm^2/(V·s)[72,194]. When its surface flatness suppresses graphene wrinkling, this substrate seems more excellent. Meanwhile, the BN has well electrical properties, which can be used for the gate electrode of graphene devices without loss of functionality of graphene. Furthermore, the optical phonon mode of BN surface is three times more than the energy of silicon oxide, which means that graphene with BN substrate may have better performance under the high temperature and field.

Table 1-2 The crystallographic information of h-BN and graphite[193]

Material	Crystal structure	Nearest neighbour distance	Latties parameters	Inter-layer spacing
h-BN	hexagonal	0.144 nm	a: 0.250 nm c: 0.666 nm	0.333 nm
graphite	hexagonal	0.142 nm	a: 0.246 nm c: 0.670 nm	0.335 nm

In several experimental preparations of graphene/h-BN, the lattice constants of graphene and h-BN has 1.84 % difference, and there is a certain angle of rotation between the two hexagonal-hole sheet layers, making graphene/h-BN heterostructures form Murray stripes (Figure 1-33)[70,195]. This formation of super-lattice, results in a lot of new properties, in addition to the zero band gap, Fermi velocity changes, states of jitter of local density, and the super-lattice features of local charge density[196].

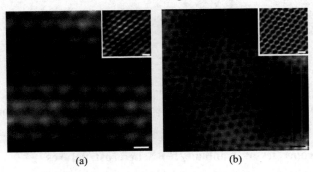

(a) (b)

Figure 1-33 Real space and Fourier transforms of Moiré patterns with different ratios in (a) and (b)[195]

Giovannetti et al.[72] in 2007 proposed the introduction of a band gap (Figure 1-34) by depositing the monolayer graphene on h-BN substrate. With a different stacked manner, graphene/h-BN heterostructures band gap varied, and different distance of layers generated variation of band gap. Hüser et al.[197] by calculating obtained the electronic properties of graphene/h-BN heterostructures, and found that in band, some small band gap can be opened, as shown in the band. Wang et al.[198] used an angular resolution electron spectroscopy, for the first time to directly measure original and second-level best Dirac band structures (Figure 1-35) of graphene/h-BN heterostructure, and in the original and second-level Dirac cones found at up to 100 meV and 160 meV energy gaps respectively, which revealed the importance of space inversion symmetry in a heterogeneous structure and the band modulation capacity of the van der Waals heterostructure.

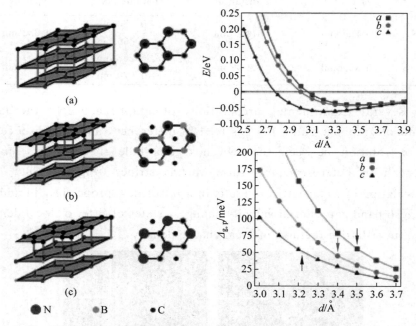

Figure 1-34　The three inequivalent orientations of single-layer graphene on a h-BN surface and total energy E of graphene on h-BN surface for the three manners[71]

3. Magnetic properties of graphene/h-BN heterostructures

Theoretical research on magnetic C and h-BN composite system is prior to

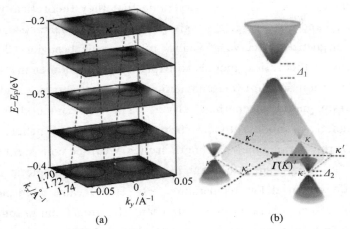

Figure 1-35 (a) Stacking of constant-energy maps of EDC curvature to show the conical dispersion at the two κ points. d1~d5, dispersions along cuts d1~d5 shown in a1. The white dashed lines are guides for the evolution of SDCs dispersions. The black arrow indicates the dispersion from the original Dirac cone; (b) Schematic drawing of the band structure in the graphene/h-BN heterostructure, showing the gapped original Dirac point with gapped SDPs at three out of six corners of the SBZ (κ points)[198]

the experimental study[199-202]. Studies found that C-BN system appears itinerant electron ferromagnetic and antiferromagnetic. Ramasubramaniam et al.[201] in 2011 studied in 2D h-BN graphene nano-islands with mosaic Zigzag edges appeared antiferromagnetic by using density functional theory, and this antiferromagnetic Nair temperature increased with the density of graphene unit area in h-BN mosaic rises. Berseneva et al.[202] also used this method to calculate the carbon nano-islands, and found a triangular inlay in h-BN will generate spin 1/2 magnetic moment. These are two models of graphene/h-BN heterostructures composites, and their research provided a theoretical approach for the later research on magnetic properties of graphene/h-BN heterostructures composite system.

Ding et al.[203] in 2011 specifically studied the magnetic properties of graphene/h-BN composites, particularly for abnormal paramagnetic behavior of graphene/h-BN composites at low-temperature, by proposed-analysis of low-temperature magnetization curve, and found that paramagnetic samples (<20 K), mainly from the point defects in the sample and the angular momentum quantum number $J = 1/2$. It is the first time that they theoretically explained why $J = 1 / 2$, and also the first to give $J = 1/2$ in graphene system. For low-temperature magnetization curve anomalies paramagnetic behavior (20~50 K), they used several theoretical models to explain this phenomenon, such as the RKKY interaction in antiferromagnetic, spin-glass state, and itinerant electron paramagnetic strengthen effect. The experimental results and antiferromagnetic spin-glass theory have a degree of conflict, and only itinerant electron paramagnetic strengthen theory can be consistent with all of our results. So, they speculated that the abnormal paramagnetic behavior of graphene/h-BN system at low temperature (20~50 K) came from paramagnetic strengthening effect generated by the itinerant electrons from graphene and h-BN interface or near the interface between (Figure 1-36).

1.4.5 Potential applications of graphene/h-BN heterostructure

The introduction of h-BN as substrate is to solve the graphene in traditional silicon material strain imposed on the substrate and doping adverse problems. At the same time, the formation of van der Waals heterostructure of graphene/h-BN is for increasing the mobility of graphene and making nanometer functional devices for potential application, which have the practical significance.

Field effect transistor (FET): In 2010, Dean et al.[204] prepared the FET of graphene/h-BN heterostructure and had the device performance test for the first time (Figure 1-37). The used mechanical stripping method for single crystal of h-BN on single and double graphene field-effect tubes. The result of the test showed the mobility of current carrier reached 60,000 $cm^2/(V \cdot s)$, more than three times the mobility of graphene similar to calculation method of SiO_2 substrate, the uniformity of charge is less than 7×10^{10} cm^{-2}, 3 times smaller than silica basal graphene at room temperature. When the temperature is 15 K, the uniformity of charge is 1×10^9 cm^{-2}, similar to suspended graphene. In addition to an order of magnitude higher than that of the test results, the roughness, intrinsic doping level and chemical reactivity are lower.

From then on, more and more the similar works have made. Decker et al.[70], Lee et al.[205-206], Iqbal et al.[207], and Song et al.[181] prepared the

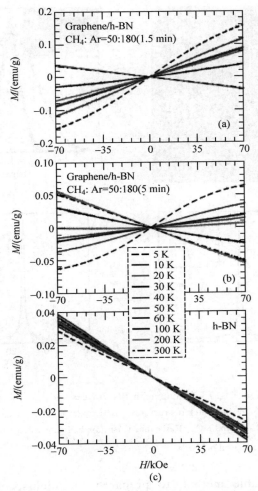

Figure 1-36 The measured mass magnetization Mas functions of magnetic fields H for (a) gra/h-BN sample 1; (b) gra/h-BN sample 2, and (c) h-BN substrate, at different temperatures T = 5 K, 10 K, 20 K, 30 K, 40 K, 50 K, 60 K, 100 K, 200 K, and 300 K[203]

FET with graphene/h-BN, which obtained the satisfied results, respectively. All of above, witness the progress of graphene/h-BN van der Waals heterostructure on nanoelectronics applications.

Quantum tunneling transistor: In 2012, Britnell et al. demonstrated a quantum tunneling transistor, made of graphene/h-BN heterostructrue. In this experiment, they made quantum tunneling from graphene electrode through h-BN, for insulating barrier. From the result, they concluded that the tunneling

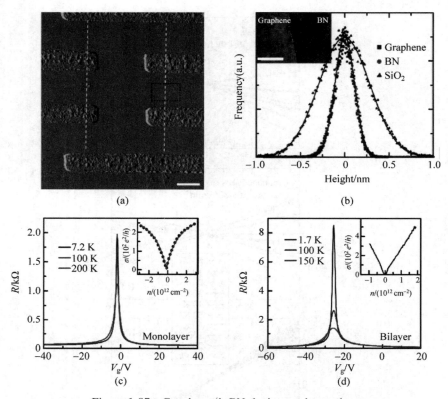

Figure 1-37 Graphene/h-BN devices and test chart
(a) AFM image of graphene/h-BN heterostructure; (b) Histogram of the height distribution measured by AFM for graphene, h-BN and SiO_2; Resistance versus applied gate voltage for monolayer graphene (c) and bilayer graphene (d) on h-BN[204]

devices offer a viable method to prepared the high-speed graphene-based analogue electronics (Figure 1-38)[182].

Thermoelectric device: In 2014, Chen et al. studied the thermal interface conductance across a graphene/h-BN heterojunction[208]. Based on the study, they prepared graphene/h-BN heterostructures to measure the thermoelectric transport[209]. They can separate the thermoelectric contribution to the I-V property of the device structures by using AC lock-in technique, and by using Raman spectroscopy, they measured the temperature gradient, so they can study the thermoelectric transport produced at interfaces of heterostructures. The Figure 1-39 shows us thermoelectric voltage and temperature gradient, they ascertained the Seebeck coefficient is 99.3 μV/K to the device.

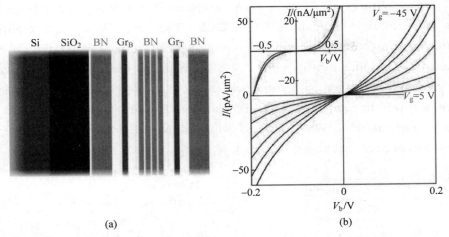

Figure 1-38 Schematic structure of field-effect tunneling transistor (a), and (b) tunneling characteristics for a graphene/h-BN device with 6±1 layers of h-BN as the tunnel barrier[182]

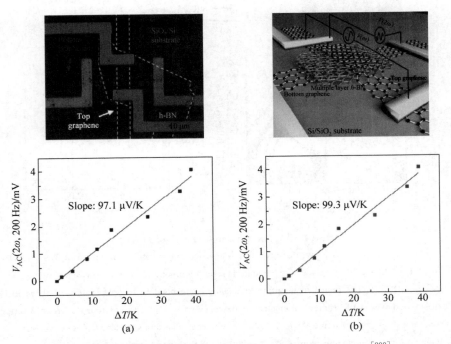

Figure 1-39 Schematic figure and test results of the device[209]

LED: Withers et al.[210] made graphene/h-BN/WSe$_2$/h-BN/graphene heterostructures to prepare LED in 2015. They used WSe$_2$ light-emitting tunneling transistors with enhanced brightness at room temperature. They cleaved and exfoliated h-BN onto a Si/SiO$_2$ substrate, then, a graphene flake is peeled onto h-BN crystal, h-BN/graphene/h-BN/WX$_2$/h-BN heterostructures have been prepared to completed the LED structure. Figure 1-40 shows the EQE T-dependence for three WSe$_2$ devices, which shows the characteristic increase with temperature reaching 5 % at 300 K.

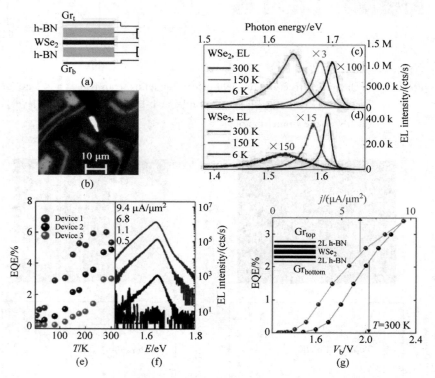

Figure 1-40　The flow and test figure of LED

(a) Schematic of the device architecture; (b) Magnification monochrome image of WSe$_2$ LEQW device. The change of photon energy under different temperature curves for WSe$_2$ in (c) and MoSe$_2$ in (d); (e) The image of quantum efficiency corresponding to temperature of three LED devices based WSe$_2$; (f) Individual electroluminescence spectra plotted for four different injection current densities for a sample device; (g) The external quantum efficiency plotted against bias voltage and injection current density at T = 300 K for the same sample device in (f)[210]

Wang et al.[211] prepared graphene/h-BN van der Waals light-emitting diodes in 2015. IN the two experiments, the heterostructures are mainly used for LED, which provided direction for the preparation of nano-optoelectronic device.

Solar cell: In 2015, Li et al.[212] designed a graphene/h-BN/GaAs sandwich diode as solar cell and photodetector. From Figure 1-41 we can know that the barrier height of graphene/GaAs heterojunction could be increased from 0.88 eV to 1.02 eV, when h-BN was been inserted. H-BN enhanced Fermi level tuning effect, power conversion efficiency (PCE) of 10.18 % could be achieved for graphene/h-BN/GaAs, which compared with 8.63 % of graphene/GaAs structures. The performance of structure based photodetector was also improved with on/off ratio increased.

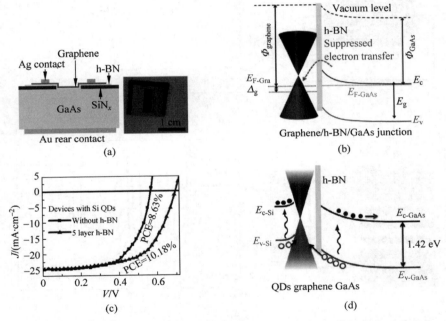

Figure 1-41 Graphene/h-BN/GaAs sandwich diode as solar cell and photodetector
(a) Schematic of the graphene/h-BN/GaAs sandwich device; (b) Electronic band structure of graphene/h-BN/GaAs heterostructure; (c) J-V curves of solar cells based on graphene/GaAs and graphene/h-BN/GaAs heterostructure with Si QDs introduced photo-induced doping; (d) Electronic band structure of the graphene/h-BN/GaAs solar cell with Si QDs covering on graphene under illumination[212]

IPPS and TPS: In 2011, Xu et al.[213] constructed an in-plane pressure sensor (IPPS) and a tunneling pressure sensor (TPS), made of graphene/h-BN heterostructures. They modeled the responses as a function of external pressure. The result of test shows that the current varies by 3 orders of magnitude as pressure increases from 0 to 5 nN/nm^2. The IPPS current is negatively correlated to pressure, whereas TPS current exhibits positive correlation to pressure. IPPS and TPS both have a pressure range of about 5 GPa, bigger than conventional MEMS pressure sensors (Figure 1-42). The result points to the direction towards realizing viable, atomic scale graphene-based sensors for pressure measurements.

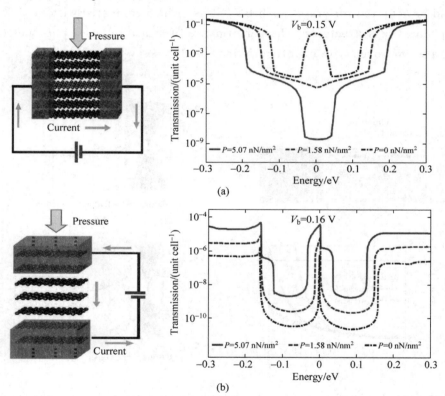

Figure 1-42　The schematic structures of (a) an in-plane pressure sensor and (b) a tunnelling pressure sensor. Transmission spectra of (c) the in-plane sensor and (d) the tunnelling sensor as functions of electron energy with different external pressures[213]

Combined BN with graphene is multi-dimensional in many fields: from the perspective of transport doping, and from the horizontal/vertical components. People can conduct researches on physics based on BN excellent substrate effect, and also can do research on electronics to use BN wide band gap. Therefore, there is still a hope to see a good research on the aspects of mixing graphite and BN.

1.5 Summary and outlook

For logic devices of graphene/h-BN heterostructures, although a lot of theories and experiments proved graphene based on h-BN has excellent optical characteristics, because of the zero-band-gap properties of graphene, it is difficult to prepare a high-performance logic circuit as a directly FET. Therefore, the band of the 2D carbon-based material has a very important role in the regulation of optoelectronic device applications. For example, continuous regulation research of graphene/h-BN heterostructures in the far infrared, the visible region, and the ultraviolet region, provide a basis for the preparation of applications of the photovoltaic devices. Account to the different absorptions of graphene and h-BN in light regions, it is certain that composite system of graphene/h-BN heterostructures has a great use in the visible region, such as photoluminescence. On the other hand, the regulation of the energy-band engineering can be achieved by designing 2D heterostructure materials combination and dielectric shielding. How well between the graphene layer of graphene/h-BN heterostructure and the h-BN layer are bonded, and how much is the force between the electrons, will be revealed in the future. We believe that practical applications of heterostructure graphene/h-BN will greatly expand because of unique optoelectronic properties of this structure.

For graphene, due to its characteristics of optoelectronic devices, energy storage, and other aspects of mechanical reinforcement, the basic applied research of the new 2D materials has made great progress. If combined with other new 2D materials, by using 2D layered materials with different chemical bonds, and the formation of different physical and chemical properties of the new lattice structure, it should have greater applications. For example: biomedical devices, photo detectors, chemical catalysis, clean energy generation/storage, surface-enhanced Raman scattering and so on.

Perhaps the most meaningful thing of studying graphene is that graphene caused people to explore other 2D materials, which also broke what people have studied of 3D materials for a long time. At least in the history of science in materials, before graphene occurred, there was no one focusing on 2D materials. Therefore, the heuristic value of graphene is far more than the value of its excellent performance. For instance, after the combination of the same or similar 2D structure of several different materials, the new composite material has a lot of characteristics making the scientific community excited, and the occurrence of 2D composite materials make properties of the materials more diversified.

For colored figures please scan the QR code.

References

[1] GEIM A K, NOVOSELOV K S. The rise of graphene[J]. Nature Materials, 2007, 6(3): 183-191.
[2] OHTA T, BOSTWICK A, SEYLLER T, et al. Controlling the electronic structure of bilayer graphene[J]. Science, 2006, 313(5789): 951-954.
[3] NOVOSELOV K S, GEIM A K, MOROZOV S V, et al. Electric field effect in atomically thin carbon films[J]. Science, 2004, 306(5696): 666-669.
[4] CHEN J, BADIOLI M, ALONSO-GONZALEZ P, et al. Optical nano-imaging of gate-tunable graphene plasmons[J]. Nature, 2012, 487(7405): 77-81.
[5] GENG X, NIU L, XING Z, et al. Aqueous-processable noncovalent chemically converted graphene-quantum dot composites for flexible and transparent optoelectronic films[J]. Advanced Materials, 2010, 22(5): 638-642.
[6] GEIM A K. Random walk to graphene (nobel lecture)[J]. Angewandte Chemie International Edition, 2011, 50(31): 6966-6985.
[7] AVOURIS P, CHEN Z, PEREBEINOS V. Carbon-based electronics[M]. Nanoscience and Technology: A Collection of Reviews from Nature Journals, 2010: 174-184.
[8] TANG Y B, LEE C S, CHEN Z H, et al. High-quality graphenes via a facile quenching method for field-effect transistors[J]. Nano Letters, 2009, 9(4): 1374-1377.
[9] LEE C, WEI X, KYSAR J W, et al. Measurement of the elastic properties and intrinsic strength of monolayer graphene[J]. Science, 2008, 321(5887): 385-388.
[10] BOLOTIN K I, SIKES K J, JIANG Z, et al. Ultrahigh electron mobility in suspended graphene[J]. Solid State Communications, 2008, 146(9/10): 351-355.
[11] KIM K S, ZHAO Y, JANG H, et al. Large-scale pattern growth of graphene films for stretchable transparent electrodes[J]. Nature, 2009, 457(7230): 706-710.
[12] BALANDIN A A. Thermal properties of graphene and nanostructured carbon materials[J]. Nature Materials, 2011, 10(8): 569-581.

[13] FALKOVSKY L A. Symmetry constraints on phonon dispersion in graphene[J]. Physics Letters A, 2008, 372(31): 5189-5192.

[14] IIJIMA S. Helical microtubules of graphitic carbon[J]. Nature, 1991, 354(6348): 56-58.

[15] BETHUNE D S, KIANG C H, DE VRIES M S, et al. Cobalt-catalysed growth of carbon nanotubes with single-atomic-layer walls[J]. Nature, 1993, 363(6430): 605-607.

[16] IIJIMA S, ICHIHASHI T. Single-shell carbon nanotubes of 1-nm diameter[J]. Nature, 1993, 363(6430): 603-605.

[17] GUAN L, SUENAGA K, IIJIMA S. Smallest carbon nanotube assigned with atomic resolution accuracy[J]. Nano Letters, 2008, 8(2): 459-462.

[18] WEN Q, ZHANG R, QIAN W, et al. Growing 20 cm long DWNTs/TWNTs at a rapid growth rate of 80-90 μm/s[J]. Chemistry of Materials, 2010, 22(4): 1294-1296.

[19] IIJIMA S, AJAYAN P M, ICHIHASHI T. Growth model for carbon nanotubes[J]. Physical Review Letters, 1992, 69(21): 3100-3103.

[20] GALPAYA D, WANG M, GEORGE G, et al. Preparation of graphene oxide/epoxy nanocomposites with significantly improved mechanical properties[J]. Journal of Applied Physics, 2014, 116(5): 053518.

[21] SAITO R, FUJITA M, DRESSELHAUS G, et al. Electronic structure of chiral graphene tubules[J]. Applied Physics Letters, 1992, 60(18): 2204-2206.

[22] HAMADA N, SAWADA S, OSHIYAMA A. New one-dimensional conductors: graphitic microtubules[J]. Physical Review Letters, 1992, 68(10): 1579-1581.

[23] TREACY M M J, EBBESEN T W, GIBSON J M. Exceptionally high Young's modulus observed for individual carbon nanotubes[J]. Nature, 1996, 381(6584): 678-680.

[24] LOURIE O, COX D M, WAGNER H D. Buckling and collapse of embedded carbon nanotubes[J]. Physical Review Letters, 1998, 81(8): 1638-1641.

[25] RITSCHEL M, UHLEMANN M, GUTFLEISCH O, et al. Hydrogen storage in different carbon nanostructures[J]. Applied Physics Letters, 2002, 80(16): 2985-2987.

[26] KONG J, FRANKLIN N R, ZHOU C, et al. Nanotube molecular wires as chemical sensors[J]. Science, 2000, 287(5453): 622-625.

[27] MCEUEN P L, FUHRER M S, PARK H. Single-walled carbon nanotube electronics [J]. IEEE transactions on nanotechnology, 2002, 1(1): 78-85.

[28] APPENZELLER J, MARTEL R, DERYCKE V, et al. Carbon nanotubes as potential building blocks for future nanoelectronics[J]. Microelectronic Engineering, 2002, 64(1-4): 391-397.

[29] DEHEER W A, BACSA W S, CHATELAIN A, et al. Aligned carbon nanotube films: production and optical and electronic properties[J]. Science, 1995, 268(5212): 845-847.

[30] BOMMELI F, DEGIORGI L, WACHTER P, et al. Evidence of anisotropic metallic behaviour in the optical properties of carbon nanotubes [J]. Solid State Communications, 1996, 99(7): 513-517.

[31] GAO X, ZHOU Z, ZHAO Y, et al. Comparative study of carbon and BN nanographenes: ground electronic states and energy gap engineering[J]. The Journal of Physical Chemistry C, 2008, 112(33): 12677-12682.

[32] WHEELOCK P B, COOK B C, HARRINGA J L, et al. Phase changes induced in hexagonal boron nitride by high energy mechanical milling[J]. Journal of Materials Science, 2004, 39(1): 343-347.

[33] LORRETTE C, WEISBECHER P, JACQUES S, et al. Deposition and characterization of hex-BN coating on carbon fibres using tris (dimethylamino) borane precursor[J]. Journal of the European Ceramic Society, 2007, 27(7): 2737-2743.

[34] LEI G E, JIAN Y, TAI Q. Study progress of preparation methods of hexagonal boron nitride[J]. Electronic Components and Materials, 2008, 27: 22-29.

[35] WATANABE K, TANIGUCHI T, KANDA H. Direct-bandgap properties and evidence for ultraviolet lasing of hexagonal boron nitride single crystal[J]. Nature Materials, 2004, 3(6): 404-409.

[36] KUBOTA Y, WATANABE K, TSUDA O, et al. Deep ultraviolet light-emitting hexagonal boron nitride synthesized at atmospheric pressure [J]. Science, 2007, 317(5840): 932-934.

[37] GOLBERG D, BANDO Y, HUANG Y, et al. Boron nitride nanotubes and nanosheets[J]. ACS Nano, 2010, 4(6): 2979-2993.

[38] PACILE D, MEYER J C, GIRIT Ç Ö, et al. The two-dimensional phase of boron nitride: Few-atomic-layer sheets and suspended membranes [J]. Applied Physics Letters, 2008, 92(13): 133107.

[39] LEE C, LI Q, KALB W, et al. Frictional characteristics of atomically thin sheets[J]. Science, 2010, 328(5974): 76-80.

[40] NAGASHIMA A, TEJIMA N, GAMOU Y, et al. Electronic structure of monolayer hexagonal boron nitride physisorbed on metal surfaces[J]. Physical Review Letters, 1995, 75(21): 3918-3921.

[41] NOVOSELOV K S, JIANG D, SCHEDIN F, et al. Two-dimensional atomic crystals [J]. Proceedings of the National Academy of Sciences, 2005, 102(30): 10451-10453.

[42] CRANE T P, COWAN B P. Magnetic relaxation properties of helium-3 adsorbed on hexagonal boron nitride[J]. Physical Review B, 2000, 62(17): 11359-11362.

[43] MILLER M, OWENS F J. Tuning the electronic and magnetic properties of boron nitride nanotubes[J]. Solid State Communications, 2011, 151(14/15): 1001-1003.

[44] SICHEL E K, MILLER R E, ABRAHAMS M S, et al. Heat capacity and thermal conductivity of hexagonal pyrolytic boron nitride [J]. Physical Review B, 1976, 13(10): 4607-4611.

[45] HENAGER C H, PAWLEWICZ W T. Thermal conductivities of thin, sputtered

optical films[J]. Applied Optics, 1993, 32(1): 91-101.

[46] RAVICHANDRAN J, MANOJ A G, LIU J, et al. A novel polymer nanotube composite for photovoltaic packaging applications[J]. Nanotechnology, 2008, 19(8): 085712.

[47] HAO S, ZHOU G, DUAN W, et al. Tremendous spin-splitting effects in open boron nitride nanotubes: application to nanoscale spintronic devices[J]. Journal of the American Chemical Society, 2006, 128(26): 8453-8458.

[48] ZHANG G J, ANDO M, OHJI T, et al. High-performance boron nitride-containing composites by reaction synthesis for the applications in the steel industry [J]. International Journal of Applied Ceramic Technology, 2005, 2(2): 162-171.

[49] JIN M S, KIM N O. Photoluminescence of hexagonal boron nitride (h-BN) film[J]. Journal of Electrical Engineering and Technology, 2010, 5(4): 637-639.

[50] ZHANG Y, HE X, HAN J, et al. Combustion synthesis of hexagonal boron-nitride-based ceramics[J]. Journal of Materials Processing Technology, 2001, 116(2/3): 161-164.

[51] HARRISON C, WEAVER S, BERTELSEN C, et al. Polyethylene/boron nitride composites for space radiation shielding[J]. Journal of applied polymer science, 2008, 109(4): 2529-2538.

[52] HANIGOFSKY J A, MORE K L, LACKEY W J, et al. Composition and microstructure of chemically vapor-deposited boron nitride, aluminum nitride, and boron nitride + aluminum nitride composites[J]. Journal of the American Ceramic Society, 1991, 74(2): 301-305.

[53] NASLAIN R, DUGNE O, GUETTE A, et al. Boron nitride interphase in ceramic-matrix composites[J]. Journal of the American Ceramic Society, 1991, 74(10): 2482-2488.

[54] MASHOFF T, PRATZER M, GERINGER V, et al. Bistability and oscillatory motion of natural nanomembranes appearing within monolayer graphene on silicon dioxide[J]. Nano Letters, 2010, 10(2): 461-465.

[55] HAMILTON J C, BLAKELY J M. Carbon segregation to single crystal surfaces of Pt, Pd and Co[J]. Surface Science, 1980, 91(1): 199-217.

[56] SHIKIN A M, PRUDNIKOVA G V, ADAMCHUK V K, et al. Surface intercalation of gold underneath a graphite monolayer on Ni (111) studied by angle-resolved photoemission and high-resolution electron-energy-loss spectroscopy [J]. Physical Review B, 2000, 62(19): 13202-13208.

[57] ROSEI R, DE CRESCENZI M, SETTE F, et al. Structure of graphitic carbon on Ni (111): a surface extended-energy-loss fine-structure study[J]. Physical Review B, 1983, 28(2): 1161-1164.

[58] SHIKIN A M, FARIAS D, RIEDER K H. Phonon stiffening induced by copper intercalation in monolayer graphite on Ni (111)[J]. Europhysics Letters, 1998, 44(1): 44-49.

[59] DEDKOV Y S, SHIKIN A M, ADAMCHUK V K, et al. Intercalation of copper underneath a monolayer of graphite on Ni (111)[J]. Physical Review B, 2001, 64(3): 035405.

[60] BRUGGER T, GUNTHER S, WANG B, et al. Comparison of electronic structure and template function of single-layer graphene and a hexagonal boron nitride nanomesh on Ru (0001)[J]. Physical Review B, 2009, 79(4): 045407.

[61] MORITZ W, WANG B, BOCQUET M L, et al. Structure determination of the coincidence phase of graphene on Ru (0001)[J]. Physical Review Letters, 2010, 104(13): 136102.

[62] LAND T A, MICHELY T, BEHM R J, et al. STM investigation of single layer graphite structures produced on Pt (111) by hydrocarbon decomposition[J]. Surface Science, 1992, 264(3): 261-270.

[63] STARR D E, PAZHETNOV E M, STADNICHENKO A I, et al. Carbon films grown on Pt (1 1 1) as supports for model gold catalysts[J]. Surface Science, 2006, 600(13): 2688-2695.

[64] CHEN J H, JANG C, XIAO S, et al. Intrinsic and extrinsic performance limits of graphene devices on SiO_2[J]. Nature Nanotechnology, 2008, 3(4): 206-209.

[65] LEE D S, RIEDL C, KRAUSS B, et al. Raman spectra of epitaxial graphene on SiC and of epitaxial graphene transferred to SiO_2 [J]. Nano Letters, 2008, 8(12): 4320-4325.

[66] GERINGER V, LIEBMANN M, ECHTERMEYER T, et al. Intrinsic and extrinsic corrugation of monolayer graphene deposited on SiO_2 [J]. Physical Review Letters, 2009, 102(7): 076102.

[67] FORBEAUX I, THEMLIN J M, DEBEVER J M. Heteroepitaxial graphite on 6H-SiC (0001): interface formation through conduction-band electronic structure[J]. Physical Review B, 1998, 58(24): 16396.

[68] MENDES-DE-SA T G, GONCALVES A M B, MATOS M J S, et al. Correlation between (in) commensurate domains of multilayer epitaxial graphene grown on SiC (0001) and single layer electronic behavior [J]. Nanotechnology, 2012, 23(47): 475602.

[69] HASS J, VARCHON F, MILLAN-OTOYA J E, et al. Why multilayer graphene on 4H-SiC (0001) behaves like a single sheet of graphene[J]. Physical Review Letters, 2008, 100(12): 125504.

[70] DECKER R, WANG Y, BRAR V W, et al. Local electronic properties of graphene on a BN substrate via scanning tunneling microscopy[J]. Nano Letters, 2011, 11(6): 2291-2295.

[71] PONOMARENKO L A, GEIM A K, ZHUKOV A A, et al. Tunable metal-insulator transition in double-layer graphene heterostructures[J]. Nature Physics, 2011, 7(12): 958-961.

[72] GIOVANNETTI G, KHOMYAKOV P A, BROCKS G, et al. Publisher's note:

substrate-induced band gap in graphene on hexagonal boron nitride: ab initio density functional calculations [J]. Physical Review B, 2007, 76(7): 079902.

[73] DEAN C R, YOUNG A F, MERIC I, et al. Boron nitride substrates for high-quality graphene electronics[J]. Nature Nanotechnology, 2010, 5(10): 722-726.

[74] BRITNELL L, GORBACHEV R V, JALIL R, et al. Field-effect tunneling transistor based on vertical graphene heterostructures[J]. Science, 2012, 335(6071): 947-950.

[75] HAIGH S J, GHOLINIA A, JALIL R, et al. Cross-sectional imaging of individual layers and buried interfaces of graphene-based heterostructures and superlattices[J]. Nature Materials, 2012, 11(9): 764-767.

[76] DEAN C, YOUNG A F, WANG L, et al. Graphene based heterostructures[J]. Solid State Communications, 2012, 152(15): 1275-1282.

[77] PONOMARENKO L A, GORBACHEV R V, YU G L, et al. Cloning of Dirac fermions in graphene superlattices[J]. Nature, 2013, 497(7451): 594-597.

[78] DEAN C R, WANG L, MAHER P, et al. Hofstadter's butterfly and the fractal quantum Hall effect in moiré superlattices[J]. Nature, 2013, 497(7451): 598-602.

[79] PARTOENS B, PEETERS F M. From graphene to graphite: electronic structure around the K point[J]. Physical Review B, 2006, 74(7): 075404.

[80] MA P, JIN Z, GUO J N, et al. Top-gated graphene field-effect transistors on SiC substrates[J]. Chinese Science Bulletin, 2012, 57(19): 2401-2403.

[81] WU M, WU X, GAO Y, et al. Materials design of half-metallic graphene and graphene nanoribbons[J]. Applied Physics Letters, 2009, 94(22): 223111.

[82] BIEL B, TRIOZON F, BLASE X, et al. Chemically induced mobility gaps in graphene nanoribbons: a route for upscaling device performances[J]. Nano Letters, 2009, 9(7): 2725-2729.

[83] JIAO L, ZHANG L, WANG X, et al. Narrow graphene nanoribbons from carbon nanotubes[J]. Nature, 2009, 458(7240): 877-880.

[84] SUTTER P W, FLEGE J I, SUTTER E A. Epitaxial graphene on ruthenium[J]. Nature Materials, 2008, 7(5): 406-411.

[85] BERGER C, SONG Z, LI X, et al. Electronic confinement and coherence in patterned epitaxial graphene[J]. Science, 2006, 312(5777): 1191-1196.

[86] WANG H, YU G. Direct CVD graphene growth on semiconductors and dielectrics for transfer-free device fabrication[J]. Advanced Materials, 2016, 28(25): 4956-4975.

[87] GÓMEZ-NAVARRO C, BURGHARD M, KERN K. Elastic properties of chemically derived single graphene sheets[J]. Nano Letters, 2008, 8(7): 2045-2049.

[88] POOT M, VAN DER ZANT H S J. Nanomechanical properties of few-layer graphene membranes[J]. Applied Physics Letters, 2008, 92(6): 063111.

[89] YOON D, SON Y W, CHEONG H. Negative thermal expansion coefficient of graphene measured by Raman spectroscopy [J]. Nano Letters, 2011, 11(8): 3227-3231.

[90] SEOL J H, JO I, MOORE A L, et al. Two-dimensional phonon transport in supported

graphene[J]. Science, 2010, 328(5975): 213-216.

[91] XU Y, CHEN X, GU B L, et al. Intrinsic anisotropy of thermal conductance in graphene nanoribbons[J]. Applied Physics Letters, 2009, 95(23): 233116.

[92] NI Z H, WANG H M, KASIM J, et al. Graphene thickness determination using reflection and contrast spectroscopy[J]. Nano Letters, 2007, 7(9): 2758-2763.

[93] NAIR R R, BLAKE P, GRIGORENKO A N, et al. Fine structure constant defines visual transparency of graphene[J]. Science, 2008, 320(5881): 1308.

[94] KUZMENKO A B, VAN HEUMEN E, CARBONE F, et al. Universal optical conductance of graphite[J]. Physical Review Letters, 2008, 100(11): 117401.

[95] SALVETAT J P, BONARD J M, BACSA R, et al. Physical properties of carbon nanotubes[C]//AIP Conference Proceedings. American Institute of Physics, 1998, 442(1): 467-480.

[96] CASTRO N A H, GUINEA F, PERES N M R, et al. The electronic properties of graphene[J]. RvMP, 2009, 81(1): 109-162.

[97] WU M, CAO C, JIANG J Z. Light non-metallic atom (B, N, O and F)-doped graphene: a first-principles study[J]. Nanotechnology, 2010, 21(50): 505202.

[98] LHERBIER A, BLASE X, NIQUET Y M, et al. Charge transport in chemically doped 2D graphene[J]. Physical Review Letters, 2008, 101(3): 036808.

[99] VARYKHALOV A, SCHOLZ M R, KIM T K, et al. Effect of noble-metal contacts on doping and band gap of graphene[J]. Physical Review B, 2010, 82(12): 121101.

[100] PEREIRA V M, NETO A H C, PERES N M R. Tight-binding approach to uniaxial strain in graphene[J]. Physical Review B, 2009, 80(4): 045401.

[101] CASTRO E V, NOVOSELOV K S, MOROZOV S V, et al. Biased bilayer graphene: semiconductor with a gap tunable by the electric field effect[J]. Physical Review Letters, 2007, 99(21): 216802.

[102] ZHOU S Y, GWEON G H, FEDOROV A V, et al. Substrate-induced bandgap opening in epitaxial graphene[J]. Nature Materials, 2007, 6(10): 770-775.

[103] MATTAUSCH A, PANKRATOV O. Ab initio study of graphene on SiC [J]. Physical Review Letters, 2007, 99(7): 076802.

[104] SHEMELLA P, NAYAK S K. Electronic structure and band-gap modulation of graphene via substrate surface chemistry [J]. Applied Physics Letters, 2009, 94(3): 032101.

[105] YANG H X, HALLAL A, TERRADE D, et al. Proximity effects induced in graphene by magnetic insulators: First-principles calculations on spin filtering and exchange-splitting gaps[J]. Physical Review Letters, 2013, 110(4): 046603.

[106] NOVOSELOV K S, GEIM A K, MOROZOV S V, et al. Two-dimensional gas of massless Dirac fermions in graphene[J]. Nature, 2005, 438(7065): 197-200.

[107] ZHANG Y, TAN Y W, STORMER H L, et al. Experimental observation of the quantum Hall effect and Berry's phase in graphene[J]. Nature, 2005, 438(7065): 201-204.

[108] TOMBROS N, JOZSA C, POPINCIUC M, et al. Electronic spin transport and spin precession in single graphene layers at room temperature[J]. Nature, 2007, 448(7153): 571-574.

[109] CHO S, CHEN Y F, FUHRER M S. Gate-tunable graphene spin valve[J]. Applied Physics Letters, 2007, 91(12): 123105.

[110] WANG Y, HUANG Y, SONG Y, et al. Room-temperature ferromagnetism of graphene[J]. Nano Letters, 2009, 9(1): 220-224.

[111] SEPIONI M, NAIR R R, RABLEN S, et al. Limits on intrinsic magnetism in graphene[J]. Physical Review Letters, 2010, 105(20): 207205.

[112] GOMEZ-SANTOS G, STAUBER T. Measurable lattice effects on the charge and magnetic response in graphene[J]. Physical Review Letters, 2011, 106(4): 045504.

[113] FERRARI A C, MEYER J C, SCARDACI V, et al. Raman spectrum of graphene and graphene layers[J]. Physical Review Letters, 2006, 97(18): 187401.

[114] DRESSELHAUS M S, DRESSELHAUS G, HOFMANN M. Raman spectroscopy as a probe of graphene and carbon nanotubes[J]. Philosophical Transactions of the Royal Society A: Mathematical, Physical and Engineering Sciences, 2008, 366(1863): 231-236.

[115] BERCIAUD S, RYU S, BRUS L E, et al. Probing the intrinsic properties of exfoliated graphene: Raman spectroscopy of free-standing monolayers[J]. Nano Letters, 2009, 9(1): 346-352.

[116] WANG Y, ALSMEYER D C, MCCREERY R L. Raman spectroscopy of carbon materials: structural basis of observed spectra[J]. Chemistry of Materials, 1990, 2(5): 557-563.

[117] LIU Z, HE D, WANG Y, et al. Improving photovoltaic properties by incorporating both SPF Graphene and functionalized multiwalled carbon nanotubes[J]. Solar Energy Materials and Solar Cells, 2010, 94(12): 2148-2153.

[118] TOPSAKAL M, AKTÜRK E, CIRACI S. First-principles study of two-and one-dimensional honeycomb structures of boron nitride[J]. Physical Review B, 2009, 79(11): 115442.

[119] BHATTACHARYA A, BHATTACHARYA S, DAS G P. Band gap engineering by functionalization of BN sheet[J]. Physical Review B, 2012, 85(3): 035415.

[120] LE M Q. Size effects in mechanical properties of boron nitride nanoribbons[J]. Journal of Mechanical Science and Technology, 2014, 28(10): 4173-4178.

[121] MORTAZAVI B, CUNIBERTI G. Mechanical properties of polycrystalline boron-nitride nanosheets[J]. RSC Advances, 2014, 4(37): 19137-19143.

[122] CORSO M, AUWÄRTER W, MUNTWILER M, et al. Boron nitride nanomesh[J]. Science, 2004, 303(5655): 217-220.

[123] LASKOWSKI R, BLAHA P, GALLAUNER T, et al. Single-layer model of the hexagonal boron nitride nanomesh on the Rh (111) surface[J]. Physical Review Letters, 2007, 98(10): 106802.

[124] GRAD G B, BLAHA P, SCHWARZ K, et al. Density functional theory investigation of the geometric and spintronic structure of h-BN/Ni (111) in view of photoemission and STM experiments[J]. Physical Review B, 2003, 68(8): 085404.

[125] TONKIKH A A, VOLOSHINA E N, WERNER P, et al. Structural and electronic properties of epitaxial multilayer h-BN on Ni (111) for spintronics applications[J]. Scientific Reports, 2016, 6: 23547.

[126] AUWÄRTER W, MUNTWILER M, OSTERWALDER J, et al. Defect lines and two-domain structure of hexagonal boron nitride films on Ni (1 1 1)[J]. Surface science, 2003, 545(1/2): L735-L740.

[127] HUDA M N, KLEINMAN L. h-BN monolayer adsorption on the Ni (111) surface: a density functional study[J]. Physical Review B, 2006, 74(7): 075418.

[128] NAGASHIMA A, TEJIMA N, GAMOU Y, et al. Electronic dispersion relations of monolayer hexagonal boron nitride formed on the Ni (111) surface[J]. Physical Review B, 1995, 51(7): 4606-4613.

[129] HAN W Q, WU L, ZHU Y, et al. Structure of chemically derived mono-and few-atomic-layer boron nitride sheets[J]. Applied Physics Letters, 2008, 93(22): 223103.

[130] JIN C, LIN F, SUENAGA K, et al. Fabrication of a freestanding boron nitride single layer and its defect assignments [J]. Physical Review Letters, 2009, 102(19): 195505.

[131] GAO R, YIN L, WANG C, et al. High-yield synthesis of boron nitride nanosheets with strong ultraviolet cathodoluminescence emission[J]. The Journal of Physical Chemistry C, 2009, 113(34): 15160-15165.

[132] MEYER J C, CHUVILIN A, ALGARA-SILLER G, et al. Selective sputtering and atomic resolution imaging of atomically thin boron nitride membranes[J]. Nano Letters, 2009, 9(7): 2683-2689.

[133] SONG L, CI L, LU H, et al. Large scale growth and characterization of atomic hexagonal boron nitride layers[J]. Nano Letters, 2010, 10(8): 3209-3215.

[134] NAG A, RAIDONGIA K, HEMBRAM K P S S, et al. Graphene analogues of BN: novel synthesis and properties[J]. ACS Nano, 2010, 4(3): 1539-1544.

[135] LIN Y, WILLIAMS T V, XU T B, et al. Aqueous dispersions of few-layered and monolayered hexagonal boron nitride nanosheets from sonication-assisted hydrolysis: critical role of water[J]. The Journal of Physical Chemistry C, 2011, 115(6): 2679-2685.

[136] WARNER J H, RUMMELI M H, BACHMATIUK A, et al. Atomic resolution imaging and topography of boron nitride sheets produced by chemical exfoliation[J]. ACS Nano, 2010, 4(3): 1299-1304.

[137] LIN Y, WILLIAMS T V, CONNELL J W. Soluble, exfoliated hexagonal boron nitride nanosheets[J]. The Journal of Physical Chemistry Letters, 2010, 1(1): 277-283.

[138] ZHI C, BANDO Y, TANG C, et al. Large-scale fabrication of boron nitride

nanosheets and their utilization in polymeric composites with improved thermal and mechanical properties[J]. Advanced Materials, 2009, 21(28): 2889-2893.

[139] GORIACHKO A, HE Y, KNAPP M, et al. Self-assembly of a hexagonal boron nitride nanomesh on Ru (0001)[J]. Langmuir, 2007, 23(6): 2928-2931.

[140] BOLDRIN L, SCARPA F, CHOWDHURY R, et al. Effective mechanical properties of hexagonal boron nitride nanosheets[J]. Nanotechnology, 2011, 22(50): 505702.

[141] ANDREW R C, MAPASHA R E, UKPONG A M, et al. Mechanical properties of graphene and boronitrene[J]. Physical Review B, 2012, 85(12): 125428.

[142] ZHOU H, ZHU J, LIU Z, et al. High thermal conductivity of suspended few-layer hexagonal boron nitride sheets[J]. Nano Research, 2014, 7(8): 1232-1240.

[143] JO I, PETTES M T, KIM J, et al. Thermal conductivity and phonon transport in suspended few-layer hexagonal boron nitride [J]. Nano Letters, 2013, 13 (2): 550-554.

[144] CHEN C C, LI Z, SHI L, et al. Thermoelectric transport across graphene/hexagonal boron nitride/graphene heterostructures[J]. Nano Research, 2015, 8(2): 666-672.

[145] ZHOU W, QI S, AN Q, et al. Thermal conductivity of boron nitride reinforced polyethylene composites[J]. Materials Research Bulletin, 2007, 42(10): 1863-1873.

[146] BARONE V, PERALTA J E. Magnetic boron nitride nanoribbons with tunable electronic properties[J]. Nano Letters, 2008, 8(8): 2210-2214.

[147] WU R Q, LIU L, PENG G W, et al. Magnetism in BN nanotubes induced by carbon doping[J]. Applied Physics Letters, 2005, 86(12): 122510-122513.

[148] WU R Q, PENG G W, LIU L, et al. Possible graphitic-boron-nitride-based metal-free molecular magnets from first principles study[J]. Journal of Physics: Condensed Matter, 2005, 18(2): 569-576.

[149] SI M S, XUE D S. Magnetic properties of vacancies in a graphitic boron nitride sheet by first-principles pseudopotential calculations[J]. Physical Review B, 2007, 75(19): 193409.

[150] LAI L, LU J, WANG L, et al. Magnetic properties of fully bare and half-bare boron nitride nanoribbons [J]. The Journal of Physical Chemistry C, 2009, 113 (6): 2273-2276.

[151] ZHENG F, ZHOU G, LIU Z, et al. Half metallicity along the edge of zigzag boron nitride nanoribbons[J]. Physical Review B, 2008, 78(20): 205415.

[152] GORBACHEV R V, RIAZ I, NAIR R R, et al. Hunting for monolayer boron nitride: optical and Raman signatures[J]. Small, 2011, 7(4): 465-468.

[153] KATSNELSON M I, NOVOSELOV K S, GEIM A K. Chiral tunnelling and the Klein paradox in graphene[J]. Nature Physics, 2006, 2(9): 620-625.

[154] HUARD B, SULPIZIO J A, STANDER N, et al. Transport measurements across a tunable potential barrier in graphene [J]. Physical Review Letters, 2007, 98(23): 236803.

[155] ELIAS D C, NAIR R R, MOHIUDDIN T M G, et al. Control of graphene's properties by reversible hydrogenation: evidence for graphane[J]. Science, 2009, 323(5914): 610-613.

[156] LIU L, SHEN Z. Bandgap engineering of graphene: a density functional theory study[J]. Applied Physics Letters, 2009, 95(25): 252104.

[157] OH D H, SHIN B G, AHN J R. Band engineering of bilayer graphene by metal atoms: First-principles calculations [J]. Applied Physics Letters, 2010, 96(23): 231916.

[158] ZHANG Y, TANG T T, GIRIT C, et al. Direct observation of a widely tunable bandgap in bilayer graphene[J]. Nature, 2009, 459(7248): 820-823.

[159] GUO Y, GUO W, CHEN C. Tuning field-induced energy gap of bilayer graphene via interlayer spacing[J]. Applied Physics Letters, 2008, 92(24): 243101.

[160] OSHIMA C, ITOH A, ROKUTA E, et al. A hetero-epitaxial-double-atomic-layer system of monolayer graphene/monolayer h-BN on Ni (111) [J]. Solid State Communications, 2000, 116(1): 37-40.

[161] BRUMFIELD B E, STEWART J T, WIDICUS W S L, et al. A quantum cascade laser cw cavity ringdown spectrometer coupled to a supersonic expansion source[J]. Review of Scientific Instruments, 2010, 81(6): 063102.

[162] BJELKEVIG C, MI Z, XIAO J, et al. Electronic structure of a graphene/hexagonal-BN heterostructure grown on Ru (0001) by chemical vapor deposition and atomic layer deposition: extrinsically doped graphene[J]. Journal of Physics: Condensed Matter, 2010, 22(30): 302002.

[163] TANG S, DING G, XIE X, et al. Nucleation and growth of single crystal graphene on hexagonal boron nitride[J]. Carbon, 2012, 50(1): 329-331.

[164] TANG S, WANG H, WANG H S, et al. Silane-catalysed fast growth of large single-crystalline graphene on hexagonal boron nitride[J]. Nature Communications, 2015, 6(1): 1-7.

[165] ZHONG X, YAP Y K, PANDEY R, et al. First-principles study of strain-induced modulation of energy gaps of graphene/BN and BN bilayers[J]. Physical Review B, 2011, 83(19): 193403.

[166] FAN Y, ZHAO M, WANG Z, et al. Tunable electronic structures of graphene/boron nitride heterobilayers[J]. Applied Physics Letters, 2011, 98(8): 083103.

[167] KAN E, REN H, WU F, et al. Why the band gap of graphene is tunable on hexagonal boron nitride[J]. The Journal of Physical Chemistry C, 2012, 116(4): 3142-3146.

[168] KRESSE G, HAFNER J. Ab initio molecular dynamics for liquid metals[J]. Physical Review B, 1993, 47(1): 558-561.

[169] GAO G, GAO W, CANNUCCIA E, et al. Artificially stacked atomic layers: toward new van der Waals solids[J]. Nano Letters, 2012, 12(7): 3518-3525.

[170] ZOMER P J, DASH S P, TOMBROS N, et al. A transfer technique for high

mobility graphene devices on commercially available hexagonal boron nitride[J]. Applied Physics Letters, 2011, 99(23): 232104.

[171] LEON J A, MAMANI N C, RAHIM A, et al. Transferring few-layer graphene sheets on hexagonal boron nitride substrates for fabrication of graphene devices[J]. Graphene, 2014, 3: 25-35.

[172] WANG M, JANG S K, JANG W J, et al. A platform for large-scale graphene electronics-CVD growth of single-layer graphene on CVD-grown hexagonal boron nitride[J]. Advanced Materials, 2013, 25(19): 2746-2752.

[173] WANG M, KIM M, ODKHUU D, et al. Catalytic transparency of hexagonal boron nitride on copper for chemical vapor deposition growth of large-area and high-quality graphene[J]. ACS Nano, 2014, 8(6): 5478-5483.

[174] KIM S M, HSU A, ARAUJO P T, et al. Synthesis of patched or stacked graphene and h-BN flakes: a route to hybrid structure discovery[J]. Nano Letters, 2013, 13(3): 933-941.

[175] GAO T, SONG X, DU H, et al. Temperature-triggered chemical switching growth of in-plane and vertically stacked graphene-boron nitride heterostructures[J]. Nature Communications, 2015, 6(1): 1-8.

[176] GANNETT W, REGAN W, WATANABE K, et al. Boron nitride substrates for high mobility chemical vapor deposited graphene[J]. Applied Physics Letters, 2011, 98(24): 242105.

[177] LIU Z, SONG L, ZHAO S, et al. Direct growth of graphene/hexagonal boron nitride stacked layers[J]. Nano Letters, 2011, 11(5): 2032-2037.

[178] YAN Z, PENG Z, SUN Z, et al. Growth of bilayer graphene on insulating substrates[J]. ACS Nano, 2011, 5(10): 8187-8192.

[179] TANG S, DING G, XIE X, et al. Nucleation and growth of single crystal graphene on hexagonal boron nitride[J]. Carbon, 2012, 50(1): 329-331.

[180] TANG S, WANG H, ZHANG Y, et al. Precisely aligned graphene grown on hexagonal boron nitride by catalyst free chemical vapor deposition[J]. Scientific Reports, 2013, 3: 2666.

[181] SONG X, GAO T, NIE Y, et al. Seed-assisted growth of single-crystalline patterned graphene domains on hexagonal boron nitride by chemical vapor deposition[J]. Nano Letters, 2016, 16(10): 6109-6116.

[182] BRITNELL L, GORBACHEV R V, JALIL R, et al. Field-effect tunneling transistor based on vertical graphene heterostructures[J]. Science, 2012, 335(6071): 947-950.

[183] DEAN C, YOUNG A F, WANG L, et al. Graphene based heterostructures[J]. Solid State Communications, 2012, 152(15): 1275-1282.

[184] PONOMARENKO L A, GORBACHEV R V, YU G L, et al. Cloning of Dirac fermions in graphene superlattices[J]. Nature, 2013, 497(7451): 594-597.

[185] DEAN C R, WANG L, MAHER P, et al. Hofstadter's butterfly and the fractal

quantum Hall effect in moiré superlattices[J]. Nature, 2013, 497(7451): 598-602.
[186] HUNT B, SANCHEZ-YAMAGISHI J D, YOUNG A F, et al. Massive Dirac fermions and Hofstadter butterfly in a van der Waals heterostructure[J]. Science, 2013, 340(6139): 1427-1430.
[187] YANG W, CHEN G, SHI Z, et al. Epitaxial growth of single-domain graphene on hexagonal boron nitride[J]. Nature Materials, 2013, 12(9): 792-797.
[188] ZHANG C, ZHAO S, JIN C, et al. Direct growth of large-area graphene and boron nitride heterostructures by a co-segregation method[J]. Nature Communications, 2015, 6(1): 1-8.
[189] WATANABE K, TANIGUCHI T, KURODA T, et al. Band-edge luminescence of deformed hexagonal boron nitride single crystals[J]. Diamond and Related Materials, 2006, 15(11/12): 1891-1893.
[190] WATANABE K, TANIGUCHI T, KURODA T, et al. Effects of deformation on band-edge luminescence of hexagonal boron nitride single crystals[J]. Applied Physics Letters, 2006, 89(14): 141902.
[191] WANG J, CAO S, SUN P, et al. Optical advantages of graphene on the boron nitride in visible and SW-NIR regions [J]. RSC advances, 2016, 6 (112): 111345-111349.
[192] YANG X, ZHAI F, HU H, et al. Far-field spectroscopy and near-field optical imaging of coupled plasmon-phonon polaritons in 2D van der Waals heterostructures [J]. Advanced Materials, 2016, 28(15): 2931-2938.
[193] PAKDEL A, ZHI C, BANDO Y, et al. Low-dimensional boron nitride nanomaterials[J]. Materials Today, 2012, 15(6): 256-265.
[194] MAYOROV A S, GORBACHEV R V, MOROZOV S V, et al. Micrometer-scale ballistic transport in encapsulated graphene at room temperature[J]. Nano Letters, 2011, 11(6): 2396-2399.
[195] XUE J, SANCHEZ-YAMAGISHI J, BULMASH D, et al. Scanning tunnelling microscopy and spectroscopy of ultra-flat graphene on hexagonal boron nitride[J]. Nature Materials, 2011, 10(4): 282-285.
[196] YANKOWITZ M, XUE J, CORMODE D, et al. Emergence of superlattice Dirac points in graphene on hexagonal boron nitride[J]. Nature Physics, 2012, 8(5): 382-386.
[197] HÜSER F, OLSEN T, THYGESEN K S. How dielectric screening in two-dimensional crystals affects the convergence of excited-state calculations: monolayer MoS_2 [J]. Physical Review B, 2013, 88(24): 245309.
[198] WANG E, LU X, DING S, et al. Gaps induced by inversion symmetry breaking and second-generation Dirac cones in graphene/hexagonal boron nitride [J]. Nature Physics, 2016, 12(12): 1111-1115.
[199] OKADA S, OSHIYAMA A. Magnetic ordering in hexagonally bonded sheets with first-row elements[J]. Physical Review Letters, 2001, 87(14): 146803.

[200] CHOI J, KIM Y H, CHANG K J, et al. Itinerant ferromagnetism in heterostructured C/BN nanotubes[J]. Physical Review B, 2003, 67(12): 125421.

[201] RAMASUBRAMANIAM A, NAVEH D, TOWE E. Tunable band gaps in bilayer transition-metal dichalcogenides[J]. Physical Review B, 2011, 84(20): 205325.

[202] BERSENEVA N, KRASHENINNIKOV A V, NIEMINEN R M. Mechanisms of postsynthesis doping of boron nitride nanostructures with carbon from first-principles simulations[J]. Physical Review Letters, 2011, 107(3): 035501.

[203] DING X, SUN H, XIE X, et al. Anomalous paramagnetism in graphene on hexagonal boron nitride substrates[J]. Physical Review B, 2011, 84(17): 174417.

[204] DEAN C R, YOUNG A F, MERIC I, et al. Boron nitride substrates for high-quality graphene electronics[J]. Nature Nanotechnology, 2010, 5(10): 722-726.

[205] LEE K H, SHIN H J, LEE J, et al. Large-scale synthesis of high-quality hexagonal boron nitride nanosheets for large-area graphene electronics[J]. Nano Letters, 2012, 12(2): 714-718.

[206] LEE J, HA T J, PARRISH K N, et al. High-performance current saturating graphene field-effect transistor with hexagonal boron nitride dielectric on flexible polymeric substrates[J]. IEEE electron device letters, 2013, 34(2): 172-174.

[207] IQBAL M W, IQBAL M Z, JIN X, et al. Superior characteristics of graphene field effect transistor enclosed by chemical-vapor-deposition-grown hexagonal boron nitride [J]. Journal of Materials Chemistry C, 2014, 2(37): 7776-7784.

[208] CHEN C C, LI Z, SHI L, et al. Thermal interface conductance across a graphene/hexagonal boron nitride heterojunction [J]. Applied Physics Letters, 2014, 104(8): 081908.

[209] CHEN C C, LI Z, SHI L, et al. Thermoelectric transport across graphene/hexagonal boron nitride/graphene heterostructures[J]. Nano Research, 2015, 8(2): 666-672.

[210] WITHERS F, DEL POZO-ZAMUDIO O, SCHWARZ S, et al. WSe_2 light-emitting tunneling transistors with enhanced brightness at room temperature [J]. Nano Letters, 2015, 15(12): 8223-8228.

[211] WANG X, XIA F. Van der Waals heterostructures: stacked 2D materials shed light [J]. Nature Materials, 2015, 14(3): 264-265.

[212] LI X, LIN S, LIN X, et al. Graphene/h-BN/GaAs sandwich diode as solar cell and photodetector[J]. Optics Express, 2016, 24(1): 134-145.

[213] XU Y, GUO Z, CHEN H, et al. In-plane and tunneling pressure sensors based on graphene/hexagonal boron nitride heterostructures [J]. Applied Physics Letters, 2011, 99(13): 133109.

Chapter 2

Electrical Properties and Recent Electrical Applications of Graphene, h-BN, Graphene/h-BN Heterostructures

2.1 Graphene

Graphene is a planar film composed of carbon atoms with sp^2 hybrid orbital, and a honeycomb-like lattice. It is a 2 dimensional(2D) material with only one atomic layer thickness; graphene is fullerene (0 dimensional, 0D), carbon nanotubes (1 dimensional, 1D), graphite (3 dimensional, 3D) of the basic constituent elements can be regarded as infinite aromatic molecules (Figure 2-1). Since graphene was discovered in 2004, by virtue of its unique mechanical, thermal, electrical, optical and other aspects of outstanding performances, it has become a scientific research focus[1-8].

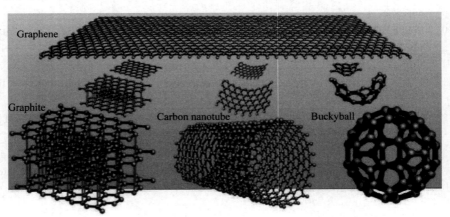

Figure 2-1 Graphene is the basic unit of various carbon compounds[4]

2.1.1 The structure of 2D graphene

Ideal graphene is a 2D honeycomb lattice structure composed of carbon six membered rings. This planar hexagonal lattice structure, and can be regarded as a layer of graphite exfoliated. Each carbon atom is sp^2 hybrid, and connected each other by strong bond to ensure the stability of material structure and excellent mechanical properties.

The structure of graphene is shown in Figure 2-2(a), a_1 and a_2 are the lattice vectors of the unit cell; δ_1, δ_2 and δ_3 are the mutually adjacent vectors; the distance of A and B is $a_{\text{C-C}} = 1.42$ Å. There are two points K and K' at the honeycomb edge corner in Figure 2-2(b), in which the Brillouin zone corresponds to the unit cell. The points are the Dirac points of graphene lattice, where b_1 and b_2 are the reciprocal-lattice vectors of the unit cell. The bond between electrons in graphene is composed of the σ bond of the sp^2 structure in the same plane and the π bond perpendicular to the plane[9].

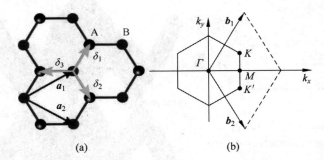

Figure 2-2 (a) Honeycomb structure of graphene[3]; (b) Brillouin zone corresponding to each graphene unit[3,9]

2.1.2 The electronic structure of graphene

Graphene is a zero-energy semiconductor. In the perfect graphene, the atoms are divided into two categories: A atoms and B atoms, as shown in Figure 2-2(a), carbon-carbon bond length is 1.42 Å, lattice constant is 2.460 Å, then the lattice can be written as: $a_1 = (1,0)$, $a_2 = \left(\dfrac{1}{2}, \dfrac{\sqrt{3}}{2}\right)$. The Fourier transform can be obtained in the inverted space of the inverted vector (Figure 2-2(b)): $b_1 = \dfrac{4\pi}{\sqrt{3}\,a}(1,0)$, $b_2 = \dfrac{4\pi}{\sqrt{3}\,a}\left(\dfrac{\sqrt{3}}{2}, -\dfrac{1}{2}\right)$.

Using the tightly constrained approximation method, we can describe the dispersion relation of these π electrons and the Hamiltonian are expressed as[2-3]:

$$\varepsilon^{\pm}(k_x, k_y) = \pm \gamma_0 \sqrt{1 + 4\cos\dfrac{\sqrt{3}\,k_x a}{2}\cos\dfrac{k_y a}{2} + 4\cos^2\dfrac{k_y a}{2}}, \quad (2\text{-}1)$$

$$\hat{H} = \hbar v_F \begin{pmatrix} 0 & k_x - ik_y \\ k_x + ik_y & 0 \end{pmatrix} = \hbar v_F \boldsymbol{\sigma} \cdot \boldsymbol{k} \quad (2\text{-}2)$$

The band structure of the graphene is shown in Figure 2-3.

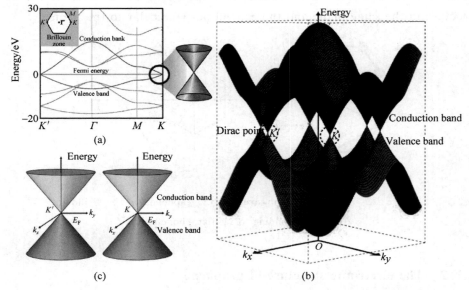

Figure 2-3 The band structure near the Fermi level of graphene
The 2D schematic diagram in (a)[9], the 3D schematic diagram in (b)[10], respectively; (c) is the Dirac cone of K and K', which is corresponding to the Fermi level of (b)[10]

2.1.3 The electronic property of graphene

There are three σ bonds in each lattice of graphene. The p orbitals of all carbon atoms are perpendicular to the hybridization plane of sp^2 and form a delocalized π bond in a shoulder-by-side fashion, which runs through the whole graphene[11-14], π electrons are free to move in the plane which give graphene the room temperature semi-integer quantum Hall effect, bipolar electric field effect, superconductivity, high carrier rate and excellent electrical properties because its good conductive unique structure. Its carrier mobility at room temperature can reach 15,000 cm^2/(V·s), carrier over the bandgap into the energy of the higher empty, empty band in the presence of electrons conduction band, the lack of an electron results in after the formation of a positively charged space, a hole. The electrons in the conduction band and the holes in the valence band are collectively referred to as electron-hole pairs, and the electrons and holes are free to move into free carriers. They produce directional motion under the action of external electric field to form macroscopic currents, which are electron conduction and hole conduction[15-31].

Each unit lattice of graphene has two carbon atoms that cause two equal equivalent tapered intersections (K and K') points in each Brillouin zone, and the energy at the intersection point is linearly related to the wave vector[3,10]. Thus, the effective mass of electrons and holes in graphene is zero, and all electrons and holes are called Dirac fermions. The intersection point is Dirac's point, the energy near it is zero, and the (band gap) of the graphene is zero. The unique carrier properties of the graphene and the massless Dirac-Fermi characteristics allow it to observe the Hall effect and the anomalous half-integer quantum Hall effect at room temperature. When the current flows perpendicularly to the outer magnetic field through the conductor, the potential difference between the two end faces is perpendicular to the magnetic field and the current direction, indicating its unique carrier properties and excellent electrical properties[32-48].

Klein paradox and chiral tunneling: According to the quantum mechanics point of view, regardless of whether the particle energy is higher than the barrier, there is a certain probability through the barrier, while the particles also have a certain probability pass the barrier. However, the electrons in the graphene are tunneled through the barrier at a probability of 100 %. In this phenomenon, the electrons of the graphene exhibit the properties of the massless particles and are not bounced or reflected by the presence of the barrier. Figure 2-4 shows the Klein tunneling effect of electrons in graphene.

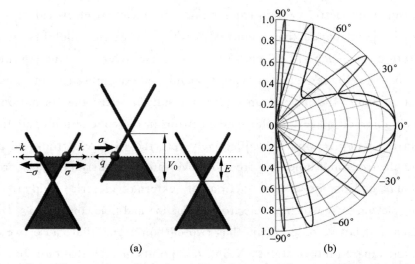

Figure 2-4 The spectrum of quasiparticles and tran smission probability of graphene
(a) Schematic diagrams of the spectrum of quasiparticles in single-layer graphene;
(b) Transmission probability T through a 100 nm-wide barrier as a function of the incident angle for single-layer graphene[27]

Ballistic transport: The electrons in graphene hardly encounter scattering during transport. Since the electrons in the graphene can move like particles of zero mass, the electrons propagate at any rate faster than any known conductor at room temperature. When the electrons of the graphene are moved in the orbit, they do not scatter due to lattice defects or foreign electrons, so the average free path of electrons in graphene can reach 1 μm, and the average free path of electrons in the general material is in nanometer level[32-33]. Figure 2-5 shows the ballistic transport of graphene.

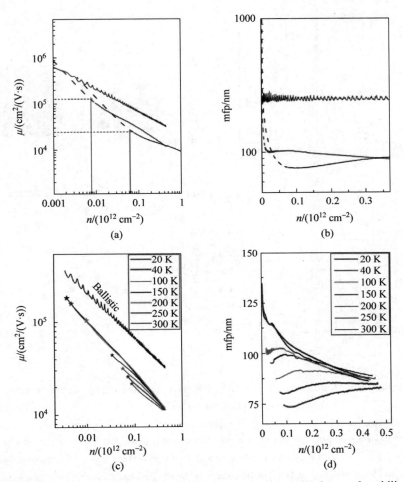

Figure 2-5 The ballistic predictions of carrier density dependence of mobility
(a) The ballistic predictions of carrier density dependence of mobility in NSG (black curve) and SG (red curve) at 100 K; (b) Ballistic simulation of carrier density with the change of mean free path in NSG and SG at 100 K; (c) Map of the density and mobility outside the puddle regime at different temperatures; (d) Map of the density and mean free path outside the puddle regime at different temperatures[32]

Ambipolar electric field effects: The conduction band and valence band of graphene have symmetry. There are two kinds of carriers of electrons and holes. Figure 2-6(a) shows the ambipolar electric field effects of single layer graphene[2].

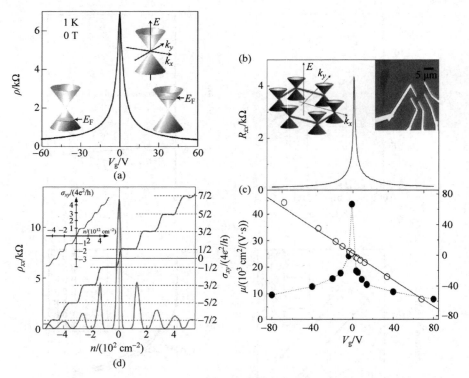

Figure 2-6   Ambipolar electric field effects, resistance, charge carrier density and mobility, and Hall conductance

(a) Ambipolar electric field effects of monolayer graphene[2]; (b) The change of resistance of graphene devices with gate voltage; (c) Charge carrier density and mobility of graphene as a function of gate voltage[23]; (d) Longitudinal resistance and Hall conductance versus carrier concentration. Illustration of the variation of Hall conductance with carrier concentration in bilayer graphene[22]

Minimum quantum conductivity and high mobility: In the vicinity of the Dirac point, although the carrier density is zero, graphene exhibits a small electrical conductivity, about 4 e^2/h. From 10 K to 100 K, the mobility is almost independent of temperature[22,47-48]. At room temperature and carrier density of 10^{12} cm^{-2}, due to the scattering of phonon scattering in graphene, the upper bound of the mobility is constrained to be 200,000 cm^2/(V·s), and the resistivity corresponding to this value is about 10^6 Ω·cm. Figures 2-6(b) and (c) show the resistivity ρ_{xx}, density n_s and carrier mobility μ as a function of gate voltage in the single layer graphene field-effect transistor[23].

Anomalous quantum Hall effects: The quantum Hall effect is usually found only in 2D conductors, and can be observed only in very clean Si or GaAs at very low temperatures and strong magnetic fields. However, graphene can present the quantum Hall effect on room temperature, and the anomalous quantum Hall effects to occur to low temperature (below 4 K), just as in Figure 2-6(d)[22,45]. The Hall conductivity at room temperature is $2\ e^2/h$, $6\ e^2/h$, $10\ e^2/h$, ⋯ for quantum conductance ($2\ e^2/h$) odd times, this behavior has been interpreted as *electron in graphene with relativistic quantum mechanics*, no static quality[45-46]. When the magnetic field is in the external, the quantum behavior of the conductivity of graphene exhibits anomalous characteristics[45]:

$$\sigma_{xy} = \pm (4e^2/h)(n+1/2) \quad (2\text{-}3)$$

The Landau level index N and the original difference in $1/2$, of which the dual energy valley and double spin are generated by the multiplication factor of 4[49-56].

2.1.4 The recent application of graphene in electronic property

Excellent electrical properties of graphene can be used in electronic transport devices, solar or lithium-ion batteries, supercapacitors and other nano electronics. Meanwhile the electrons in graphene are standard Dirac fermions, which makes graphene can be a good physical experimental platform for the study of quantum electrodynamics[57-82].

Graphene FET: The field effect transistor based on carbon nanotubes has been a hot spot on recent ten years[57-64]. Graphene has many advantages, its carrier mobility is much higher than the carbon nanotubes (2×10^4 cm^2/(V·s))[22-23,63], the very large critical current density (1×10^8 A/cm^2)[1], it does not need to assemble large parallel arrays of nanotubes to achieve high current. Since the first FET device is assembled based on graphene, the research craze has raised on electrical properties of graphene graphite in the academic circles[1,64-65].

Using a high-κ gate dielectric without the bandgap engineering, Meric et al. designed a top-gated graphene FET (GFET)[65]. In this device, they obtained current saturation and high transconductances, which are well-suited for analogue applications. Figure 2-7 shows the test results of the device.

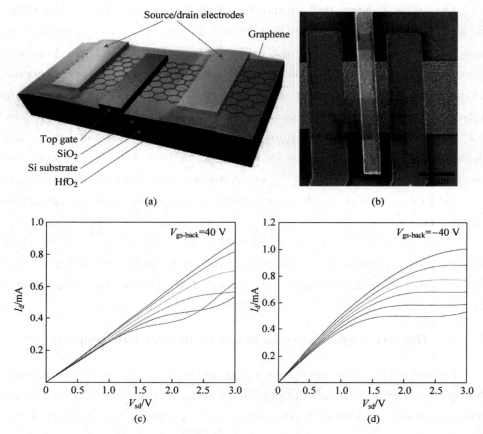

Figure 2-7 Graphene field-effect transistor and current-voltage characteristics
Schematic diagram of graphene field-effect transistor (a) and SEM micrograph (b). The current-voltage characteristics of graphene field-effect device for different $V_{\text{gs-top}}$ in (c) and (d)[65]

Logic inverters: Logic gates are part of the basic components of digital electronics, cascaded between them in order to achieve more complex logic functions[66-70].

Rizzi et al. designed the graphene complementary inverter with the same input and output logic voltage levels, so as to achieve cascade[71]. Figure 2-8(a) shows the schematic diagram of inverter, Figure 2-8(b) is the integration of two inverter for series connection. Figure 2-8(c) shows the waveforms for $V_{\text{DD}} = 2.5$ V and $f = 50$ kHz, which is belong to the input/output. Figures 2-8(d)~(f) show the DC characteristics of the inverter for ambient conditions.

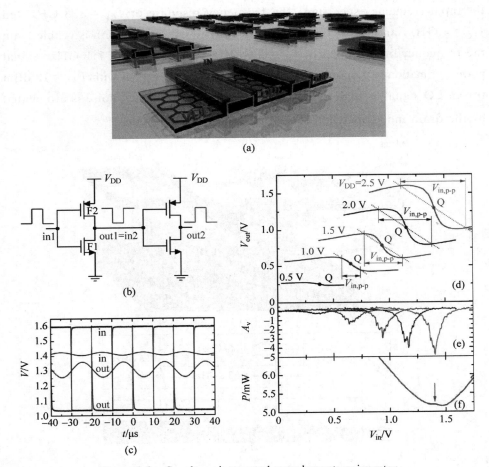

Figure 2-8 Graphene integrated complementary inverter
(a) Schematic diagram; (b) The circuit is made of two inverters; (c) The test results of graphene inverter; (d)~(f) DC characteristics of inverter with ambient conditions[71]

Graphene inverters can find their way of realistic applications where high-speed operation is desired, but power consumption is not a problem, similar to emitter-coupled logic.

RF frequency mixer and integrated circuit: The first graphene-monolithic-integrated circuit was reported on 2011[72]. The bipolar behavior of graphene based mixing has been successfully demonstrated, similar to the transistor's key applications including RF signal amplification[72-76].

Figure 2-9(a) shows the mixer to integrate circuit. Figure 2-9(b) shows the input frequency spectrum, the frequency response of $f_{RF} = 3.8$ GHz and $f_{LO} = 4$ GHz, and the drain bias with 2 V. The mixing function is visible from the frequency of two tones of $f_{IF} = 200$ MHz and $f_{RF} + f_{LO}$ of 7.8 GHz. Signal power estimation is proportional to that of the input RF signal with $P_{RF} = 12$ dBm for an LO signal as high as $P_{LO} = 20$ dBm. High frequency noise is attenuated by the drain inductance and therefore has a low amplitude[72].

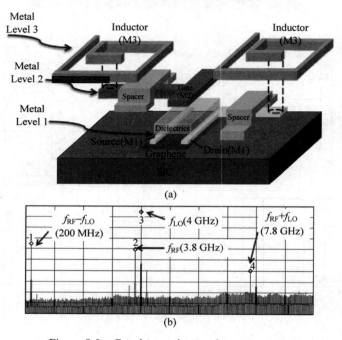

Figure 2-9 Graphene mixer and out spectrum
(a) Schematic diagram of graphene mixer for integrated circuit; (b) Out spectrum of mixer for different frequency[72]

Graphene ring oscillators: The ring oscillator is one of the most important circuits used to evaluate the performance limits of digital technology. However, the ring oscillator based on low-dimensional nanomaterials exhibits limited performance due to low current driving or parasitic phenomena[77].

Guerriero et al. made a ring oscillator by graphene[77], which is made by the method of chemical vapor deposition. The maximum oscillation frequency is 1.28 GHz, the maximum output voltage swing is 0.57 V. These two characteristics are still limited by the parasitic capacitance in the circuit, and

Chapter 2 Electrical Properties and Recent Electrical Applications of Graphene, h-BN, Graphene/h-BN Heterostructures

are not limited by the intrinsic properties of the graphene transistor assembly, the assembly of ring oscillators in any low-dimensional nanomaterials can now be achieved quickly and is the most sensitive to the power supply voltage fluctuations. Figure 2-10 shows the circuit diagram and test results in the device.

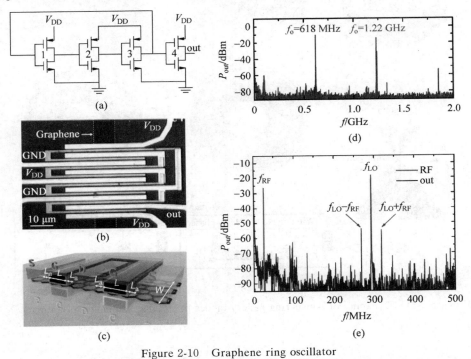

Figure 2-10 Graphene ring oscillator
(a) Circuit diagram, (b) microscope image and (c) schematic of a inverters. (d) Power spectrum of medium and small ring oscillators. (e) The power spectrum of graphene mixer of RF and out signals[77]

Circular graphene resonators: Joseph A. Stroscio et al. have found that the spectral characteristics induced by Berry phase is significantly increased from the ring-shaped graphene P-N junction resonator when a relatively small critical magnetic field is reached. This phenomenon is achieved by opening the π Berry phase associated with the topological properties of the Dirac fermions in graphene[78].

The Berry phase can be switched on/off by a very small magnetic field in the order of 10 mV · T, which can be used for a range of optoelectronic graphene equipment applications. Figures 2-11(a)~(e) show the test results of the graphene resonators.

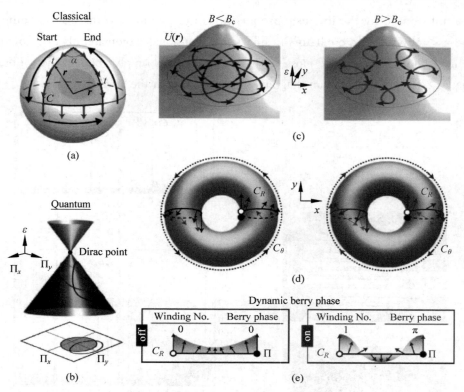

Figure 2-11　Dynamics of whispering gallery modes in circular graphene resonators[78]

2.2　h-BN

Boron nitride (BN) is a new type of wide-band-gap nano materials with excellent properties and potential applications (Figure 2-12)[82-87]. It is a typical Ⅲ-Ⅴ group compound, which is composed of nitrogen atoms and boron atoms, and the two are combined according to different hybridization modes to form BN with different phase structures[87-90]. The h-BN is the only one in the nature of the BN phase, belonging to the hexagonal system, "white"[91]. The h-BN material has a layered structure similar to that of graphite, which has the same crystal structure and parameters as graphene. Graphene is a zero-band-gap semiconductor, while h-BN single-layer material is a wide-band-gap semiconductor materials, so called *white graphene*[92-93].

Chapter 2 Electrical Properties and Recent Electrical Applications of Graphene, h-BN, Graphene/h-BN Heterostructures

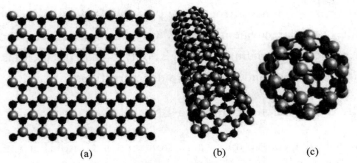

Figure 2-12 Boron nitride nanomaterials
(a) 2D h-BN nanosheet; (b) 1D nanotube; (c) 0D fullerene[83]

2.2.1 The structure of 2D h-BN

Each of the six BN atoms are an infinite extension of the hexagonal grid composed of B atoms and N atoms, which are arranged in the direction of ABAB in the c axis[92-95]. In each layer, B and N atoms are bonded by a strong covalent bond of sp^2, the bond length is $a = b = 0.2504$ nm (Figure 2-13); the adjacent layers are bonded by the weak van deer Waals force, the bond length is 0.6661 nm and the density is 2.28 g/cm^3. As a result, the bond strength of h-BN along the c axis is small, and the distance between the layers is easy to slipping[96-100].

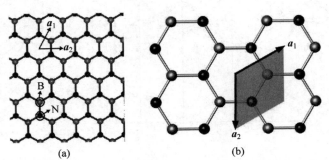

Figure 2-13 Honeycomb structure and unit cell
(a) The honeycomb structure of 2D h-BN with Bravais lattice vectors[94]; (b) Monolayer h-BN with highlighted unit cell[95]

In the inner layer of boron and nitrogen atoms of h-BN sp^2 hybrid, the formation of honeycomb structure, and the electronegativity of boron and nitrogen atoms is different, nitrogen atom charge distribution bias the angle of six-ring plane, a polar covalent bond is very strong; between the layers of atoms and no direct bonding, there is only weak interactions, such as electrostatic interactions and van der Waals interactions.

2.2.2 The electronic structure of h-BN

For h-BN, each cell contains a B atom and a N atom, a total of eight electrons, sp^2 hybridization between B and N atoms, each B (N) atom and the nearest three N (B) atoms form between the three σ bonds, there are six electrons filled to the σ orbit, the remaining two electrons in the $2p_z$ state, forming a vertical plane of the π bond. Figure 2-14 shows the band structures of h-BN with different states. From the figure, we can know that 2D h-BN is a direct band gap, the top of the valence band and the bottom of the rewind are at the high symmetry point K[94-95,101].

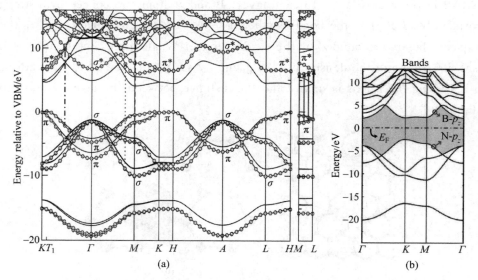

Figure 2-14　Band structure of bulk and monolayer h-BN
(a) The band structure of bulk h-BN[101]; (b) The band structure of monolayer h-BN[94]

In order to study the interaction between disordered and localized electrons, the Hamiltonian corresponding to h-BN[95] is as follows:

$$H = H_0 + \sum_{i a \sigma} V_{i\sigma}^a n_{i\sigma}^a + U \sum_{ia} n_{i\uparrow}^a n_{i\downarrow}^a, \qquad (2\text{-}4)$$

where, H_0 is the Hamiltonian of single particle, $V_{i\sigma}^a$ is disorder potential, $n_{i\sigma}^a$ is number operator, σ is spin indices, i is site, and α is orbital. They are three terms that describe the interaction of a single particle, a disorder, and an electron[95].

In Figure 2-14 (a), the calculation results in the LDA for the implementation and GW for the fine solid line and the open loop, in which the band structure is plotted along the high symmetry direction[101-104].

As can be seen from Figure 2-14(b), the calculated results of the energy band similar to that of the BN crystals are calculated as the h-BN[94]. The π band and π^* band of graphene intersect with the K and K^* points, and at this point, the energy band is opened in the BN as the bonding and anti-bonding combination of N-p_z and B-p_z orbitals. The contribution to the N-p_z orbital is the full band of the valence band edge. The calculated indirect band gap is 4.64 eV[94].

2.2.3 The electronic property of h-BN

Although the h-BN structure is similar to graphite, it is very different with graphene, because of its structure contains two kinds of B and N atoms, such as graphene is half metallic materials, while the h-BN is an insulator at room temperature, the resistivity of the two show anisotropic characteristics.

Indirect bandgap semiconductor: The h-BN is a kind of wide-band-gap semiconductor with high thermal stability and chemical stability. Over the past few years, h-BN has attracted much more attention as a result of the performance of the van der Waals structure, which is formed by the other 2D materials——h-BN. Even so, the band-gap properties for h-BN are rarely concerned[83-84].

Cassabois et al. made high-purity h-BN crystals, using optical experiments to show that the h-BN has a strong emission response to 215 nm, and further reveals the potential of h-BN as a deep UV radiation material[84]. The indirect band gap of 5.955 eV is proved by optical spectrum. The experimental results also show the existence of the phonon-assisted optical transition, and the binding energy of the exciton is about 130 meV. In the simplest case of phonon scattering, the MK phonon wave vector completely determines the momentum conservation, as shown in Figure 2-15(b), MK corresponds to the green arrow in the Brillouin zone; the vertical arrow reveal the position of the double fold line in the photoluminescence spectrum, which corresponds to 22 meV, 64 meV, 95 meV, 162 meV and 188 meV, respectively. There are the two-photon excitation spectrum of h-BN and at 5.86 eV as a function of twice the excitation energy.

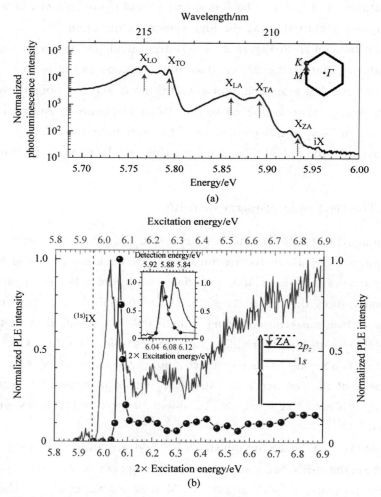

Figure 2-15 Phonon modes and two-photon excitation spectroscopy
(a) Identification of the phonon modes involved in the phonon-assisted recombination lines in h-BN[85]; (b) Two-photon excitation spectroscopy in h-BN[84]

Dielectric screening of 2D h-BN: The h-BN is an excellent dielectric substrate for graphene, molybdenum disulfide, and other 2D nanomaterials, optimizing the performance of these 2D materials. And the h-BN as the substrate material shows an outstanding performance of its dielectric shielding properties[105-107].

Li et al. produced the h-BN nanosheets using the method of Scotch tape exfoliation[108]. They verified the theory of dielectric shielding by experiment, and obtained the same results of the simulation experiment. They created 108 atomic models to simulate the multilayer BN system (Figure 2-16). At the same time, the distribution of the electric potential for the plane perpendicular to the plane of BN is calculated by introducing the external electric field at the $k_B T = 21$ meV electron temperature with the dispersion relation of Dirac-Fermi distribution[108]. The electric microscope and the interaction of the first-principles calculation with van der Waals, consider screening research of different thickness calculation and nonlinear Thomas-Fermi theory model of BN nanosheets medium[109-110].

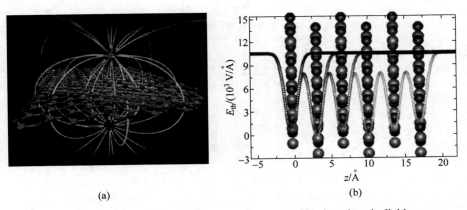

(a)　　　　　　　　　　　　　(b)

Figure 2-16　Dielectric screening effect and effective electric field
(a) Simulation results of dielectric screening effect of boron nitride nanosheets; (b) Effective electric field variation of boron nitride nanosheets with different layers[109]

Effect of uniaxial plane strain on the electronic properties: Figure 2-17 shows the change in band-gap width corresponding to the change in L_x and L_y[111]. From the figure, the spatial structural properties of h-BN were changed under uniaxial strain. Through the study of h-BN can always change in the system under different strain and electron localization function diagram, it was found that the addition of large strain in the direction perpendicular to the B-N bond, when $L_x \leqslant 0.3388$ nm, the system is in simple orthorhombic structure; when $L_x \geqslant 0.3488$ nm, the system is in a simple rectangular structure; when $L_x > 0.6$ nm finally tends to isolated BN chain. At the same time, the electronic structure and properties of the system and the isolated BN chain tend to be consistent. Put a big strain in the direction parallel to the B-N key system, change the chain structure of BN staggered side by side directly from the original simple orthorhombic structure, no rectangular structure, when $L_y > 0.571$ nm, the final system tends to the chain structure of BN isolated, the nature of electronic structure and properties of the system with isolated BN chains tend to be consistent[111-114].

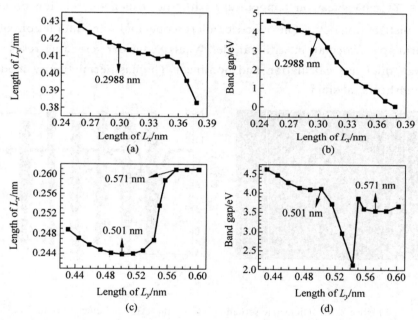

Figure 2-17 The change L_y and bandgap width as a function of L_x, and also the change L_x and bandgap width as a function of L_y
(a) The change in L_y as a function of L_x; (b) The change in bandgap width as a function of L_x for h-BN with system with asymmetrical strain distributions perpendicular to B-N bonds; (c) The change in L_x as a function of L_y; (d) The change in bandgap width as a function of L_y for h-BN with system with asymmetrical strain distributions perpendicular to B-N bonds[111]

The research results show that in plane strain on the action of six-dimensional (6D) BN nanosheets structure, electronic properties have rich change, and can be simple and effective on the nature of the modulation. It provides a theory of understanding and regulation according to the 6D angle of BN optoelectronic properties.

Effect of element doping on conductivity: The h-BN and graphene also belong to the sp^2 hybrid layered structure, graphene is a good conductor, and the h-BN has excellent insulation and high-breakdown electric field (35 kV/mm) because of its large band-gap width. In fact, the electrical properties of h-BN can be controlled by doping, which can be N type doping or P type doping[115-119].

At present, there are few reports about h-BN doping. The results show that in situ Mg and Zn doped h-BN are P type, and P type h-BN thin films with a mobility of 27 cm^2/(V · s) and can be obtained by rapid thermal annealing after Be ion implantation[118]. Figure 2-18 shows the curves of P-type resistivity change with temperature in h-BN doped with Mg[118]. From the figure, we can see a sample of magnesium-doped BN at a temperature of 300 K, the resistivity value of 12 Ω and a comparison of the E_A estimates based on resistivity versus temperature. The E_A value is lower than the measured value of 150~300 meV[111,119-121], and the Hall-effect shows that the free-hole concentration of p is about 1.1×10^{18} cm^3 and the carrier mobility μ is about 0.5 cm^2/(V · s).

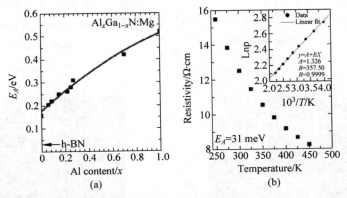

Figure 2-18 The curves of p-type resistivity change with temperature in h-BN doped with Mg in (a) and (b)[118]

Table 2-1 summarizes the experiments and examples of the influence of doping on the electronic properties of 2D BN in recent years[115-119].

Table 2-1 The experiments and examples of the influence of doping on the electronic properties

Doping element	Doping concentration/ (10^{19} cm^{-3})	Doping mode	Activation energy /meV	Resistivity/ $\Omega \cdot$ cm	Mobility/ (cm^2/(V·s))	References
Be	3.00	Ion implantation	210	43000	27	[115]
Mg	0.189	In-situ doping	300	0.128	35.2	[116-117]
	0.163			0.132	29.1	
Zn	1.00	In-situ doping	31	12	0.5	[118]

Insulated gate dielectric: The surface of the six BN sheet is smooth, the lattice structure is similar to that of graphite, and has a large optical phonon mode, a wide band gap, no dangling bonds on the surface and electron trapping[122-127].

Figures 2-19(a) and (b) show the I-V characteristic curves corresponding to the device[128] made of graphite-BN-graphite and Au-BN-Au. As can be seen

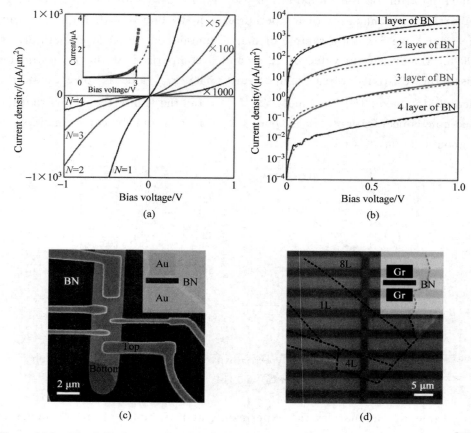

Figure 2-19 The I-V characteristic curves (a) of device (c) and (b) correspond to (d)[128]

from the figure, the *I-V* characteristic curves of the two devices have a linear relationship near the zero-bias voltage, and the *I-V* characteristic curve depends on the higher bias voltage. The results show that the atomic thickness of h-BN as a high-breakdown electric field has no defect in the medium, revealing the tunneling current characteristics of h-BN with different thicknesses. The results of these tests show that h-BN has great potential as insulating layer and dielectric layer of high-carrier FET.

2.2.4 The recent electrical application of h-BN

The dielectric properties of BN are very significant in the application of electrical energy as the insulator of wide band gap. In addition, the physical and chemical properties of h-BN with the thickness of the atomic layer are relatively stable, which makes it very useful in the preparation of nano electronic devices.

The h-BN as substrate material: Recently, h-BN has been regarded as an excellent substrate material and tunneling dielectric. The focused study is mainly due to the fact that h-BN has an atomically flat surface, and there are no dangling bonds or charge traps on the surface. On h-BN substrate, graphene/h-BN heterojunction structures exhibit high mobility, and almost no intrinsic doping, the switch has significantly improved than the traditional graphene/SiO_2 substrate. In addition, due to the large effect of the substrate and graphene, graphene/h-BN heterostructure can also be used as a material to study the intrinsic mechanical properties of graphene[129-133].

Lee et al. used CVD to prepare h-BN nanosheets and used them as substrates to prepare thin-film transistors with graphene[135]. As shown in Figure 2-20, in the presence of h-BN, the electron mobility and drain current switching ratios of the transistors are three times (573 cm^2/V at $-2 \times 10^{11}/cm^2$) and h-BN times higher than those without the substrate.

Counter electrode in dye-sensitized solar cells: The lattice of 2D h-BN has the same lattice structure as graphite, and has been used as a counter electrode for dye-sensitized solar cell (DSC) catalytic materials.

Xu et al. used X-ray diffraction (XRD) to confirm the crystal structure of h-BN, using scanning electron microscopy (SEM) and electrochemical characterization of BN film counter electrode the electrochemical impedance

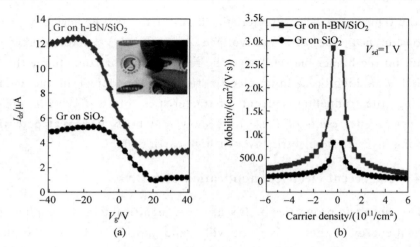

Figure 2-20 The I_{ds}-V_g characteristic curves and relationship between electron mobility and electron carrier density

(a) The I_{ds}-V_g characteristic curves of graphene TFTs on h-BN nanosheets and on SiO_2. Illustration is a schematic diagram of TFTs. (b) Relationship between electron mobility and electron carrier density in graphene[134]

spectroscopy (EIS) and the corresponding impedance parameters[135].

The high-charge transfer resistance (R_{ct}) of the h-BN electrode induces the low efficiency (η) of BN based on the dye-sensitized solar cell electrode. This means that it is difficult to be reduced by three iodide ions (I_3^-) to catalytic activity. Figure 2-21 shows the result of the test[135]. We can see that the R_{ct} of h-BN counter electrode is higher than that of the graphite.

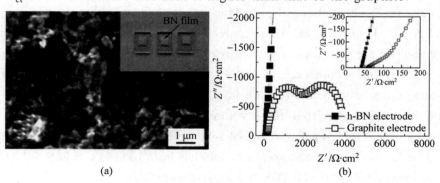

Figure 2-21 SEM picture and Nyquist plots

(a) The SEM picture of h-BN. The electrode of h-BN is shown in insert. (b) Nyquist plots of graphite and h-BN. The insert shows the expanded range of the ordinate and abscissa in high frequency region[135]

MSM detector of h-BN: The h-BN has excellent physical properties of high-temperature stability and corrosion resistance, as well as large optical absorption and neutron capture interface, large negative electron affinity. The layered structure and lattice constants is similar with graphene, high thermal conductivity and band-gap plane, so these properties is known: the BN graphene electronic and optoelectronic ideal template, the gate dielectric layer, the electrical insulation and the thermal conductor[136-138].

Figure 2-22(a) shows the band-gap edge fluorescence emission spectra of the BN epitaxial layer at temperature 10 K. The shape and configuration of the photoluminescence emission spectral line along the crystallographic x-axis ($E_{emi}/\!/c$) are similar to the shape and configuration of the vertical x-axis ($E_{emi}\perp c$)[136]. Figure 2-22(b) shows the relative spectral response of the h-BN MSM photodetector at different bias voltages[138]. Illustration is an example of the h-BN MSM photodetector. It can be seen from the line that the photodetector has a sharp peak at 220 nm, and there is a sharp cutoff wavelength around 230 nm, which corresponds to the PL of the edge band at 5.48 eV (or 227 nm). Figure 2-22(c) is the I-V characteristic curve of MSM photodetector based on h-BN thin film[137]. The MSM photodetector shows a weak dark current of $\sim 10^{-10}$ A/cm^2 at a bias voltage of 100 V. Figure 2-22(d) shows the c-direction of the h-BN epitaxial layer (out-of-plane) of the I-V characteristic curve, which shows the breakdown occurs around the 810 V[137]. This implies that the $E_B \sim 4.5$ MV/cm is significantly lower than that of the ultrathin layer of micron scale cross-section separated from the powder crystal h-BN.

The FETs based on h-BN: The h-BN acts as a very important role in packaging other 2D materials.

Chuang et al. used a variety of transition metal sulfides and BN to assemble a 2D low-resistance ohmic contact, they used the van der Waals component to replace the doped transition metal sulfides as source/drain contacts[139-144]. Figure 2-23(a) shows the assembly structure diagram of TET, which are made of WSe$_2$ and h-BN with 2D/2D contacts. From Figure 2-23(b) we can see that h-BN is used at the top and the bottom of the device,

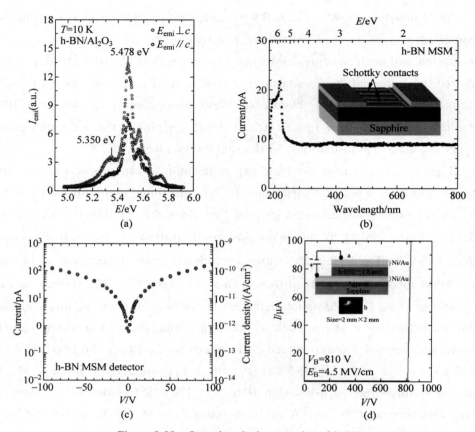

Figure 2-22 Optoelectrical propetries of h-BN
(a) The polarization-resolved band-edge PL spectra of h-BN at 10 K[136]; (b) The relative spectral responses of h-BN MSM photodetectors measured at a bias of 30 V[138]; (c) The I-V characteristic of h-BN MSM photodetector epilayers[137]; (d) Out-of-plane (vertical) I-V characteristics of a released h-BN epilayer, which has a cross section area of about 4 mm^2[137]

respectively. For WSe$_2$, h-BN encapsulates the 2D materials, just as dielectric layer. From Figures 2-23(c) and (d), the FETs show low contact resistances about 0.3 kΩ·μm, high on/off ratios above 10^9, and drive currents above 320 μA/μm[144].

The low ohmic contact of 2D/2D represents a new device, which overcomes the bottleneck of TMDS preparation, and also can be used for the preparation of various 2D materials.

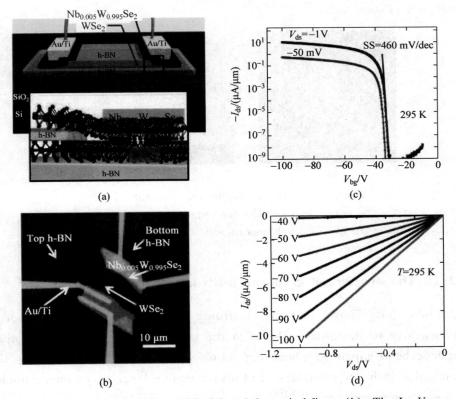

Figure 2-23　The scheme of WSe$_2$ FETs (a) and the optical figure (b); The I_{ds}-V_{bg} characteristics curve (c) and the I_{ds}-V_{ds} characteristics (d) of the device[144]

2.3　Graphene/h-BN heterostructures

　　In order to improve the mobility of graphene and graphene band-gap opening angle for scientific workers to study the interaction between substrate and graphite and graphene, different substrates are proposed: SiO$_2$[145], Al$_2$O$_3$[146], h-BN[147], SiC[148], etc. In all of the substrate materials, BN has an atomically flat surface, the surface roughness is very low, no dangling bonds, and graphene is van der Waals force weak, on the transport properties of graphene carrier transport with minimal impact, plus it with the graphene lattice mismatch is only 1.7%, no the doping effect on graphene[82-87], making six-BN graphene device substrate become one of the powerful candidates (Figure 2-24)[134,147,149-150].

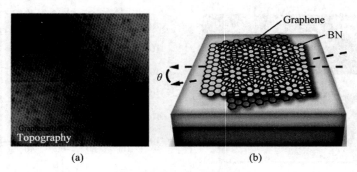

(a)　　　　　　　　　　　(b)

Figure 2-24 STM topography and mismatch angle θ
(a) The STM topography of Graphene/h-BN with 60nm×60nm size[149]; (b) The scheme of graphene/h-BN with mismatch angle θ[150]

2.3.1　The structure of graphene/h-BN heterostructures

Studies have shown that the electronic structure of graphene bilayer is very sensitive to the distance between the layers, and in the study of open graphene band gap in graphene deposition we can open a small gap in the Dirac point in h-BN substrates, and also a double layer of graphene higher than electron transfer rate[151-164].

We have introduced the C—C bond length a_{C-C} = 0.246 nm in 2D graphene lattice, the bond length of B-N is a_{B-N} = 0.2504 nm in h-BN crystal lattice, it is because of the difference 1.7% in length of two kinds of 2D crystal honeycomb lattices, always appear in the moiré pattern in the heterostructure[156]. The results show that the distance between the two kinds of crystal structures is about d_{g-h-BN} = 0.322 nm[163].

From Figure 2-25 (a), the B and N atoms in BN have different correspondence with C atoms in graphene. The main correspondence is shown in Figure 2-26. The configurations I and II are Bernal arrangement (AB and AB′), and the C atom of a sublattice is just above the center of the h-BN, and another C atom of the sublattice is located above the B and N atoms. The configuration III is AA type, that is, all the C atoms of graphene and B and N atoms of h-BN are strict one-to-one correspondence. In addition to these three ways, there are other irregular correspondence[156-164].

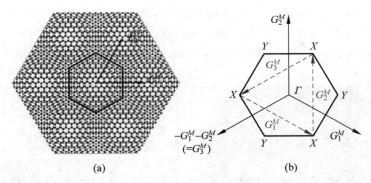

Figure 2-25 The mismatch angle and Brillouin zone of graphene/h-BN superlattice (a) The graphene/h-BN heterostructures with mismatch angle $\theta = 0$. The red hexagon is shown unit cell. (b) The Brillouin zone of graphene/h-BN superlattice with reciprocal lattice vectors G_i^M [156].

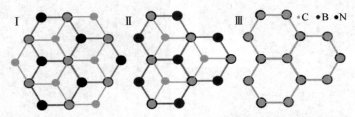

Figure 2-26 The corresponding configuration I, II, III of C atom and B, N atom in graphene/h-BN superlattice[161]

The 3D atomic structure of a suspended graphene/h-BN van der Waals heterostructure: Recently, the researchers studied the highly oriented graphene films floating on monolayer h-BN by transmission electron microscopy (TEM) and scanning transmission electron microscopy (STEM), and developed a method capable of recording electron beams under scanning transmission electron microscopy scattering direction detection method. This method is extremely sensitive to locally deposited atoms[165].

By comparing the experimental data with the model simulation values, they found that the heterostructure was significantly curved in the direction perpendicular to the surface of the sample, forming a wave-like structure consistent with the moiré wavelength. The formation of this wave-like structure is due to the interaction between the layers and the layers, and the structure makes the strain in the layer also change (Figure 2-27). We can see that the 3D atomic structure of the graphene/h-BN van der Waals heterostructure, shows, the self-warping of the heterogeneous structure in the van der Waals heterogeneous structure, which is considered to be smooth (Figure 2-28)[165].

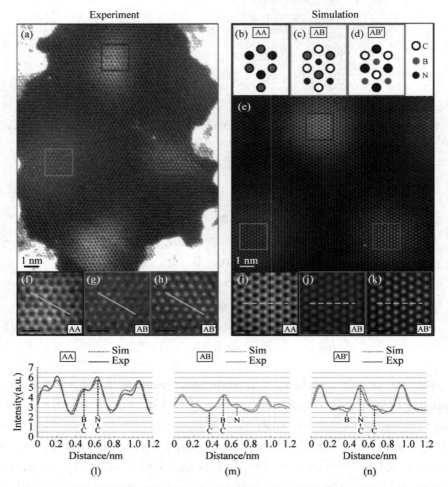

Figure 2-27 The highly oriented graphene films floating on monolayer hexagonal boron nitride

(a) A heterogeneous structure of the atomic resolution of the MAADF image. The angle scattering contrast makes the high-symmetry region brighter. The structural model of the uppermost high-symmetry region is shown in Figures (b)~(d). (e) A scanning transmission electron microscope (MAADF) simulation of the region of interest of the heterogeneous structure. and (e) a scanning electron microscopy (TEM). (l)~(n) in the graph (f)~(k) along the yellow line area experiment (solid line) and simulation (dotted line) gray value distribution. The scale of (f)~(k) is 0.5 nm[165]

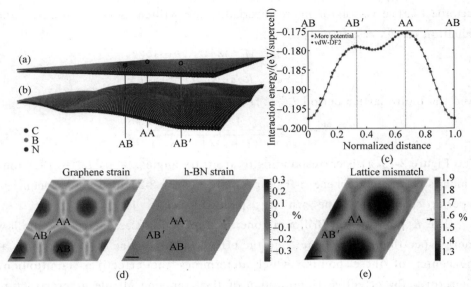

Figure 2-28 physical properties of graphene/hexagonal boron nitride heter ostructures (a) Rigid structural model of the graphene/h-BN bilayer before complete relaxation. (b) Structural model of graphene/h-BN bilayer after complete relaxation. The relaxation model has a visible twist in the vertical sample plane. (c) The interlayer interaction energy curve for each super monomer (four carbons, two borons, and two nitrogen atoms). Among them, the blue dot shows the value of the different stack structure calculated by the density functional theory, the red dot indicates the Maussian potential energy, and its parameters are obtained by fitting the numerical point of the density functional theory. (d) in-plane strain maps of graphene (left) and h-BN (right). (e) Lattice mismatch. The black arrow next to the color bar indicates the initial crystal mismatch of the two crystals before relaxation

The 3D atomic structure of the heterostructures of the suspended graphene/h-BN van der Waals was revealed by projection electron microscopy and scanning electron microscopy. It was found that there were lattice mismatch and stack orientation difference in the van der Waals heterogeneous structure, and the van der Waals interlayer force determines the strain in the uniaxial plane and the deformation of the vertical monolayer direction.

2.3.2 The electronic structure of graphene/h-BN heterostructures

The red regular hexagon region in Figure 2-25(a) is defined as a basic unit of the superlattice, and the center point coordinates are $O(0,0)$, $\boldsymbol{a}_1 = a(1,0)$ and $\boldsymbol{a}_2 = a(1/2, \sqrt{3}/2)$ is defined as the lattice vector of graphene, h-BN lattice vector $\tilde{\boldsymbol{a}}_i = \boldsymbol{MR}\boldsymbol{a}_i$, $(i = 1, 2)$. $\boldsymbol{R}$ is rotation matrix by θ, $\boldsymbol{M} = (1 + \varepsilon)\boldsymbol{1}$ is isotropic expansion, $(1 + \varepsilon) = a_{\text{h-BN}}/a \approx 1.018$, so the primitive lattice vector and reciprocal lattice vectors of moiré superlattice are written as Equation (2-5) and Equation (2-6), respectively[156]:

$$L_i^M = (1 - \mathbf{R}^{-1}\mathbf{M}^{-1})\mathbf{a}_i, \quad (i = 1, 2), \tag{2-5}$$

$$G_i^M = (1 - \mathbf{M}^{-1}\mathbf{R})\mathbf{a}_i^*, \quad (i = 1, 2). \tag{2-6}$$

And the moire lattice period $L^M = |L_1^M| = |L_2^M|^{[158]}$, and

$$L^M = \frac{1+\varepsilon}{\sqrt{\varepsilon^2 + 2(1+\varepsilon)(1-\cos\theta)}} a. \tag{2-7}$$

The Figure 2-29(a) is corresponding to arbitrary angle $\theta = 0$, $L^M = 13.8$ nm, which is belong to graphene/h-BN superlattice. Figure 2-25(b) is corresponding to superlattice Brillouin zone, which is spanned by G_i^M.

The K point of the Brillouin zone in graphene lattice determines the low energy spectrum of graphene, and the effective Dirac cone of h-BN resides at the center of these points, which determines the effective Hamiltonian. Therefore, the effective Hamiltonian of the graphene nitride heterostructure is the 2×2 matrix form of the Hamiltonian and the BN hamiltonian. So, the Hamiltonian of graphene/h-BN superlattice near K point is shown as[156]:

$$H_{\text{graphene/h-BN}} = \begin{pmatrix} H_{\text{graphene}} & U^\dagger \\ U & H_{\text{h-BN}} \end{pmatrix}. \tag{2-8}$$

Figures 2-29(a) and (b) are the band structures of graphene/h-BN system, which are calculated by the tight-binding model and effective continuum mode with arbitrary angle $\theta = 0$. Figure 2-29(c) is the first and second electron and hole bands of K valley calculated by continuum model. Figure 2-29(d) is superposition of the cone dispersion energy at K point in graphene/h-BN superlattice, Figure 2-29(e) is the 3D band structure of graphene/h-BN heterostructure, respectively[156].

In the different stacking modes of graphene and BN, different configurations have different equilibrium positions and interlayer energy. In the case of the same interlayer spacing, the stability of the configurations is as follows: configuration Ⅰ > configuration Ⅱ > configuration Ⅲ[163]. This is due to the interaction between the π electrons of graphene and BN single layer cations and the mutual exclusion of anions. Just above the N anion closer to the graphene six-lattice center, the position of π electron density was the lowest, while the B cation is above love graphene six-lattice center so as to enhance the attraction effect.

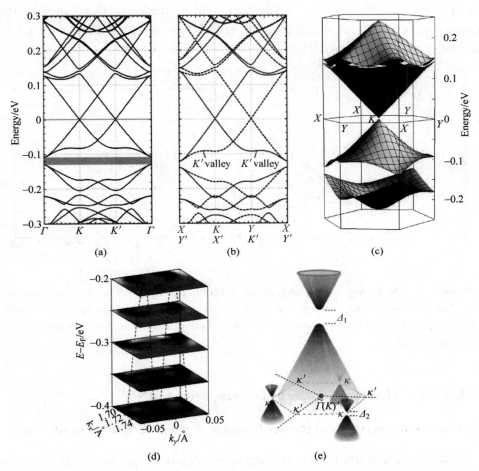

Figure 2-29 The first and second electron and hole bands of K valley

(a) and (b) are the band structure of graphene/h-BN heterostructure, which calculated by tight-binding modele and effective continuum model[158]; (c) The energy dispersion of graphene/h-BN superlattice of K-valley[158]; (d) Angle resolved photoemission spectroscopy for direct measurement of second stage Dirac cone[168]; (e) Schematic diagram of the energy band of the original Dirac cone and the second stage of the Dirac cone[168]

Heterostructural band changes with distance: Figure 2-30(a) is an image of the variation of the distance between the layers and the binding energy with the distance[165]. As can be seen from the figure, AB stack corresponds to the most stable. Figure 2-30(a) is an image of the variation of the distance between the layers and the binding energy with the distance. And from the figure, AB stack corresponds to the most stable configuration, at this time the layer spacing $d = 3.30$ Å.[163] In other studies, the calculated values of the binding energies of graphene and BN by the van der wa function were also observed (Figure 2-30(b))[159-164].

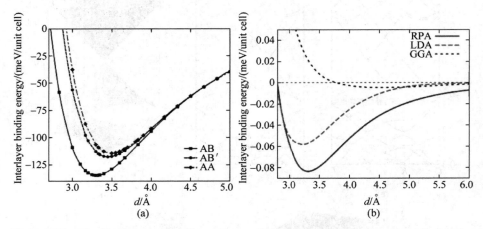

Figure 2-30 The binding energies of graphene and boron nitride by the van der walls function (a) The variation of binding energy of different structures of graphene nitride boron nitride with interlayer spacing[163]; (b) The relationship between band energy and interlayer spacing calculated by different methods[167]

2.3.3 The electronic properties of graphene/h-BN

1. Vibrational properties and phonon dispersion of graphene/h-BN heterostructure

Graphene and BN form a heterogeneous structure, the lattice dynamics of the heterostructure, especially the low frequency plane mode, which is influenced by the adjacent layers. Moreover, it can be seen from many studies that the phonon dispersion spectra of heterostructures are obviously different from those of the graphene and BN[11,83,163,168].

Figure 2-31 (a) is the graphene/h-BN phonon dispersion spectrum calculated by DFPT method, the model is a single-layer graphene and single-layer h-BN AA stack. Due to the mismatch between the graphene and the h-BN lattices, the phonon dispersion mode of the strained graphene/h-BN heterostructure is obtained after compensating for the lattice deformation caused by the strain. BN plane phonon branch K_6 acoustic mode (ZA) moved to higher-frequency position, infrasound learning mode can be assigned to the peak in 47 mV (P_3). And the acoustic Γ_{6-} mode (ZA) of graphene plane moved to the frequency position of 36 meV (P_2). Graphene phonon M_{2+} (ZO) becomes strong, moving to 86 meV, close to the 86 meV position (P_5). The LA-phonon branch of graphene becomes weaker, which makes the K_5 phonon close to 140 meV (P_8)[168].

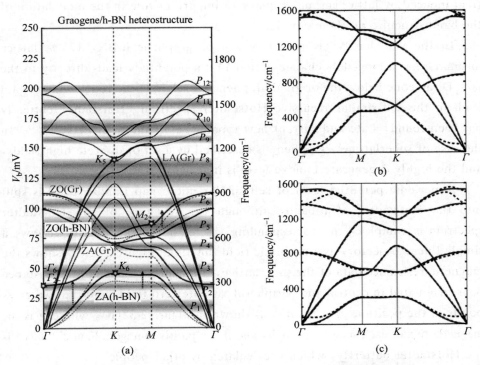

Figure 2-31　Graphene/h-BN phonon dispersion spectrum
(a) The phonon dispersion of graphene/h-BN heterostructure[168], (b) monolayer graphene[163], (c) monolayer h-BN[163]

On the other hand, Figure 2-31 shows the phonon dispersion relation of a single-layer graphene (Figure 2-31(b)), a single-layer BN (Figure 2-31(c)) and graphene/h-BN heterostructure (Figure 2-31(a)). It can be assumed that the two components of the atlas (b) and (c) coincide into a map (a). Each model is equivalent to the acoustic modes around the localized Γ points. These patterns are almost flat at M and K points. It is noted that the lowest two BN bifurcation with the z direction polarization caused by the non-perturbative anharmonic effect is called the Fermi resonance. This also leads to a new phenomenon, that is, the energy transfer between the upper branches of the split and the low-frequency dissipation and the two modes.

2. Hofstadter's butterfly

In the van der Waals heterostructure, the superlattice periodic potential field induced by lattice mismatch plays an important role in the modulation of the band structure of 2D materials.

In the van der Waals model system of graphene/h-BN, 1.7% lattice mismatch has a long-slow change period 14 nm potential, leads directly to the new Dirac cone below the original graphene (level 2) Dirac cone, which again leads to the self-similar recursive Hofstadter spectrum (Hofstadter butterfly has been found) state and current new topological quantum effects[169-173]. In the case of superlattices, graphene is described by a set of discrete high fields, and the highly degenerate Landau level is indexed by an integer N.

A periodic potential level of self-similar micro gap Landau band is split into the Hofstadter Mini-band, although the structure of the Hofstadter spectrum is complex, the corresponding fractal micro-gap density follows a simple linear trajectory as a magnetic field function. Figure 2-32(a) shows the magnetic resistance data of the superlattice, it includes a Landau fan resistance peak originated in central and peripheral satellite zero field, and the two cross points in the position ($\phi = \phi_0/q$), as shown in Figure 2-32(b), where ϕ is the magnetic flux of each superlattice unit, and q is a positive integer. Figure 2-32(c) is the Hofstadter butterfly, which is calculated by other people.

Different researchers designed both the experimental device and test Hofstadter butterfly spectrum differences, but can produce consistent mechanism of spectrum, namely the graphene/h-BN heterostructure lattice layer between the super spatial variation results in moiré pattern superlattice, broking the symmetry of the C lattice. The appearance of the Landau energy level is also one of the reasons why Hofstadter is observed in the high field.

3. The fractal quantum Hall effect

Hofstadter spectroscopy provides a unique tunable system and the basis for the study of topological order in strong interactions. In the previous study, only the Landau energy level was used to fill the low Bloch band. It is found

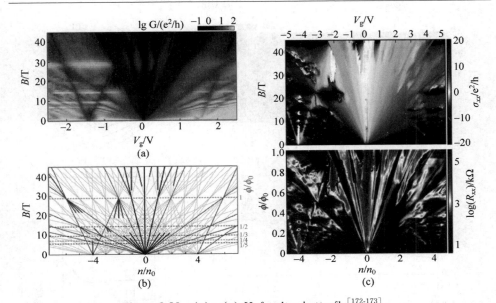

Figure 2-32 (a)~(c) Hofstadter butterfly[172-173]
(a) The magnetic resistance data of the superlattice; (b) The two cross points in the position ($\phi = \phi_0/q$); (c) Hofstadter butterfly

that the coexistence of the traditional fractional quantum Hall effect and the quantum Hall effect is related to the energy spectrum of the branch. Especially in the large magnetic field, a series of other states appear in the fractional Bloch filling index. These Bloch band quantum Hall effects can be calculated and observed in the experiment.

Figure 2-33(a) is a spectrum of the state of the quantum Hall effect that can be applied to a finite vertical magnetic field. More details are shown in the diagram, which corresponds to the region of the longitudinal resistance selected in Figure 2-32(c), where the longitudinal resistance is in opposition to the magnetic field on the vertical axis, and the Landau filling fraction is along the horizontal axis[172]. From this figure, by choosing the locus of the spectral line, corresponding to the different filling fraction in the constant magnetic field, as shown in Figure 2-33(c)[172], according to the longitudinal resistance and Hall resistance, they measured the Hall conductance. Figure 2-33(b) is the Landau levels of graphene/h-BN[174].

4. As a tunable hyperbolic metamaterial

The h-BN, as a kind of natural hyperbolic material, has the same

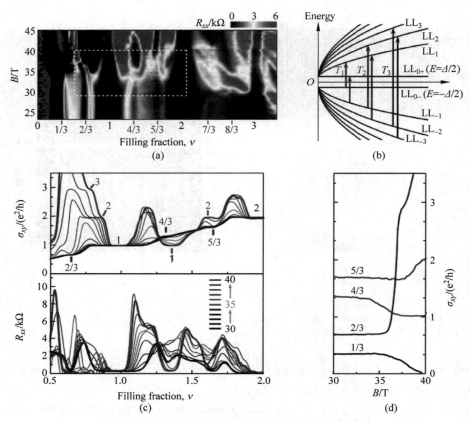

Figure 2-33 Electric properties of graphene/h-BN
(a) High field area data maps of longitudinal resistance and Landau filling fraction corresponding to figure 32(c)[172]; (b) The landau levels of graphene/h-BN superlattice[174]; (c) Hall conductivity (top) and longitudinal resistance (bottom) corresponding to horizontal line cuts within the dashed region in (a) with different B[172]; (d) The relationship between the Hall conductivity and the magnetic field in the fixed filling fraction[172]

dielectric constant in the direction of the basal plane ($\varepsilon^t = \varepsilon^x = \varepsilon^y$) and the opposite signs in the normalplane direction ($\varepsilon^t \varepsilon^z < 0$). Based on the characteristics of the whole, the BN has a limited planar structure as the propagation mode of the multimode waveguide, which is originated from the coupling of photons and phonons. However, the BN lattice determines the electrical properties of the material, so that the control of these hyperbolic phonon polariton modes still faces challenges[175-181].

The recent success of nano-infrared imaging revealed when the graphene/h-BN heterostructure formation, can effectively control the hyperbolic model, so the eigen mode heterostructure is hyperbolic plasma-phonon polaritons[182-184]. Due to the low ohmic loss of the hyperbolic plasmon-phonon polaritons in the graphene/h-BN heterostructure, the propagation length is about 1.5~2 times higher than that of the hyperbolic phonon polaritons in h-BN (Figure 2-34 is the property of hyperbolic materials based on graphene/h-BN)[185]. Therefore, the hyperbolic plasmon-phonon polaritons has the advantages of the combination of the surface plasmon polaritons and the h-BN hyperbolic phonon polaritons. Therefore, heterostructure the materials of graphene/h-BN can be classified as electromagnetic metamaterials that are not present in the properties of their constituent elements.

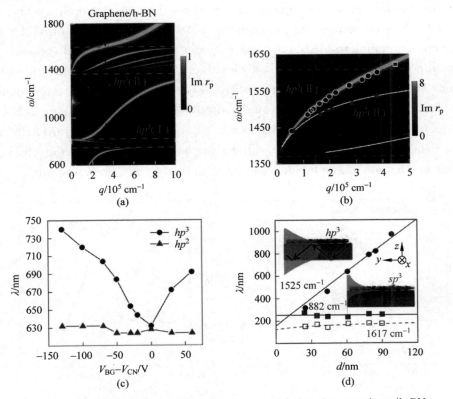

Figure 2-34 The property of hyperbolic materials based on graphene/h-BN (a) Calculated dispersion of graphene/h-BN superlattice, which is corresponding to hyperbolic plasmon-phonon polarions wih $E_F = 0.37$ eV. (The line hp^3 belongs to hyperbolic plasmon-phonon polaritons, the line sp^3 belongs to surface plasmon-phonon polaritons.) (b) Experimental dispersion relation of type II HP^2 in h-BN (red triangles), hp^3 (blue circles) and sp^3 in graphene/h-BN (pink square) with a Fermi energy of $E_F = 0.37$ eV. (c)~(d) Tuning of the graphene/h-BN polariton wavelength by electrostatic gating and varying the meta-structure thickness[185]

5. Field emission characteristics

In the 2.2.4 section point 1 of this book, BN has been used as the base material of graphene, and the carrier mobility of graphene has been improved a lot. Many researchers have tried to make use of the high electron mobility of graphene/h-BN heterostructures as electron emission materials[134,186-187].

Yamada et al. prepared graphene/h-BN/Si and h-BN/Si heterostructures, and also calculated the field emission characteristics of them, in order to study the characteristics of graphene/h-BN heterostructure[187]. Figure 2-35(a) shows the variation of the emission current with threshold voltage of graphene/h-BN heterostructure. From the figure, we can know that the threshold voltages at 1.6×10^3 V is corresponding to graphene/h-BN heterostructure, and 2.1×10^3 V is corresponding to h-BN. Obviously, under the same threshold voltage, the emission current of the former is higher than that of the latter. On the other hand, they also investigated the electronic structure using UPS and Fowler-Nordheim plot to discuss the property of the field emission. Figure 2-35(b) shows the Fowler-Nordheim plot, which is obtained from the Figure 2-35(a). The two graphs are in the low voltage region, indicating that the observed electron emission characteristics can be explained by the tunneling line[187].

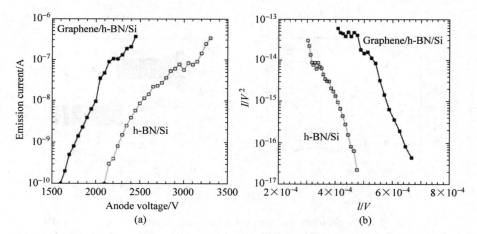

Figure 2-35 Experimental emission current and theoretical simulations
(a) The variation of the emission current with threshold voltage of graphene/h-BN/Si and h-BN/Si;
(b) Fowler-Nordheim plot calculated from (a)[187]

The results show the modification of graphene in h-BN transfer function, thus changing the field emission mechanism. It is possible to control the bandgap engineering of graphene, which provides the basis for the vacuum electronic devices based on graphene.

6. Ballistic transport

In the 1.3.3 section of this book, the ballistic transport properties of suspended graphene are introduced in detail. However, the suspended graphene is easily affected by the environment, external doping and many other factors, which affect the mobility of carriers in graphene[188-196].

Recently, a sandwich structure has been fabricated (h-BN/graphene/h-BN)[188], where the graphene is encapsulated by h-BN. When the carrier concentration is $n \sim 10^{11}$ cm^{-2}, the mobility of the carriers reaches 100,000 cm^2/(V·s)[191], and the mean free path is up to $l = 1$ μm. Figure 36(a) shows bend resistance (R_B) at different temperatures from 2 K to 250 K. As can be seen from the figure, R_B is obviously dependent on the temperature, and the temperature of 250 K below the negative resistance phenomenon. Figure 2-36(b) is the variation of R_B with carrier concentration (n) at 2 K. At the beginning of the 10^{11} cm^{-2} there is a negative resistance. The illustration shows the Hall crosses in the heterostructure. Figure 2-36(c) shows that, at the $V_g = 3$ V, R_B at different temperature varies with the magnetic field strength $B(T)$. It can be seen clearly that with the increase of magnetic field strength, the change of R_B is very obvious, and the results are in good agreement with the characteristics of ballistic transport in previous literatures. The calculation of Hall cross in inset is using the billiard-ball model. Figure 2-36(d) shows the spectral lines of Hall resistance R_H with magnetic field strength $B(T)$ at 50 K and 250 K. It can be seen that the Hall resistance is less affected by temperature[188].

From the above results we can see that in the package in h-BN, graphene exhibits negative bending resistance and obvious anomalous quantum Hall effect is a direct consequence of different carrier concentration at room temperature at the micron scale ballistic transport. The h-BN acts as an ultra-thin gate dielectric, and graphene is virtually unaffected by the environment[190,194-195].

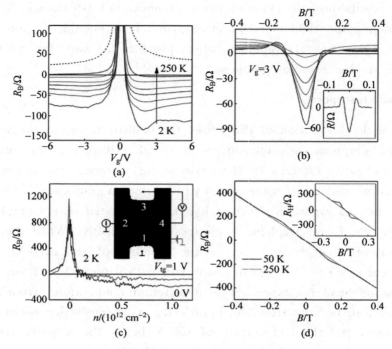

Figure 2-36 Electric properties

(a) The variation of bending resistance with V_g at different temperatures; (b) The variation of bending resistance with carrier concentration at 50 K and 250 K temperature; (c)~(d) Ballistic transport in magnetic field of h-BN/graphene/h-BN heterostructure[188]

7. Conversion between metal and insulator

At low temperature, the resistance of the disordered conductor is higher than that of the quantum resistance h/e^2, and the behavior of the insulation is usually manifested. This universal phenomenon is known as Anderson localization. Among the numerous materials, including graphene with defects and disordered chemical derivatives, although the resistivity of graphene is close to h/e^2 at neutral point, the strong localization phenomenon can't be observed. A recent study demonstrates the Anderson localization of the transition between metal and insulator in high-quality graphene[192,196-200].

Figure 2-37(a) is a schematic diagram of the experimental setup. The device is a sandwich mechanism using experimental method of dry physical transfer preparation, a graphene/h-BN/graphene/h-BN heterostructure, wherein

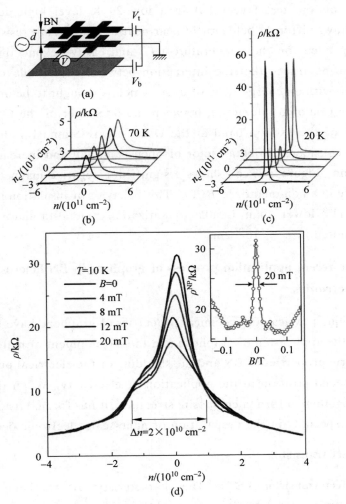

Figure 2-37 The Anderson localization of the transition between metal and insulator in high quality graphene

(a) The schematic of device (graphene/h-BN heterostructure); The test of the device in (b) and (c) with different temperature; (d) The test of the device with different B[200]

the graphene top exposed as control layer, and the encapsulation of graphene layers in h-BN as the test layer, h-BN as insulation medium and substrate material. Figures 2-37 (b) and (c) are the device at 70 K and 20 K temperatures, Hall resistance ρ_{xy} with carrier concentration changes in the image. See from the figure, at the temperature of 70 K, the control layer has little effect on the test layer; but in a low 20 K level high doping case, graphene shows different behavior in near neutral point, Hall resistance ρ_{xy} is strongly dependent on the temperature (T) and easily beyond the threshold h/e^2. It can be argued that the interaction between adjacent layers, namely the Coulomb with an energy scale of $e/\varepsilon d$ which is thought to be an important factor in the Coulomb interaction between the interlayer of the Coulomb and the opening of the exciton band at the Dirac point. Similarly, the magnetic field is also considered to be a factor of open band gap and increasing carrier concentration. Figure 2-37(d) shows a significant negative magnetoresistance of the device in a non-quantized $B(T)$. This behavior of insulation is limited in $B^* \approx 10$ mT, lower than Landau quantization, metal-insulator transition (MIT) is limited in $n \leqslant 10^{10}$ cm^{-2}.

2.3.4 The recent application progress of graphene/h-BN heterostructure in electronics

The unique physical and chemical properties of graphene have made great progress in the application of graphene and the development of materials. The high-dielectric properties of BN and the shielding of the electrical properties of BN has a special position in the application of electricity. When the two are combined together to form a composite structure, it has formed its own unique properties, especially in the preparation of nano-scale electronic devices.

1. Field effect transistor

Field effect transistor (FET): The characteristics of graphene in neutral point conductivity and Klein tunneling electron tunneling transport through the potential barrier, hindering the performance of FETs based on graphene, such as limit switch conversion ratio of $\sim 10^3$, and the temperature is higher than that of 10 K when it is difficult to achieve. After the special electrical properties of graphene/h-BN heterostructures have been discovered, the nano-scale transistors based on heterostructures have been tested and recognized, which is expected to solve the above problems[201-213].

Figure 2-38(a) is a schematic diagram of a number of FET based on graphene/h-BN heterostructures in recent years[201]. Illustration is an optical image of the FET. Figure 2-38(b) is the Raman spectrum of the modified material, the characteristic peaks correspond to the single-layer graphene and the single-layer h-BN in the atomic vibration characteristic of the heterostructure. Figure 2-38(c) is an image of graphene at room temperature, which varies of conductance (σ) with the back-gate voltage on SiO_2 substrate and h-BN substrate, respectively. It can be seen from the figure, for h-BN substrate, the conductivity of graphene in the back gate voltage (V_{BG}) changes small, there is a strong response, which was significantly higher than that of graphene based on SiO_2 substrate. Figure 2-38(d) is the plot of extraction of the mobility with the hole and electron and the intrinsic doping level. It can be seen from the diagram that the mobility of carriers based on h-BN is much higher than that of graphene carriers on SiO_2.

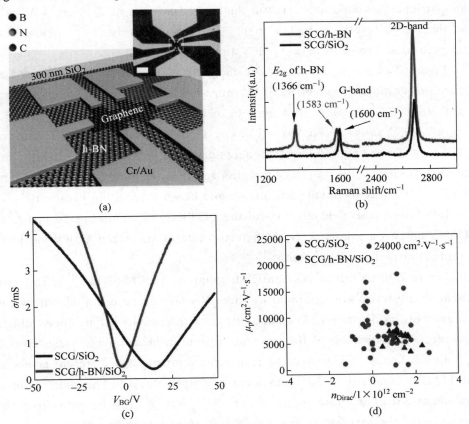

Figure 2-38 Field effect transistor and characterization
(a) The field effect transistor based on graphene/h-BN heterostructure. The inset is the optical map of the device; (b) The Raman spectrum of the device. (c) and (d) are the results of the test with the device[201]

Based on the BN substrate, the average intrinsic doping level of graphene crystals was significantly lower than that on the silica substrate. This may be due to the lower dangling bonds on the BN substrate. This finding is also used to measure the carrier mobility of graphene single crystals.

Thinfilm transistors (TFTs): In today's era of science and technology, the development of printing electronic products in various industries has become very important, many studies focus on the new technology nanomaterials, including organic nanoparticles, inorganic nanoparticles and nanotubes/nanowires composite structure. However, high operating voltage (> 50 V), low mobility (< 10 cm^2/(V · s)), and low current injection are still challenging the organic TFT[214-215]. The combination of inorganic nanoparticles and nanotubes leads to the carrier mobility of >10 cm^2/(V · s), and the on：off ratio is higher than $> 10^6$, respectively[216-217]. However, there are some problems in scalability and integration.

Figure 2-39(a) is the schematic diagram of the TFT, h-BN is used as separator, graphene as electrode (400 nm thick) and WSe$_2$ as channel ($t \sim 1$ μm, $L = 200$ μm, $\omega = 15.6$ mm) to spray h-BN (400 nm thick) on WSe$_2$. Figure 2-39(c) shows the test result of the TFT, which have on：off ratios of >25, $g_m = 22\mu S$[203]. Although it is necessary to improve the on：off ratio and the on：off speed, it is necessary to optimize the ionic liquid, but this new type of TFT still shows its unique advantages and broad application prospects.

Interlayer tunnel field effect transistors (ITFETs): Potential performance of ITFETs based on graphene/h-BN heterostructures transcend CMOS devices, making it attract much attention[219-221].

Figure 2-40(a) is a schematic diagram of the ITFETs[204]. The two graphene substrates are separated by three BN layers and encapsulated by the top and bottom BN layers. The back gate is by the heavily n-type doped silicon structure. The thickness of the top and bottom graphene layers varies from 2 to 5 atomic layers. Figure 2-40(b) is at room temperature, different layers of the ITFETs standard I-V characteristics with a device. The illustration is equivalent circuit diagram to the Figure 2-40(a). As can be seen from the figure, with the increase in the number of graphene layers, the greater the slope of the I_{INT}-V_{TL} characteristic curve, the more sensitive the current and voltage response. Figure 2-40(c) shows the variation of the differential

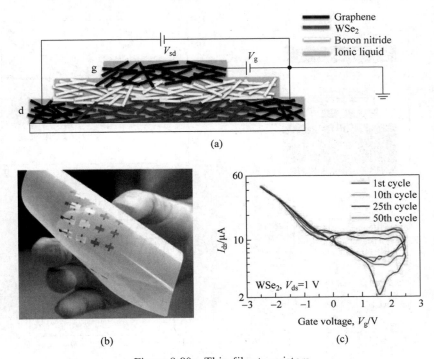

Figure 2-39 Thin film transistors

(a) Schematic diagram of TFT; (b) Fabrication of TFT arrays; (c) Test results for TFT[203]

conductance at the top of the bias voltage (V_{TL}). Similar to Figure 2-40(d), when the back-gate voltage is $V_{BG} = -40$ V, the variation of the differential conductance (dI_{INT}/dV_{TL}) of graphene with five layers is stronger with the change of the bias voltage. This is due to the resonant tunneling of the graphene band in the interlayer alignment mode, while the graphene layer increases, due to the narrower span and higher strength, making the conductance peak clearer. The width of the first resonance peak of the corresponding device extracted by Lorentzian fitting is inversely proportional to the thickness of the graphene layers. When the number of graphene layers changed from 2 to 5, the corresponding peak width (Γ) also changed from 295 mV to 55 mV, and became sharper.

Flexible graphene field-effect transistors (GFETs): A single layer of graphene and a single layer of BN constitute a heterogeneous structure, and in the preparation of electronic devices have good flexibility. The assembly of the FET has been verified by the use of BN encapsulated graphene[222-229].

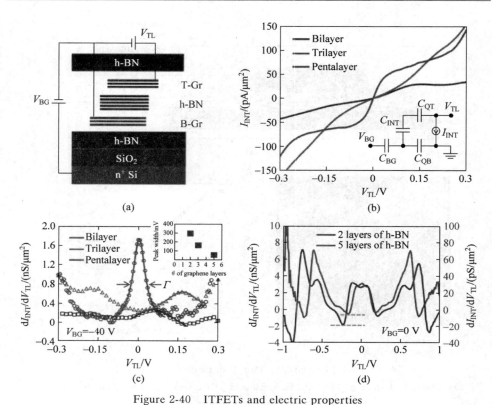

Figure 2-40 ITFETs and electric properties
(a) The schematic of ITFETs; (b)~(d) The test results of the device with different layers[207]

Figure 2-41 shows the results of a flexible GFET and the resulting image[216]. This flexible GFETs assembly with a 2 μm channel length exhibits excellent performances: extraordinary room temperature carrier mobility (μ_{FE} = 1.0×10^4 cm^2/(V·s)), strong saturation current characteristics (r_0 = 2,000 Ω), and high mechanical flexibility (strain limits of 1 %). These values are unprecedented. On the other hand, a flexible radio frequency FET/(RF-FET) with a 375 nm channel length exhibits excellent performance: μ_{FE} = 1.0×10^4 cm^2/(V·s), r_0 = 132.5 Ω. The unity gain frequency f_T and the unity power gain frequency f_{max} reached 12 GHz and 10.6 GHz, respectively. The ratio of the cutoff frequency to 0.5 (f_{max}/f_T = 0.5) corresponds to a record of the flexible GFETS[215].

Based on the previous results, we can conclude that the electrical properties of graphene carriers are attributed to the dielectric properties of BN.

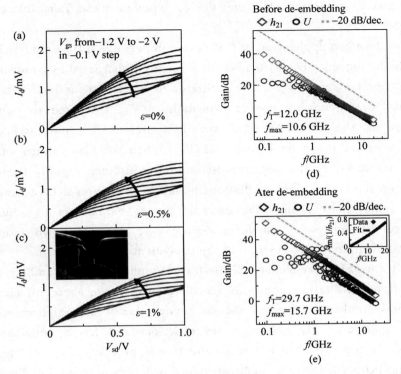

Figure 2-41 Flexible graphene field-effect transistor and the resulting image
(a)~(c) The I-V characteristics of GFETs. The insert is the physical of the device;
(d)~(e) Radio frequency characteristics of a flexible graphene RF-FET[213]

They are the electrical properties of BN itself, as well as the lattice alignment of the heterogeneous structure, which ensures that the electrical properties of graphene have been fully reflected. This provides a powerful driving force for the development of graphene-based flexible electronics.

2. Oscillator

Radio frequency oscillator based on a resonant tunneling transistor: With the in-depth study, in the regulation of the electronic heterostructures reached unprecedented achievements, not only through the selection of materials, packing mode, but also by adjusting their internal strain and the relative position of the component layers to achieve the purpose of controlling the electronic properties. When a wide range of nanoscale graphene-based FETs are fabricated, other nanoelectronic devices based on these transistors are also combined.

A well-known application of this device in negative differential resistance is as a high-frequency oscillator, which is usually constructed by connecting an external resonant circuit[229]. The oscillator in the picture is implemented in this way: the inherent parasitic capacitance (C_{tot}) is used to build the oscillator as the capacitance of the LC circuit (L, inducatance; C, capacitance), as shown in Figure 2-42(b). When the bias voltage and gate voltage are tuned to the negative differential resistance region, the device performs a stable sine wave oscillation, as shown in Figures 2-42(a) and (c). At the same time, change the external circuit parameters to adjust the oscillation frequency, as shown in Figure 2-42(b). The two insets are structural diagram of the device and the equivalent circuit diagram.

Split closed-loop resonator: Single-layer graphene has a familiar lattice structure, but the lattice constants are different. In the aspect of electrical properties, graphene is a zero-band-gap semiconductor with high mobility at room temperature. 2D BN as a dielectric material has a wide-band-gap structure. If the crystal structure of the two is precisely aligned, then the electronic properties of the heterostructure are very interesting. Especially with the improvement of the preparation method of the heterogeneous structure, the preparation technology is becoming more and more mature. Graphene heterostructures with various shapes can be prepared on the BN substrate. Using this technique, a split ring resonator as a bandpass filter has been fabricated recently.

Figure 2-43(a) shows a test of closed-loop resonator based on graphene/h-BN heterostructures, h-BN as dielectric substrate of graphene materials, this unique oscillator is the plane structure[230]. The illustration is the optical image of the device, which is a planar split closed loop resonator. The 1.95 GHz is the resonating frequency, which is close to the value of copper microstrips with similar geometries. Figures 2-43(b) and (c) are the radio-frequency measurements of the resonator. The blue point line is the calculated results, and the other two-point line is the experimental results at 350 ℃ and room temperature, respectively.

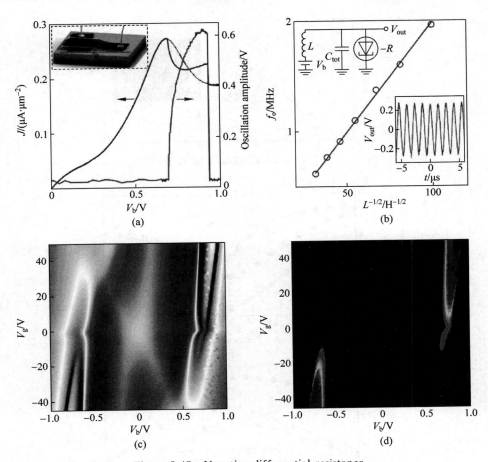

Figure 2-42 Negative differential resistance

(a) $J(V_b)$ characteristics with and without an external LC circuit; (b) Resonant frequency of the oscillator versus the inductance of the LC circuit; (c) A dI/dV_b map measured with a 330μH inductance; (d) Corresponding amplitude map[229]

3. Nanocapacitor

It is generally believed that it is a feasible method to reduce the size of the dielectric material in order to increase the capacitance of the capacitor. However, the quantum capacitance and the dead layer effect tend to reduce the capacitance of the micro nanostructures, which is in sharp contrast to the traditional electrostatic expectation. When the properties of graphene/h-BN heterostructures have been gradually recognized and utilized, they can be used

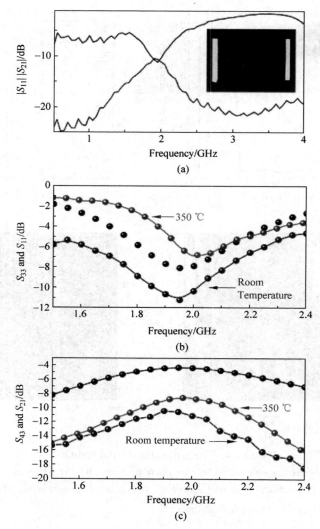

Figure 2-43 RF measurements of closed-loop resonator[230]

to achieve excellent capacitance performance[231-232].

Figure 2-44 is the use of graphene/h-BN heterostructures prepared by the test of the capacitor. The most basic unit contains only BN and graphene. Figures 2-44(a) and (b) show that the dielectric constant of h-BN thin films is different under different stacking structures. As can be seen from the figure, when the capacitance is below 5 nm, the capacitance value increases significantly, more than 100 %. It is shown that the anomalous increase in capacitance is due to the negative capacitance of this special material. From BN-1 to BN-7, there is no graphene in the structure of the GBNG-10, when the structure GBN-8 is used to the structure of graphene/h-BN heterostructure. As the thickness of h-BN film changed from 26.7 nm to 0.8 nm, which are corresponding to BN-1 and GBNG-10, respectively. And the number of h-BN layers in the layers of the capacitors also decreased to 2, while the capacitance loss increased from 0.139 pf to 11.948 pf. It is obvious that the excellent performance of the graphene/h-BN heterostructure in the nanocapacitor is extraordinary.

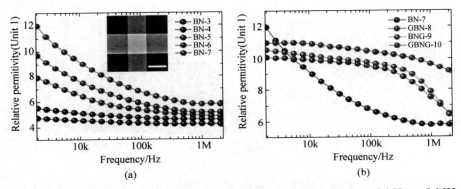

Figure 2-44 The extended measurements of relative permittivity from 2 kHz to 2 MHz (a) Measured relative permittivity of the Au-h-BN-Au stacking groups from 2 kHz to 2 MHz. The insert is optical map of nanocapacitor; (b) Comparison between different stacking structures in the same range[231]

Nonlinear quantum capacitors (The core of a quantum computer): Graphene and quantum computers are the hallmarks of the next-generation human technology innovation, and recent research has shown that the development of these two most exciting technologies may be combined. Researchers have developed a graphene-based quantum capacitor that can be compatible with the low-temperature conditions of the superconducting circuit. When connected to a circuit, the capacitor has the potential to produce stable quantum bits and is easier to manufacture than other known non-linear cryogenic devices, with less sensitivity to electromagnetic interference[234].

It consists of insulating BN sandwiched between two graphene sheets (Figure 2-45). Due to the unusual nature of this sandwich structure and graphene, the incoming charge is proportional to the generated voltage. This nonlinearity is a necessary step in the process of generating quantum bits. The device can significantly improve the processing of quantum information, but there are other potential applications. It can be used to create very nonlinear high-frequency circuits—up to the terahertz mode—or for super-coupling between the mixer, the amplifier and the photon. The use of graphene in the quantum capacitor design of the special performance, will enable us to better understand how to create a practical quantum computer.

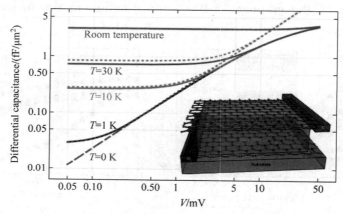

Figure 2-45 Test results of nonlinear quantum capacitors at different temperatures
The illustration is a schematic diagram of the structure of a nonlinear quantum capacitor, which is made of graphene/h-BN/graphene heterostructure[234]

4. Photodetector

Recently, the researchers based on copper-nickel alloy substrate growth of high-quality h-BN and graphene film research basis, by depositing h-BN single-crystal growth of graphene, successfully prepared high-quality graphene/h-BN plane quality knot. The deposition time of graphene on copper-nickel alloy is very fast, and the shorter graphene deposition time reduces the destruction of h-BN film during the growth of graphene film. At the same time, due to the excellent catalytic ability of Cu-Ni alloy, the crystallization of graphene is eliminated while increasing the crystallization quality of BN single crystal, so that the graphene crystal domain is only at the vertex angle of the triangular h-BN single-crystal domain. The nuclei are grown along the h-BN orientation, Figure 2-46 shows the optical images of the device and test results[235].

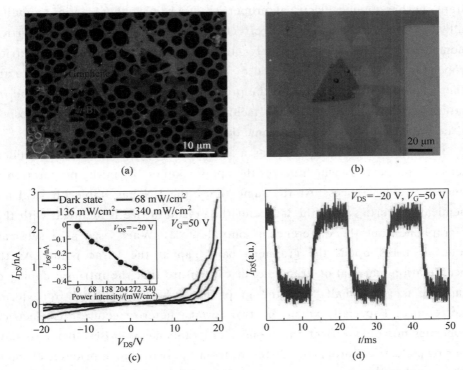

Figure 2-46 The optical images of the device and experimental results
(a) and (b) are optical images of the device. (c) I-V characteristics of the device in the dark and under different illumination intensities with a gate bias set at 50 V. The inset shows the detected current as a function of illumination power intensity for $V_{DS} = -20$ V. (d) Time-resolved photoresponse[235]

On the basis of high-quality graphene/h-BN planar heterojunction, the WSe_2/MoS_2 2D photodetector was prepared by using graphene as the contact electrode and h-BN as the insulating substrate, and the graphene/h-BN plane heterojunction quality and electrical properties, based on the heterogeneous junction material platform to carry out basic research and 2D logic integrated circuit applications to provide a basis for exploration.

2.4 Summary and outlook

With the increasing understanding of the properties of graphene, more and more people focus on the applications and great potential of graphene.

Graphene has unique electrical properties, such as higher carrier mobility, ballistic transport, bipolar field effect, massless Dirac fermion properties, anomalous quantum Holzer effect, etc.. At the same time, the electrical properties of graphene are affected by the shape, size, doping, substrate and other related factors. Thanks to the preparation methods of graphene and the maturity of nano scale processing technology, graphene has a broad prospect in the preparation of nano electronic devices[236-238].

On the other hand, h-BN, as a wide-band-gap insulator, has excellent dielectric properties, which makes the application of BN in the preparation of nano electronic devices. At the same time, the BN has stable physical and chemical properties, and the lattice of the crystal lattice is identical with that of graphene, but the difference is only 1.7 %. When used as a substrate material, h-BN opens the graphene band gap at the Dirac point. As the encapsulation material of graphene, it can ensure that the intrinsic doping of graphene is very small, and the graphene is hardly affected by external interference. Especially when the two form a heterostructure, its electrical properties not only reflect the monolayer graphene or h-BN, but also have their own electrical properties different from the two, which prides itself on h-BN more dangling bonds. A large number of experiments and research show that graphene, BN, and graphene/h-BN heterostructures have broad prospects in the construction of nanoscale electronic devices and optoelectronic devices, especially in the development of great characterization and characterization methods, improving the research of nanometer materials and control processing. Various types of transistors, oscillators, capacitors, etc., based on graphene/h-BN heterostructures, once again tell people that heterogeneous structures made of different materials will be available in the near future.

References

[1] NOVOSELOV K S, GEIM A K, MOROZOV S V, et al. Electric field effect in atomically thin carbon films[J]. Science, 2004, 306(5696): 666-669.
[2] GEIM A K, NOVOSELOV K S. The rise of graphene[J]. Nature Materials, 2009, 6(3): 183-191.
[3] CASTRO N A H, GUINEA F, PERES N M R, et al. The electronic properties of graphene[J]. RvMP, 2009, 81(1): 109-162.

[4]   GEIM A K, KIM P. Carbon wonderland[J]. Scientific American, 2008, 298(4): 90-97.
[5]   NAIR R R, BLAKE P, GRIGORENKO A N, et al. Fine structure constant defines visual transparency of graphene[J]. Science, 2008, 320(5881): 1308-1308.
[6]   LEE C, WEI X, KYSAR J W, et al. Measurement of the elastic properties and intrinsic strength of monolayer graphene[J]. Science, 2008, 321(5887): 385-388.
[7]   BALANDIN A A. Thermal properties of graphene and nanostructured carbon materials [J]. Nature Materials, 2011, 10(8): 569-581.
[8]   BALANDIN A A, GHOSH S, TEWELDEBRHAN D, et al. Extremely high thermal conductivity of graphene: prospects for thermal management applications in silicon nanoelectronics[C]. 2008 IEEE Silicon Nanoelectronics Workshop. IEEE, 2008: 1-2.
[9]   ABBOTT'S I E. Graphene: exploring carbon flatland[J]. Phys. Today, 2007, 60(8): 35.
[10] ANDO T. The electronic properties of graphene and carbon nanotubes[J]. NPG Asia Materials, 2009, 1(1): 17-21.
[11] DRESSELHAUS M S, JORIO A, SAITO R. Characterizing graphene, graphite, and carbon nanotubes by Raman spectroscopy[J]. Annu. Rev. Condens. Matter Phys., 2010, 1(1): 89-108.
[12] WANG L, MERIC I, HUANG P Y, et al. One-dimensional electrical contact to a two-dimensional material[J]. Science, 2013, 342(6158): 614-617.
[13] REICH S, THOMSEN C. Raman spectroscopy of graphite[J]. Philosophical Transactions of the Royal Society of London. Series A: Mathematical, Physical and Engineering Sciences, 2004, 362(1824): 2271-2288.
[14] YAN J A, RUAN W Y, CHOU M Y. Electron-phonon interactions for optical-phonon modes in few-layer graphene: first-principles calculations[J]. Physical Review B, 2009, 79(11): 115443.
[15] SAHA S K, WAGHMARE U V, KRISHNAMURTHY H R, et al. Phonons in few-layer graphene and interplanar interaction: a first-principles study[J]. Physical Review B, 2008, 78(16): 165421.
[16] PARK C H, GIUSTINO F, SPATARU C D, et al. Angle-resolved photoemission spectra of graphene from first-principles calculations[J]. Nano Letters, 2009, 9(12): 4234-4239.
[17] PARK C H, GIUSTINO F, COHEN M L, et al. Electron-phonon interactions in graphene, bilayer graphene, and graphite[J]. Nano Letters, 2008, 8(12): 4229-4233.
[18] CHARLIER J C, EKLUND P C, ZHU J, et al. Electron and phonon properties of graphene: their relationship with carbon nanotubes[M]//Carbon nanotubes. Berlin: Springer, 2007: 673-709.
[19] SLONCZEWSKI J C, WEISS P R. Band structure of graphite[J]. Physical Review, 1958, 109(2): 272.
[20] SEMENOFF G W. Condensed-matter simulation of a three-dimensional anomaly[J].

Physical Review Letters, 1984, 53(26): 2449.
[21] HALDANE F D M. Model for a quantum Hall effect without Landau levels: condensed-matter realization of the "parity anomaly"[J]. Physical Review Letters, 1988, 61(18): 2015.
[22] NOVOSELOV K S, GEIM A K, MOROZOV S V, et al. Two-dimensional gas of massless Dirac fermions in graphene[J]. Nature, 2005, 438(7065): 197-200.
[23] ZHANG Y, TAN Y W, STORMER H L, et al. Experimental observation of the quantum Hall effect and Berry's phase in graphene[J]. Nature, 2005, 438(7065): 201-204.
[24] BOYANOVSKY D, BLANKENBECLER R, YAHALOM R. Physical origin of topological mass in 2+1 dimensions[J]. Nuclear Physics B, 1986, 270: 483-505.
[25] NOVOSELOV K S, MCCANN E, MOROZOV S V, et al. Unconventional quantum Hall effect and Berry's phase of 2π in bilayer graphene[J]. Nature Physics, 2006, 2(3): 177-180.
[26] MCCANN E, FAL'KO V I. Landau-level degeneracy and quantum Hall effect in a graphite bilayer[J]. Physical Review Letters, 2006, 96(8): 086805.
[27] KATSNELSON M I, NOVOSELOV K S, GEIM A K. Chiral tunnelling and the Klein paradox in graphene[J]. Nature Physics, 2006, 2(9): 620-625.
[28] DOMBEY N, CALOGERACOS A. Seventy years of the Klein paradox[J]. Physics Reports, 1999, 315(1/2/3): 41-58.
[29] HUARD B, SULPIZIO J A, STANDER N, et al. Transport measurements across a tunable potential barrier in graphene[J]. Physical Review Letters, 2007, 98(23): 236803.
[30] BARBIER M, PEETERS F M, VASILOPOULOS P, et al. Dirac and Klein-Gordon particles in one-dimensional periodic potentials[J]. Physical Review B, 2008, 77(11): 115446.
[31] ALLAIN P E, FUCHS J N. Klein tunneling in graphene: optics with massless electrons[J]. The European Physical Journal B, 2011, 83(3): 301-317.
[32] DU X, SKACHKO I, BARKER A, et al. Approaching ballistic transport in suspended graphene[J]. Nature Nanotechnology, 2008, 3(8): 491-495.
[33] MIAO F, WIJERATNE S, ZHANG Y, et al. Phase-coherent transport in graphene quantum billiards[J]. Science, 2007, 317(5844): 1530-1533.
[34] CUEVAS J C, YEYATI A L. Subharmonic gap structure in short ballistic graphene junctions[J]. Physical Review B, 2006, 74(18): 180501.
[35] BEENAKKER C W J. Specular Andreev reflection in graphene[J]. Physical Review Letters, 2006, 97(6): 067007.
[36] TWORZYDLO J, TRAUZETTEL B, TITOV M, et al. Sub-Poissonian shot noise in graphene[J]. Physical Review Letters, 2006, 96(24): 246802.
[37] DICARLO L, WILLIAMS J R, ZHANG Y, et al. Shot noise in graphene[J]. Physical Review Letters, 2008, 100(15): 156801.
[38] DU X, SKACHKO I, ANDREI E Y. Josephson current and multiple Andreev

reflections in graphene SNS junctions[J]. Physical Review B, 2008, 77(18): 184507.
[39] LI G, ANDREI E Y. Observation of Landau levels of Dirac fermions in graphite[J]. Nature Physics, 2007, 3(9): 623-627.
[40] ANDO T, FOWLER A B, STERN F. Electronic properties of two-dimensional systems[J]. Reviews of Modern Physics, 1982, 54(2): 437.
[41] MILLER D L, KUBISTA K D, RUTTER G M, et al. Observing the quantization of zero mass carriers in graphene[J]. Science, 2009, 324(5929): 924-927.
[42] JIANG Z, HENRIKSEN E A, TUNG L C, et al. Infrared spectroscopy of Landau levels of graphene[J]. Physical Review Letters, 2007, 98(19): 197403.
[43] HENRIKSEN E A, CADDEN-ZIMANSKY P, JIANG Z, et al. Interaction-induced shift of the cyclotron resonance of graphene using infrared spectroscopy[J]. Physical Review Letters, 2010, 104(6): 067404.
[44] SHIZUYA K. Many-body corrections to cyclotron resonance in monolayer and bilayer graphene[J]. Physical Review B, 2010, 81(7): 075407.
[45] NOVOSELOV K S, JIANG Z, ZHANG Y, et al. Room-temperature quantum Hall effect in graphene[J]. Science, 2007, 315(5817): 1379-1379.
[46] ZHANG Y, TAN Y W, STORMER H L, et al. Experimental observation of the quantum Hall effect and Berry's phase in graphene[J]. Nature, 2005, 438(7065): 201-204.
[47] MOROZOV S V, NOVOSELOV K S, KATSNELSON M I, et al. Giant intrinsic carrier mobilities in graphene and its bilayer[J]. Physical Review Letters, 2008, 100(1): 016602.
[48] CHEN J H, JANG C, XIAO S, et al. Intrinsic and extrinsic performance limits of graphene devices on SiO_2[J]. Nature Nanotechnology, 2008, 3(4): 206-209.
[49] BERGER C, SONG Z, LI X, et al. Electronic confinement and coherence in patterned epitaxial graphene[J]. Science, 2006, 312(5777): 1191-1196.
[50] WALLACE P R. The band theory of graphite[J]. Physical Review, 1947, 71(9): 622.
[51] DAMLE K, SACHDEV S. Nonzero-temperature transport near quantum critical points[J]. Physical Review B, 1997, 56(14): 8714.
[52] KOVTUN P K, SON D T, STARINETS A O. Viscosity in strongly interacting quantum field theories from black hole physics[J]. Physical Review Letters, 2005, 94(11): 111601.
[53] SON D T. Vanishing bulk viscosities and conformal invariance of the unitary Fermi gas[J]. Physical Review Letters, 2007, 98(2): 020604.
[54] KARSCH F, KHARZEEV D, TUCHIN K. Universal properties of bulk viscosity near the QCD phase transition[J]. Physics Letters B, 2008, 663(3): 217-221.
[55] LEVITOV L, FALKOVICH G. Electron viscosity, current vortices and negative nonlocal resistance in graphene[J]. Nature Physics, 2016, 12(7): 672-676.
[56] YOO M J, FULTON T A, HESS H F, et al. Scanning single-electron transistor microscopy: imaging individual charges[J]. Science, 1997, 276(5312): 579-582.

[57] JAVEY A, KIM H, BRINK M, et al. High-κ dielectrics for advanced carbon-nanotube transistors and logic gates[J]. Nature Materials, 2002, 1(4): 241-246.

[58] JAVEY A, GUO J, FARMER D B, et al. Self-aligned ballistic molecular transistors and electrically parallel nanotube arrays[J]. Nano Letters, 2004, 4(7): 1319-1322.

[59] KLINKE C, CHEN J, AFZALI A, et al. Charge transfer induced polarity switching in carbon nanotube transistors[J]. Nano Letters, 2005, 5(3): 555-558.

[60] KANG S J, KOCABAS C, OZEL T, et al. High-performance electronics using dense, perfectly aligned arrays of single-walled carbon nanotubes[J]. Nature Nanotechnology, 2007, 2(4): 230-236.

[61] AVOURIS P, CHEN Z, PEREBEINOS V. Carbon-based electronics[M]//Nanoscience and Technology: A Collection of Reviews from Nature Journals, 2010: 174-184.

[62] AKINWANDE D, CLOSE G F, WONG H S P. Analysis of the frequency response of carbon nanotube transistors[J]. IEEE Transactions on Nanotechnology, 2006, 5(5): 599-605.

[63] SCHEDIN F, GEIM A K, MOROZOV S V, et al. Detection of individual gas molecules adsorbed on graphene[J]. Nature Materials, 2007, 6(9): 652-655.

[64] LI S L, MIYAZAKI H, KUMATANI A, et al. Low operating bias and matched input-output characteristics in graphene logic inverters[J]. Nano Letters, 2010, 10(7): 2357-2362.

[65] MERIC I, HAN M Y, YOUNG A F, et al. Current saturation in zero-bandgap, top-gated graphene field-effect transistors[J]. Nature Nanotechnology, 2008, 3(11): 654-659.

[66] SORDAN R, TRAVERSI F, RUSSO V. Logic gates with a single graphene transistor [J]. Applied Physics Letters, 2009, 94(7): 51.

[67] GUERRIERO E, POLLONI L, RIZZI L G, et al. Graphene audio voltage amplifier [J]. Small, 2012, 8(3): 357-361.

[68] KIM W, RIIKONEN J, LI C, et al. Highly tunable local gate controlled complementary graphene device performing as inverter and voltage controlled resistor [J]. Nanotechnology, 2013, 24(39): 395202.

[69] LIAO A D, WU J Z, WANG X, et al. Thermally limited current carrying ability of graphene nanoribbons[J]. Physical Review Letters, 2011, 106(25): 256801.

[70] LIU L, RYU S, TOMASIK M R, et al. Graphene oxidation: thickness-dependent etching and strong chemical doping[J]. Nano Letters, 2008, 8(7): 1965-1970.

[71] RIZZI L G, BIANCHI M, BEHNAM A, et al. Cascading wafer-scale integrated graphene complementary inverters under ambient conditions[J]. Nano Letters, 2012, 12(8): 3948-3953.

[72] LIN Y M, VALDES-GARCIA A, HAN S J, et al. Wafer-scale graphene integrated circuit[J]. Science, 2011, 332(6035): 1294-1297.

[73] YANG X, LIU G, BALANDIN A A, et al. Triple-mode single-transistor graphene amplifier and its applications[J]. ACS Nano, 2010, 4(10): 5532-5538.

[74] XIA F, PEREBEINOS V, LIN Y, et al. The origins and limits of metal-graphene junction resistance[J]. Nature Nanotechnology, 2011, 6(3): 179-184.

[75] WANG Q H, HERSAM M C. Room-temperature molecular-resolution characterization of self-assembled organic monolayers on epitaxial graphene[J]. Nature Chemistry, 2009, 1(3): 206.

[76] EMTSEV K V, BOSTWICK A, HORN K, et al. Towards wafer-size graphene layers by atmospheric pressure graphitization of silicon carbide[J]. Nature Materials, 2009, 8(3): 203-207.

[77] WOOD J D, SCHMUCKER S W, LYONS A S, et al. Effects of polycrystalline Cu substrate on graphene growth by chemical vapor deposition[J]. Nano Letters, 2011, 11(11): 4547-4554.

[78] GHAHARI F, WALKUP D, GUTIÉRREZ C, et al. An on/off Berry phase switch in circular graphene resonators[J]. Science, 2017, 356(6340): 845-849.

[79] CASIRAGHI C, PISANA S, NOVOSELOV K S, et al. Raman fingerprint of charged impurities in graphene[J]. Applied Physics Letters, 2007, 91(23): 233108.

[80] CHEN F, XIA J, TAO N. Ionic screening of charged-impurity scattering in graphene [J]. Nano Letters, 2009, 9(4): 1621-1625.

[81] LIN Y M, DIMITRAKOPOULOS C, JENKINS K A, et al. 100-GHz transistors from wafer-scale epitaxial graphene[J]. Science, 2010, 327(5966): 662-662.

[82] PAKDEL A, ZHI C, BANDO Y, et al. Low-dimensional boron nitride nanomaterials [J]. Materials Today, 2012, 15(6): 256-265.

[83] FURTHMÜLLER J, HAFNER J, KRESSE G. Ab initio calculation of the structural and electronic properties of carbon and boron nitride using ultrasoft pseudopotentials [J]. Physical Review B, 1994, 50(21): 15606.

[84] YU W J, LAU W M, CHAN S P, et al. Ab initio study of phase transformations in boron nitride[J]. Physical Review B, 2003, 67(1): 014108.

[85] ZHANG Y, SUN H, CHEN C. Structural deformation, strength, and instability of cubic BN compared to diamond: A first-principles study[J]. Physical Review B, 2006, 73(14): 144115.

[86] XU Y N, CHING W Y. Calculation of ground-state and optical properties of boron nitrides in the hexagonal, cubic, and wurtzite structures[J]. Physical Review B, 1991, 44(15): 7787.

[87] KNITTLE E, KANER R B, JEANLOZ R, et al. High-pressure synthesis, characterization, and equation of state of cubic C-BN solid solutions[J]. Physical Review B, 1995, 51(18): 12149.

[88] PAKDEL A, ZHI C, BANDO Y, et al. Low-dimensional boron nitride nanomaterials [J]. Materials Today, 2012, 15(6): 256-265.

[89] LIU J, VOHRA Y K, TARVIN J T, et al. Cubic-to-rhombohedral transformation in boron nitride induced by laser heating: in situ Raman-spectroscopy studies[J]. Physical Review B, 1995, 51(13): 8591.

[90] BLASE X, RUBIO A, LOUIE S G, et al. Quasiparticle band structure of bulk hexagonal boron nitride and related systems [J]. Physical Review B, 1995, 51(11): 6868.

[91] ZUPAN J. Energy bands in boron nitride and graphite[J]. Physical Review B, 1972, 6(6): 2477.

[92] OOI N, RAIRKAR A, LINDSLEY L, et al. Electronic structure and bonding in hexagonal boron nitride[J]. Journal of Physics: Condensed Matter, 2005, 18(1): 97.

[93] OOI N, RAJAN V, GOTTLIEB J, et al. Structural properties of hexagonal boron nitride[J]. Modelling and Simulation in Materials Science and Engineering, 2006, 14(3): 515.

[94] TOPSAKAL M, AKTÜRK E, CIRACI S. First-principles study of two-and one-dimensional honeycomb structures of boron nitride[J]. Physical Review B, 2009, 79(11): 115442.

[95] EKUMA C E, DOBROSAVLJEVIĆ V, GUNLYCKE D. First-principles-based method for electron localization: application to monolayer hexagonal boron nitride [J]. Physical Review Letters, 2017, 118(10): 106404.

[96] MEYER J C, CHUVILIN A, ALGARA-SILLER G, et al. Selective sputtering and atomic resolution imaging of atomically thin boron nitride membranes [J]. Nano Letters, 2009, 9(7): 2683-2689.

[97] HAN W Q, WU L, ZHU Y, et al. Structure of chemically derived mono-and few-atomic-layer boron nitride sheets[J]. Applied Physics Letters, 2008, 93(22): 223103.

[98] WATANABE K, TANIGUCHI T, KANDA H. Direct-bandgap properties and evidence for ultraviolet lasing of hexagonal boron nitride single crystal[J]. Nature Materials, 2004, 3(6): 404-409.

[99] GAO Y, REN W, MA T, et al. Repeated and controlled growth of monolayer, bilayer and few-layer hexagonal boron nitride on Pt foils[J]. ACS Nano, 2013, 7(6): 5199-5206.

[100] KIM K K, HSU A, JIA X, et al. Synthesis of monolayer hexagonal boron nitride on Cu foil using chemical vapor deposition[J]. Nano Letters, 2012, 12(1): 161-166.

[101] ARNAUD B, LEBÈGUE S, RABILLER P, et al. Huge excitonic effects in layered hexagonal boron nitride[J]. Physical Review Letters, 2006, 96(2): 026402.

[102] BLASE X, RUBIO A, LOUIE S G, et al. Quasiparticle band structure of bulk hexagonal boron nitride and related systems [J]. Physical Review B, 1995, 51(11): 6868.

[103] FURTHMÜLLER J, HAFNER J, KRESSE G. Ab initio calculation of the structural and electronic properties of carbon and boron nitride using ultrasoft pseudopotentials [J]. Physical Review B, 1994, 50(21): 15606.

[104] XU Y N, CHING W Y. Calculation of ground-state and optical properties of boron nitrides in the hexagonal, cubic, and wurtzite structures[J]. Physical Review B, 1991, 44(15): 7787.

[105] JANG C, ADAM S, CHEN J H, et al. Tuning the effective fine structure constant in graphene: opposing effects of dielectric screening on short-and long-range potential scattering[J]. Physical Review Letters, 2008, 101(14): 146805.

[106] KIM K K, HSU A, JIA X, et al. Synthesis and characterization of hexagonal boron nitride film as a dielectric layer for graphene devices[J]. ACS Nano, 2012, 6(10): 8583-8590.

[107] DATTA S S, STRACHAN D R, MELE E J, et al. Surface potentials and layer charge distributions in few-layer graphene films[J]. Nano Letters, 2009, 9(1): 7-11.

[108] CASTELLANOS-GOMEZ A, CAPPELLUTI E, ROLDÁN R, et al. Electric-field screening in atomically thin layers of MoS_2: the role of interlayer coupling[J]. Advanced Materials, 2013, 25(6): 899-903.

[109] LI L H, SANTOS E J G, XING T, et al. Dielectric screening in atomically thin boron nitride nanosheets[J]. Nano Letters, 2015, 15(1): 218-223.

[110] SANTOS E J G, KAXIRAS E. Electric-field dependence of the effective dielectric constant in graphene[J]. Nano Letters, 2013, 13(3): 898-902.

[111] LI J, GUI G, ZHONG J. Tunable bandgap structures of two-dimensional boron nitride[J]. Journal of Applied Physics, 2008, 104(9): 094311.

[112] PARK C H, LOUIE S G. Energy gaps and Stark effect in boron nitride nanoribbons [J]. Nano Letters, 2008, 8(8): 2200-2203.

[113] ZHENG F, ZHOU G, LIU Z, et al. Half metallicity along the edge of zigzag boron nitride nanoribbons[J]. Physical Review B, 2008, 78(20): 205415.

[114] SLOTMAN G J, FASOLINO A. Structure, stability and defects of single layer hexagonal BN in comparison to graphene[J]. Journal of Physics: Condensed Matter, 2012, 25(4): 045009.

[115] HE B, ZHANG W J, YAO Z Q, et al. P-type conduction in beryllium-implanted hexagonal boron nitride films[J]. Applied Physics Letters, 2009, 95(25): 252106.

[116] LU M, BOUSETTA A, BENSAOULA A, et al. Electrical properties of boron nitride thin films grown by neutralized nitrogen ion assisted vapor deposition[J]. Applied Physics Letters, 1996, 68(5): 622-624.

[117] DAHAL R, LI J, MAJETY S, et al. Epitaxially grown semiconducting hexagonal boron nitride as a deep ultraviolet photonic material[J]. Applied Physics Letters, 2011, 98(21): 211110.

[118] NOSE K, OBA H, YOSHIDA T. Electric conductivity of boron nitride thin films enhanced by in situ doping of zinc[J]. Applied Physics Letters, 2006, 89(11): 112124.

[119] XUE Y, LIU Q, HE G, et al. Excellent electrical conductivity of the exfoliated and fluorinated hexagonal boron nitride nanosheets[J]. Nanoscale Research Letters, 2013, 8(1): 49.

[120] WATANABE K, TANIGUCHI T, KANDA H. Direct-bandgap properties and evidence for ultraviolet lasing of hexagonal boron nitride single crystal[J]. Nature Materials, 2004, 3(6): 404-409.

[121] EVANS D A, MCGLYNN A G, TOWLSON B M, et al. Determination of the optical band-gap energy of cubic and hexagonal boron nitride using luminescence excitation spectroscopy [J]. Journal of Physics: Condensed Matter, 2008, 20(7): 075233.

[122] PONOMARENKO L A, GEIM A K, ZHUKOV A A, et al. Tunable metal-insulator transition in double-layer graphene heterostructures [J]. Nature Physics, 2011, 7(12): 958-961.

[123] AMET F, WILLIAMS J R, GARCIA A G F, et al. Tunneling spectroscopy of graphene-boron-nitride heterostructures[J]. Physical Review B, 2012, 85(7): 073405.

[124] BRITNELL L, GORBACHEV R V, JALIL R, et al. Field-effect tunneling transistor based on vertical graphene heterostructures[J]. Science, 2012, 335(6071): 947-950.

[125] SIMMONS J G. Generalized formula for the electric tunnel effect between similar electrodes separated by a thin insulating film[J]. Journal of applied physics, 1963, 34(6): 1793-1803.

[126] KHARCHE N, NAYAK S K. Quasiparticle band gap engineering of graphene and graphone on hexagonal boron nitride substrate[J]. Nano Letters, 2011, 11(12): 5274-5278.

[127] LEE G H, YU Y J, LEE C, et al. Electron tunneling through atomically flat and ultrathin hexagonal boron nitride[J]. Applied Physics Letters, 2011, 99(24): 243114.

[128] BRITNELL L, GORBACHEV R V, JALIL R, et al. Electron tunneling through ultrathin boron nitride crystalline barriers [J]. Nano Letters, 2012, 12(3): 1707-1710.

[129] KUBOTA Y, WATANABE K, TSUDA O, et al. Deep ultraviolet light-emitting hexagonal boron nitride synthesized at atmospheric pressure [J]. Science, 2007, 317(5840): 932-934.

[130] WATANABE K, TANIGUCHI T, NIIYAMA T, et al. Far-ultraviolet plane-emission handheld device based on hexagonal boron nitride[J]. Nature Photonics, 2009, 3(10): 591-594.

[131] LEE C, LI Q, KALB W, et al. Frictional characteristics of atomically thin sheets [J]. Science, 2010, 328(5974): 76-80.

[132] DEAN C R, YOUNG A F, MERIC I, et al. Boron nitride substrates for high-quality graphene electronics[J]. Nature Nanotechnology, 2010, 5(10): 722-726.

[133] LI X, ZHU Y, CAI W, et al. Transfer of large-area graphene films for high-performance transparent conductive electrodes[J]. Nano Letters, 2009, 9(12): 4359-4363.

[134] LEE K H, SHIN H J, LEE J, et al. Large-scale synthesis of high-quality hexagonal boron nitride nanosheets for large-area graphene electronics[J]. Nano Letters, 2012, 12(2): 714-718.

[135] XU S J, LUO Y F, ZHONG W, et al. Investigation of hexagonal boron nitride for

application as counter electrode in dye-sensitized solar cells[J]. Advanced Materials Research, 2012, 512-515: 242-245.

[136] MAJETY S, CAO X K, LI J, et al. Band-edge transitions in hexagonal boron nitride epilayers[J]. Applied Physics Letters, 2012, 101(5): 051110.

[137] LI J, MAJETY S, DAHAL R, et al. Dielectric strength, optical absorption, and deep ultraviolet detectors of hexagonal boron nitride epilayers[J]. Applied Physics Letters, 2012, 101(17): 171112.

[138] CAO X K, MAJETY S, LI J, et al. Optoelectronic properties of hexagonal boron nitride epilayers[J]. 2013, 8631: 863128.

[139] PODZOROV V, GERSHENSON M E, KLOC C, et al. High-mobility field-effect transistors based on transition metal dichalcogenides[J]. Applied Physics Letters, 2004, 84(17): 3301-3303.

[140] RADISAVLJEVIC B, RADENOVIC A, BRIVIO J, et al. Single-layer MoS_2 transistors[J]. Nature Nanotechnology, 2011, 6(3): 147-150.

[141] YANG L, MAJUMDAR K, LIU H, et al. Chloride molecular doping technique on 2D materials: WS_2 and MoS_2[J]. Nano Letters, 2014, 14(11): 6275-6280.

[142] LEE C H, LEE G H, VAN DER ZANDE A M, et al. Atomically thin p-n junctions with van der Waals heterointerfaces[J]. Nature Nanotechnology, 2014, 9(9): 676.

[143] HOWELL S L, JARIWALA D, WU C C, et al. Investigation of band-offsets at monolayer-multilayer MoS_2 junctions by scanning photocurrent microscopy[J]. Nano Letters, 2015, 15(4): 2278-2284.

[144] CHUANG H J, CHAMLAGAIN B, KOEHLER M, et al. Low-resistance 2D/2D ohmic contacts: a universal approach to high-performance WSe_2, MoS_2, and $MoSe_2$ transistors[J]. Nano Letters, 2016, 16(3): 1896-1902.

[145] BLAKE P, HILL E W, CASTRO NETO A H, et al. Making graphene visible[J]. Applied Physics Letters, 2007, 91(6): 063124.

[146] BALANDIN A A, GHOSH S, BAO W, et al. Superior thermal conductivity of single-layer graphene[J]. Nano Letters, 2008, 8(3): 902-907.

[147] DEAN C R, YOUNG A F, MERIC I, et al. Boron nitride substrates for high-quality graphene electronics[J]. Nature Nanotechnology, 2010, 5(10): 722-726.

[148] WANG X, XU J B, WANG C, et al. High-performance graphene devices on SiO_2/Si substrate modified by highly ordered self-assembled monolayers[J]. Advanced Materials, 2011, 23(21): 2464-2468.

[149] DECKER R, WANG Y, BRAR V W, et al. Local electronic properties of graphene on a BN substrate via scanning tunneling microscopy[J]. Nano Letters, 2011, 11(6): 2291-2295.

[150] DEAN C R, WANG L, MAHER P, et al. Hofstadter's butterfly and the fractal quantum Hall effect in moiré superlattices[J]. Nature, 2013, 497(7451): 598-602.

[151] GUO Y, GUO W, CHEN C. Tuning field-induced energy gap of bilayer graphene via interlayer spacing[J]. Applied Physics Letters, 2008, 92(24): 243101.

[152] ZHOU S Y, GWEON G H, FEDOROV A V, et al. Substrate-induced bandgap opening in epitaxial graphene[J]. Nature Materials, 2007, 6(10): 770-775.

[153] GIOVANNETTI G, KHOMYAKOV P A, BROCKS G, et al. Substrate-induced band gap in graphene on hexagonal boron nitride: ab initio density functional calculations[J]. Physical Review B, 2007, 76(7): 073103.

[154] SLAWIŃSKA J, ZASADA I, KOSIŃSKI P, et al. Reversible modifications of linear dispersion: graphene between boron nitride monolayers[J]. Physical Review B, 2010, 82(8): 085431.

[155] LU Y H, HE P M, FENG Y P. Asymmetric spin gap opening of graphene on cubic boron nitride (111) substrate[J]. The Journal of Physical Chemistry C, 2008, 112(33): 12683-12686.

[156] MOON P, KOSHINO M. Electronic properties of graphene/hexagonal-boron-nitride moiré superlattice[J]. Physical Review B, 2014, 90(15): 155406.

[157] ZHOU S, HAN J, DAI S, et al. Van der Waals bilayer energetics: generalized stacking-fault energy of graphene, boron nitride, and graphene/boron nitride bilayers [J]. Physical Review B, Condensed Matter and Materials Physics, 2015, 92(15): 155438.1-155438.13.

[158] HUNT B, SANCHEZ-YAMAGISHI J D, YOUNG A F, et al. Massive Dirac fermions and Hofstadter butterfly in a van der Waals heterostructure[J]. Science, 2013, 340(6139): 1427-1430.

[159] ZHONG X, YAP Y K, PANDEY R, et al. First-principles study of strain-induced modulation of energy gaps of graphene/BN and BN bilayers[J]. Physical Review B, 2011, 83(19): 193403.

[160] LEBEDEVA I V, LEBEDEV A V, POPOV A M, et al. Dislocations in stacking and commensurate-incommensurate phase transition in bilayer graphene and hexagonal boron nitride[J]. Physical Review B, 2016, 93(23): 235414.e.

[161] SLOTMAN G J, DE WIJS G A, FASOLINO A, et al. Phonons and electron-phonon coupling in graphene-h-BN heterostructures[J]. Annalen Der Physik, 2014, 526(9/10): 381-386.

[162] SACHS B, WEHLING T O, KATSNELSON M I, et al. Adhesion and electronic structure of graphene on hexagonal boron nitride substrates[J]. Physical Review B, 2011, 84(19): 195414.

[163] GIOVANNETTI G, KHOMYAKOV P A, BROCKS G, et al. Substrate-induced band gap in graphene on hexagonal boron nitride: ab initio density functional calculations[J]. Physical Review B, 2007, 76(7): 073103.

[164] FAN Y, ZHAO M, WANG Z, et al. Tunable electronic structures of graphene/boron nitride heterobilayers[J]. Applied Physics Letters, 2011, 98(8): 083103.

[165] ARGENTERO G, MITTELBERGER A, REZA A M M, et al. Unraveling the 3D atomic structure of a suspended graphene/h-BN van der Waals heterostructure[J]. Nano Letters, 2017, 17(3): 1409-1416.

[166] WANG E, LU X, DING S, et al. Gaps induced by inversion symmetry breaking and second-generation Dirac cones in graphene/hexagonal boron nitride[J]. Nature Physics, 2016, 12(12): 1111-1115.

[167] SACHS B, WEHLING T O, KATSNELSON M I, et al. Adhesion and electronic structure of graphene on hexagonal boron nitride substrates[J]. Physical Review B, 2011, 84(19): 195414.

[168] JUNG S, PARK M, PARK J, et al. Vibrational properties of h-BN and h-BN-graphene heterostructures probed by inelastic electron tunneling spectroscopy[J]. Scientific Reports, 2015, 5(1): 1-9.

[169] DEAN C R, WANG L, MAHER P, et al. Hofstadter's butterfly and the fractal quantum Hall effect in moiré superlattices[J]. Nature, 2013, 497(7451): 598-602.

[170] HOFSTADTER D R. Energy levels and wave functions of Bloch electrons in rational and irrational magnetic fields[J]. Physical Review B, 1976, 14(6): 2239.

[171] ALBRECHT C, SMET J H, VON KLITZING K, et al. Evidence of Hofstadter's fractal energy spectrum in the quantized Hall conductance[J]. Physical Review Letters, 2001, 86(1): 147.

[172] WANG L, GAO Y, WEN B, et al. Evidence for a fractional fractal quantum Hall effect in graphene superlattices[J]. Science, 2015, 350(6265): 1231-1234.

[173] HUNT B, SANCHEZ-YAMAGISHI J D, YOUNG A F, et al. Massive Dirac fermions and Hofstadter butterfly in a van der Waals heterostructure[J]. Science, 2013, 340(6139): 1427-1430.

[174] CHEN Z G, SHI Z, YANG W, et al. Observation of an intrinsic bandgap and Landau level renormalization in graphene/boron-nitride heterostructures[J]. Nature Communications, 2014, 5: 4461.

[175] DAI S, FEI Z, MA Q, et al. Tunable phonon polaritons in atomically thin van der Waals crystals of boron nitride[J]. Science, 2014, 343(6175): 1125-1129.

[176] CALDWELL J D, KRETININ A V, CHEN Y, et al. Sub-diffractional, volume-confined polaritons in a natural hyperbolic material: hexagonal boron nitride (Presentation Recording)[C]//Metamaterials, Metadevices, and Metasystems 2015. International Society for Optics and Photonics, 2015, 9544: 95440R.

[177] PODDUBNY A, IORSH I, BELOV P, et al. Hyperbolic metamaterials[J]. Nature Photonics, 2013, 7(12): 948-957.

[178] DAI S, MA Q, ANDERSEN T, et al. Subdiffractional focusing and guiding of polaritonic rays in a natural hyperbolic material[J]. Nature Communications, 2015, 6(1): 1-7.

[179] FEI Z, RODIN A S, ANDREEV G O, et al. Gate-tuning of graphene plasmons revealed by infrared nano-imaging[J]. Nature, 2012, 487(7405): 82-85.

[180] WANG Y, FANG Z, SCHLATHER A, et al. Tunable absorption enhancement with graphene nanodisk arrays[J]. APS, 2014, 2014: C1. 347.

[181] YAN H, LOW T, ZHU W, et al. Damping pathways of mid-infrared plasmons in

graphene nanostructures[J]. Nature Photonics, 2013, 7(5): 394-399.
[182] LIU Z, LEE H, XIONG Y, et al. Far-field optical hyperlens magnifying sub-diffraction-limited objects[J]. Science, 2007, 315(5819): 1686-1686.
[183] VAKIL A, ENGHETA N. Transformation optics using graphene[J]. Science, 2011, 332(6035): 1291-1294.
[184] IORSH I V, MUKHIN I S, SHADRIVOV I V, et al. Hyperbolic metamaterials based on multilayer graphene structures[J]. Physical Review B, 2013, 87(7): 075416.
[185] DAI S, MA Q, LIU M K, et al. Graphene on hexagonal boron nitride as a tunable hyperbolic metamaterial[J]. Nature Nanotechnology, 2015, 10(8): 682-686.
[186] SUGINO T, KIMURA C, YAMAMOTO T. Electron field emission from boron-nitride nanofilms[J]. Applied Physics Letters, 2002, 80(19): 3602-3604.
[187] YAMADA T, MASUZAWA T, EBISUDANI T, et al. Field emission characteristics from graphene on hexagonal boron nitride[J]. Applied Physics Letters, 2014, 104(22): 221603.
[188] MAYOROV A S, GORBACHEV R V, MOROZOV S V, et al. Micrometer-scale ballistic transport in encapsulated graphene at room temperature[J]. Nano Letters, 2011, 11(6): 2396-2399.
[189] SANDNER A, PREIS T, SCHELL C, et al. Ballistic transport in graphene antidot lattices[J]. Nano Letters, 2015, 15(12): 8402-8406.
[190] BANSZERUS L, SCHMITZ M, ENGELS S, et al. Ballistic transport exceeding 28 μm in CVD grown graphene[J]. Nano Letters, 2016, 16(2): 1387-1391.
[191] DEAN C R, YOUNG A F, MERIC I, et al. Boron nitride substrates for high-quality graphene electronics[J]. Nature Nanotechnology, 2010, 5(10): 722-726.
[192] WEINGART S, BOCK C, KUNZE U, et al. Low-temperature ballistic transport in nanoscale epitaxial graphene cross junctions[J]. Applied Physics Letters, 2009, 95(26): 262101.
[193] CASTRO E V, OCHOA H, KATSNELSON M I, et al. Limits on charge carrier mobility in suspended graphene due to flexural phonons[J]. Physical Review Letters, 2010, 105(26): 266601.
[194] BOLOTIN K I, SIKES K J, HONE J, et al. Temperature-dependent transport in suspended graphene[J]. Physical Review Letters, 2008, 101(9): 096802.
[195] DU X, SKACHKO I, BARKER A, et al. Approaching ballistic transport in suspended graphene[J]. Nature Nanotechnology, 2008, 3(8): 491-495.
[196] GÓMEZ-NAVARRO C, WEITZ R T, BITTNER A M, et al. Electronic transport properties of individual chemically reduced graphene oxide sheets[J]. Nano Letters, 2007, 7(11): 3499-3503.
[197] BOSTWICK A, MCCHESNEY J L, EMTSEV K V, et al. Quasiparticle transformation during a metal-insulator transition in graphene[J]. Physical Review Letters, 2009, 103(5): 056404.
[198] TIKHONENKO F V, KOZIKOV A A, SAVCHENKO A K, et al. Transition

between electron localization and antilocalization in graphene[J]. Physical Review Letters, 2009, 103(22): 226801.

[199] BANSZERUS L, SCHMITZ M, ENGELS S, et al. Ballistic transport exceeding 28 μm in CVD grown graphene[J]. Nano Letters, 2016, 16(2): 1387-1391.

[200] PONOMARENKO L A, GEIM A K, ZHUKOV A A, et al. Tunable metal-insulator transition in double-layer graphene heterostructures [J]. Nature Physics, 2011, 7(12): 958-961.

[201] KIM S M, HSU A, PARK M H, et al. Synthesis of large-area multilayer hexagonal boron nitride for high material performance[J]. Nature Communications, 2015, 6(1): 1-11.

[202] BRITNELL L, GORBACHEV R V, JALIL R, et al. Field-effect tunneling transistor based on vertical graphene heterostructures[J]. Science, 2012, 335(6071): 947-950.

[203] KELLY A G, HALLAM T, BACKES C, et al. All-printed thin-film transistors from networks of liquid-exfoliated nanosheets[J]. Science, 2017, 356(6333): 69-73.

[204] KANG S, PRASAD N, MOVVA H C P, et al. Effects of electrode layer band structure on the performance of multilayer graphene-hBN-graphene interlayer tunnel field effect transistors[J]. Nano Letters, 2016, 16(8): 4975-4981.

[205] CHARI T, MERIC I, DEAN C, et al. Properties of self-aligned short-channel graphene field-effect transistors based on boron-nitride-dielectric encapsulation and edge contacts[J]. IEEE Transactions on Electron Devices, 2015, 62(12): 4322-4326.

[206] LEE G H, YU Y J, CUI X, et al. Flexible and transparent MoS_2 field-effect transistors on hexagonal boron nitride-graphene heterostructures [J]. ACS Nano, 2013, 7(9): 7931-7936.

[207] STOLYAROV M A, LIU G, RUMYANTSEV S L, et al. Suppression of 1/f noise in near-ballistic h-BN-graphene-h-BN heterostructure field-effect transistors[J]. Applied Physics Letters, 2015, 107(2): 023106.

[208] WANG H, TAYCHATANAPAT T, HSU A, et al. BN/graphene/BN transistors for RF applications[J]. IEEE Electron Device Letters, 2011, 32(9): 1209-1211.

[209] KAYYALHA M, CHEN Y P. Observation of reduced 1/f noise in graphene field effect transistors on boron nitride substrates[J]. Applied Physics Letters, 2015, 107(11): 113101.

[210] ROY T, TOSUN M, KANG J S, et al. Field-effect transistors built from all two-dimensional material components[J]. ACS Nano, 2014, 8(6): 6259-6264.

[211] PARK N, KANG H, PARK J, et al. Ferroelectric single-crystal gated graphene/hexagonal-BN/ferroelectric field-effect transistor [J]. ACS Nano, 2015, 9 (11): 10729-10736.

[212] SONG X, GAO T, NIE Y, et al. Seed-assisted growth of single-crystalline patterned graphene domains on hexagonal boron nitride by chemical vapor deposition[J]. Nano Letters, 2016, 16(10): 6109-6116.

[213] PETRONE N, CHARI T, MERIC I, et al. Flexible graphene field-effect transistors

encapsulated in hexagonal boron nitride[J]. ACS Nano, 2015, 9(9): 8953-8959.

[214] SIRRINGHAUS H. 25th anniversary article: organic field-effect transistors: the path beyond amorphous silicon[J]. Advanced Materials, 2014, 26(9): 1319-1335.

[215] COROPCEANU V, CORNIL J, DA SILVA FILHO D A, et al. Charge transport in organic semiconductors[J]. Chemical Reviews, 2007, 107(4): 926-952.

[216] YANG J, CHOI M K, KIM D H, et al. Designed assembly and integration of colloidal nanocrystals for device applications[J]. Advanced Materials, 2016, 28(6): 1176-1207.

[217] ZAUMSEIL J. Single-walled carbon nanotube networks for flexible and printed electronics[J]. Semiconductor Science and Technology, 2015, 30(7): 074001.

[218] KIM K, YANKOWITZ M, FALLAHAZAD B, et al. Van der Waals heterostructures with high accuracy rotational alignment[J]. Nano Letters, 2016, 16(3): 1989-1995.

[219] GUPTA A, CHEN G, JOSHI P, et al. Raman scattering from high-frequency phonons in supported n-graphene layer films[J]. Nano Letters, 2006, 6(12): 2667-2673.

[220] LUI C H, LI Z, CHEN Z, et al. Imaging stacking order in few-layer graphene[J]. Nano Letters, 2011, 11(1): 164-169.

[221] WANG L, MERIC I, HUANG P Y, et al. One-dimensional electrical contact to a two-dimensional material[J]. Science, 2013, 342(6158): 614-617.

[222] SIRE C, ARDIACA F, LEPILLIET S, et al. Flexible gigahertz transistors derived from solution-based single-layer graphene [J]. Nano Letters, 2012, 12(3): 1184-1188.

[223] PETRONE N, MERIC I, HONE J, et al. Graphene field-effect transistors with gigahertz-frequency power gain on flexible substrates[J]. Nano Letters, 2013, 13(1): 121-125.

[224] WU Y, JENKINS K A, VALDES-GARCIA A, et al. State-of-the-art graphene high-frequency electronics[J]. Nano Letters, 2012, 12(6): 3062-3067.

[225] LEE S K, JANG H Y, JANG S, et al. All graphene-based thin film transistors on flexible plastic substrates[J]. Nano Letters, 2012, 12(7): 3472-3476.

[226] LU C C, LIN Y C, YEH C H, et al. High mobility flexible graphene field-effect transistors with self-healing gate dielectrics[J]. ACS Nano, 2012, 6(5): 4469-4474.

[227] KIM B J, JANG H, LEE S K, et al. High-performance flexible graphene field effect transistors with ion gel gate dielectrics[J]. Nano Letters, 2010, 10(9): 3464-3466.

[228] CHENG R, JIANG S, CHEN Y, et al. Few-layer molybdenum disulfide transistors and circuits for high-speed flexible electronics[J]. Nature Communications, 2014, 5(1): 1-9.

[229] MISHCHENKO A, TU J S, CAO Y, et al. Twist-controlled resonant tunnelling in graphene/boron nitride/graphene heterostructures [J]. Nature Nanotechnology, 2014, 9(10): 808-813.

[230] LIU Z, MA L, SHI G, et al. In-plane heterostructures of graphene and hexagonal

boron nitride with controlled domain sizes[J]. Nature Nanotechnology, 2013, 8(2): 119-124.

[231] SHI G, HANLUMYUANG Y, LIU Z, et al. Boron nitride-graphene nanocapacitor and the origins of anomalous size-dependent increase of capacitance[J]. Nano Letters, 2014, 14(4): 1739-1744.

[232] OEZCELIK V O, CIRACI S. High-performance planar nanoscale dielectric capacitors [J]. Physical Review B, 2015, 91(19): 195445.

[233] ÖZÇELI K, ONGUN V, CIRACI S. Size dependence in the stabilities and electronic properties of α-graphyne and its boron nitride analogue[J]. The Journal of Physical Chemistry C, 2013, 117(5): 2175-2182.

[234] KHORASANI S, KOOTTANDAVIDA A. Nonlinear graphene quantum capacitors for electro-optics[J]. npj 2D Materials and Applications, 2017, 1: 1.

[235] LU G, WU T, YANG P, et al. Synthesis of high-quality graphene and hexagonal boron nitride monolayer in-plane heterostructure on Cu-Ni alloy[J]. Advanced Science, 2017, 4(9): 1.

[236] WANG J, MA F, SUN M. Graphene, hexagonal boron nitride, and their heterostructures: properties and applications[J]. RSC Advances, 2017, 7.

[237] WANG J, CAO S, SUN P, et al. Optical advantages of graphene on the boron nitride in visible and SW-NIR regions[J]. Rsc Advances, 2016, 6(112): 111345-111349.

[238] WANG J, CAO S, DING Y, et al. Theoretical investigations of optical origins of fluorescent graphene quantum dots[J]. Entific Reports, 2016, 6: 24850.

Chapter 3

Optical, Photonic and Optoelectronic Properties of Graphene, h-NB and Their Hybrid Materials

3.1 Introduction to graphene

Graphene is the first two dimensional (2D) atomic crystal prepared by scientists[1-4], which constitutes the basic unit of the other carbon material (Figure 3-1)[5]. In recent years, graphene research has made many breakthroughs. Also, large amount of graphene preparation made significant progress. This carbon material with an atomic thickness has excellent mechanical properties, electrical conductivity, optical properties, thermal conductivity and impermeability, making it immensely attractive for many applications[6-12].

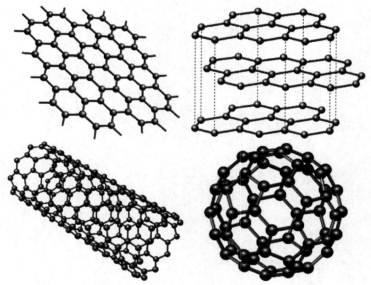

Figure 3-1 Graphene: the basic unit of the other carbon material[5]

3.1.1 Graphene's structure, electronic band

Ideal graphene is a single layer of 2D atomic crystal with orthohexagonal lattice structure[13]. The length of C—C bond is around 0.142 nm, and the thickness of the layer is 0.35 nm.

The bonding between electrons and electrons in the graphene consists of the σ bond of the sp^2 structure and the π bond vertical to the same plane. The adjacent π bonds are coupling with each other to form the basis of graphene conduction[14]. Figure 3-2(a) shows the positional distribution of the atoms A and B in a single graphene cell and the energy-momentum relationship of electrons and thus behave as massless Dirac fermions[4,15-16]. There are A and B sub-lattices in each single cell of graphene, and the spins of the electrons on the A

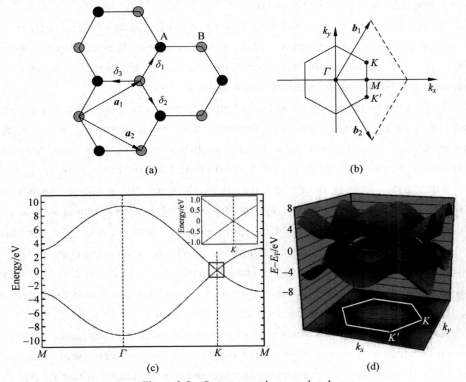

Figure 3-2 Structure and energy band
(a) The structure of graphene lattice[14]; (b) The Brillouin zone of a unit graphene[14]; (c) The band structure of a single graphene layer along M, Γ, K, M[13]; (d) The band structure (top) and Brillouin zone (bottom) of graphene

and B sub-lattices have chiral features with each other. Graphene is a semiconductor, which has zero band gap, and at six points of high symmetry in the Brillouin zone (Figure 3-2(b)), the conduction band and the valence band intersect at one point, called the Dirac point (Figures 3-2(c) and (d))[13,17]. Near the Dirac point, the energy and momentum show a linear dispersion relationship $E = \hbar \nu_F \mathbf{K}$[14,17], where $\nu_F = 10^6$ m/s, the linear dispersion relationship makes the effective mass of electrons in graphene equal zero, the electron in graphene is a type of massless Dirac fermion[14,17].

3.1.2 Electronic properties of graphene, which impact the optical properties

Unique crystal structure and electronic structure make the material have a lot of new and noteworthy properties.

At room temperature, graphene is able to exhibit quantum Hall effect, so it appears anomalous quantum Hall effect at low temperatures (below 4 K)[18-19]. This behavior has been interpreted as *electrons in relativistic quantum mechanics in graphene with no quiescent mass*.

The conduction band and valence band of graphene are symmetrical, and there are two charge carriers: electrons and holes. At the same time, confirmed by experimental results the mobility of electrons and holes in graphene is same[20], which is called Ambipolar Electric Field Effects.

The carrier density near the Dirac point is zero, but graphene shows the existence of minimum conductivity, on the order of 4 e^2/h, between 10 K and 100 K, and its mobility is almost independent of temperature[18,21-22], because of Berrys phase[4]. Electrons in graphene have the characteristics of Klein tunnelling, and pass through the barrier with the probability of 100 %[14]. In this phenomenon, the properties of massless particles are not rebounded or scattered by the presence of barrier during the motion.

At room temperature, the speed of conduction electrons is faster than any known conductor, and when graphene electrons move in orbit, scattering will not occur due to lattice defects or external electrons. Therefore, the average free path of the electrons in the graphene can reach 1 μm[23], while the average material in the electronic free path is in the nanometer scale. People often call it ballistic transport.

As a zero band-gap semiconductor, the unique carrier characteristics and excellent electrical properties of graphene make it have very special optical properties, because the optical properties of matter are related to electrons state density, they directly affects all the properties of matter optics[4,15-16].

Ideal graphene is only a single atomic thickness, making it have optical visibility[24-25], its transmittance (T) can be indicated according to the parameters of the fine structure[26]. Dirac electrons' linear dispersion makes applications of graphene in broadband implements possible; as a result of the Pauli blocking[27-28]; we can observe its saturation absorption; the unbalanced carrier leads to hot luminescence[29-30], and the graphene chemical and physical properties cause it to luminescence[31-34], making graphene the ideal material for optoelectronic devices and optoelectronics.

3.1.3 Optical properties of graphene

1. Linear optical absorption

Graphene absorbs 2.3 % of incident light, even for one atomic thickness, and the frequency of incident light has nothing to do with the absorption intensity (Figure 3-3)[26]. These unique optical properties are generated by electrons conical band and holes at the Dirac point.

Graphene on top of a Si/SiO$_2$ substrate can be identified by the usage of the optical image contrast[24]. A research team of Geim A. K. studied the image that when SiO$_2$ used as a substrate representing the relationship between multilayer graphene interference[26]. They converst the incident light and substrate's thickness to maximize the contrast. According to the Fresnel equation[35], the photoconductivity of the suspended monolayer graphene $G_0 = e^2/4\hbar$ gives that the transmissivity $T_{opt} = (1 + 2\pi G/c)^{-2} \approx 1 - \pi\alpha \approx 0.977$. This is because the valence and conduction bands of graphene intersect near Dirac point and the photoconductivity of graphene in the photonic bandgap over a wide $G_1(w) = e^2$ is independent of the frequency, but depends on the fine structure constant α, unrelated to the frequency. In the visible light region, the reflectance of each layer $A = 1 - T \approx \pi\alpha \approx 2.3$ % as shown in Figure 3-3(a). At the atomic scale, graphene exhibits a strong broadband absorption (wavelength range of 300~2,500 nm), which is about 50 times that

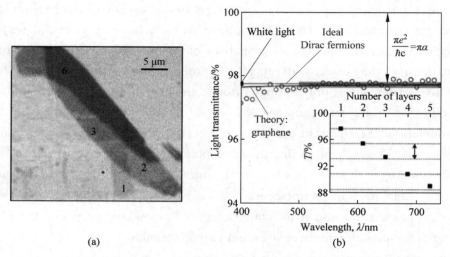

Figure 3-3 Optical image and transmittance spectrum of single-layer graphene
(a) Schematic of optical micrograph for different layers[24]; (b) Transmittance spectrum of single-layer graphene[26]

of GaAs with the same thickness.

Electrons in graphene, as non-mass 2D particles, result in a very important non-wavenumber absorption ($\pi\alpha \approx 2.3\%$) for normal incident light below 3 eV[35]. In addition, when the light energy is less than twice the Fermi level, the monolayer and multilayer graphene become completely transparent due to the Pauli barrier effect. These properties are mainly attributed to the density of graphene in electron transport of the van der Hove singular point and the electron transition between the bands, which are suitable for use in many controllable photonic devices.

2. Photoluminescence

Graphene has special characters and potential applications in photoluminescence and electromagnetic transport, which have aroused a wide range of research interests[36-47]. Graphene fragments and graphene quantum dots show unique photoluminescence properties in the preparation process; the physical and chemical treatment of graphene to reduce the coherence between π electrons is another way to make graphene photoluminescence. After a slight oxidation plasma treatment, a single graphene sheet can bring light[31], and at this time photoluminescence has a large area of uniform (Figure 3-4(a)). Likewise, bulk graphene and dispersions also exhibit extensive photoluminescence[47-49]. By etching the top layer, the lower layer intact, this makes the preparation of the mixture possible. Based on graphene materials photoluminescence being routinely prepared, the combination of this conductive layer and photoluminescence can be applied to sandwich-type light-emitting diodes(LEDs), a wavelength range from infrared (IR) to blue spectral[31,49].

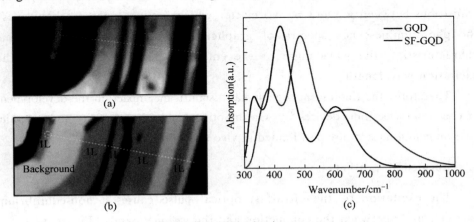

Figure 3-4 Photoluminescence

(a),(b) Photoluminescence and elastic scattering image[31]; (c) Comparison of optical absorption of graphene quantum dots with NH_2 modified graphene quantum dots[50]

Someone once prepared a graphene quantum dot that can emit blue light by hydrothermal method[40]. Li et al. prepared a green-emitting quantum dot by electrochemical method, which can be made as the electron acceptor material for photovoltaic devices[41]. Eala et al. found that the graphene oxide can emit ultraviolet(UV) and blue light when excited by UV light. Sun et al. used photoluminescence of oxidized graphene to image living cells in the near-infrared(NIR) region[33]; Wang et al. studied the photoluminescence mechanism of graphene quantum dots[50], which revealed the excited state transfer between electrons and holes and the influence of edge effect light luminescence (Figure 3-4(b)), made in visible photoluminescence devices within the scope of the preparation of a theoretical basis.

As the size of quantum dots decreases, graphene exhibits quantum confinement effects and unique edge effects[36,50-52]. Some teams interpret the photoluminescence of graphene oxide as radiative recombination of e-h pairs of localization states of sp^2 cluster[32]. The energy gap between the π^* and π states is defined by the size or the conjugate length of the sp^2 cluster[53-54], which is more likely to be explained by the graphene edge effect and the defects associated with oxidation. Some people have explained that the photoluminescence of GQD is derived from radiative recombination of electrons in the free Zigzag sites of graphene edges[35], rather than the general transitions between π^* and π. No matter which luminescence mechanism is, the photoluminescence spectra of graphene quantum dots have the same characteristics, the wavelength of the emission spectrum changes with the excitation wavelength.

Therefore, the fluorescent organic has a significant impact on the development of cheap optoelectronic devices[55]. On the other hand, their toxicity and potential environmental hazards are also limited in vivo and other applications[56-57].

3. Saturated absorption

The excitation of the ultra-fast optical pulse causes a non-equilibrium carrier cluster between the conduction and the valence bands (Figure 3-5)[28]. In the time-resolved experiments[58], two relaxation time scales can be observed: one is the fast relaxation corresponding to the mutual collision and phonon scattering electrons between the conduction and the valence bands, which is about 100 fs, and the other is the slower relaxation corresponding to electron band relaxation and thermophonon cooling[59-60].

On the picosecond scale on, the linear dispersion relation of Dirac indicates that there is always resonance between electrons and holes for any excitation. The electrons and holes are solved quantitatively using the kinetic equation and the distribution function of holes. ($f_e(p)$ and $f_h(p)$), where p is the momentum of the Dirac point[28]. When the pulse time is greater than the relaxation time, the electrons reach a quiescent state during the pulse time. At the effective temperature, the electron and the hole reach the thermal equilibrium by collision. On the other hand, the number of electrons and holes determines their density of states and the total energy density[28]. Due to Paul blocking, by the factor $1 + \Delta\alpha_1/\alpha_2 = [1 - f_e(p)][1 - f_h(p)]$, the decrease of photon absorption for each layer by given laser energy can be determined by

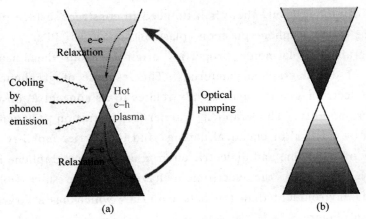

Figure 3-5 Optical excitation and relaxation in graphene

(a) Schematic diagram of optical excitation in graphene; (b) Low band relaxation[28]

the populations.

Similarly, because of the Dirac points, linear dispersion relation, the collision of the paired carriers does not cause relaxation between the conduction and the valence bands, which protects the total number of electrons and holes, respectively[28,61]. When the energy of electrons and holes approaches the Dirac point, the phonon emission causes interband relaxation, and spontaneous emission of hot electrons and hole groups also occurs[29-30].

Theoretically, to certain amount of material, a decoupling SLG is able to provide higher saturable absorption than others, because of the two-order dispersion of the graphene sheet and the collision of the sub carriers, the relaxation of the band gap is produced[28].

4. Graphene plasmon

Surface plasmon polaritons (SPPs) are the collective oscillations of free electrons induced by external electromagnetic fields (such as photons or electrons), and have surface electromagnetic field propagation properties: the electric field intensity has the maximum value at the interface between the metal and the medium. When increasing the vertical distance from the metal surface, the field strength exponentially decreases[62]. Recently graphene has been confirmed to be a plasma-derived waveguide material in the IR frequency range[63], also known as terahertz metamaterials. Graphene has similar properties to metals in conductivity and can transport SPPs. Since it has only a thin layer of atomic level, there is little need to construct surface properties, so it can be called graphene plasmon (plasmon polaritons, PPs).

The graphene plasmonic properties originate from the linear optical properties of doped graphene materials. The response of graphene to high-frequency electromagnetic waves mainly relates to its concentration of carrier and carrier mobility. The graphene carrier concentration can generally be modulated by classical or chemical doping, and the carrier mobility relates to the quality of grapheme and dielectric environment. Since graphene is a single atomic layer material, the electronic behavior is more susceptible to the surrounding environment than the bulk material, which has a larger specific surface area (Figure 3-6)[64-68].

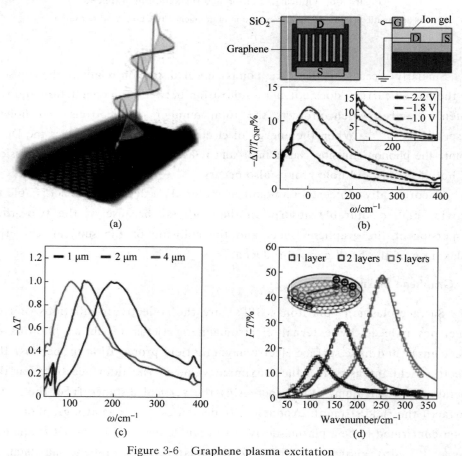

Figure 3-6　Graphene plasma excitation

(a) Schematic diagram of graphene plasma excitation; (b) Schematic diagram and test results of plasma resonance test of graphene; (c)~(d) Test patterns of different widths and layers[65]

The chemical potential μ and the Fermi level E_F determine the performance of graphene. By adjusting chemical doping or photoelectric reflection, the transition from electrolyte to metal can be achieved. The μ between the conduction and valence band gap and the in-band conductivity and graphene are relevant to the frequency ω of the incident light. When the chemical potential $\mu < \hbar\omega/2$, the band-gap transition plays a dominant role and controls the dynamic high-frequency conductivity; when the chemical potential $\mu > \hbar\omega/2$, the transition in the band plays a dominant role and controls the megahertz range of conductivity. Under these conditions, the momentum of PPs is enhanced, so PPs can propagate in graphene. The competitive relationship between the transition of light in the band and the transition of the band gap makes it possible to couple the tunable optical response and the polarization selective coupling.

3.1.4 The application of photonics and optoelectronics

1. Flexible electronic devices

Electronic products widely use conductive coatings, such as touch screen[69], electronic paper, organic photovoltaic cells[70], and organic LED[71], and the need for low surface resistance and high transmittance of special applications. Figure 3-7 by way of comparison the transparent conductive layer based on graphene is different from other properties of photoelectric materials, showsclearly the photoelectric properties of them[72-73].

Graphene meets the needs of electronic and optical equipment, single-layer transmittance of up to 97.7%, but in the past people thought that indium tin oxide (ITO) performance would be better[74]. However, considering the annual increase in the quality of graphene, price of ITO will be higher, and the cost of deposition method preparation of ITO will be higher as well. So, graphene will certainly get a larger market share. Excellent flexibility and corrosion resistance of graphene are the most important properties of flexible electronic materials equipment, but ITO cannot achieve in this respect.

The electrical properties required for different applications of the different motors are not the same (such as surface resistance). Due to the different production methods, there will be a variety of conductive coatings. Therefore,

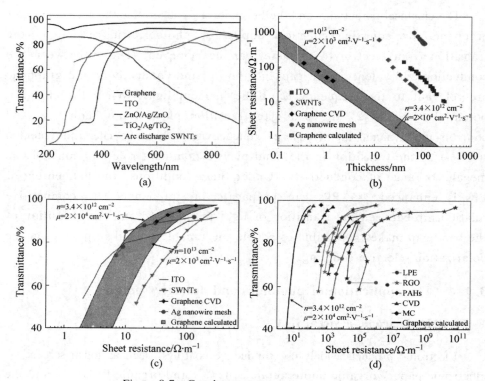

Figure 3-7 Graphene as transparent conductor

(a) Transmittance for different transparent conductors: GTCFs[72], single-walled carbon nanotubes (SWNTs)[73], ITO[74], ZnO/Ag/ZnO and TiO$_2$/Ag/TiO$_2$[75-76]. (b) Thickness dependence of the sheet resistance. The blue rhombuses show roll-to-roll GTCFs based on CVD-grown graphene[72]; red squares, ITO[74]; grey dots, metal nanowires[74]; green rhombuses, SWNTs[73]. Two limiting lines for GTCFs are also plotted (enclosing the shaded area). (c) Transmittance versus sheet resistance for different transparent conductors: blue rhombuses, roll-to-roll GTCFs based on CVD-grown graphene[72]; red line, ITO[74]; grey dots, metal nanowires[74]; green triangles, SWNTs[73]. Shaded area enclosed by limiting lines for GTCFs calculated using n and μ as in (b). (d) Transmittance versus sheet resistance for GTCFs grouped according to production strategies: triangles, CVD[10, 72, 77-78]; blue rhombuses, micromechanical cleavage (MC)[79]; red rhombuses, organic synthesis from polyaromatic hydrocarbons (PAHs)[80]; dots, liquid-phase exfoliation (LPE) of pristine graphene[79, 81-83]; and stars, reduced graphene oxide (RGO)[84-88]

the electrodes of the contact screen (products requiring the CVD method) have a relatively high surface conductance on the basis of 90% light transmittance[72]. Graphene electrodes used in contact panels have the advantage that graphene has greater stability. In addition, the graphene has a tenfold higher fracture strain than ITO, which means that graphene can be used in collapsible and bendable equipment.

Flexible electronic paper is very attractive. Its bending radius is in the range of 5 mm to 10 mm. This requirement is very easy to achieve for graphene. And graphene can absorb visible light, which for the color of electronic paper is very important. However, the graphene electrode contact resistance and metal loop are still a big problem.

Despite the relatively high resistance of graphene films, the advantages of flexibility and high mechanical strength of graphene ensure that the graphene equipment can have more flexible applications.

2. Photodetector

One of the most widely researched optoelectronic devices, graphene photodetectors are and can be used in the wide-band spectral region between UV and IR. The ultra-wide working bandwidth is an advantage of graphene photodetector, allowing graphene photodetectors to be used in high-speed data communications. The high carrier activity of graphene provides a carrier for rapid extraction of the picture, as it allows for higher bandwidth operation. At the reported velocity of saturated carriers, the bandwidth of the graphene photodetector due to time constraint is expected to reach 1.5 THz[89-90]. In fact, the graphene photodetector has the maximum bandwidth 640 GHz due to the delay of the capacitor rather than the delay of the transfer time[89].

At present, the graphene photodetector uses a local potential change near the surface of the metal-graphene to extract the photodetector carrier[91-92]. The optical response rate can reach 40 GHz; the detector-operating rate can reach 10 GHz. However, the maximum response rate is relatively low due to the limitation of the smaller effective detection area and the thin graphene to the absorption rate.

There are many solutions to increase a graphene photodetector sensitivity, for example, using nanostructured plasma to enhance the local optical electric field or to increase the photo-graphene interaction length by combining with a waveguide[93-94].

Without the band gap of graphene low response and low optical gain and limits the applications of graphene-based photoelectric detector, even the superfast and broadband optical response of graphene. Figure 3-8 shows a photodetector based on graphene-Bi_2Te_3 heterostructure which overcomes these difficulties[95]. Owing to topological insulators family with small hexagonal symmetry structure like to that of graphene with zero-gap materials, device of the photocurrent can be effectively enhanced and fail to detect a decline in spectral width. The result shows that the graphene-Bi_2Te_3 photodetector has higher optical response and higher sensitivity than the pure monolayer graphene apparatus, and the detection wavelength range of the device is extended to the near IR region (980 nm), the communication band (1,550 nm).

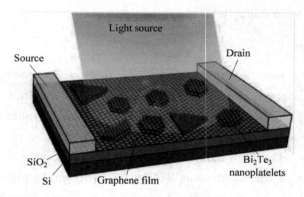

Figure 3-8　Photodetector based on graphene-Bi_2Te_3 heterostructure[95]

3. Optical modulator

An excellent optical modulator is able to achieve the performance through stripping the resulting graphene, which absorbs a small amount of incident light from a broad spectrum of light and is able to respond quickly. In order to achieve these properties, in a single graphene layer[96], the inter-band transition photoelectrons are modulated by a driving voltage across a wide band to obtain an optical modulator with a bandwidth in the near-IR region of more than 1 GHz[97]. The use of mutually constrained double graphene can reduce the delay in the RC delay by providing some structure change, providing an area that can reach hundreds of gigabits. Theoretically operating light modulators with bandwidth in excess of 50 GHz cannot be achieved[94]. Graphene is a latent material for megahertz wireless communications because the loss of light in graphene is much less than in precious metals.

Liu et al. assembled and fabricated an electro-absorption modulator based on graphene waveguide integration for the first time[97], in which actively tuning the Fermi level of a monolayer graphene sheet achieves modulation (Figure 3-9). The gigahertz graphene modulator demonstrates a strong electroabsorption modulation of 0.1 dB/μm and operates over a broad range of wavelengths, from 1.35 μm to 1.6 μm.

Figure 3-9　Schematic diagram of graphene optical modulator[97]

4. Mode-locked laser/THz generator

Ultrafast passive mode-locked lasers have been used in spectroscopy, micro-materials processing[98], biomedical[99], and safety applications. They are often used as a saturable absorber, by selecting high-intensity-light transmission to cause the modulation of light intensity. Compared to the semiconductor saturable absorber[100], graphene monolayer absorbance is very high, which in the low light intensity of the wide band area can be saturated[28,101]. Controllable modulation depth, high thermal conductivity, wide frequency tuning, relaxation time and high damage threshold of the ultrafast carrier are all the advantages of the graphene-saturated absorber[26,102-103].

Scientists have devoted much of their research to fiber and solid-state lasers[104]. However, people also can use graphene-saturated absorbers in semiconductor laser technology. Wavelength multiplexing eliminates the need for optical interconnects with a series of lasers of different wavelengths. Using a single laser in different longitudinal modes can generate different wavelengths such as mode-locked lasers[105]. Active mode-locked silicon hybrid lasers have been studied to meet the needs of laser technology[106], but graphene-saturated absorbers can provide simple passive mode-locked semiconductor lasers for operation and processing.

Wan et al. prepared high-quality graphene by CVD method on the copper[107]. It is the first time to implement the use of high-quality single-graphene as a saturable absorber (SA) to prepare the stability of the diode-pumped passive mode-locked Tm: YAP laser (Figure 3-10). The absorption pump power is 7.67 W, and the output power of the laser can reach 256 mW at 1,988.5 nm, which is the center wavelength. When the pulse frequency is 62.38 MHz, the pulse energy can reach 4.1 nJ[107]. Studies show that using graphene to create low-cost and ultrafast laser is promising.

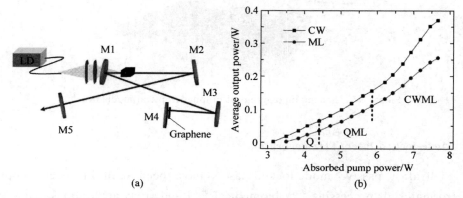

Figure 3-10 (a) Assembly diagram of mode-locked laser based on graphene;
(b) Test pattern of graphene mode-locked laser[107]

3.2 Introduce of h-BN

Hexagonal boron nitride (h-BN) is the only phase structure in all phases of boron nitrides(BNs)[108-111], which is hexagonal, white and similar to layer structural features of graphene, also known as "white graphene". Each atomic layer is composed by B and N atoms alternately arranged in an infinite extension of the hexagonal grid (Figure 3-11), the atomic layer along the c axis direction in accordance with the ABAB order. In each layer, B and N atoms are strongly bonded by sp^2 covalent bond, with bond length $a = b = 0.2504$ nm. Weak van der Waals force bonds layers and layers. The bond length $c = 0.6661$ nm and density of 2.28 g/cm^3. Therefore, h-BN along the c axis direction of the bonding force is very small; space of atomic layer is large; it is easy to slide between layers; and it is a good lubricant. Hexagonal boron nitride has a wide range of applications due to its excellent physicochemical properties[112-119]. It can be used as oxidizing additives for refractory materials, insulating films in electronic devices, transparent insulators for electroluminescent devices due to its transparentness in X-ray and visible regions, and in the manufacture of sub-nanometer VLSIs protective film and so on.

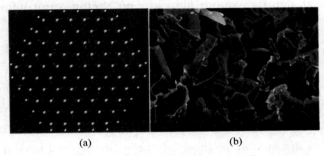

(a) (b)

Figure 3-11 (a) The planar structure of 2D h-BN[111]; (b) SEM surface photographs of the h-BN film[119]

3.2.1 The electronic band structure of 2D h-BN

Graphene and 2D h-BN has similar honeycomb structure. In 2D h-BN, each cell contains a B and an N atom, with total of eight electrons; between B atoms and N atoms performs sp^2 hybrid; each B (N) atoms and the most immediate of the three N (B) atoms generate three σ bonds; there are six electrons filled into the σ orbit; the remaining two electrons in $2p_z$ state, formats the vertical plane π bond[120].

2D h-BN is directly band-gap, top of valence band and bottom of rewinding are all at the high-symmetry point K. HOMO and LUMO of the system are determined by the π and π^* that state on the N and the B atoms, respectively. The bonding combination of B-sp^2 and N-sp^2 orbitals forms the bond between nearest B and N atoms. Electron transfers from B atom to N atom, because of the electronegativity difference between B atom and N atom. Between B and N, the bond shows ionic properties, opposite to pure covalent bonds in graphene. The properties of h-BN are mainly due to the charge transfer between nitrogen and boron atoms, hence the complementary structure and honeycomb structure of graphene with h-BN[112-113].

Figure 3-12 (a) shows the structure and charge of atomic, charge transferring between nitrogen and boron atoms, and Figure 3-12(b) is the density of states of 2D h-BN electronic structure. High density around N atoms is represented by contour plots of total charge. The charge density of 2D h-BN minus charge densities of free B and N atoms equals to the difference charge density, i.e. $\Delta \rho = \rho_{BN} - \rho_B - \rho_N$. Charge transfer from B to N atoms can be indicated by high-density contour plots around N atoms protruding toward the B-N bonds[114], 2D h-BN is a semiconductor. The calculations of electronic energy bands and h-BN crystal are similar[115]. As bonding and anti-bonding combinations of N-p_z and B-p_z orbitals, the π and π^* bands of graphene crossing at the K and K points of the BZ open a gap in the 2D h-BN. N-p_z makes significant contribution for the filled band at the edge of valence band. The calculation of band gap is not direct, E_G = 4.64 eV. TDOS and partial density of states are similar to those of the h-BN layered crystal as shown in Figure 3-12.

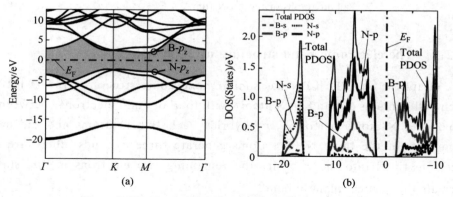

Figure 3-12 Energy bands and density of state of h-BN
(a) Optimized atomic structure, energy bands of 2D h-BN; (b) Density of state of 2D h-BN[120]

3.2.2 The optical properties of h-BN

BN is the lightest, strongest ionicity group Ⅲ-Ⅴ material, whose optical properties have drawn people's close attention. In the short-wave LEDs, semiconductor lasers, and photodetectors, BN has a wide range of applications. Especially, in recent years, h-BN emission spectra have been found to be very strong excitonic emission peaks in the UV region[116,118,121], which can be applied to deep UV lasers. The optical properties of BN nanostructures also draw great interest[122-126].

1. The IR characteristics of h-BN

As we know Fourier transform infrared (FTIR) have been used to examined new prepared products, which is the effective method to characteristically determine the structure of new experimental products, so some researchers used FTIR to the structural nature of h-BN[127-130].

Some researchers have got the data of BN vibrations in h-BN in the state of single crystal. The peak value is 1,365 cm^{-1}[128]. The phonon modes of turbostratic h-BN is 792 cm^{-1} and 1,384 cm^{-1}. From some literatures, we can know that the phonon modes have shift to 800 cm^{-1} and 1,372 cm^{-1} when h-BN is in multilayer state. On the other hand, the phonon modes of h-BN tubes shift to 811 cm^{-1} and 1,377 cm^{-1}, when h-BN is in polycrystalline state[129]. From Figure 3-13, we can see that, there are two peaks at 1,374 cm^{-1} and 818 cm^{-1} from FTIR spectrum, due to the strong vibrations of BN in h-BN nanosheets. With the aid of FTIR, the structural properties and phase composition of experimental h-BN products had been revealed. We can also know that the peak at 1,374 cm^{-1} belongs to the transverse optical modes of the sp^2-bonded h-BN when B-N is in plane. When nitrogen-boron-nitrogen plane is out of plane, there are a peak at 818 cm^{-1} appeared, which could belong to bending vibration of B-N-B[127,130]. However, there are no any absorption peak belonging to the raw materials or any impurity emerged in visible region.

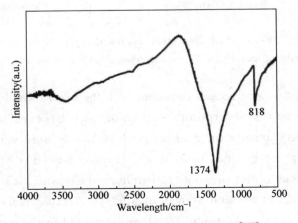

Figure 3-13　Infrared spectra of h-BN[127]

2. The UV characteristics of h-BN

With the deepening of the research, the theoretical calculation and experimental verification has been shown the UV absorption spectrum of the 2D h-BN materials[131-134].

By the simulation of ideal 2D h-BN nanometre materials, Wang et al. calculated the single layer h-BN's absorption spectrum by using DFT and DT-DFT methods, the spectrum of 2D single-layer h-BN materials revealed the optical absorption property of h-BN[131]. From Figure 3-14(a), we can see three strong absorption peaks in the spectrum. On the other hand, there is a shoulder around 210 nm in the deep UV region. The three absorption peaks are around at 197 nm, 203 nm and 208 nm, respectively. By using the charge different density (CDD) method, the absorption peaks are due to strong charge transfer and the vibration of the BN.

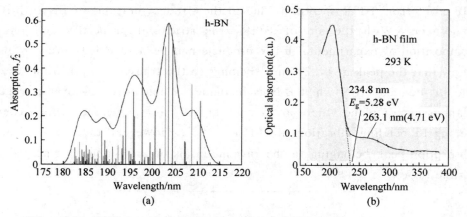

Figure 3-14 UV-vis optical absorption spectra of 2D h-BN nanometer material
(a) Simulation model[131]; (b) The experimental h-BN nanosheets at 293 K[132]

The theoretical calculation is consistent with the experimental verification. Jin et al. got the deep UV absorption spectrum of the h-BN experiment materials. There is a strong absorption peak around at 234.8 nm, which the optical energy band gap corresponds to 5.28 eV (Figure 3-14(b))[132]. This data is basically the same as the previous researchers, just as 5.2 eV and 5.14 eV, made by Hoffman et al. and Stenzel et al.[133-134] respectively. On the other hand, an absorption peak around 263.1 nm appeared in the optical absorption spectrum, corresponding to optical energy band gap with the value of 4.71 eV. They made sure that, the defect level of the BN nanosheets resulted this absorption peak, where the absorption peak energy corresponds to the difference between 5.28 eV (corresponding to the optical energy band gap) and 0.63 eV (corresponding to the activation energy of the donor level). By using the TSC method, the observing result can prove the influence of the donor level, which leads to the emergence of the absorption peak at 263.1 nm.

Both the results of the experiment and simulation are proved once again, that 2D h-BN has strong optical response in deep UV area.

3. Luminescence property of h-BN

As one of semiconductor materials, the bandgap energy properties and band structure of h-BN materials have been studied for some years. With the acceleration development of research, both direct bandgap and indirect bandgap have been revealed little by little. Luminescence is one of unique optical features of h-BN[132,135-139].

Cathodoluminescence (CL) property: Watanabe et al. studied the optical and electrical properties of h-BN crystal[136]. There is a CL spectrum of experimental h-BN materials at room temperature in Figure 3-15(a). From the Figure 3-15(a), a strong peak in the spectrum of the h-BN single layer crystal had been successfully made by them. The peak appeared at 215 nm with the bandgap is 5.76 eV at the room temperature. On the other hand, there is a shoulder structures appeared at 300 nm, corresponding with the bandgap is 4 eV. By researching this shoulder, either the structure of h-BN defects or the impurities in h-BN might cause this result, just as oxygen and carbon[135]. The ratio of two peaks of about 100 times, all in the deep UV region.

Photostimulated luminescence (PSL) of h-BN: As described above, CL spectroscopy had been thought the existence corresponding to near-bandgap structure, which is in the deep UV region. The PSL is thought the result, to capture the free carrier at the distal lattice point, the visible light makes the carriers recombine in these traps[137,140-142].

Though the research of Figure 3-15(b)[137], PSL is revealed that the state depends on the infinite lifetime, which the depopulation is triggered with incident light. When the energy of incident light beyond 5.5 eV, PSL

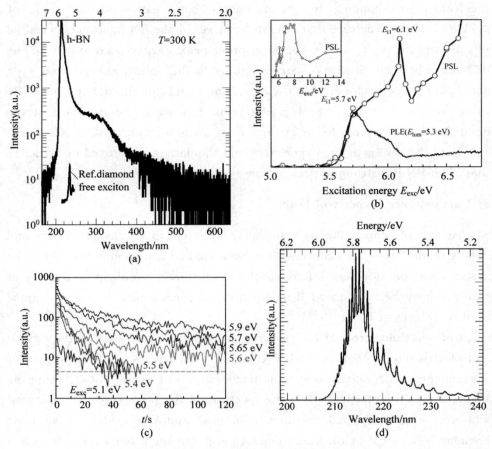

Figure 3-15 Photostimulated luminescence

(a) Cathodoluminescence spectrum of h-BN[136]; (b) The PSL excitation spectrum and PLE spectrum of polycrystalline h-BN[137]; (c) Phosphorescence decay curves of different photon excitation energy[137]; (d) Laser emission spectrum of h-BN[136]

spectrum appeared. On the other hand, the quality of sample or defects may destroy the crystalline symmetry, which will make the hypothetic dark state get accessible. There is no free exciton emission in the figure of the case, so people considered that the strong coupling of the singlet with triplet excitons leads to the emergence of the results and a shortened lifetime. Accordingly, there is no exciton state of long lifetime in the case. In Figure 3-15(b), by using open circles, a small part of PSL excitation spectrum has been represented corresponding to the whole spectrum, the whole measured spectrum is made as the inset in Figure 3-15(b). The range of energy from 5 eV to 7eV is considered the PSL excitation energy. The spectrum to be index increase, then it reached the maximum at 5.7 eV. After this, the growth of the image become slower until it reaches the next maximum at 6.1 eV. After brief drop, the image adds progressively and reaches the final at 7 eV. Purple line represents the PLE spectrum, which is the near band-edge luminescence with the value is 5.3 eV.

Again, we cannot measure the bandgap energy by using the PSL excitation. But we can get the valuable information of exciton or band-edge energy position by the combination with the PLE spectrum.

Phosphorescence property of h-BN: From Figure 3-15(c)[137], the phosphorescence decay curves can be seen of different photon excitation energy, the range from 5.4 eV to 5.9 eV. When the energy at $E_{exc} \leqslant 5.5$ eV, the phosphorescence decay curves can be characterized by monoexponential decay during the characteristic time of 5.8 s. The PM dark noise is shown in the figure by the dotted line. When the energy is at $E_{exc} \geqslant 5.6$ eV, the long-lived part of low intensity arises, it is the higher part of the figure just like a plateau. The long component of the recombination at large distance tends to infinity, which belongs to the charges. When the excitation energy added from 5.1 eV to 5.9 eV, the intensity of phosphorescence increased about 20 times.

Some researchers studied the phosphorescence spectrum, they considered it is related to the dipole forbidden, that is dark exciton states. Watanabe et al. considered the coupling between the dark exciton states and the white exciton states[136], which makes the phosphorescence spectra arise in the absorption spectrum of h-BN monocrystals.

Laser emission: Figure 3-15(d) is the laser emission spectrum of h-BN, measured by Watanabe et al.[136] When light transmitted through the experiment, the sharp fringes of the Figure 3-15(d) is correspondence with fringes of absorption spectrum, the fringes' wavelengths from 213 nm to 224 nm at low temperature 8 K. This fringe mode spacing can be expressed with a formula as follows: $\lambda^2/2D(n - \lambda dn/d\lambda)$, D expresses the thickness of h-BN, n means the index of refraction and λ denotes the wavelength, severally.

The spectrum of laser emission spectrum reveals that the change of emission wavelength range is at the beginning around at 208 nm and at the end around at 240 nm. The fringes rise sharply from 208 nm to 215 nm and achieve maximum intensity at 215 nm, where the strong UV luminescence happens with the energy 5.76 eV. From 215 nm to 240 nm, the fringes sharply decline. On the other hand, when the wavelength is at 240 nm, the emission intensity close to zero. When $dn/d\lambda$ changes, the dispersion term can lead to the mode spacing change. Similar to the exciton polariton effect near the exciton energy level, the exciton effect determines the abnormal dispersion.

4. Phonon polaritons on h-BN

When optical phonons are coupling to the photons in polar crystals, the collective modes are called phonon polaritons. Those materials show the optical phonons, similar to some polar van der Waals solids, such property might be usually called polaritonic effects. And the strong phonon resonances, let h-BN become a good polaritonic effect materials.

Dai et al.[143] studied this character of h-BN by using s-SNOM. They adjusted the intensity of incident light and changed the number of layers to achieve the realization of the phonon polariton. Figure 3-16(a) is the dispersion relation of phonon polaritons of h-BN. From the image we can know that AFM tip alter the range of momenta was supporting phonon polaritons of h-BN. The polaritons propagate rapidly on the surface of BN. With the change of the incident wavelength, the dispersion relation of the phonon polariton is expressed in Figure 3-16(a). Figure 3-16(b) shows the images of different layers of h-BN and different lengths of edge, Figures 3-16(c) and (d) are experimental measurement polariton wavelengths and the data predicted according to the principle, which are in good agreement.

From their study, the phonon polaritons can confine and tunable electromagnetic energy on the nanometer scale. By altering the layers' number, the wavelength and confinement of phonon polaritons can be designed.

3.2.3 Potential applications of h-BN

BN has such optical and photoelectric properties that have broad application prospects in the optical field, especially in the deep UV region.

1. MSM photodetector

H-BN epilayers have some features, just as optical absorption and dielectric strength, as a result, it has potential as deep UV detector materials for application.

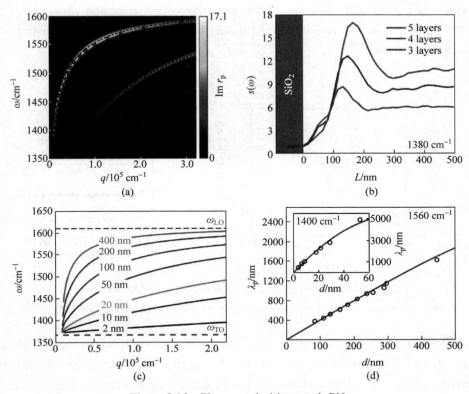

Figure 3-16　Phonon polaritions on h-BN

(a) Sketch map of phonon polaritons of h-BN[143]; The test spectrum line of h-BN with different layers, width in (b)~(d)

Li et al. synthesized h-BN epilayers by using CVD method[144]. Analogous to the optical absorption property of graphene, they tested the absorption coefficient, about 7×10^5 cm of band edge in h-BN. Both the dielectric strength and the absorption coefficient are better than that of wurtzite AlN. Figure 3-17 shows the typical I-V performances of MSM detectors, based on h-BN epilayer. When the bias voltage is at 100 V, the low dark current is 200 pA and current density is 10^{-10} A/cm^2, severally. There are other data of spectral responses corresponding to the different bias voltages in the figure. From Figure 3-17(b), there is a sharp peak of optical responsivity at 220 nm and a cutoff wavelength at 230 nm, respectively. It is in good agreement with the fluorescence emission peak of the band-edge of h-BN, which is at 227 nm (5.48 eV). On the other hand, there are no detectable responses in low-frequency region up to 800 nm. However, the visible rejection ratio of h-BN MSM detectors in UV region is 2~3 orders of magnitude shorter than that of the AlN detectors[145-146].

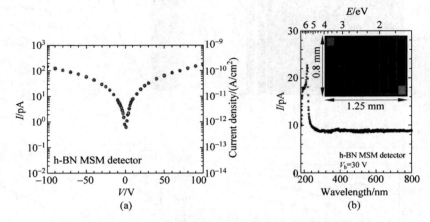

Figure 3-17 The typical I-V performance of MSM detectors
(a) I-V characteristics of photodetector; (b) The relative spectral response of detector[144]

Based on the above study, they summed up several advantages such as high/deep UV to visible rejection ratio, due to the high absorption of the thin active layers, it is suitable for oxidation resistance and chemical inertness and high dielectric strength.

2. Deep UV emitters

In the p-type conductivity demonstration progress of the epitaxial growth of h-BN method of epitaxial layer represents the p-layer revolutionary and overcome in the Al rich AlGaN deep purple problems inherent low p-type conductivity of a special opportunity deep UV device application. However, the epitaxial growth h-BN on AlGaN is prerequisite for the formation of p-type AlGaN deep UB device structure[147-150].

Majety et al. used the method above-mentioned assembled a heterostructure, which prepared for p-type h-BN and AlGaN deep UV device structure[150]. They used CVD method to prepare h-BN the AlGaN/AlN/Al_2O_3 templates. The epitaxial layers of h-BN are also highly insulated in the AlN and N type AlGaN templates, trying to prove that h-BN/AlGaN p-n junction is grown by Mg doping. By testing the diode performance of heterostructure, they demonstrated the feasibility of p-type h-BN as electron blocking of its highly conduction, which can be made as p-contact layer of deep UV emitters. The characteristics of I-V corresponding to p-n structure, which doped Mg in buffer layer, were annealed at the temperature 1,020 ℃. The test values under different temperatures are shown in Figure 3-18(b). The test results showed that these p-n heterostructures have wide application prospects in the high-efficiency deep UV photovoltaic devices.

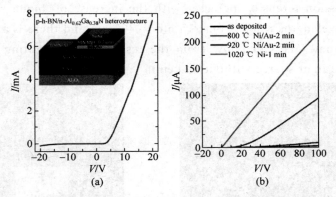

Figure 3-18 h-BN/AlGaN p-n junction and I-V characteristics
(a) I-V characteristics of a p-BN: Mg/n-Al$_{0.62}$Ga$_{0.38}$N/ AlN structure; (b) I-V characteristics of emitters at different times and temperatures[150]

3. Far-UV(FUV) plane-emission device (PED)

Some scholars have been hoping to replace the traditional FUV lamps with high-efficiency solid-state device with long service life. But, due to the low efficiency of UV, the research is still continuing to develop FUV devices with high performance[151-153].

Watanabe et al. studied the potential application of h-BN as the FUV fluorescent materials[153]. The height of the luminescent of h-BN, FUV emission planar compact device is equipped with a field emission array as excitation source, proves its stable operation and output power of 0.2 MW at 225 nm. Due to its low current consumption during operation, the device can be driven by a dry battery. This convenient FUV device may prove useful in chemical and biotechnological applications such as chemical modification, photo-catalysis and sterilization. The operating state of a FUV plane-emitting device consisting of a field emitter as a cathode, the power of h-BN on the screen, a vacuum chamber and a wire electrode around the screen is shown in Figure 3-19(a), which is driven by dry battery. Figure 3-19(b) shows a FUV plane-emission device. Its output power is shown in Figure 3-19(c), from the Figure 3-19(c), we can know that, with the increase of the acceleration voltage from 3 kV to 8 kV, the output power increase nonlinearly from 0.02 mV to 0.25 mV. The output spectra of this device are shown in Figure 3-19(d). There are three spectral lines in the figure corresponding to different anode voltages, and the emission intensity increases with the increase of anode voltage. The emission wavelengths range from 215 nm to 400 nm and there is a strong peak around 227 nm. We can know from the test, the cold-lighting device as a conventional light source without makeup.

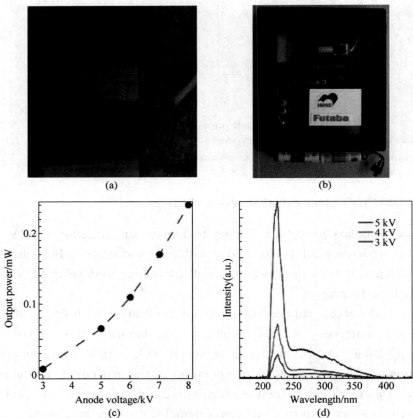

Figure 3-19 FUV fluorescent materials
(a)~(b) The photograph of different far-ultraviolet (FUV) plane-emission device in operation; (b) Photograph of a battery-driven FUV plane-emission device in operation; (c)~(d) The test spectrum line of FUV[153]

3.3 The introduce of graphene/h-BN van der Waals heterostructure

Different van der Waals heterostructures can be formed by different 2D materials with van der Waals forces, so they can show the properties that the 2D materials do not possess. As the substrate of graphene, the performance of h-BN is the best. They formed heterostructure (Figure 3-20), which became one of research hot spots[154-158]. The emergence of this artificial heterojunction provides researchers with a great deal of flexibility for the design of different structures and devices[156].

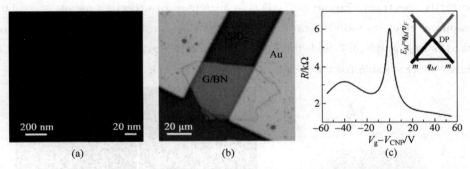

Figure 3-20 Graphene/BN heterostructure and typical transport property
(a) Atomic force microscopy image shows a high coverage of monolayer graphene alone with a small portion of bilayer graphene (bright area ~0.3%) and bare BN (dark area ~3%). The inset shows a high-resolution AFM image of the graphene/BN moiré superlattice with a period of 151 nm;
(b) Optical micrograph of a two-terminal field-effect graphene/BN device on a SiO_2/Si substrate;
(c) Gate-dependent resistance of a typical graphene/BN device at room temperature[154]

3.3.1 The structure of graphene/h-BN van der Waals heterostructure

Graphene/h-BN is one of the most typical representatives[159-165]. Graphene and h-BN combine together; the graphene surface will appear Moire' stripes (Figure 3-21(a)); Moire' fringe cycle are closely related to the angle between the two. This Moire' fringe can be regarded as a modulation of the periodic potential of graphene on a BN substrate, leading to the reconstruction of graphene bands, such as the generation of self-similar superlattice subbands and the opening of the graphene gap. The lattice orientation of graphene is the same as that of BN, and its energy is the lowest and most stable.

As an insulating material with a bandgap of 5.97 eV, h-BN has an atomically flat surface, no dangling bonds, weak doping effect, etc. The surface roughness is extremely low, shows weak van der Waals force to graphite, and has minimal impact on the graphite carrier transport properties, and the mismatch degree of it and the graphite thin lattice is only 1.7%, with no doping effect to graphite. Therefore, intrinsic physical properties of the graphene can be maximally maintained (Figures 3-21(b) and (c)). More importantly, a 2D superlattice structure formed by graphene on h-BN can control the graphene band structure and form an additional Dirac point[166-169]. In order to explore a series of new physical phenomena such as Hofstadter butterfly spectrum (Figure 3-21(d)), it provides an effective means[165]. The graphene electron mobility on the h-BN substrate can be comparable to that of free-floating graphene, and the substrate-supported graphene/h-BN structure is clearly more suitable for the design of graphene-based electronic devices[170-171].

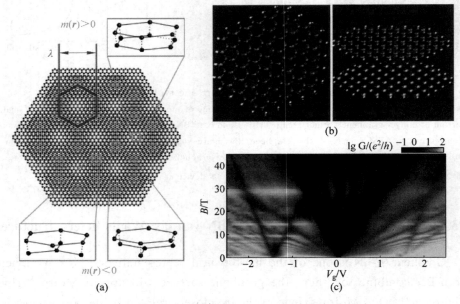

Figure 3-21 Graphene/h-BN heterostructure and devices

(a) Schematic of the moiré pattern for graphene (gray) on h-BN (red and blue)[155] graphene/h-BN heterostructure schematic diagram; (b) Top view and side view[137]; (c) Two-terminal magnetoconductance of device A1 up to 45 T[155]

3.3.2 The energy bandgap structure of graphene/h-BN van der Waals heterostructure

The quantum nature of electrons generates a band structure that determines the conduction of electrons and the optical properties of the material. Long-range superlattices, on the other hand, can produce microstructures that are dispersed in a better energy scale crossing a reduced Brillouin zone, which produces negative differential conductance and Bloch oscillation, and so on.

With the development of research, the arrival of van der Waals heterojunction of high-quality graphene/h-BN (deviation angle of less than 1°) has significantly changed the status quo[157,168,172]. In this heterostructure, the periodic potential of the electrons in the graphene is exploited by the hexagonal moiré fringes generated by the inconsistence between the two crystals. The microstructures of Dirac electrons have also been demonstrated by scanning tunneling, capacitance and optical microscopy. These studies illustrate the electronic structure of the so-called Hofstadter butterfly that appears in a quantized magnetic field.

Figure 3-22 shows the key features of the micro-band energy dynamics in the moiré superlattice[172], and points out the direction of new transport effects to further explore. At the technical level, a clear demonstration of such a micro-strip conductance implies that graphene/h-BN is a practical platform based on micro-band energy physical devices. The high efficiency of the photocurrent generated at the edge of the graphene superlattice in the magnetic field may be caused by the transition orbit observed by the researchers; in addition, Hertz devices such as a Bloch oscillator can benefit from longer scattering times in the system.

Wang et al. reported direct experimental results on the dispersion of SDCs in 0°-aligned graphene/h-BN heterostructures by using angle-resolved photoemission spectroscopy. It reveals SDCs at the corners of the superlattice Brillouin zone, and at the only one of the two superlattice valleys[173]. They also made the gaps of approximately 100 meV and approximately 160 meV, which observed at the SDCs and the original graphene Dirac cone, respectively (Figure 3-23). Their research result brings out the important role of the potential of a strong inversion-symmetry-breaking perturbation in the physics of graphene/h-BN,

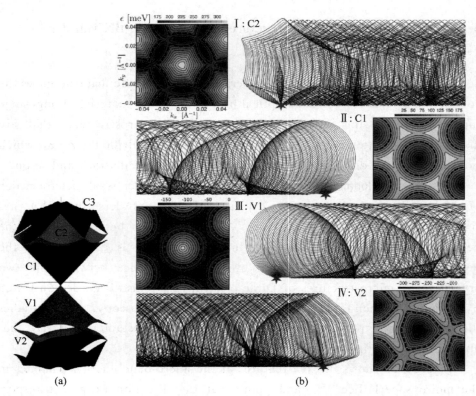

Figure 3-22 Key features of the micro-band energy dynamics in the moiré superlattice (a) Calculated miniband structure of the graphene/h-BN superlattice; (b) Representative ensembles of simulated skipping orbits emanating from an emitter (red star) at the boundary of the graphene/h-BN superlattice possessing the miniband dispersion[170]

and bridges gaps of critical knowledge in the band structure engineering of Dirac fermions thought to be a superlattice potential.

3.3.3 The optical and photoelectric properties of graphene/h-BN van der Waals heterostructure

2.3% of incident light is absorbed by graphene, despite the fact that it has only one atomic thickness, and the absorption intensity is irrelevant to the frequency of the incident light. The absorption intensity corresponds linearly with the number of graphene layers. Graphene has broadband absorption characteristic, especially in visible region, where the response to the optical is obvious. Aditionally, h-BN has an absorption peak in the UV region. In other

Chapter 3 Optical, Photonic and Optoelectronic Properties of Graphene, h-NB and Their Hybrid Materials

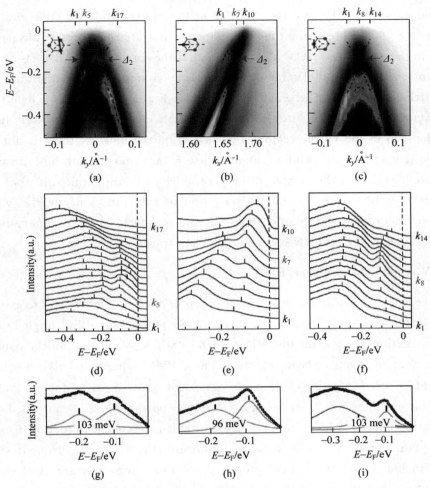

Figure 3-23 Observation of gap opening at second-generation Dirac points of Graphene/h-BN[173]

(a)~(c) ARPES data through the SDPs along different directions; (d)~(f) EDCs between the momenta indicated in (a)~(c); (g)~(i) Fitting results of the EDCs across the SDPs in (a)~(c) with two ((g),(h)) or three (i) Lorentzian peaks

words, the graphene/h-BN van der Waals heterostructure may have peculiar characteristics in the correlation response[175-180].

1. Optical advantages of graphene/h-BN in visible and SW-NIR(near infraved) regions

Wang et al. calculated the optical properties of this structure in the entire optical region and the electron transfer in the excited state by DFT and TD-DFT[137]. The results show that the graphene has a strong optical absorption peak in the visible and NUV regions, whereas the single-layer h-BN has optical absorption in the deep UV region. When they compose heterostructures, BN has little absorption effect on graphene; from the situation of single-layer graphene and heterostructures in the excited state electron-hole transfer, it can also be seen in the visible region, between graphene and BN there is not any charge transfer, but inside graphene itself the charge and hole transfer (Figure 3-24(a)). The graph of Figure 3-24(b) also confirms that the influence of BN on the graphene as a graphene substrate is negligible, which further theoretically explains that for the production of graphene-based optoelectronic devices, BN is a very good substrate.

2. UV features of graphene/h-BN

Recently, Liu Zhongfan research group of Peking University successfully completed the preparation of patterned graphene/h-BN single-crystal heterojunctions[175]. Prior to CVD growth, PMMA was coated on the copper-foil surface by electron beam etching. These PMMA particles, like *seed*, act as centers of nucleation, allowing patterned growth of graphene during preparation of CVD. Meanwhile, PMMA particles also controlled the nucleation density and grain size, and provided a possible direction for the high-quality graphene/h-BN heterojunction growth. In the experiment, they also concluded that the optical properties of the heterostructure, UV-visible absorption spectra showed that there are graphene and h-BN absorption peaks at 270 nm and 200 nm; graphene/h-BN van der Waals heterojunction has high visible-light transparency at 550 nm, which provides a possibility for heterojunction to be used in optoelectronic devices (Figure 3-25).

3. Negative refraction effect

Recently, the theoretical and experimental studies on the hybridization between the graphene plasmon polariton (PP) and h-BN phonon polaritons have been extensively studied. In this regard, the study of the stacked structure of graphene and h-BN as the multilayer structure of the crystal is less[181-184].

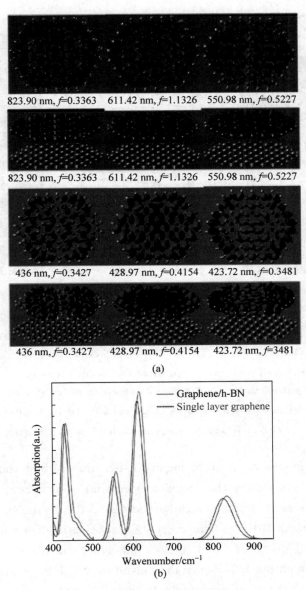

Figure 3-24

(a) Charge-transfer densities for the strong electronic transitions in single-layer on single-layer h-BN substrate with top and side views, where the green and the red stand for the hole and electrons, respectively; (b) The calculated absorption spectra of single-layer graphene and graphene/h-BN[137]

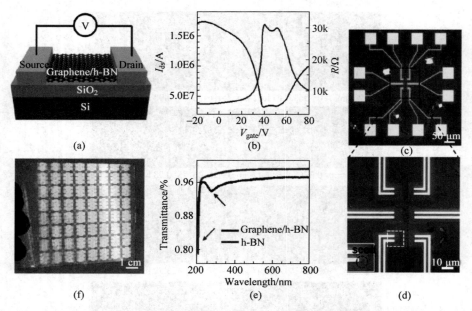

Figure 3-25 Electrical and optical properties of patterned graphene/h-BN
(a) Schematic depiction of FET devices based on graphene/h-BN stacks; (b) Source-drain current (Ids) and channel resistance (R) as a function of back gate voltage (V_{gate}) measured at room temperature; (c) Optical image of patterned FET devices, fabricated using transferred graphene/h-BN patterns on 300 nm SiO_2/Si; (d) Close-view of (c) showing 3 × 2 devices; (e) UV-vis transmittance of graphene/h-BN (blue) and h-BN (red); (f) Photograph of the transparent graphene FET arrays fabricated using graphene/h-BN stacks on the quartz glass[175]

As a tunable materials, graphene and h-BN stack of hyperbolic crystal can be adjusted by controlling the chemical potential of graphene to give more freedom. Sayem et al. studied on the properties of the crystal structure stacked by graphene and h-BN[184]. There are two schematic diagrams of negative refraction in Figure 3-26(b), h-BN and graphene/h-BN hyper-crystals, respectively, which the hyper-crystals stacked by single layer graphene and h-BN. By matching the wave-vector ($\theta_{rk,hc}$) with the pointing vector ($\theta_{rs,hc}$), a function corresponds to the incidence angle at wavenumber 810 cm^{-1}, which again corresponding to different from graphene chemical potential value. The 2D map of refraction angles is shown in Figure 3-26(c), for $\mu_c = 0.25$ eV. Figure 3-26(d) shows us the reflection for different μ_c. From 725 nm to 850 nm, the reflection increases slowly with the increase of wave number.

When the wave number increases at 800 cm^{-1} this reflex is stopped. For the wave numbers in the range of 800 ~ 820 cm^{-1}, the reflection decreases sharply. After 820 cm^{-1}, the reflection increased sharply, reaching a maximum value near 830 cm^{-1}, and then the reflection decreased slowly, but the small μ_c refraction still high.

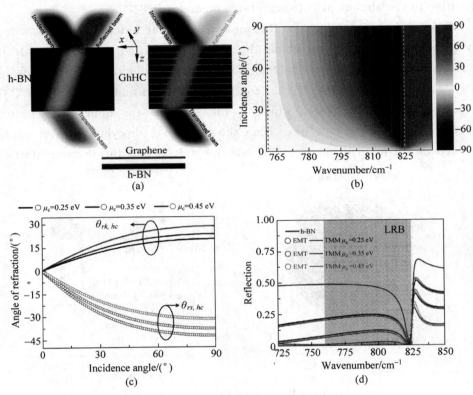

Figure 3-26 Negative refraction effect

(a) Schematic representation of negative refraction in bare h-BN and GhHC; (b) Refraction angles for the pointing vector (symbols "o") and wave-vector (solid lines) for GhHC as a function of incidence angle at wavenumber 800 cm^{-1} for different values of chemical potential of graphene; (c) 2D map of the refraction angles for the pointing vector as a function of wavenumber and incidence angle for μ_c = 0.25 eV; (d) Reflection of h-BN and GhHC as a function of wavenumber with incidence angle 30°[184]

All-angle negative refraction of the hyper crystal much higher than that of bare h-BN. On the other hand, graphene can completely control the negative refraction of the optical properties of BN and the transmission characteristics of the BN structure without hindering the negative refraction.

4. The Raman character of graphene/h-BN superlattices

As a fast and nondestructive detection technique, Raman spectroscopy is widely used in the qualitative study of some materials. In particular, it is much sensitive to the alignment between graphene and h-BN[185-187].

Eckmann et al. studied the Raman spectra of the heterosturcture stacked by graphene on h-BN, in which the lattice of two kinds of materials is perfectly aligned[188]. Figure 3-27(a) is the schematic diagram for graphene/h-BN. Due to the difference of the length of B-N bond and the length of C-C bond, graphene and h-BN, which is the same as the honeycomb lattice structure of the six party, have been shown from a superlattice after being stacked into a heterostructure. The wavelength λ_M and the low superlattice

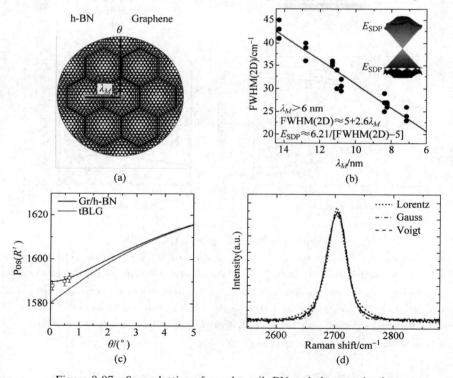

Figure 3-27 Super lattice of graphene/h-BN and characterization
(a) Superlattice potential of graphene/h-BN; (b) The test spectrum of FWHM and λ_M as the function; (c) The test spectral line of different heterosturcture by θ and R' as the function; (d) The Raman shift of h-BN/graphene superlattice with different calculated method[188]

Dirac point (SPD) energy E_{SDP} corresponding to the stacking angle θ between graphene and h-BN. The linear dependence of FWHM (2D peak) and moiré wavelength for twist angles below 2°. They make the function FWHM (2D) $\approx$ $5 + 2.6\lambda_M$. The comparing spectral lines between graphene/h-BN and tBLG (twisted bilayer graphene) of peck R' as a function of the mismatch angle in Figure 3-27(c). From it, we can know that, the position of peak R' increases with the increase of the lattice alignment angle of graphene and h-BN. The theoretical calculation of the two structures is basically the same. Figure 3-27(d) is the Raman spectrum, which calculated corresponding to Lorantz, Gauss and Voigt, respectively.

The Raman spectra of these superlattices reveal the relationship between the energy and the angle of the superlattice, which provides a reliable theoretical basis for the structure of the optoelectronic devices.

5. Plasmon-phonon coupling and plasmon delocalization, high pressure and low loss of plasmons in graphene/h-BN hetero structure

High-quality graphene/h-BN heterostructures including other 2D materials are considered to have broad prospects in applications[189-192].

Figures 3-28(a) and (b) show the SINS (synchrotron infrared nanospectroscopy) spectra of the graphene/SiO_2 and graphene/h-BN. Figure 3-28(a), the wavenumber at 765 cm^{-1} belongs to a small additional band of G/SiO_2, which is the enhancement corresponding to the band at 1,120 cm^{-1} (red point line). To Figure 3-28(b), the wavenumber at 817 cm^{-1} belongs to the longitudinal optical (LO) phonon of h-BN (green point line); the wavenumber at 1,365 cm^{-1} belongs to the transverse optical (TO) phonon of h-BN (green point line). On the other hand, there is a strength enhancement at 817 cm^{-1}, the same strength at 1,365 cm^{-1}, respectively. The enhancement belongs to the transverse surface plasmons of graphene with the LO phonon band of h-BN. But, the coupling of plasmons-phonon polaritons have no effect on TO modes[192].

Compared with other metal materials, graphene has the unique property of strong field confinement and relatively low transport decay in the plasmons transport capacity[193-195]. Woessner et al. studied the propagating plasmons of high quality in graphene with h-BN by near-field microscopy and also do a

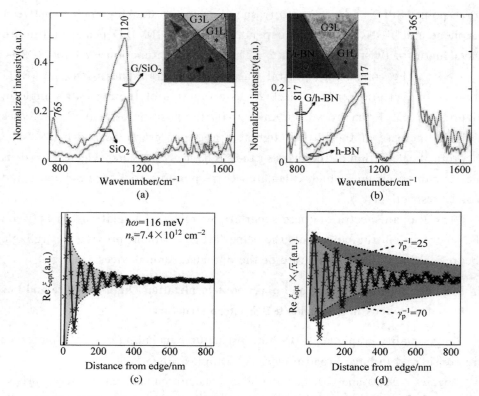

Figure 3-28　SINS spectra of graphene/SiO$_2$ in (a) and graphene/h-BN in (b)[192]; (c) and (d) is the spectrum line of plasmons damping in graphene/h-BN[196]

quantitative study on the plasmon damping, which corresponding to the spectrum in Figure 3-28(c) and (d)[196]. From the figure, with the increase of the distance from the boundary of the heterostructure, the plasma damping decreases nonlinearly. The complex parameters ξ_{opt}, being made as ordinate Figures 3-28(c) and (d). The oscillating signal is fitted with

$$\xi_{opt} = A \frac{e^{i2q_p x}}{\sqrt{x}} + B \frac{e^{iq_p x}}{x^a}. \qquad (3-1)$$

The attenuation of fringes away from the edge is caused by the combination of the damping and the geometric diffusion of the circular wave. The results show that the main damping is due to the intrinsic thermal phonons scattering in graphene and the dielectric loss in h-BN.

3.3.4 Potential applications of graphene/h-BN heterostructures in optical property

Graphene/h-BN heterostructures compared with the single-layer graphene and h-BN, the optical properties and photoelectric properties has shown excellent performance, which has broad prospects in the application of optical devices and optoelectronic devices.

1. Light-harvesting devices—The potential applications in solar cell and photodetector

Under the limitation of the current conversion efficiency standards, such as Shockley-Queisser limit, the absorption of a single photon can only excite an electron in a conventional light harvesting device[197].

A device about h-BN/graphene/h-BN heterostructures was prepared by Wu et al.[197] as shown in Figure 3-29(a). Through experiments, the collection of multiple hot carriers in the six-molar BN semiconductor superlattice was investigated.

Figures 3-29(a) and (c) show the results and optical images of the photonic collection device. Figure 3-29(b) is a moiré pattern formed by graphene stack on the h-BN. Figure 3-29(d) is the change of the resistance of the Dirac point and secondary Dirac point in the excitation field with the base graphite as the back gate and the longitudinal resistance as a function of 50 mT. When the back-gate voltage is -4.7 V and 4.4 V, there are corresponding to two peaks, corresponding to the sPDs. A strong peak at the back gate voltage of -0.1 V is the main Dirac point. Figure 3-29(e) is an image of the photocurrent variation corresponding to Figure 3-29(d). Near the back-gate voltage -4 V and 4 V, there is a phenomenon of photocurrent enhancement, corresponding to the electron and hole of secondary Dirac point position. The excitation laser is 660 nm. Figure 3-29(f) shows the device's scanning photocurrent image.

The use of graphene has been demonstrated by the photo-Nernst effect (which has been shown to absorb at least one of the five carriers of a photon), so that the zero-bias optical response is recorded at a 0.3 A/W. This phenomenon is due to the enhancement of the energy coefficient of the Lifshtiz transition at the low-energy van Hove singularities. This electronic

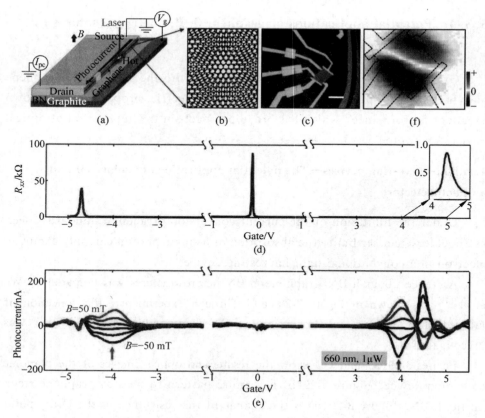

Figure 3-29 Anomalous photo-Nernst effect in graphene/BN superlattices[197]

collection device provides an effective and flexible means for the fabrication of optoelectronic resonators and LED, based on the flexible optoelectronic devices of van der Waals heterostructure.

2. Light-emitting diodes (LED)

Withers et al. contrasted to the conventional epitaxial heterostructures, stacked in the vertical direction by a mechanical transfer method[198-200], interacting weakly with each other by the van der Waals force. Figure 3-30(a) is a simulation diagram of LED heterostructure, heterogeneous structure is different from the traditional containing only the semiconductor, they used three different 2D materials: h-BN insulator, graphene metal, transition metal sulfides with direct band gap (MoS_2). Figure 3-30(b) is a schematic diagram of the LED, corresponding to the Figure 3-30(a) of figure TMDC, in which the electrons and holes pass from the graphene layer tunnel through the h-BN dielectric layer to TMDC at the external bias. The inset is a physical image of LED corresponding to Figure 3-30(a). Figure 3-30(c) is the comparison between PL MoS_2 monolayer and EL spectra, it can be seen from the figure, the PL frequency is closed to the frequency of EL around $V_b = 2.4$ V. After analysis, this phenomenon is the radiative recombination of X^- corresponding to the EL spectral line overlap.

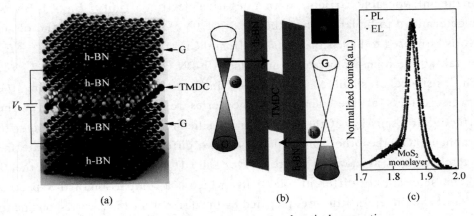

Figure 3-30 LED heterostructure and optical properties

(a) The schematic of LED[199]. (b) The working principle diagram of LED[199]. The inset is the physical optical image. (c) The comparison diagram of PL and EL in single layer MoS_2[198]

In addition, the authors also show that the external emission efficiency of multilayer TMDCs LED is close to 10%, which is comparable to that of modern LED based on organic materials. Although the wavelength of LED heterostructure in the work of the single semiconductor from TMDCs range from 600 nm to 700 nm based, it has the potential to extend the communication range of emission wavelength, and could reach the mid IR band.

In fact, the authors attempt to make thin, transparent and flexible polyester film LED. The main factors restricting the development of this technology are the lack of a scalable synthesis strategy for heterostructures. However, it is encouraging because simple van der Waals heterostructures have been proven to grow and the mechanics of 2D layers of mechanical transfer have been made, suggesting that this barrier may eventually be overcome.

3. Solar cell

As the application of low-cost photovoltaic, the structure of heterostructure based on graphene has become an excellent candidate for it[201-203].

Meng et al. prepared the heterostructure by using CVD method[203]. Figure 3-31(a) is the graphic illustration of graphene/h-BN/Si. Figure 3-31(b) shows that h-BN is employed to an effective electron blocking layer, inset the graphene and n-Si. Owning to the wide band gap, h-BN can reduce recombination of the unfavourable carriers. With a negative electron affinity, E_C of h-BN/Si is determined to be larger than that of Si (4.05 eV). The valence band offset E_V is estimated to be less than 0.63 eV from the formula $\Delta E_V = E_g^{hBN} - E_g^{Si} - \Delta E_C$ by taking the room-temperature band gaps of h-BN (5.80 eV) and Si (1.12 eV). The existence of small ΔE_V is also an important problem in avoiding the adverse effects of the increase of the series resistance of the cell in the effective transmission of the hole from the silicon layer through h-BN layer to graphene. On the other hand, they used two different methods to prepare the heterostructures, as shown in Figure 3-31(c) and (d). Figure 3-31(e) shows the J-V characteristics of different solar cells, which is a comparison of the spectral lines of the different structures prepared by the different methods corresponding to Figures 3-31 (c) and (d). From the comparison, we can know that the heterostructure of one-step methods is better than that of two-step method. The PCE (power conversion efficiency) of solar cell reaches 10.93 %.

The layered h-BN is also used to improve the performance of the device. It is because of its unique physical and chemical properties and the appropriate band structure, h-BN as an effective electronic barrier, hole transport, thus inhibiting the interface complex, so that the open circuit voltage significantly improved. On the other hand, based on h-BN, the carrier conductivity increases significantly and the resistance of the solar cell is reduced, leading to the increase of the short-circuit current density. By improving the preparation method of the heterogeneous, the graphene/h-BN heterostructure can be directly grown, the interface defect and contamination can be avoided, and the performance of the device can be greatly improved.

4. Nanoresonators based on graphene/h-BN/SiO$_2$ heterostructures

The optical properties of graphene and h-BN have been changed obviously

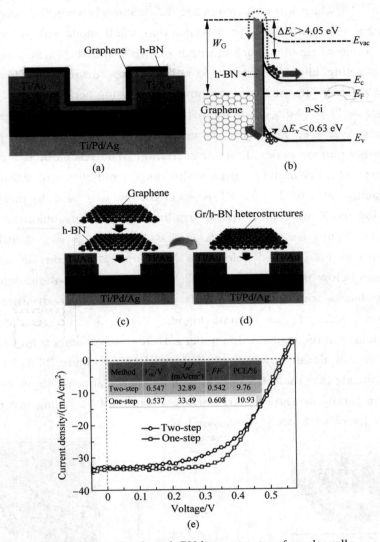

Figure 3-31 Graphene-h-BN heterostructure for solar cell
(a) Schematic diagram of solar cells; (b) Schematic diagram of working principle of solar cells; Schematic diagram of solar cells prepared by one step; (c) and two step method (d); (e) Comparison of two methods for solar cell performance test $J-V$ characteristics[203]

due to graphene/h-BN heterosturctures. In particular, the coupling between the surface plasmons of the graphene and the phonons of h-BN leads to a new frequency-wavevector dispersion relation[189-192,204].

Brar et al. assembled graphene nanoresonators on a single-layer h-BN nanosheets[205], which was used to measure the coupling between the plasmons and the excitons near graphene. They found that small mode volume and high-modulus graphene plasma shock strength of h-BN and the formation of two-phonon coupling allow for strong coupling, thus forming the two clearly separated the hybridization pattern shows anticrossing behaviour.

Figure 3-32(a) is the schematic of the nanoresonator based on graphene/h-BN heterostructure. The figure shows the motion of the graphene plasmons and the h-BN phonon under the laser excitation. The resonator is etched into an electron rod array pattern with a width range from 30 nm to 300 nm and a width/spacing ratio of 1 : 2. FTIR spectroscopy was used to measure the transmission mode with the vertical irradiation of nanoresonators, the test result is in Figure 3-32(b), with $1.0 \times 10^{13}/cm^2$ carrier density. The wavenumber at 1,370 cm^{-1} belongs to h-BN optical phonon, in which the spectra lines below the Figure 3-32(b). With the increase of the width, the frequency has a significant red shift and the strength of transmission modulation ($-\Delta T/T_{CNP}$) become higher. Figure 3-32(c) is the carrier-induced change in the transmission under different wavenumbers and energy of the same carrier density, which is corresponding to Figure 3-32(b). These features indicate that the h-BN optical phonon energy appears above and below the experimentally measured characteristics, and shows a strong inverse cross behavior pattern with strong correspondence.

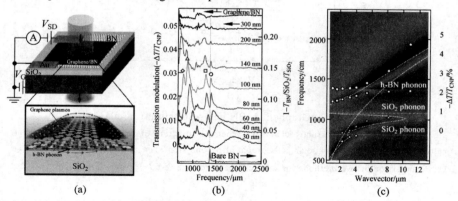

Figure 3-32 Graphene/h-BN/SiO$_2$ heterostructure for sensor
(a) Schematic of graphene nanoresonators; (b) Experimental device physical test line of nanoresonators; (c) The theoretical calculation spectrum corresponding to (b)[205]

This work is based on graphene/h-BN heterostructure nanoresonator to test the graphene plasmons and h-BN phonon coupling results. However, the fabrication of the nanoresonator shows that the heterostructure has a broad prospect in the application of optoelectronic devices.

5. Photodetector and autocorrelator

Recently, 2D materials have shown a number of new properties, such as optical emission, parametric nonlinearity, broadband ultrafast optoelectronic detection, saturable absorption and on-chip electro-optic modulation. These unique properties make them an alternative to the optoelectronic platform for semiconductor devices[206-209].

A heterostructure consists of a single-layer graphene encapsulated in a six-party BN, with high efficiency of photoelectric detection and ultrafast metrology[208]. Figure 3-33(a) is the structure of the photodetector. The illustration of the red circle is the side simulation of the heterogeneous structure formed by the encapsulation of single-layer graphene with h-BN. Figure 3-33(b) is the result of an electrical spectrum analyzer obtained by a high-speed RF probe, with measured power Δf ($V_{DS} = V_{GS} = 0$). The measurement results show that the cutoff frequency of 42 GHz is 3 dB, and the results match the maximum velocity of graphene photodetector. At the same time, the optical response of the zero-drain source bias is also observed, a typical distinction between graphene and semiconductor high speed photodiode. Figure 3-33(c) is the delay time and the intensity of the optical current as the coordinates of the different input power, the change of the photocurrent image. From the figure, at $\Delta t = 0$, the spectrum of relaxation time and width corresponds to the obvious sink.

The results of the measurement of photocurrent by using the gate voltage and drain source voltage as a function show that the thermal electrons mediate the photoelectric response. At the same time, the observed saturation photocurrent corresponds to the super collisional cooling mechanism of electron phonon. This nonlinear optical response can be tuned to picosecond scale timing resolution on the optical on-chip autocorrelation measurements and extremely low peak power[209].

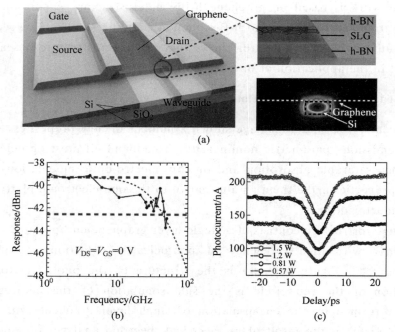

Figure 3-33 Photodetector

(a) schematic of the h-BN/SLG/h-BN photodetector on a buried silicon waveguide. The inset shows the electric field energy density of the fundamental TE waveguide mode, obtained by finite-element simulation. (b) High-speed response of the graphene photodetector. (c) Autocorrelation traces of the graphene-based autocorrelator with different input average power of the laser pulses[209]

6. High-speed electro-optic modulator

Efficient, nanoscale, electro-optic modulator is an important part of optical interconnects. Due to the weak photoelectric effect of silicon, the silicon become is a technical bottleneck in the modulator. As a substitute for silicon, graphene has special optical and electronic properties, which makes it an alternative to silicon-based materials in optoelectronic applications[210-212].

Based on the graphene/h-BN heterostructure and the silicon photonic crystal resonant cavity, Gao et al. integrated a high-speed graphene electro-optic modulator. There is a strong interaction between the light and the material in the submicron cavity, which makes the reflection of the resonator with high efficiency. After testing, the modulation depth is 3.2 dB, and the cutoff frequency is 1.2 GHz. Figure 3-34(a) is a schematic diagram of the

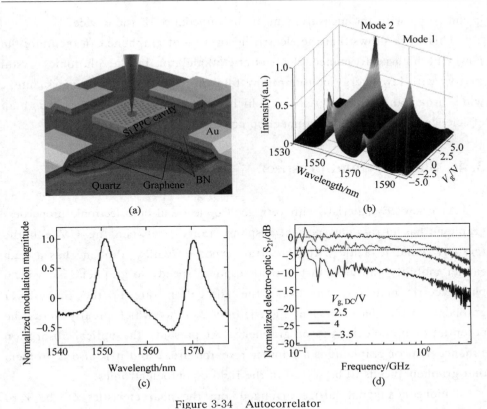

Figure 3-34 Autocorrelator

(a) Schematic of modulator based on graphene/h-BN; (b) Cavity reflection spectrum as a function of V_G and wavelength λ; (c)~(d) The test spectrum of modulator with different parameter[212]

structure of the modulator, its preparation is based on the graphene/h-BN heterostructure. Double-layer graphene capacitors on the quartz substrate coupled with a planar photonic crystal cavity. The gate voltage (V_g) is scanned slowly, and the optical response of the modulator is measured by observing the reflection spectrum of the cavity, as shown in Figure 3-34(b). During the gate voltage increased at the range from 2.5 V to 6.7 V, the carrier density of two graphene layers increased slowly. When the incident wavelength is at 1,551 nm and 1,570 nm, there are two peaks corresponding to the carrier density, moreover, the intensity becomes stronger suddenly at $V_g = \pm 5$ V. Figure 3-34(d) shows the high-speed response characteristics of the low-pass filter with 3 dB cutoff frequency of 1.2 GHz. The RC time constant of the bilayer graphene capacitors limits the cutoff frequency, which is consistent with the measurement of the impedance of the device.

This work shows that the electric adsorption of graphene can facilitate the powerful high-performance electro-optic modulator in the photonic crystal cavity, with low energy consumption, told the wavelength scale mark, etc., wide broadband characteristics, which makes the equipment with high efficiency in photonic interconnection conversion, and stability.

3.4 Summary and prospect

As a new 2D material, with very good optical and optoelectronic properties, graphene has ultra-wide spectral response range, between the UV and the terahertz band to achieve full spectral response; while graphene has a high carrier mobility and ultra-fast optical response speed, so graphene is an ideal photoelectric material. However, the absorption rate of the single-layer graphene on the space incident light is only 2.3 %, which greatly limits the potential of optoelectronics of graphene. At present, the optical absorption enhancement of graphene is a popular research area, and it is also a problem that graphene needs to be solved in the field of optoelectronics.

Unique hexagonal lattice arranged and the characteristics of the zero band-gap allow the electrons on the graphene to move freely, which makes graphene have many extraordinary features. The 2D h-BN has a similar structure to the graphene lattice: wide band-gap electronic features make its chemical properties different from graphene, electrons will not be easily flow in the BN atoms, which basically is an insulator. The graphene has unique optical properties: single-layer graphene is approximately transparent, and has a broadband optical response throughout the light region, especially in the visible region, all of which make graphene suitable for use in next-generation optoelectronic devices and optical communication systems. Its excellent electrical conductivity and light transmission make it especially suitable for high-performance photoelectric detection equipment.

The h-BN has a zero optical response in the visible region; only in the deep UV region has an absorption peak. It is that two optical properties with almost different 2D materials, when the graphene interlayer lattice aligned BN lattice, a *super-lattice* was born, and the superlattice is what people pursue with the structure of high-efficiency photoelectric properties. This feature is due to quantum mechanics and some unknown rules dominate the known interaction between particles, especially in the super lattice between the van Hough singular point detection of special quantum region. These high-electron-density regions are not available between the graphene layers or between the BN layers. When high-energy photons are directed to the superlattice, a high-energy photon can be converted into electrons in the van Hove singularity region, and these electrons, if collected by the electrodes, form a current. With this finding, researchers can create a more efficient device that allows more electronic feedback when photons are acquired. Future work will explore how to fuse the excited electrons into current for optimal energy conversion efficiency, and remove some of their superlattice-annoying problems, such as the need for magnetic fields. We remain convinced that this efficient process between photons and electrons will bring significant progress.

References

[1] NOVOSELOV K S, GEIM A K, MOROZOV S V, et al. Electric field effect in atomically thin carbon films[J]. Science, 2004, 306(5696): 666-669.

[2] TAKEDA K, SHIRAISHI K. Theoretical possibility of stage corrugation in Si and Ge analogs of graphite[J]. Physical Review B, 1994, 50(20): 14916.

[3] CAHANGIROV S, TOPSAKAL M, AKTÜRK E, et al. Two-and one-dimensional honeycomb structures of silicon and germanium[J]. Physical Review Letters, 2009, 102(23): 236804.

[4] GEIM A K, NOVOSELOV K S. The rise of graphene [M]//Nanoscience and technology: a collection of reviews from nature journals, 2010: 11-19.

[5] NETO A C, GUINEA F, PERES N M. Drawing conclusions from graphene[J]. Physics World, 2006, 19(11): 33.

[6] GEIM A K. Random walk to graphene (Nobel lecture)[J]. Angewandte Chemie, 2011, 50 (31): 6966-6985.

[7] TANG Y B, LEE C S, CHEN Z H, et al. High-quality graphenes via a facile quenching method for field-effect transistors [J]. Nano Letters, 2009, 9 (4): 1374-1377.

[8] LEE C, WEI X, KYSAR J W, et al. Measurement of the elastic properties and intrinsic strength of monolayer graphene[J]. Science, 2008, 321(5887): 385-388.

[9] BOLOTIN K I, SIKES K J, JIANG Z, et al. Ultrahigh electron mobility in suspended

graphene[J]. Solid State Communications, 2008, 146(9/10): 351-355.
[10] KIM K S, ZHAO Y, JANG H, et al. Large-scale pattern growth of graphene films for stretchable transparent electrodes[J]. Nature, 2009, 457(7230): 706-710.
[11] BALANDIN A A, GHOSH S, BAO W, et al. Superior thermal conductivity of single-layer graphene[J]. Nano Letters, 2008, 8(3): 902-907.
[12] FALKOVSKY L A. Symmetry constraints on phonon dispersion in graphene[J]. Physics Letters A, 2008, 372(31): 5189-5192.
[13] PARTOENS B, PEETERS F M. From graphene to graphite: electronic structure around the K point[J]. Physical Review B, 2006, 74(7): 075404.
[14] CASTRO NETO A H, GUINEA F, PERES N M R, et al. The electronic properties of graphene[J]. RvMP, 2009, 81(1): 109-162.
[15] CHARLIER J C, EKLUND P C, ZHU J, et al. Electron and phonon properties of graphene: their relationship with carbon nanotubes[M]//Carbon nanotubes. Berlin: Springer, 2007: 673-709.
[16] WALLACE P R. The band theory of graphite[J]. Physical Review, 1947, 71(9): 622.
[17] AVOURIS P, CHEN Z, PEREBEINOS V. Carbon-based electronics [M]// Nanoscience and Technology: A Collection of Reviews from Nature Journals, 2010: 174-184.
[18] NOVOSELOV K S, GEIM A K, MOROZOV S V, et al. Two-dimensional gas of massless Dirac fermions in graphene[J]. Nature, 2005, 438(7065): 197-200.
[19] ZHANG Y, TAN Y W, STORMER H L, et al. Experimental observation of the quantum Hall effect and Berry's phase in graphene[J]. Nature, 2005, 438(7065): 201-204.
[20] BERGER C, SONG Z, LI X, et al. Electronic confinement and coherence in patterned epitaxial graphene[J]. Science, 2006, 312(5777): 1191-1196.
[21] MOROZOV S V, NOVOSELOV K S, KATSNELSON M I, et al. Giant intrinsic carrier mobilities in graphene and its bilayer[J]. Physical Review Letters, 2008, 100(1): 016602.
[22] CHEN J H, JANG C, XIAO S, et al. Intrinsic and extrinsic performance limits of graphene devices on SiO_2[J]. Nature Nanotechnology, 2008, 3(4): 206-209.
[23] DU X, SKACHKO I, BARKER A, et al. Approaching ballistic transport in suspended graphene[J]. Nature Nanotechnology, 2008, 3(8): 491-495.
[24] CASIRAGHI C, HARTSCHUH A, LIDORIKIS E, et al. Rayleigh imaging of graphene and graphene layers[J]. Nano Letters, 2007, 7(9): 2711-2717.
[25] BLAKE P, HILL E W, CASTRO NETO A H, et al. Making graphene visible[J]. Applied Physics Letters, 2007, 91(6): 063124.
[26] NAIR R R, BLAKE P, GRIGORENKO A N, et al. Fine structure constant defines visual transparency of graphene[J]. Science, 2008, 320(5881): 1308-1308.
[27] HASAN T, SUN Z, WANG F, et al. Nanotube-polymer composites for ultrafast photonics[J]. Advanced Materials, 2009, 21(38/39): 3874-3899.

[28] SUN Z, HASAN T, TORRISI F, et al. Graphene mode-locked ultrafast laser[J]. ACS Nano, 2010, 4(2): 803-810.

[29] STÖHR R J, KOLESOV R, PFLAUM J, et al. Fluorescence of laser-created electron-hole plasma in graphene[J]. Physical Review B, 2010, 82(12): 121408.

[30] LUI C H, MAK K F, SHAN J, et al. Ultrafast photoluminescence from graphene[J]. Physical Review Letters, 2010, 105(12): 127404.

[31] GOKUS T, NAIR R R, BONETTI A, et al. Making graphene luminescent by oxygen plasma treatment[J]. ACS Nano, 2009, 3(12): 3963-3968.

[32] EDA G, LIN Y Y, MATTEVI C, et al. Blue photoluminescence from chemically derived graphene oxide[J]. Advanced Materials, 2010, 22(4): 505-509.

[33] SUN X, LIU Z, WELSHER K, et al. Nano-graphene oxide for cellular imaging and drug delivery[J]. Nano Research, 2008, 1(3): 203-212.

[34] LUO Z, VORA P M, MELE E J, et al. Photoluminescence and band gap modulation in graphene oxide[J]. Applied Physics Letters, 2009, 94(11): 111909.

[35] KUZMENKO A B, VAN HEUMEN E, CARBONE F, et al. Universal optical conductance of graphite[J]. Physical Review Letters, 2008, 100(11): 117401.

[36] PONOMARENKO L A, SCHEDIN F, KATSNELSON M I, et al. Chaotic Dirac billiard in graphene quantum dots[J]. Science, 2008, 320(5874): 356-358.

[37] WANG Z, CHONG Y, JOANNOPOULOS J D, et al. Observation of unidirectional backscattering-immune topological electromagnetic states[J]. Nature, 2009, 461(7265): 772-775.

[38] RADOVIC L R, BOCKRATH B. On the chemical nature of graphene edges: origin of stability and potential for magnetism in carbon materials[J]. Journal of the American Chemical Society, 2005, 127(16): 5917-5927.

[39] GENG X, NIU L, XING Z, et al. Aqueous-processable noncovalent chemically converted graphene-quantum dot composites for flexible and transparent optoelectronic films[J]. Advanced Materials, 2010, 22(5): 638-642.

[40] PAN D, ZHANG J, LI Z, et al. Hydrothermal route for cutting graphene sheets into blue-luminescent graphene quantum dots[J]. Advanced Materials, 2010, 22(6): 734-738.

[41] LI Y, HU Y, ZHAO Y, et al. An electrochemical avenue to green-luminescent graphene quantum dots as potential electron-acceptors for photovoltaics[J]. Advanced Materials, 2011, 23(6): 776-780.

[42] SHEN J, ZHU Y, CHEN C, et al. Facile preparation and upconversion luminescence of graphene quantum dots[J]. Chemical Communications, 2011, 47(9): 2580-2582.

[43] GUO C X, YANG H B, SHENG Z M, et al. Layered graphene/quantum dots for photovoltaic devices[J]. Angewandte Chemie International Edition, 2010, 49(17): 3014-3017.

[44] SUN Y P, ZHOU B, LIN Y, et al. Quantum-sized carbon dots for bright and colorful photoluminescence[J]. Journal of the American Chemical Society, 2006, 128(24):

7756-7757.
[45] GUPTA V, CHAUDHARY N, SRIVASTAVA R, et al. Luminscent graphene quantum dots for organic photovoltaic devices[J]. Journal of the American Chemical Society, 2011, 133(26): 9960-9963.
[46] LIU R, WU D, FENG X, et al. Bottom-up fabrication of photoluminescent graphene quantum dots with uniform morphology[J]. Journal of the American Chemical Society, 2011, 133(39): 15221-15223.
[47] EDA G, LIN Y Y, MATTEVI C, et al. Blue photoluminescence from chemically derived graphene oxide[J]. Advanced Materials, 2010, 22(4): 505-509.
[48] SUN X, LIU Z, WELSHER K, et al. Nano-graphene oxide for cellular imaging and drug delivery[J]. Nano Research, 2008, 1(3): 203-212.
[49] LU J, YANG J, WANG J, et al. One-pot synthesis of fluorescent carbon nanoribbons, nanoparticles, and graphene by the exfoliation of graphite in ionic liquids[J]. ACS Nano, 2009, 3(8): 2367-2375.
[50] WANG J, CAO S, DING Y, et al. Theoretical investigations of optical origins of fluorescent graphene quantum dots[J]. Scientific Reports, 2016, 6: 24850.
[51] LI X, WANG X, ZHANG L, et al. Chemically derived, ultrasmooth graphene nanoribbon semiconductors[J]. Science, 2008, 319(5867): 1229-1232.
[52] SON Y W, COHEN M L, LOUIE S G. Energy gaps in graphene nanoribbons[J]. Physical Review Letters, 2006, 97(21): 216803.
[53] ROBERTSON J, O'REILLY E P. Electronic and atomic structure of amorphous carbon[J]. Physical Review B, 1987, 35(6): 2946.
[54] MOSER J, GRAETZEL M. Light-induced electron transfer in colloidal semiconductor dispersions: single vs. dielectronic reduction of acceptors by conduction-band electrons [J]. Journal of the American Chemical Society, 1983, 105(22): 6547-6555.
[55] SHEATS J R, ANTONIADIS H, HUESCHEN M, et al. Organic electroluminescent devices[J]. Science, 1996, 273(5277): 884-888.
[56] ROTHBERG L J, LOVINGER A J. Status of and prospects for organic electroluminescence [J]. Journal of Materials Research, 1996, 11(12): 3174-3187.
[57] FRANGIONI J V. In vivo near-infrared fluorescence imaging[J]. Current Opinion in Chemical Biology, 2003, 7(5): 626-634.
[58] BREUSING M, ROPERS C, ELSAESSER T. Ultrafast carrier dynamics in graphite [J]. Physical Review Letters, 2009, 102(8): 086809.
[59] KAMPFRATH T, PERFETTI L, SCHAPPER F, et al. Strongly coupled optical phonons in the ultrafast dynamics of the electronic energy and current relaxation in graphite[J]. Physical Review Letters, 2005, 95(18): 187403.
[60] LAZZERI M, PISCANEC S, MAURI F, et al. Electron transport and hot phonons in carbon nanotubes[J]. Physical Review Letters, 2005, 95(23): 236802.
[61] GONZÁLEZ J, GUINEA F, VOZMEDIANO M A H. Unconventional quasiparticle lifetime in graphite[J]. Physical Review Letters, 1996, 77(17): 3589.

[62] RAETHER H. Surface plasmons on smooth surfaces [M]//Surface Plasmons on Smooth and Rough Surfaces and on Gratings. Berlin: Springer, 1988: 4-39.
[63] VAKIL A, ENGHETA N. Transformation optics using graphene[J]. Science, 2011, 332(6035): 1291-1294.
[64] BAO Q, LOH K P. Graphene photonics, plasmonics, and broadband optoelectronic devices[J]. ACS Nano, 2012, 6(5): 3677-3694.
[65] GRIGORENKO A N, POLINI M, NOVOSELOV K S. Graphene plasmonics[J]. Nature Photonics, 2012, 6(11): 749-758.
[66] GARCIA DE ABAJO F J. Graphene plasmonics: challenges and opportunities[J]. ACS Photonics, 2014, 1(3): 135-152.
[67] LOW T, AVOURIS P. Graphene plasmonics for terahertz to mid-infrared applications[J]. ACS Nano, 2014, 8(2): 1086-1101.
[68] JU L, GENG B, HORNG J, et al. Graphene plasmonics for tunable terahertz metamaterials[J]. Nature Nanotechnology, 2011, 6(10): 630-634.
[69] BONACCORSO F, SUN Z, HASAN T A, et al. Graphene photonics and optoelectronics [J]. Nature Photonics, 2010, 4(9): 611.
[70] TONG S W, WANG Y, ZHENG Y, et al. Graphene intermediate layer in tandem organic photovoltaic cells [J]. Advanced Functional Materials, 2011, 21 (23): 4430-4435.
[71] WU J, AGRAWAL M, BECERRIL H A, et al. Organic light-emitting diodes on solution-processed graphene transparent electrodes[J]. ACS Nano, 2010, 4(1): 43-48.
[72] BAE S, KIM H, LEE Y, et al. Roll-to-roll production of 30-inch graphene films for transparent electrodes[J]. Nature Nanotechnology, 2010, 5(8): 574.
[73] GENG H Z, KIM K K, SO K P, et al. Effect of acid treatment on carbon nanotube-based flexible transparent conducting films[J]. Journal of the American Chemical Society, 2007, 129(25): 7758-7759.
[74] LEE J Y, CONNOR S T, CUI Y, et al. Solution-processed metal nanowire mesh transparent electrodes[J]. Nano Letters, 2008, 8(2): 689-692.
[75] SAHU D R, LIN S Y, HUANG J L. ZnO/Ag/ZnO multilayer films for the application of a very low resistance transparent electrode[J]. Applied Surface Science, 2006, 252(20): 7509-7514.
[76] HAMBERG I, GRANQVIST C G. Evaporated Sn-doped In2O3 films: basic optical properties and applications to energy-efficient windows [J]. Journal of Applied Physics, 1986, 60(11): R123-R160.
[77] REINA A, JIA X, HO J, et al. Large area, few-layer graphene films on arbitrary substrates by chemical vapor deposition[J]. Nano Letters, 2009, 9(1): 30-35.
[78] GOMEZ D A L, ZHANG Y, SCHLENKER C W, et al. Continuous, highly flexible, and transparent graphene films by chemical vapor deposition for organic photovoltaics [J]. ACS Nano, 2010, 4(5): 2865-2873.
[79] BLAKE P, BRIMICOMBE P D, NAIR R R, et al. Graphene-based liquid crystal

device[J]. Nano Letters, 2008, 8(6): 1704-1708.
[80] WANG X, ZHI L, TSAO N, et al. Transparent carbon films as electrodes in organic solar cells[J]. Angewandte Chemie International Edition, 2008, 47(16): 2990-2992.
[81] HERNANDEZ Y, NICOLOSI V, LOTYA M, et al. High-yield production of graphene by liquid-phase exfoliation of graphite[J]. Nature Nanotechnology, 2008, 3(9): 563-568.
[82] LOTYA M, HERNANDEZ Y, KING P J, et al. Liquid phase production of graphene by exfoliation of graphite in surfactant/water solutions[J]. Journal of the American Chemical Society, 2009, 131(10): 3611-3620.
[83] GREEN A A, HERSAM M C. Solution phase production of graphene with controlled thickness via density differentiation[J]. Nano Letters, 2009, 9(12): 4031-4036.
[84] MATTEVI C, EDA G, AGNOLI S, et al. Evolution of electrical, chemical, and structural properties of transparent and conducting chemically derived graphene thin films[J]. Advanced Functional Materials, 2009, 19(16): 2577-2583.
[85] EDA G, FANCHINI G, CHHOWALLA M. Large-area ultrathin films of reduced graphene oxide as a transparent and flexible electronic material [J]. Nature Nanotechnology, 2008, 3(5): 270-274.
[86] WANG X, ZHI L, MÜLLEN K. Transparent, conductive graphene electrodes for dye-sensitized solar cells[J]. Nano Letters, 2008, 8(1): 323-327.
[87] BECERRIL H A, MAO J, LIU Z, et al. Evaluation of solution-processed reduced graphene oxide films as transparent conductors[J]. ACS Nano, 2008, 2(3): 463-470.
[88] WU J, BECERRIL H A, BAO Z, et al. Organic solar cells with solution-processed graphene transparent electrodes[J]. Applied Physics Letters, 2008, 92(26): 237.
[89] XIA F, MUELLER T, LIN Y, et al. Ultrafast graphene photodetector[J]. Nature Nanotechnology, 2009, 4(12): 839-843.
[90] MERIC I, HAN M Y, YOUNG A F, et al. Current saturation in zero-bandgap, top-gated graphene field-effect transistors[J]. Nature Nanotechnology, 2008, 3(11): 654-659.
[91] XIA F, MUELLER T, GOLIZADEH-MOJARAD R, et al. Photocurrent imaging and efficient photon detection in a graphene transistor[J]. Nano Letters, 2009, 9(3): 1039-1044.
[92] MUELLER T, XIA F, AVOURIS P. Graphene photodetectors for high-speed optical communications[J]. Nature Photonics, 2010, 4(5): 297-301.
[93] ECHTERMEYER T J, BRITNELL L, JASNOS P K, et al. Strong plasmonic enhancement of photovoltage in graphene[J]. Nature Communications, 2011, 2(1): 1-5.
[94] KIM K, CHOI J Y, KIM T, et al. A role for graphene in silicon-based semiconductor devices[J]. Nature, 2011, 479(7373): 338-344.
[95] QIAO H, YUAN J, XU Z, et al. Broadband photodetectors based on graphene-Bi2Te3 heterostructure[J]. ACS Nano, 2015, 9(2): 1886-1894.

[96] WANG F, ZHANG Y, TIAN C, et al. Gate-variable optical transitions in graphene [J]. Science, 2008, 320(5873): 206-209.

[97] LIU M, YIN X, ULIN-AVILA E, et al. A graphene-based broadband optical modulator[J]. Nature, 2011, 474(7349): 64-67.

[98] LIU X, DU D, MOUROU G. Laser ablation and micromachining with ultrashort laser pulses[J]. IEEE Journal of Quantum Electronics, 1997, 33(10): 1706-1716.

[99] DREXLER W, MORGNER U, KÄRTNER F X, et al. In vivo ultrahigh-resolution optical coherence tomography[J]. Optics Letters, 1999, 24(17): 1221-1223.

[100] KELLER U, WEINGARTEN K J, KARTNER F X, et al. Semiconductor saturable absorber mirrors (SESAM's) for femtosecond to nanosecond pulse generation in solid-state lasers[J]. IEEE Journal of Selected Topics in Quantum Electronics, 1996, 2(3): 435-453.

[101] BAO Q, ZHANG H, WANG Y, et al. Atomic-layer graphene as a saturable absorber for ultrafast pulsed lasers [J]. Advanced Functional Materials, 2009, 19(19): 3077-3083.

[102] ZHANG H, TANG D, KNIZE R J, et al. Graphene mode locked, wavelength-tunable, dissipative soliton fiber laser [J]. Applied Physics Letters, 2010, 96(11): 111112.

[103] XU J L, LI X L, HE J L, et al. Performance of large-area few-layer graphene saturable absorber in femtosecond bulk laser[J]. Applied Physics Letters, 2011, 99(26): 261107.

[104] TAN W D, SU C Y, KNIZE R J, et al. Mode locking of ceramic Nd: yttrium aluminum garnet with graphene as a saturable absorber[J]. Applied Physics Letters, 2010, 96(3): 031106.

[105] DE SOUZA E A, NUSS M C, KNOX W H, et al. Wavelength-division multiplexing with femtosecond pulses[J]. Optics Letters, 1995, 20(10): 1166-1168.

[106] KOCH B R, FANG A W, LIVELY E, et al. Mode locked and distributed feedback silicon evanescent lasers[J]. Laser & Photonics Reviews, 2009, 3(4): 355-369.

[107] WAN H, CAI W, WANG F, et al. High-quality monolayer graphene for bulk laser mode-locking near 2 μm[J]. Optical and Quantum Electronics, 2016, 48(1): 11.

[108] POUCH J J, ALTEROVITZ S A. A review of synthesis and properties of boron nitride [J]. Material and Manufacturing Process, 1991, 6(2): 373-374.

[109] HAUBNER R, WILHELM M, WEISSENBACHER R, et al. Boron nitrides—properties, synthesis and applications[M]//High Performance Non-Oxide Ceramics II. Berlin: Springer, 2002: 1-45.

[110] GAO X, ZHOU Z, ZHAO Y, et al. Comparative study of carbon and BN nanographenes: ground electronic states and energy gap engineering[J]. The Journal of Physical Chemistry C, 2008, 112(33): 12677-12682.

[111] WANG J, CAO S, SUN P, et al. Optical advantages of graphene on the boron nitride in visible and SW-NIR regions[J]. RSC Advances, 2016, 6(112): 111345-111349.

[112] BLASE X, RUBIO A, LOUIE S G, et al. Quasiparticle band structure of bulk hexagonal boron nitride and related systems [J]. Physical Review B, 1995, 51(11): 6868.

[113] ZUPAN J. Energy bands in boron nitride and graphite[J]. Physical Review B, 1972, 6(6): 2477.

[114] OOI N, RAIRKAR A, LINDSLEY L, et al. Electronic structure and bonding in hexagonal boron nitride[J]. Journal of Physics: Condensed Matter, 2005, 18(1): 97.

[115] OOI N, RAJAN V, GOTTLIEB J, et al. Structural properties of hexagonal boron nitride[J]. Modelling and Simulation in Materials Science and Engineering, 2006, 14(3): 515.

[116] WATANABE K, TANIGUCHI T, KANDA H. Direct-bandgap properties and evidence for ultraviolet lasing of hexagonal boron nitride single crystal[J]. Nature Materials, 2004, 3(6): 404-409.

[117] WATANABE K, TANIGUCHI T, KURODA T, et al. Band-edge luminescence of deformed hexagonal boron nitride single crystals[J]. Diamond and Related Materials, 2006, 15(11/12): 1891-1893.

[118] WATANABE K, TANIGUCHI T, KURODA T, et al. Effects of deformation on band-edge luminescence of hexagonal boron nitride single crystals[J]. Applied Physics Letters, 2006, 89(14): 141902.

[119] JIN M S, KIM N O. Photoluminescence of hexagonal boron nitride (h-BN) film[J]. Journal of Electrical Engineering and Technology, 2010, 5(4): 637-639.

[120] TOPSAKAL M, AKTÜRK E, CIRACI S. First-principles study of two-and one-dimensional honeycomb structures of boron nitride[J]. Physical Review B, 2009, 79(11): 115442.

[121] ARNAUD B, LEBÈGUE S, RABILLER P, et al. Huge excitonic effects in layered hexagonal boron nitride[J]. Physical Review Letters, 2006, 96(2): 026402.

[122] GUO G Y, LIN J C. Second-harmonic generation and linear electro-optical coefficients of BN nanotubes[J]. Physical Review B, 2005, 72(7): 075416.

[123] PARK C H, SPATARU C D, LOUIE S G. Excitons and many-electron effects in the optical response of single-walled boron nitride nanotubes [J]. Physical Review Letters, 2006, 96(12): 126105.

[124] LAURET J S, ARENAL R, DUCASTELLE F, et al. Optical transitions in single-wall boron nitride nanotubes[J]. Physical Review Letters, 2005, 94(3): 037405.

[125] MARINOPOULOS A G, WIRTZ L, MARINI A, et al. Optical absorption and electron energy loss spectra of carbon and boron nitride nanotubes: a first-principles approach[J]. Applied Physics A, 2004, 78(8): 1157-1167.

[126] WIRTZ L, MARINI A, RUBIO A. Excitons in boron nitride nanotubes: dimensionality effects[J]. Physical Review Letters, 2006, 96(12): 126104.

[127] GAO R, YIN L, WANG C, et al. High-yield synthesis of boron nitride nanosheets with strong ultraviolet cathodoluminescence emission[J]. The Journal of Physical

Chemistry C, 2009, 113(34): 15160-15165.

[128] KUBOTA Y, WATANABE K, TSUDA O, et al. Deep ultraviolet light-emitting hexagonal boron nitride synthesized at atmospheric pressure[J]. Science, 2007, 317(5840): 932-934.

[129] BOROWIAK-PALEN E, PICHLER T, FUENTES G G, et al. Infrared response of multiwalled boron nitride nanotubes[J]. Chemical Communications, 2003 (1): 82-83.

[130] GU Y, ZHENG M, LIU Y, et al. Low-temperature synthesis and growth of hexagonal boron-nitride in a lithium bromide melt[J]. Journal of the American Ceramic Society, 2007, 90(5): 1589-1591.

[131] WANG J, CAO S, SUN P, et al. Optical advantages of graphene on the boron nitride in visible and SW-NIR regions [J]. RSC Advances, 2016, 6 (112): 111345-111349.

[132] JIN M S, KIM N O. Photoluminescence of hexagonal boron nitride (h-BN) film[J]. Journal of Electrical Engineering and Technology, 2010, 5(4): 637-639.

[133] HOFFMAN D M, HEINZ R E, DOLL G L, et al. Optical reflectance study of the electronic structure of acceptor-type graphite intercalation compounds[J]. Physical Review B, 1985, 32(2): 1278.

[134] WEIβMANTEL E, PFEIFER T, RICHTER F. Electron microscopic analysis of cubic boron nitride films deposited on fused silica[J]. Thin Solid Films, 2002, 408(1/2): 1-5.

[135] LOPATIN V V, KONUSOV F V. Energetic states in the boron nitride band gap[J]. Journal of Physics and Chemistry of Solids, 1992, 53(6): 847-854.

[136] WATANABE K, TANIGUCHI T, KANDA H. Direct-bandgap properties and evidence for ultraviolet lasing of hexagonal boron nitride single crystal[J]. Nature Materials, 2004, 3(6): 404-409.

[137] MUSEUR L, FELDBACH E, KANAEV A. Defect-related photoluminescence of hexagonal boron nitride[J]. Physical Review B, 2008, 78(15): 155204.

[138] KAWARADA H, MATSUYAMA H, YOKOTA Y, et al. Excitonic recombination radiation in undoped and boron-doped chemical-vapor-deposited diamonds [J]. Physical Review B, 1993, 47(7): 3633.

[139] SILLY M G, JAFFRENNOU P, BARJON J, et al. Luminescence properties of hexagonal boron nitride: cathodoluminescence and photoluminescence spectroscopy measurements[J]. Physical Review B, 2007, 75(8): 085205.

[140] ARNAUD B, LEBÈGUE S, RABILLER P, et al. Huge excitonic effects in layered hexagonal boron nitride[J]. Physical Review Letters, 2006, 96(2): 026402.

[141] MUSEUR L, KANAEV A. Near band-gap photoluminescence properties of hexagonal boron nitride[J]. Journal of Applied Physics, 2008, 103(10): 103520.

[142] ARNAUD B, LEBÈGUE S, RABILLER P, et al. Huge excitonic effects in layered hexagonal boron nitride[J]. Physical Review Letters, 2006, 96(2): 026402.

[143] DAI S, FEI Z, MA Q, et al. Tunable phonon polaritons in atomically thin van der Waals crystals of boron nitride[J]. Science, 2014, 343(6175): 1125-1129.

[144] LI J, MAJETY S, DAHAL R, et al. Dielectric strength, optical absorption, and deep ultraviolet detectors of hexagonal boron nitride epilayers[J]. Applied Physics Letters, 2012, 101(17): 171112.

[145] LI J, FAN Z Y, DAHAL R, et al. 200 nm deep ultraviolet photodetectors based on AlN[J]. Applied Physics Letters, 2006, 89(21): 213510.

[146] DAHAL R, AL TAHTAMOUNI T M, FAN Z Y, et al. Hybrid AlN-SiC deep ultraviolet Schottky barrier photodetectors[J]. Applied Physics Letters, 2007, 90(26): 263505.

[147] NAKARMI M L, KIM K H, KHIZAR M, et al. Electrical and optical properties of Mg-doped Al 0.7 Ga 0.3 N alloys[J]. Applied Physics Letters, 2005, 86(9): 092108.

[148] DAHAL R, LI J, MAJETY S, et al. Epitaxially grown semiconducting hexagonal boron nitride as a deep ultraviolet photonic material[J]. Applied Physics Letters, 2011, 98(21): 211110.

[149] NAKARMI M L, KIM K H, ZHU K, et al. Transport properties of highly conductive n-type Al-rich Al$_x$Ga$_{1-x}$N ($x \geqslant 0.7$)[J]. Applied Physics Letters, 2004, 85(17): 3769-3771.

[150] MAJETY S, LI J, CAO X K, et al. Epitaxial growth and demonstration of hexagonal BN/AlGaN p-n junctions for deep ultraviolet photonics[J]. Applied Physics Letters, 2012, 100(6): 061121.

[151] JAFFRENNOU P, BARJON J, LAURET J S, et al. Origin of the excitonic recombinations in hexagonal boron nitride by spatially resolved cathodoluminescence spectroscopy[J]. 2007, 102(11): 1.

[152] WATANABE K, TANIGUCHI T. Jahn-Teller effect on exciton states in hexagonal boron nitride single crystal[J]. Physical Review B, 2009, 79(19): 193104.

[153] WATANABE K, TANIGUCHI T, NIIYAMA T, et al. Far-ultraviolet plane-emission handheld device based on hexagonal boron nitride[J]. Nature Photonics, 2009, 3(10): 591-594.

[154] SHI Z, JIN C, YANG W, et al. Gate-dependent pseudospin mixing in graphene/boron nitride moiré superlattices[J]. Nature Physics, 2014, 10(10): 743-747.

[155] HUNT B, SANCHEZ-YAMAGISHI J D, YOUNG A F, et al. Massive Dirac fermions and Hofstadter butterfly in a van der Waals heterostructure[J]. Science, 2013, 340(6139): 1427-1430.

[156] OSHIMA C, ITOH A, ROKUTA E, et al. A hetero-epitaxial-double-atomic-layer system of monolayer graphene/monolayer h-BN on Ni(111)[J]. Solid State Communications, 2000, 116(1): 37-40.

[157] SAKAI Y, SAITO S, COHEN M L. First-principles study on graphene/hexagonal boron nitride heterostructures[J]. Journal of the Physical Society of Japan, 2015, 84(12): 121002.

[158] BJELKEVIG C, MI Z, XIAO J, et al. Electronic structure of a graphene/hexagonal-BN heterostructure grown on Ru (0001) by chemical vapor deposition and atomic layer deposition: extrinsically doped graphene[J]. Journal of Physics: Condensed Matter, 2010, 22(30): 302002.

[159] YELGEL C, SRIVASTAVA G P. Energy band gap modification of graphene deposited on a multilayer hexagonal boron nitride substrate [C]. MRS Online Proceedings Library Archive, 2012.

[160] WANG J, XU Y, CHEN H, et al. Ultraviolet dielectric hyperlens with layered graphene and boron nitride[J]. Journal of Materials Chemistry, 2012, 22 (31): 15863-15868.

[161] TANG S, DING G, XIE X, et al. Nucleation and growth of single crystal graphene on hexagonal boron nitride[J]. Carbon, 2012, 50(1): 329-331.

[162] DEAN C R, YOUNG A F, MERIC I, et al. Boron nitride substrates for high-quality graphene electronics[J]. Nature Nanotechnology, 2010, 5(10): 722-726.

[163] PONOMARENKO L A, GEIM A K, ZHUKOV A A, et al. Tunable metal-insulator transition in double-layer graphene heterostructures [J]. Nature Physics, 2011, 7(12): 958-961.

[164] HAIGH S J, GHOLINIA A, JALIL R, et al. Cross-sectional imaging of individual layers and buried interfaces of graphene-based heterostructures and superlattices[J]. Nature Materials, 2012, 11(9): 764-767.

[165] DEAN C, YOUNG A F, WANG L, et al. Graphene based heterostructures[J]. Solid State Communications, 2012, 152(15): 1275-1282.

[166] PONOMARENKO L A, GORBACHEV R V, YU G L, et al. Cloning of Dirac fermions in graphene superlattices[J]. Nature, 2013, 497(7451): 594-597.

[167] DEAN C R, WANG L, MAHER P, et al. Hofstadter's butterfly and the fractal quantum Hall effect in moiré superlattices[J]. Nature, 2013, 497(7451): 598-602.

[168] GIOVANNETTI G, KHOMYAKOV P A, BROCKS G, et al. Substrate-induced band gap in graphene on hexagonal boron nitride: ab initio density functional calculations[J]. Physical Review B, 2007, 76(7): 073103.

[169] SLAWIŃSKA J, ZASADA I, KLUSEK Z. Energy gap tuning in graphene on hexagonal boron nitride bilayer system[J]. Physical Review B, 2010, 81(15): 155433.

[170] BRITNELL L, GORBACHEV R V, JALIL R, et al. Field-effect tunneling transistor based on vertical graphene heterostructures[J]. Science, 2012, 335(6071): 947-950.

[171] WANG D, CHEN G, LI C, et al. Thermally induced graphene rotation on hexagonal boron nitride[J]. Physical Review Letters, 2016, 116(12): 126101.

[172] LEE M, WALLBANK J R, GALLAGHER P, et al. Ballistic miniband conduction in a graphene superlattice[J]. Science, 2016, 353(6307): 1526-1529.

[173] WANG E, LU X, DING S, et al. Gaps induced by inversion symmetry breaking and second-generation Dirac cones in graphene/hexagonal boron nitride [J]. Nature Physics, 2016, 12(12): 1111-1115.

[174] FAN Y, ZHAO M, WANG Z, et al. Tunable electronic structures of graphene/boron nitride heterobilayers[J]. Applied Physics Letters, 2011, 98(8): 083103.

[175] SONG X, GAO T, NIE Y, et al. Seed-assisted growth of single-crystalline patterned graphene domains on hexagonal boron nitride by chemical vapor deposition[J]. Nano Letters, 2016, 16(10): 6109-6116.

[176] GANNETT W, REGAN W, WATANABE K, et al. Boron nitride substrates for high mobility chemical vapor deposited graphene[J]. Applied Physics Letters, 2011, 98(24): 242105.

[177] LEVENDORF M P, KIM C J, BROWN L, et al. Graphene and boron nitride lateral heterostructures for atomically thin circuitry[J]. Nature, 2012, 488(7413): 627-632.

[178] WU S, WANG L, LAI Y, et al. Multiple hot-carrier collection in photo-excited graphene Moiré superlattices[J]. Science Advances, 2016, 2(5): e1600002.

[179] ZOMER P J, DASH S P, TOMBROS N, et al. A transfer technique for high mobility graphene devices on commercially available hexagonal boron nitride[J]. Applied Physics Letters, 2011, 99(23): 232104.

[180] LEE K H, SHIN H J, LEE J, et al. Large-scale synthesis of high-quality hexagonal boron nitride nanosheets for large-area graphene electronics[J]. Nano Letters, 2012, 12(2): 714-718.

[181] NARIMANOV E E. Photonic hypercrystals[J]. Physical Review X, 2014, 4(4): 041014.

[182] PODDUBNY A, IORSH I, BELOV P, et al. Hyperbolic metamaterials[J]. Nature Photonics, 2013, 7(12): 948-957.

[183] DAI S, MA Q, LIU M K, et al. Graphene on hexagonal boron nitride as a tunable hyperbolic metamaterial[J]. Nature Nanotechnology, 2015, 10(8): 682-686.

[184] AL SAYEM A, RAHMAN M M, MAHDY M R C, et al. Negative refraction with superior transmission in graphene-hexagonal boron nitride (h-BN) multilayer hyper crystal[J]. Scientific Reports, 2016, 6: 25442.

[185] ZHANG X, DAI Z, SI S, et al. Ultrasensitive SERS substrate integrated with uniform subnanometer scale "hot spots" created by a graphene spacer for the detection of mercury ions[J]. Small, 2017, 13(9): 1603347.

[186] CAROZO V, ALMEIDA C M, FERREIRA E H M, et al. Raman signature of graphene superlattices[J]. Nano Letters, 2011, 11(11): 4527-4534.

[187] CAI Q, MATETI S, YANG W, et al. Boron nitride nanosheets improve sensitivity and reusability of surface-enhanced raman spectroscopy[J]. Angewandte Chemie, 2016, 128(29): 8545-8549.

[188] ECKMANN A, PARK J, YANG H, et al. Raman fingerprint of aligned graphene/h-BN superlattices[J]. Nano Letters, 2013, 13(11): 5242-5246.

[189] HERMANN P, HOEHL A, PATOKA P, et al. Near-field imaging and nano-Fourier-transform infrared spectroscopy using broadband synchrotron radiation[J]. Optics Express, 2013, 21(3): 2913-2919.

[190] FEI Z, ANDREEV G O, BAO W, et al. Infrared nanoscopy of Dirac plasmons at the graphene-SiO_2 interface[J]. Nano Letters, 2011, 11(11): 4701-4705.

[191] AMARIE S, KEILMANN F. Broadband-infrared assessment of phonon resonance in scattering-type near-field microscopy[J]. Physical Review B, 2011, 83(4): 045404.

[192] BARCELOS I D, CADORE A R, CAMPOS L C, et al. Graphene/h-BN plasmon-phonon coupling and plasmon delocalization observed by infrared nano-spectroscopy [J]. Nanoscale, 2015, 7(27): 11620-11625.

[193] FEI Z, RODIN A S, ANDREEV G O, et al. Gate-tuning of graphene plasmons revealed by infrared nano-imaging[J]. Nature, 2012, 487(7405): 82-85.

[194] CHEN J, BADIOLI M, ALONSO-GONZÁLEZ P, et al. Optical nano-imaging of gate-tunable graphene plasmons[J]. Nature, 2012, 487(7405): 77-81.

[195] PRINCIPI A, CARREGA M, LUNDEBERG M B, et al. Plasmon losses due to electron-phonon scattering: the case of graphene encapsulated in hexagonal boron nitride[J]. Physical Review B, 2014, 90(16): 165408.

[196] WOESSNER A, LUNDEBERG M B, GAO Y, et al. Highly confined low-loss plasmons in graphene-boron nitride heterostructures[J]. Nature Materials, 2015, 14(4): 421-425.

[197] WU S, WANG L, LAI Y, et al. Multiple hot-carrier collection in photo-excited graphene Moiré superlattices[J]. Science Advances, 2016, 2(5): e1600002.

[198] WITHERS F, DEL POZO-ZAMUDIO O, MISHCHENKO A, et al. Light-emitting diodes by band-structure engineering in van der Waals heterostructures[J]. Nature Materials, 2015, 14(3): 301-306.

[199] WANG X, XIA F. Van der Waals heterostructures: stacked 2D materials shed light [J]. Nature Materials, 2015, 14(3): 264-265.

[200] WITHERS F, DEL POZO-ZAMUDIO O, SCHWARZ S, et al. WSe_2 light-emitting tunneling transistors with enhanced brightness at room temperature [J]. Nano Letters, 2015, 15(12): 8223-8228.

[201] SONG Y, LI X, MACKIN C, et al. Role of interfacial oxide in high-efficiency graphene-silicon Schottky barrier solar cells[J]. Nano Letters, 2015, 15(3): 2104-2110.

[202] LI X, LIN S, LIN X, et al. Graphene/h-BN/GaAs sandwich diode as solar cell and photodetector[J]. Optics Express, 2016, 24(1): 134-145.

[203] MENG J H, LIU X, ZHANG X W, et al. Interface engineering for highly efficient graphene-on-silicon Schottky junction solar cells by introducing a hexagonal boron nitride interlayer[J]. Nano Energy, 2016, 28: 44-50.

[204] WANG F, ZHANG Y, TIAN C, et al. Gate-variable optical transitions in graphene [J]. Science, 2008, 320(5873): 206-209.

[205] BRAR V W, JANG M S, SHERROTT M, et al. Hybrid surface-phonon-plasmon polariton modes in graphene/monolayer h-BN heterostructures[J]. Nano Letters, 2014, 14(7): 3876-3880.

[206] POSPISCHIL A, HUMER M, FURCHI M M, et al. CMOS-compatible graphene photodetector covering all optical communication bands[J]. Nature Photonics, 2013, 7(11): 892-896.

[207] GAN X, SHIUE R J, GAO Y, et al. Chip-integrated ultrafast graphene photodetector with high responsivity[J]. Nature Photonics, 2013, 7(11): 883-887.

[208] WANG X, CHENG Z, XU K, et al. High-responsivity graphene/silicon-heterostructure waveguide photodetectors[J]. Nature Photonics, 2013, 7(11): 888-891.

[209] SHIUE R J, GAO Y, WANG Y, et al. High-responsivity graphene-boron nitride photodetector and autocorrelator in a silicon photonic integrated circuit[J]. Nano Letters, 2015, 15(11): 7288-7293.

[210] LI Z Q, HENRIKSEN E A, JIANG Z, et al. Dirac charge dynamics in graphene by infrared spectroscopy[J]. Nature Physics, 2008, 4(7): 532-535.

[211] AKAHANE Y, ASANO T, SONG B S, et al. High-Q photonic nanocavity in a two-dimensional photonic crystal[J]. Nature, 2003, 425(6961): 944-947.

[212] GAO Y, SHIUE R J, GAN X, et al. High-speed electro-optic modulator integrated with graphene-boron nitride heterostructure and photonic crystal nanocavity[J]. Nano Letters, 2015, 15(3): 2001-2005.

Chapter 4

Optoelectronic Properties and Applications of Graphene-Based Hybrid Nanomaterials and van der Waals Heterostructure

4.1　Introduction

There are many problems in the development of optoelectronic devices based on semiconductors and other materials. With small size effect[1], surface effect[2-3], and quantum tunneling effect, nanomaterials have attracted much attention in the field of semiconductor. Moreover, great progress has been made in the research and preparation of photovoltaic devices. Graphene has been warmly studied as a two dimensional (2D) carbon nanomaterial[5-22]. Different dimensions of graphene's nano-materials (graphene quantum dots, nanoribbons) in addition to the characteristics of nanomaterials, graphene's nano-material itself unique optical[6-7], electronical and thermal characteristics make it in optoelectronic devices based on graphene's nano-material has been widely concerned[8-10].

Graphene is a 2D crystal with a hexagonal honeycomb lattice structure composed of C atoms in a sp^2 hybridized orbitals (Figure 4-1(b))[11-13]. Figure 4-1(a) shows the scanning electron micrographs (SEMs) of graphene. The thickness of the single-layer graphene is 0.35 nm, the length of C-C is 0.142 nm[11-12]. The protocells of graphene are defined by lattice vectors a_1 and a_2 to define two C atoms within each protocell, just as shown in Figure 4-1(c)[11]. There are three electrons in the outer shell of atom C form strong σ bonds through sp^2 hybridization, and the fourth electron forms π bonds. Figure 4-1(d) shows the graphene Brillouin zone of the unit cell. This regular hexagonal planar structure makes graphene very stable and flexible. Therefore, graphene, as the basic unit of many carbon nanomaterials, can be curled into carbon

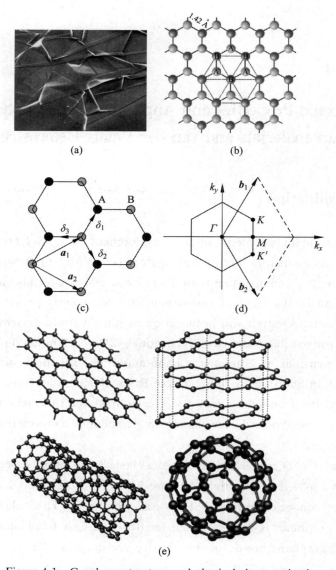

Figure 4-1 Graphene structure and physical characterizations

(a) The scanning electron micrograph of graphene(SEM)[5]; (b) The structure of graphene. A and B are equivalent atoms in a unit cell[12]; (c) A schematic diagram of the bonding of C atoms in a unit cell[11]; (d) The Brillouin zone corresponding to the unit lattice[11]; (e) Carbon-based nanomaterials with graphene as the basic unit[22]

nanotubes, stacked to form graphite, and zero-dimensional (0D) fullerenes.

Graphene has many unique mechanical, thermal, optical and electrical properties due to its unique crystal structure and novel electronic structure. The intrinsic modulus and strength of graphene are 1.1×10^{12} Pa and 1,250 Pa respectively[14]. The thermal conductivity of graphene is up to 5,000 W/(m·K) at room temperature[15]. Even more peculiar is the electron mobility of graphene at room temperature as high as 1.5×10^4 cm^2/(V·s)[16-17]. More than this, graphene has other electrical properties, such as anomalous quantum Hall effects (QHEs)[17], ambipolar electronic field effects[18-19], Klein tunneling[20] and ballistic transport[21].

4.2 The optoelectronic properties of graphene

Graphene, as a typical 2D nanometer material, has unique optoelectronic properties due to its unique band structure. In addition to the intrinsic optoelectronic properties, the hybridization or heterostructure of other 2D materials with graphene and the change of shape and configuration of graphene, which make the hybridized materials or heterostructure based on graphene nano-structure show more and more unique photoelectric properties.

4.2.1 The intrinsic optoelectronic properties of graphene nanomaterials

1. The energy band structure of zero bandgap

The conduction band and valence band corresponding to the two sublattice of A and B in the unit cell are connected to K points at the boundary of the first Brillouin zone, which proves that graphene is a semiconductor with zero band gap[11-12,23]. Near the Dirac point(DP), which at the K high symmetry point, the energy and momentum of the electron show a linear dispersion relation: $\varepsilon(k) = \hbar \nu_F |k|$, k is the wave vector of the electron wave function, ν_F is Fermi velocity ($\nu_F \approx 1.0 \times 10^6$ m/s). This linear dispersion makes the effective mass of the electrons of graphene approximately equal to zero, which means that the electrons in graphene are a massless Dirac fermion[17]. Figure 4-2(a) shows the band structure of graphene[11].

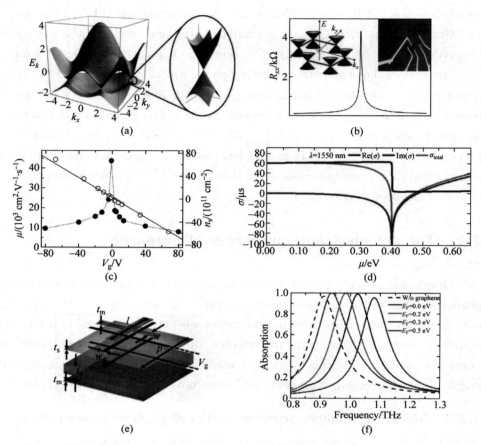

Figure 4-2 The energy band structure of zero bandgap and optoelectric properties

(a) The band structure of graphene, the inset is Dirac cone of DP K[11]; (b) The curve of the electrical conductivity of graphene[17]; (c) The relation diagram of different mobility (μ) and (V_g)[17]. (d) Carrier concentration corresponding to different Fermi levels[34]; (e) Device diagram for testing[33]; (f) The change of the Fermi level leads to the change of the optical absorption property of graphene[33]

It is the novel zero-band-gap structure of graphene that distinguishes it from other 2D materials, giving it excellent optical properties. The zero-band-gap of graphene nano-structure has been changed by various methods[24-28], for example: the preparation of graphene quantum dots of different shapes and sizes; adding substrates or forming heterogeneous structures; doped graphene with other elements; changing the external electric field and so on can change the optical response of graphene nanomaterials by changing the band gap.

2. Minimum quantum conductivity and superhigh carrier mobility

In the vicinity of the DP K, although the carrier density is 0, graphene has the minimum conductivity, which is about the order of magnitude of 4 e^2/h. Moreover, the mobility of carriers is almost independent of temperature in the range of 10 K to 100 K[17,29-30]. At room temperature, when the carrier density is $1.0 \times 10^{12}/cm^2$, due to the scattering effect of phonon, the upper constraint of carrier is 2.0×10^5 cm^2/(V·s), and the corresponding conductivity is 1.0×10^{-6} Ω·cm, Figures 4-2(b) and (c). On the other hand, the super-high carrier's mobility of graphene has also been verified in many experiments. Dean et al., measured the carrier's mobility of graphene based on h-BN substrate as high as 0.8×10^5/(cm·V·s)[28].

3. Fermi levels

Some studies have shown that the Fermi level of graphene can be adjusted by changing the bias voltage, which makes graphene have novel optical response[31-33]. Single-layer graphene can be characterized by an infinite thin conductive layer, and its surface conductivity's model can be expressed by $\sigma_s(\omega, \mu_c, \Gamma, T) = \sigma_{intra} + \sigma'_{inter} + \sigma''_{inter}$ [31] (ω is circular frequency, μ_c is the chemical potential which equals to the Fermi level E_F, Γ is the scattering efficiency). The electrical conductivity of graphene in terahertz frequency band is expressed as[31]:

$$\sigma(\omega, \mu_c, \Gamma, T) \approx \sigma_{intra} = \frac{-je^2 k_B T}{\pi \hbar^2 (\omega - j2\Gamma)} \times \left[\frac{\mu_c}{k_B T} + 2\ln(e^{\frac{-\mu_c}{k_B T}} + 1) \right]. \quad (4-1)$$

According to Equation (4-1), the conductivity of graphene can be regulated by Fermi level E_F. The relationship between Fermi level E_F and bias voltage V_g is as follows[33]:

$$E_F \approx \hbar v_F \sqrt{\frac{\pi \varepsilon_r \varepsilon_0 V_g}{e t_s}}. \quad (4-2)$$

It can be seen from the above equation that different bias voltages can be applied to regulate the Fermi level of graphene so as to regulate the optoelectronic properties of graphene. Figures 4-2 (d) and (e) show the diagram of test device and results.

On the other hand, τ_1 is relaxation time of interband, τ_2 is relaxation time of intraband, μ is the chemical potential, another representation of Equation (4-1) is[34-36]:

$$\sigma_{intra} = \frac{4\mu e^2}{2\hbar} \cdot \frac{1}{\hbar\tau_1 - i\hbar\omega}, \tag{4-3}$$

$$\sigma'_{inter} = \frac{\pi e^2}{2\hbar} \cdot \left(1 + \frac{1}{\pi}\arctan\frac{\hbar\omega - 2\mu}{\hbar\tau_2} - \frac{1}{\pi}\arctan\frac{\hbar\omega + 2\mu}{\hbar\tau_2}\right), \tag{4-4}$$

$$\sigma'_{inter} = -\frac{e^2}{4\hbar}\ln\frac{(2\mu + \hbar\omega)^2 + \hbar^2\tau_2^2}{(2\mu - \hbar\omega)^2 + \hbar^2\tau_2^2}. \tag{4-5}$$

The photoconductivity of graphene (σ) is affected by the chemical potential (μ) of graphene and the frequency of incident light. The chemical potential of graphene is determined by its carrier concentration n_0, n_0 can be expressed as $n_0 = (\mu/\hbar v)^2/\pi$. Therefore, chemical doping can also change the Fermi energy level, thus changing the optical properties of graphene. Figures 4-2(d)~(f) show the effects of chemical doping and external voltage changes on the Fermi level of graphene, thus affecting the optical properties of graphene[33-34].

4. The linear absorption

The unique band structure gives graphene novel optical properties. When light shines on the surface of graphene, electrons in the valence band are excited to the conduction band by absorbing energy of photons. The photoconductivity of single-layer graphene depends on fine structure constant α[35]:

$$\alpha = \frac{e^2}{4\pi\varepsilon_0 \hbar c} \approx \frac{1}{137}. \tag{4-6}$$

The linear transmittance of single-layer graphene is T[35]:

$$T = \frac{1}{(1 + 0.5\pi\alpha)^2} \approx 1 - \pi\alpha \approx 97.7\%. \tag{4-7}$$

As shown in Figure 2-3(a), the absorption rate of monolayer graphene is 2.3 %. In the visible region, the reflectivity of single-layer graphene is less than 0.1 %, and that of ten-layers graphene is only 2 %. Figures 4-3(a) and (b) show the optical properties of theoretical calculation results and experimental results[35]. Because of zero-bandgap structure, graphene can absorb the photonics at any wavelength[35-37]. Theoretical calculations also show that different configurations of graphene quantum dots or substrate-based graphene can be absorbed on different bands of light[38-39].

When the power of incident light is weak, the energy band near the DP of graphene is not filled, and the light absorption rate of graphene remains unchanged. When the power of the incident light is very strong, the energy band near the DP of the graphene is filled. Due to the Pauli blocking effect, the graphene cannot continue to absorb photons to achieve saturable absorption. The light absorption of graphene is no longer linear, but nonlinear absorption[40-45]. The non-linear saturable absorption coefficient of graphene is[40]:

$$\alpha^*(N) = \frac{\alpha_S^*}{1+(N/N_S)} + \alpha_{NS}^*, \qquad (4\text{-}8)$$

where α_S^* is saturated absorption coefficient, and α_{NS}^* is unsaturated absorption components, N is the photoinduced electron-hole density, N_S is saturation density. Figure 4-3(c) shows the dynamic simulation of graphene-saturated absorption.

5. Nonlinear absorption

The nonlinear effect of light is due to the interaction of electrons and photons in the electromagnet field. The formula $P = \varepsilon_0 \chi^{(1)} E$ shows the relationship between electron polarization P and the applied electric field E. $\chi^{(1)}$ is the first-order linear polarizability. When the intensity of the incident light increases, the electron polarization shows a significant nonlinear phenomenon: $P = \varepsilon_0 (\chi^{(1)} E + \chi^{(2)} E^2 + \chi^{(3)} E^3) + \cdots$ $\chi^{(2)}$ and $\chi^{(3)}$ are the second-order nonlinear susceptibility and the third order nonlinear susceptibility[41]. These nonlinear phenomena include two-photon absorption, self-phase modulation, saturable absorption and so on.

The pair excitation of free electrons and bound electrons induced by a single photon can be described by first-order susceptibility $\chi^{(1)}$, where, the real part of the susceptibility is related to the real part of the refractive index, and the imaginary part is related to the loss or gain of light. The effective linear polarization rate of a given frequency of light can be modulated by applying a direct current electric field to graphene. For second-order nonlinear susceptibility $\chi^{(2)}$, $\chi^{(2)}$ is generally considered to be 0 due to the inversion symmetry of graphene crystal cells. However, when graphene with stress, disorder or functionalization, the symmetry of its crystal cells will be destroyed, and $\chi^{(2)}$ cannot be ignored.

The nonlinear optical properties of graphene mostly depend on its third-order nonlinear susceptibility $\chi^{(3)}$. The N-order integral of surface current J_n^ν is usually used to represent the nonlinear optical response of graphene[42-44]:

$$J_n^\nu = \frac{1}{4\pi^2} \int dp\, j_n^\nu N(\varepsilon), \qquad (4\text{-}9)$$

where N is the thermal conductivity, after calculation, as follows:

$$\begin{cases} N(\varepsilon) = n_F(-\varepsilon) - n_F(\varepsilon) = \tanh(\varepsilon/2K_B T), \\ j_n^\nu = \psi^+ \hat{V}_\nu \psi, \\ \hat{V}_\nu = (\partial H)/\partial p_V, \end{cases} \qquad (4\text{-}10)$$

and the linear optical response is

$$J_1 = (e^2 E)/(4\hbar). \qquad (4\text{-}11)$$

For symmetric graphene, $\hat{v}(x) = \hat{v}(-x)$, the second-order total surface current of graphene is

$$J_2 = 0. \qquad (4\text{-}12)$$

The third-order surface current is

$$J_3 = J_3(\omega) + J_3(3\omega) = \frac{\sigma_1 e^2 v_F^2 E_0^2}{\hbar^2 \omega^4}[N_1(\omega)e^{i\omega t} + N_3(\omega)e^{3i\omega t}], \qquad (4\text{-}13)$$

where:

$$\sigma_1 = e^2/4\hbar, \quad v_F = c/300, \quad N_1(\omega) = N(\omega),$$

$$N_3(\omega) = 13N(\omega/2)/48 - N(\omega)/3 + 45N(3\omega/2)/48.$$

In Equation (4-13), $J_3(\omega) + J_3(3\omega)$ determines the nonlinearity of graphene.

6. The relationship between transition absorption and energy level

According to the different Fermi level and DP position of graphene, there are three different states of graphene energy level: eigenstates (Figure 4-3(d)), N-type (Figure 4-3(e)) and P-type (Figure 4-3(f))[45-46]. Since graphene is a 2D material with zero bandgap, the interband transition and intraband transition of electrons are competitive, and the threshold of intraband transition is $|\mu| = \hbar\omega/2$. When the electrons in the valence band absorb the energy of the photons, the electrons transition from the valence band to the conduction band by means of interband transitions, thus forming electron-hole pairs in the graphene. Figure 4-3(d) shows the processes of electrons in the valence band undergo interband transitions after the intrinsic graphene absorbs photons, and the electrons from the valence band transition to the conduction band. Figures 4-3(e) and (f) show N-type and P-type doped graphene absorbs photons, electrons excited from the valence band to the conduction band, where the energy of the photon is $\hbar\omega > 2E_F$. Figures 4-3(d)~(g) show the intraband and interband transitions of graphene.

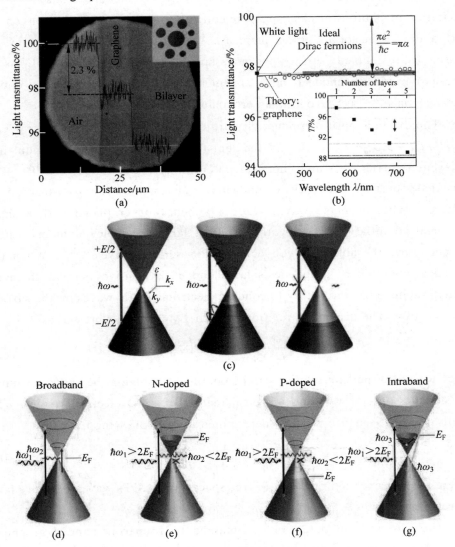

Figure 4-3 Optical properties of graphene
(a) Image of the optical absorption of graphene with different layers[35]; (b) Transmittance spectrum of single-layer graphene[35]; (c) Absorption spectra for three different samples of graphene over the range of photon energies between 0.5 eV and 1.2 eV[40]; (d)~(g) The intraband and interband transition of graphene[45-46]

7. Graphene surface plasmons (G-SPs)

Graphene's unique electronic properties make it as new infrared (IR) frequency domain plasmon waveguide and terahertz metamaterials[47-54]. By excitation of photons or electrons, the collective oscillations of electrons on the surface of a conductor are surface plasmons. When photons and graphene's surface plasmons are coupled, surface plasmons polaritons (SPPs) are formed. As an alternative to conventional metal plasmon, G-SP can be used for optical metamaterials[50], such as optical absorption and optical conversion[51-52].

The dynamic optical conductivity of graphene can be expressed as: $\sigma_s(\omega, \mu_c, \Gamma, T) = \sigma_{intra} + \sigma'_{inter} + \sigma''_{inter}$. Incident frequency ($\omega$), the relaxation time of excited carrier (τ) and chemical potential (μ) can tuned to SPPs. The imaginary part of the optical conductivity determines the transmission of different types of plane waves. Pure graphene, or graphene with a small chemical potential, has a negative σ'' in a large frequency domain, where graphene has the ability to conduct TE-SPPs surface waves[47,49]. When the chemical potential of graphene is large and σ'' is positive, similar to metal film, graphene has the ability to conduct electromagnetic waves on the surface of TM-SPPs. The dispersion relation of such TM-SPPs is expressed as[53]:

$$k_{sp} \approx \frac{\hbar^2}{4e^2 E_F}(\varepsilon + 1)\omega\left(\omega + \frac{i}{\tau_1}\right). \tag{4-14}$$

The curve of dispersion relation can be obtained by solving the above formula. k_{sp} depends on ω^2, which is a typical feature of a 2D electron system. The limit of G-SPPs surface waves is determined by the following formula:

$$k_{sp}/\lambda_0 \approx \frac{4\alpha}{\varepsilon + 1} \cdot \frac{E_F}{\hbar \omega}, \tag{4-15}$$

where, λ_0 is wavelength of light in free space, λ_{sp} is SPPs wavelength of plane wave, expressed as $\lambda_{sp} = 2\pi/\text{Re}\{k_{sp}\}$.

The experimental data can be obtained by preparing banded graphene with different chemical potentials (Figure 4-4), which is similar to traditional metal-insulator-metal waveguide[52,54]. Graphene's chemical potential can be controlled by the thickness of the dielectric layer. Another method is that the dielectric layer has the same thickness but different dielectric constants.

Graphene samples with specific shapes can be obtained by designing ingenious growth and preparation methods. The structure of graphene SPP waveguide can imitate the traditional metal-silicon waveguide structure[52].

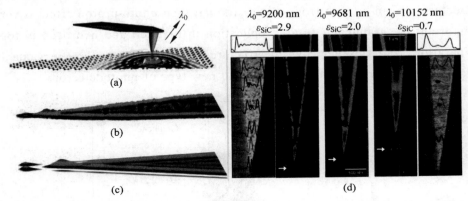

Figure 4-4   Imaging propagating and localized graphene plasmons by scattering graphene plasmons by scattering-type SNOM (scanning near-field optical microscopy) in (a)~(c); (d) Controlling the plasmon wavelength over a wide range[54]

4.2.2 The optoelectronic properties of hybrid graphene or heterostructure

1. Proton transport-huge photoelectric effect

Researchers recently discovered a new unexpected physical effect of graphene film[55-59]. Through the doping of other particles, the photoelectric effect of graphene can be greatly enhanced, which can be used to simulate photosynthesis artificially[59]. Graphene has been shown to be permeable to hot protons. The new results show that when a beam of light hits the surface of the material, the speed at which the material transfers protons increases significantly. The researchers prepared the original graphene film and modified the surface of the graphene with Pt nanoparticles (NPs). The experiment results showed that the proton transport performance of this hybrid graphene film was improved by 10 times, and the light response rate measured was about 104 A/W. Unlike conventional light detectors, graphene membranes produce hydrogen in addition to electricity. That is, for each photon, there are 104 proton gains, and there are 5,000 hydrogen molecules.

This gain is a huge number for photovoltaic devices, and also provides a new way of thinking and approaches for the new photoelectric effect and photosynthesis.

Figure 4-5 shows the experiment result of graphene/Pt NPs hybrid structure. It can be seen from the results that such photo-proton effect is very important for graphene's future application in fuel cell and hydrogen isotope separation, and this new hybrid structure has important reference value in water decomposition, photo-catalysis and new type of photodetector.

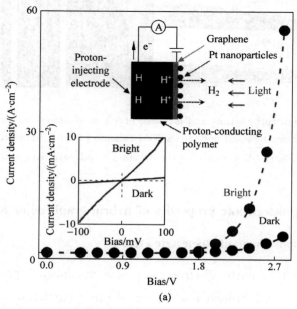

Figure 4-5 Influence of illumination on proton transport through graphene activated with Pt NPs

(a) Volt-ampere characteristics for one of devices under dark and bright conditions. The dashed curves are guides to the eye. The top inset is the schematic of our measurement setup and the bottom inset is the photo-proton effect at low biases. (b) Proton current density I as a function of illumination power P for different biases. The dashed curves are guides to the eye. The dashed black line indicates responsivity at low illumination powers. Inset, The photo-proton effect can be described by the dependence $I \propto P^{[1/4]}$. (c) Photo-proton effect observed by mass spectrometry and microsecond time response. The hydrogen flux against current density under dark and bright conditions, using biases in the range from 0 V to 3 V. The dashed line represents Faraday's law. Top inset, Example of raw data for simultaneously recorded I (black) and Φ (red) while switching illumination on and off with bias voltage, 2.3 V. Bottom inset, frequency response for the proton current measured using 1 kHz chopped illumination at 2.8 V[59]

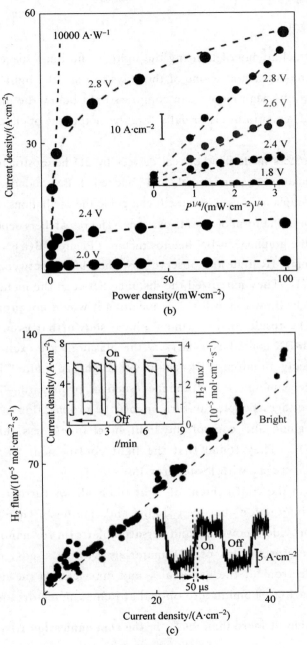

Figure 4-5(Continued)

2. Limit of light

The new technology of limiting the light in the limit space is developing, and the plasmon constraint is one of the ways to limit the light. Previous work has shown that although metals can compress light below the diffraction limit, this additional constraint is usually at the expense of additional energy loss[60-64].

The researchers made nano-optical devices by 2D heterostructures. They used graphene monolayer as semi-metal and stacked h-BN monolayer on top as insulator[65]. Graphene is used because it can guide the oscillations of light-electron-light interaction in the form of plasma excitons. The researchers deposited a series of metal rods on the graphene/h-BN heterostructure (Figure 4-6(a)), and then used the device to emit IR light and observe how the plasma between the metal and graphene travels. They narrowed the distance between the metal and graphene to see if the light limit was still valid, meaning it would not produce additional energy loss. The results of experiments have shown that even when a single monolayer of h-BN is used as a barrier, the plasmons are excited by light and can diffuse freely in atom-thick channels. Light is limited in the vertical direction between metals and graphene (such as propagating plasmons). By adjusting the voltage, people can turn on/off the propagation of the plasmons, and they can guide and control the light in channels less than 1 nm wide (Figure 4-6(b)). They found that the light confinement was more intense (Figures 4-6(c)~(e)), with less energy loss.

This shift in the confinement of light fields allows further exploration of the extreme interaction between light and matter. This heterogeneous structure can provide a powerful and versatile platform for nanophotonic. The atomic-scale heterostructure of 2D materials can be used to control both photons and electrons at the nanoscale, and may lead to the development of new devices, such as subminiature optical switches, detectors and sensors.

3. The absorption of more than 99 % in the communication frequency band of light

Graphene is an ultra-wide optical frequency response material, which can realize the full-spectrum response from the UV region to the terahertz band,

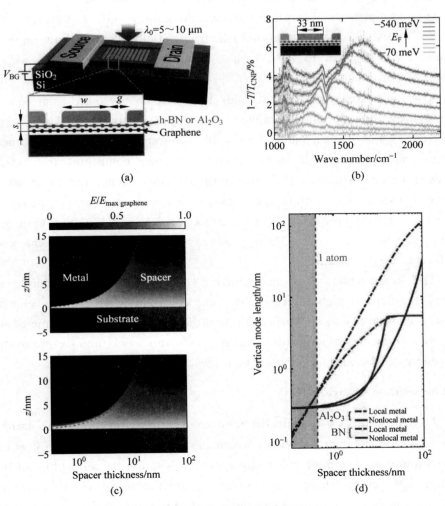

Figure 4-6　Limit of light

(a) The device made of graphene/h-BN or graphene/Al_2O_3 heterostructure. A gate voltage V_{BG} is applied between graphene and Si in order to control the Fermi energy of graphene's E_F. (b) Extinction spectra for E_F ranging from 70 meV to 540 meV. Electric field magnitude distribution of the plasmons associated to a continuous heterostructure of air/Ti/h-BN/graphene/SiO_2 as a function of h-BN thickness for (c) LMP (local metal permittivity) and (d) NMP (nonlocal metal permittivity) model limits are depicted by blue dashed lines, and graphene is located at $z = 0$. (e) Vertical field confinement for both types of dielectrics as a function of the spacer thickness for local (dash-dotted lines) and nonlocal (solid lines) metal permittivity[65]

as well as ultra-high carrier mobility and ultra-fast optical response speed. However, graphene's absorption for incident light is only 2.3 %,35 which greatly limits its use in the optoelectronic field. The enhancement of graphene's absorption to light is also an urgent problem in the area of optoelectronic field.

The heterostructure with a total thickness of less than 1 micron was prepared by using monolayer graphene[66], PMMA grating, silica layer and gold reflector, Figures 4-7(a) and (b) show the heterostructure. The heterostructure supports multiple intrinsic modes. With resonance condition, the external incident light can stimulate the mode in the structure. When the mode radiates energy outward at the same speed as the material such as graphene, the structure can fully absorb the external incident light and the highest absorption rate of graphene is 99.6 %. Figures 4-7(c)~(f) show the experiment results of device.

The heterostructure is characterized by large tolerance, simple and compact structure, high absorption rate and controllable working wavelength. It has a good application prospect in high-performance optoelectronic devices such as high-switching ratio optical modulator and high-responsiveness photodetector based on graphene and other 2D materials.

4. Photo-thermoelectric effect

To search for new optoelectronic mechanism that can break the limits of traditional semiconductor devices, which can be used to detect and collect low-energy photons. The most promising approach is to use the absorption of light by matter to generate heat, causing free electrons to move, and thus driving a current, known as the photoelectric effect. For this approach to be successful, it would require a wideband absorber capable of carrying more electrons with each other than with the photons, with a higher energy selectivity and the absorption of excess electron heat.

Using the influence of heterostructure on the properties of 2D materials, the researchers prepared graphene-based heterostructure[67]. The results showed that graphene/WSe_2/graphene/h-BN heterostructure showed potential photo-thermoelectric effects, and the light energy absorbed by graphene was

Chapter 4 Optoelectronic Properties and Applications of Graphene-Based Hybrid Nanomaterials and van der Waals Heterostructure

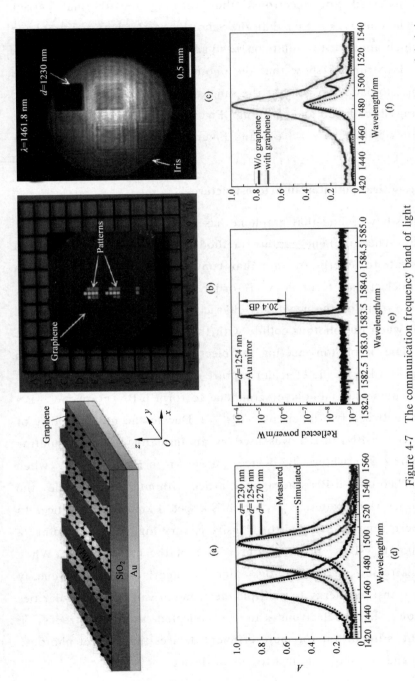

Figure 4-7 The communication frequency band of light

Schematic image (a) and optical image (b) of heterostructure, which are made of graphene, PMMA, SiO_2 and Au; (c) Reflection image of single-layer graphene-based heterostructure under the irradiations of a tunable laser with different wavelengths with $\lambda = 1461.8$ nm; (d) Measured (solid line) and simulated (dot line) absorption spectra of graphene-based heterostructure with different grating periods for TE (transverse-electric) polarization; (e) Comparisons of reflected power from a 200 thick gold layer and from graphene-based heterostructure with $d = 1,254$ nm for TE polarization under irradiations of the tunable laser with corresponding absorption peak wavelengths; (f) Measured absorption spectra of structures ($d = 1,254$ nm) with and without graphene (dot line shows the simulated spectra)[66]

effectively transferred into electrons, thus forming the thermal carrier distribution. The energy is higher than the Schottky barrier between graphene and WSe_2, which allows it to radiate outward across the barrier and produce a photocurrent. Experiments show that the photoelectric effect can detect the photons of the sub-band gap, and at the same time, it can realize large-scale electrical tuning, broadband and speeding. Figures 4-8(a)~(d) show the test results of the device in Figure 4-8(e), and Figure 4-8(f) shows the process of photoelectric effect.

5. Huge electrooptical potential in heterostructures

Researchers have found that graphene has great potential in the field of optoelectronics. The graphene capture method is more effective and has a higher success rate for energy transfer than conventional methods, which is a key step in optoelectronic technology. In order to collect the photoelectrons, it is necessary to focus on the negatively charged subatomic particles produced at the moment when the photons collide with the electrons. Energy exchange occurs at the time of photon meeting the electron, so raising the maximum transfer energy is a key to the efficient capture of the photon.

The researchers used monolayer graphene to form h-BN/graphene/h-BN heterostructure between two layers of h-BN[68]. Due to the poor mobility of electrons between h-BN, h-BN was used as an insulator for encapsulating graphene, Figure 4-9(a) shows the device. As shown in Figure 4-9(b), when graphene is stacked over h-BN's hexagonal lattice without rotation angle, the whole system forms Moiré pattern[68-70], which is a special kind of superlattice. In the moiré pattern region, the electron density is very high. At this point, a second Dirac points (sDPs) occurs at the center of the pattern[68-72]. When photons hit this region, as the photons-enter the superlattice, a high-energy photon energy transfers energy to multiple electrons in van Hove singularities (VHSs) region. These electrons can be collected using electrodes, a superlattice that is important for more efficient devices that collect photons. Figures 4-9(d) and (e) show the test results of device.

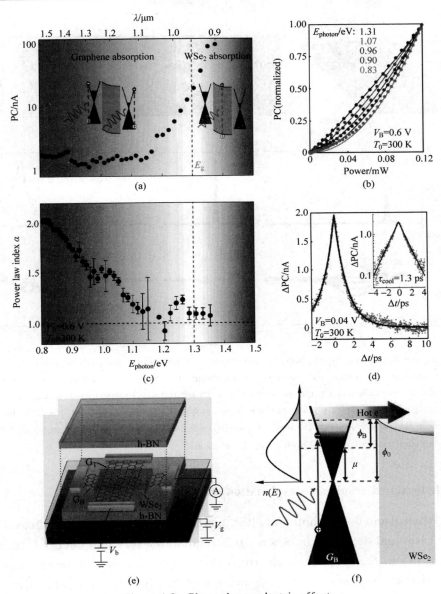

Figure 4-8 Photo-thermoelectric effect
(a) Photocurrent (PC) spectrum measured at room temperature in G/WSe$_2$/G region with laser power $P = 90\ \mu W$, $V_B = 0.6 V$ and $V_G = 0$; (b) Power dependence of the photocurrent for various values of photo energy E_{photon}; (c) Fitted power law index α versus photon energy, showing the transition from linear to superlinear power dependence; (d) Time-resolved photocurrent change $\Delta PC(\Delta t) = PC(\Delta t) - PC(\Delta t \to \infty)$, with an average power of 260 μW, $t = 30$ K and $V_B = 0.04$ V; (e) Schematic representation of the heterostructure on a 285 nm-thick SiO$_2$/Si substrate, to which a gate voltage (V_G) is applied to modify the Fermi-level (μ) of the bottom graphene; (f) Simplified band diagram of the PTI (photo-thermionic) effect at a G/WSe$_2$ interface[67]

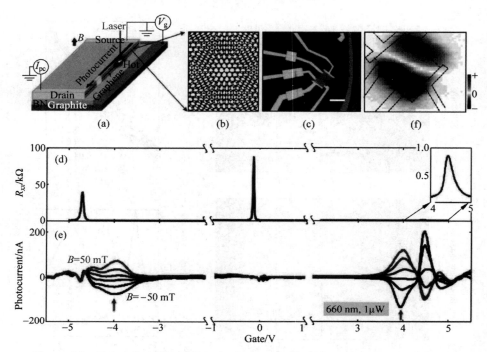

Figure 4-9 Anomalous photo-Nernst effect in h-BN/graphene/h-BN superlattices (a) Schematics of device and photocurrent measurement; (b) Cartoon depiction of a Moiré superlattice when aligning a graphene crystal with h-BN substrate; (c) Optical image of one device; (d) Longitudinal resistance R_{xx} as a function of gate at 50 mT showing one DP and two sDPs. Inset zooms in the e-sDP peak; (e) Photocurrent generation as a function of gate under a magnetic field varying from -50 mT to 50 mT with a step size of 20 mT[68]

6. Photoexcited graphene's explanation and nature

When excited by photonics, the conductivity of graphene increases in some cases and decreases in others. It is related to a way that energy from the light flows to the graphene's electrons that has been found by researchers in recent work. In other words, the process of heating the graphene electrons occurs very quickly and efficiently after the light is absorbed by the graphene[73].

The researchers prepared graphene using CVD and deposited it on a quartz substrate[73]. The ionic gate consisted of $LiClO_4$ mixed with poly-ethyl oxide in an 8∶1 ratio by mass, Figure 4-10(a) shows the device. Heavy-doped graphene has many free electrons. When the temperature of an ultra-fast electronic system rises, a large number of hot carriers are generated, and the hot carriers cause a decrease in conductivity. On the other hand, for lightly doped graphene, since the free electrons are less, an increase in the temperature of the system generates additional free electrons, resulting in an increase in conductivity. These additional carriers are the direct result of graphene's gap-free. In interstitial materials, electron heating does not result in additional free carriers. This simple case of photo-electron heating in graphene could explain many of the observed effects. Figures 4-10(b)~(d) show the experiments results. In addition to describing changes in the

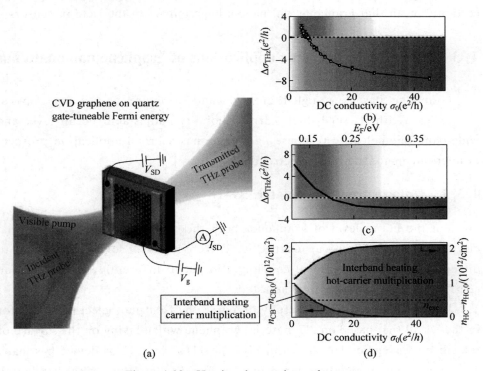

Figure 4-10 Heating the graphene electrons

(a) Illustration of optical pump-THz probe measurement technique and the gate-tunable graphene device design. By applying a gate voltage V_g to the polymer electrolyte, the researchers change the Fermi energy of graphene and thereby its DC conductivity (σ_0). (b) The measured THz photoconductivity ($\Delta\sigma_{THz}$) as function of simultaneously measured DC conductivity (σ_0). (c) Theoretical calculation of the THz photoconductivity versus DC conductivity (bottom horizontal axis) or Fermi energy (top horizontal axis) for the same parameters as in the experiment, at time $t = 300$ fs after photoexcitation. (d) The calculated carrier density in the conduction band (left vertical axis) and hot-carrier density (right vertical axis) with respect to equilibrium, for the same parameters as in the experiment[73]

conductivity of the material after photon absorption, this process can also explain the multiplier effect of the carrier. In the perspective of quantum electrodynamics, an absorbed photon can indirectly generate more than one free electron. This free electron produced by photons is responsible for the effective optical response inside the device.

Particularly, the results of this book are very accurate in understanding the process of electronic heating and can help in the design and development of graphene-based optical detection technology. In addition, this result will certainly mean that graphene will have a huge impact in the field of sensors.

4.3 Recent optoelectronic applications of graphene nanomaterials

Graphene has newfangled energy band structure and unique optical properties. With its super-high carrier mobility, adjustable Fermi level and wide-band photoelectric response, graphene has a broad application prospect in micro or nanoscale photoelectric devices.

4.3.1 Optoelectronic modulator (OM)

For the Fermi level of graphene, both the chemical doping and external electric field regulation can change it. Different Fermi levels correspond to different intensities of optical coupling. To put it differently, the absorption coefficient of graphene can be adjusted.

In 2011, Liu et al. successfully prepared the first photoelectric modulator in the world that based on 2D material graphene which laying on the surface of the silicon waveguide (as shown in Figure 4-11(a))[74]. This device has many advantages: the modulation depth reaches 0.1 dB/μm, the optical bandwidth is 1.35~1.6 μm, and the size is only 25 μm^2. This device enables CMOS (complementary metal oxide semiconductor) devices to be compatible with the ultrafast carrier mobility of graphene. It provides new ideas and technical methods for achieving highly integrated optoelectronic devices on the micro or nanoscale. Figure 4-11(a) shows the graphene OM based on the surface of the silicon waveguide. As can be seen from the figure, the modulator is composed of the bottom layer of silicon, the middle layer of Al_2O_3, and the top layer of graphene. The three layers are stacked vertically. The Fermi level of graphene can be adjusted by applying a voltage to the top graphene. The change of graphene Fermi level affects the propagation mode of the electromagnetic field in the silicon waveguide, thereby achieving the switching regulation of the optical signal.

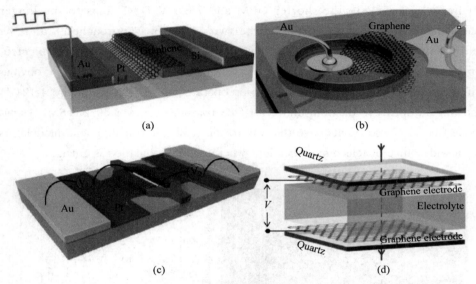

Figure 4-11 Optoelectronic modulator

(a) Schematic diagram of single-layer graphene photoelectric modulator[74]; (b) Schematic diagram of photoelectric modulator based on single-layer graphene-ring microloop structure[77]; (c) Theoretical model of graphene plasmon photoelectric modulator based on MZI structure[78]; (d) Graphene photoelectric modulator based on planar structure[79]

In the same year, Kim et al. proposed the photoelectric modulator model of ridge waveguide, which improved the dynamic response rate of the device[75]. In 2012, to overcome the effect of silicon photons on devices, Liu et al. replaced single-layer graphene with double-layer one, constructing a photoelectric modulator[75-76]. In addition to the graphene photoelectric modulator based on the direct waveguide structure, since then, a photoelectric modulator based on the micro-ring structure resonator combined with graphene has appeared (Figure 4-11(b))[77]. The other two photoelectric modulators are based on graphene combined with Mach-Zehnder Interferometer (Figure 4-11(c))[78] and graphene plane structure (Figure 4-11(d))[79].

A 10 Gbit/s graphene phase modulator (GPM) has been made by MZI configuration, which is formed with graphene-insulator-silicon capacitor[80].

The phase modulation amplitude of the device is up to 300 μm and the extinction ratio is up to 35 dB. Figure 4-12(a) shows the graphene MZI modulator, and Figure 4-12(b) shows the extinction ratio at the port at 1,550 nm. By applying 4.1 V to the shorter GPM and 7.25 V D.C. bias to the longer one, the phase difference between the two MZI arms approaches π and the extinction is maximized. On the other hand, Figure 4-12(c) shows the electro-optical S21 bandwidth of the GPM. From the image, there is an obvious inflection point at 5 GHz. The modulator operates at 10 Gbit/s, Figure 4-12(d) showing an open eye diagram and error-free transmission over 50 km SMF (sing-mode fiber). These results pave the way to the realization of graphene modulators for a wide range of telecom applications where phase modulation is crucial.

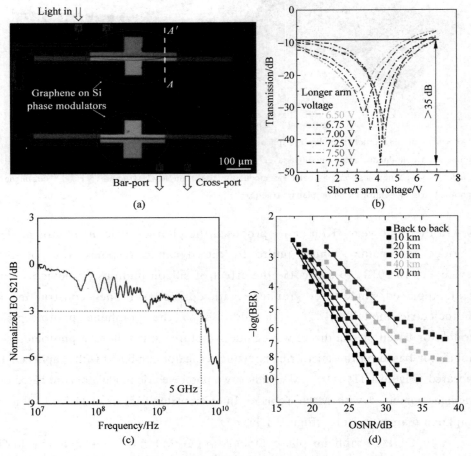

Figure 4-12　MZI phase modulator

(a) Optical micrograph of MZI phase modulator; The experiment resulted in (b), (c) and (d)[80]

Due to the photoelectric properties of graphene, the development of graphene-based photoelectric modulator has made great progress[81-97]. However, there are still many problems in the performance of devices, such as high-power consumption, slow modulation rate and low modulation depth. The biggest factor limiting the performance of the photoelectric modulator is the contact resistance of graphene to the electrode. It not only increases the energy loss of the device, but also increases the modulation voltage of the device. The modulation rate is limited by the resistance capacitance time constant. Researchers can increase the modulation rate, decrease the modulation voltage and reduce the energy loss by reducing the contact resistance. The modulation depth depends on the intensity of the interaction between graphene and light. People can increase the intensity of incident light to enhance the modulation depth. On the other hand, the single-thickness graphene has a weak interaction with light, which also limits the application potential of graphene in the region of modulators. People can change the base structure of the modulator by heterostructure with graphene and other 2D materials, such as graphene/h-BN, TMDCs/graphene or topological insulator[98-102]. Exploiting advantages of different 2D materials to meet the actual needs of the modulator and further improve the performance of the device. Table 4-1 shows the properties of graphene modulators in recent years.

Table 4-1 The properties of modulators based on graphene nanostructures

Type	Configuration	Modulation depth	Modulation width	Year	Reference
Straight	Graphene/Al_2O_3/Si	0.1 dB/μm	1.2 GHz	2011	74
Straight	Graphene/Al_2O_3/graphene	0.16/μm	1 GHz	2012	76
Straight	Graphene/ridge/graphene	5.05 dB/μm	510 GHz	2013	81
Microring	Graphene/Al_2O_3/graphene	44 dB/μm	100 GHz	2012	82
Microring	Graphene/Al_2O_3	40 %	80 GHz	2014	83
Microring	Graphene/Si_3N_4/graphene	15 dB/μm	30 GHz	2015	84
MZI	Graphene/Si/graphene	64 %	2.5 GHz	2014	85
Flat	Graphene/Ta_2O_5/Ag	5 %	154 MHz	2012	86
Flat	H-BN/graphene/h-BN/graphene/H-BN	3.2 dB/μm	1.2 GHz	2015	87
Microfiber	Graphene/microfiber	38 %	200 GHz	2014	88

Continued

Type	Configuration	Modulation depth	Modulation width	Year	Reference
Plasmon	Graphene/h-BN	50 %	100 GHz	2015	89
Straight	Al_2O_3/graphene/Al_2O_3/Graphene/Al_2O_3	3 dB/μm	35 GHz	2016	90
Straight	Graphene/SiO_2/Si	35 dB/μm	5 GHz	2017	80
Straight	Si/SiO_2/graphene/SiO_2/Si	90 %	5 GHz	2017	91
Plasmon	HfO_2/graphene/ZnS	5.6 dB/μm	29.3 GHz	2017	92
MZI	Graphene/Al_2O_3	36 dB/μm	—	2018	93
Straight	Graphene/Si_3N_4/graphene	16 dB/μm	—	2018	94
Plasmon	Graphene/ion-gel/Vg	12.1 dB/μm	—	2018	95
Plasmon	Graphene/h-BN	14~20 dB/μm	—	2018	96

4.3.2 Photodetector

As a zero-band-gap material, graphene has the advantage of ultra-wide spectrum absorption. Due to the zero-band-gap characteristics of graphene, the graphene optical detector has ultra-wide optical bandwidth. The graphene photodetector has an electrical bandwidth of 500 GHz in theory. Due to the dimmable optical response and the saturable absorption characteristics of graphene, the photoelectric response of graphene photodetector is highly tunable and nonlinear optical response.

Unlike semiconductor photodetectors, graphene has a broad spectrum range from UV to IR. At the same time, at the reported speeds of saturated carriers, graphene photodetector's bandwidth due to time constraints is expected to reach 1.5 THz. In fact, the maximum bandwidth of the graphene photodetector is limited to 640 GHz due to the delay of capacitor rather than the delay of transfer time.

1. Photodetector based on graphene/Si hybrid nanomaterials

The graphene/Si photodetector is based on the graphene/Si heterostructure to realize the photoelectric conversion to detect the light radiation. The work function of N-type Si is 4.3 eV, lower than that of graphene. When the two materials are in contact, a portion of the electrons will flow from the high-energy silicon to the low-energy graphene. At this time, the electrons in the interface layer are less and less (depleted), and the holes are accumulated at the interface. As electrons migrate, more and more electrons accumulate at the interface of graphene, forming a space charge layer. The number of carriers in the positive and negative space charge layers is very small due to the compound action. As the positive and negative charges accumulate at the interface on both sides of graphene/Si, a built-in electric field will be formed at the interface from one side of silicon to the one side of graphene. Under its action, the Fermi level of graphene and the Fermi level of silicon are gradually flattened, and the energy band bends until finally reaching equilibrium.

The valence electrons in silicon migrate to the conduction band after absorbing enough photon energy to form electron-hole pairs. The electrons are separated by the effect of electron-hole on the built electric field. The holes move in the same direction of the built electric field, and electrons move in the opposite direction of the built electric field. Electrons and holes are transmitted to the external circuit through the graphene layer and the silicon layer due to the function of the built-in electric field, respectively, to form a photo-generated current, thereby realizing a photovoltaic effect. The graphene/Si photodetector is based on this photovoltaic effect to detect light radiation.

The first graphene photodetector was developed at IBM's research laboratory in 2009[103]. The transistor-based photodetector has a bandwidth of more than 25 GHz and is then used to transmit data through the 10 Gbps optical data link. Under the irradiation of 1,550 nm, the response degree is as high as 0.5 mA/W. The detection efficiency in these devices is improved by using 0.5 mA/W asymmetric metal-graphene-metal transistor configurations. The analysis suggests that the graphene photodetector's bandwidth could eventually exceed 500 GHz. Figure 4-13(a) shows the graphene/Si photodetector. Since then, researchers have studied various types of photodetectors based on graphene, which not only greatly enhances the performance of photodetectors, but also enables a wider range of optical frequencies to be detected. Figures 4-13(b) and (c) show an optoelectronic detector with three layers of graphene and silicon[104]. As the number of graphene layers increases, the specific surface area of graphene increases, and the increase in transparent conductivity provides convenience for improving device performance. As shown in Figure 4-13(d), it is a graphene/Si Schottky junction photodetector. A layer of oxide is introduced to enhance the photoelectric response. However, the thickness of the interface oxide layer must be controlled within a certain range. An excessively thick interface oxide layer hinders the migration of photogenerated carriers, which has a negative impact on device performance[105]. Of course, Figure 4-13(e) is also a kind of reduce graphene oxide,(RGO)/N-Si photodetector[106]. A 3D graphene/Si heterojunction photodetector is shown in Figure 4-13(f). This device configuration utilizes an improved interface layer, which is to create a special microstructure that improves the effective absorption of light and the detection performance[107-108]. Figure 4-13(g) is a graphene short-wave infraved(SWIR) photodetector device that uses modified NP to enhance absorption[109].

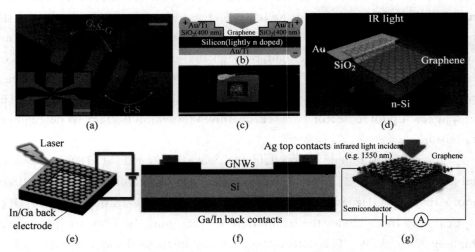

Figure 4-13 Photodetector

(a) SEM and optical (inset) images of the high-bandwidth graphene photodetectors[103]; (b) Schematic and (c) a digital photograph of a monolayer graphene (1LG)/Si heterojunction device[104]; (d) Schematic illustration of the MLG/Si Schottky junction near-infrared (NIR) photodetector[105]; (e) RGO/n-Si photodetector[106]; (f) Illustration of the GNWs-Si photodetector profile structure and band diagram of the junction region[107]; (g) Schematic diagram of the concept of graphene SWIR photodetector[109]

Among the graphene-Si composite nanomaterials[103-112], a tunneling photoconductive detector was constructed[113]. Figure 4-14(a) shows the device structure. The monolayer MoS$_2$ is inserted between the graphene and the Si substrate. Optical excited carriers from the Si can be transmitted to the graphene layer, while the introduced MoS$_2$ layer not only increases the interface state of the device, but also acts as a tunnel layer for both the graphene layer and the Si substrate. After a series of composite and interface improvements, the photoelectrical performance of the device has been greatly improved. The photoelectrical response time has reached 17 ns and the response power of the device is as high as 3×10^4 A/W under the irradiation of 16.8 nW. Compared with the performance when the MoS$_2$ layer is not inserted, various indicators of the device have been improved. Figure 4-14 (b) is the result of the performance experiment of the constant optical response of the composite photoconductive detector. Figure 4-14 (c) shows the performance test results of the device's transient light response. Figures 4-14 (d)~(f) show the optical excited carrier transmission of the photoconductive detector. This typical photoelectric detector, which has both superfast response speed and high response, has broad application prospect and value.

To understand the development process of photodetector devices, some parameters of those devices are collected in Table 4-2. This table shows the performance parameters of the photoelectric detector based partly on graphene/Si nanomaterials in recent years. It can be seen from the Table 4-2 with the improvement of the preparation process and the in-depth research. The photoelectric detection performance of the device has been greatly improved. Graphene/Si heterostructure photodetectors play a key role in both quantitative and structural performances. It is basically due to its advantages of simple structure, excellent performance and many regulation methods. With the maturity of material preparation technology and experimental methods, the improvement of preparation methods and characterization techniques, and the deepening of theoretical research, the photoelectric detector based on graphene/Si heterostructure will have a broader application prospect and value in the field of photoelectricity.

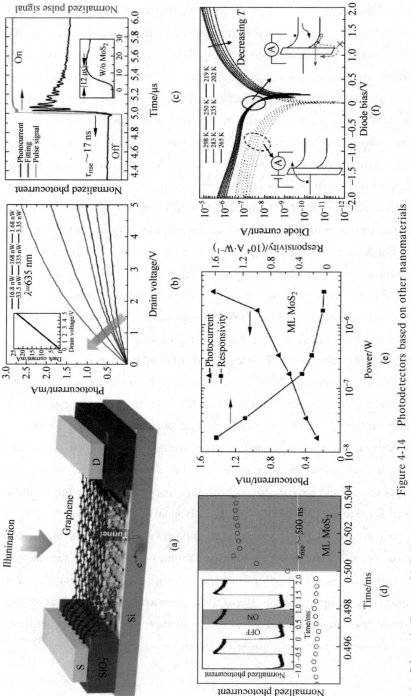

Figure 4-14 Photodetectors based on other nanomaterials

(a) Schematic diagram of hybrid graphene photoconductor. (b) Photocurrent vs. drain voltage under various light powers at 635 nm wavelength. The arrow indicates the direction of light power increases. The inset shows the dark current of the device. (c) Switching characteristics of the device. (d) Transient characteristics of the hybrid graphene photoconductor with ML MoS_2 under 635 nm illumination. (e) Photocurrent and responsivity as functions of the illumination power of the device with ML MoS_2. (f) Current vs. bias curves of the device operating in diode mode in the dark (dashed curves) and under 635 nm illumination (solid curves) at different temperatures[113]

Table 4-2 The properties of photodetectors based on graphene

Type	Configuration	Detection spectrum range/nm	Responsivity	Detectivity /(cm·Hz$^{1/2}$/W)	Year	Reference
G + Si	graphene/n-Si	400~900	225 mA/W	7.69×10^9	2013	104
	graphene/crystalline Si Schottky	850	29 mA/W	3.9×10^{11}	2013	105
	RGO/Si	445	62.92 mA/W	1.176×10^{12}	2014	106
	graphene/SiO$_2$/Si	890	0.73 A/W	5.77×10^{13}	2016	110
	graphene/(porous Si)	400~500	0.2 A/W		2014	111
	graphene/Si/TiO$_2$	420	71.9 mA/W	3.34×10^{13}	2015	108
	graphene/Si/TiO$_2$ (HNO$_3$)	420	91.9 mA/W	4.65×10^{13}	2015	108
	graphene/Si + Au	1550	83 A/W	1.0×10^8	2017	109
	graphene/Si-QDS	1013	109 A/W		2017	112

2. Photodetectors based on other nanomaterials

In fact, a large number of graphene-based nanomaterials without Si materials have been built and studied by researchers[114-119]. Among them: photoelectric detectors based on FeCl$_3$-graphene hybridization structure increased the range of photoelectric detection by 4,500 times[114]. The 3D graphene photoelectric detector, a device consisting of graphene only, reduces the interface pollution during the preparation process, thus improving the performance of photoelectric detection[115]. Based on graphene/h-BN plane heterostructure and WSe$_2$/MoS$_2$ constitute a 2D photodetector[116]. Broadband photodetectors based on graphene-Bi$_2$Te$_3$ heterostructures have higher light response and sensitivity, and a wider detection wavelength range[117]. The MoS$_2$/graphene photodetector combines MoS$_2$ with glass to provide a significant increase in carrier mobility with an optical egg response of up to 12.3 mA/W[118]. The synergistic effect of graphene and other 2D materials, or the formation of van der Waals heterostructure, or the formation of plane heterostructure, or the formation of resonant cavity, doping, etc. is essentially to improve the photoelectric response and detection sensitivity of devices.

Moreover, the range of detection is very wide, from far IR to deep UV region, different photoelectric detectors of different configurations also provide the possibility for different fields and applications.

4.3.3 Graphene-based light-emitting diodes(LEDs) and solar cells

In the photovoltaic devices, both the graphene-based nanomaterials for the photoluminescence dimethyl tube and the solar cells require the materials to have low surface resistance and high optical transparency. Graphene has good optical transparency and low surface resistance, which is consistent with the material characteristics of photovoltaic devices. The biggest drawback of graphene is the energy band structure of zero bandgap, which requires the efficient photoelectric conversion efficiency of traditional semiconductors or the band-gap opening of graphene to improve the performance of photovoltaic devices[120-134].

Lin et al. found that LEDs constructed with graphene on the surface of p-type GaN can be luminous when they connect the forward current and the reverse current[120,123-128]. Figure 4-15(a) shows the device's layered structure, with Ag NPs attached to the surface of gallium nitride, where Ag NPs are clearly visible. Figures 4-15(b) and (c) respectively show the photoluminescence curves of diodes without and with Ag NPs modification in the case of connected positive current. As can be seen from the figure, GaN-graphene-based LED modified by Ag NPs has higher brightness. At the peak of the wavelength, the device emits a yellow-green light. Figures 4-15(d) and (e) respectively show the characteristic curves of electroluminescence of diodes without and with Ag NPs in the case of connected reverse current. It can be seen from the figure that the GaN-graphene electroluminescent diode modified by Ag NPs has higher brightness. At the peak of the wavelength, the device emits pure blue light. Figure 4-15(f) compares the integrated intensity of the diodes in two cases with the change of current. It can be clearly seen from the figure that the device modified by Ag NPs gives higher luminescence.

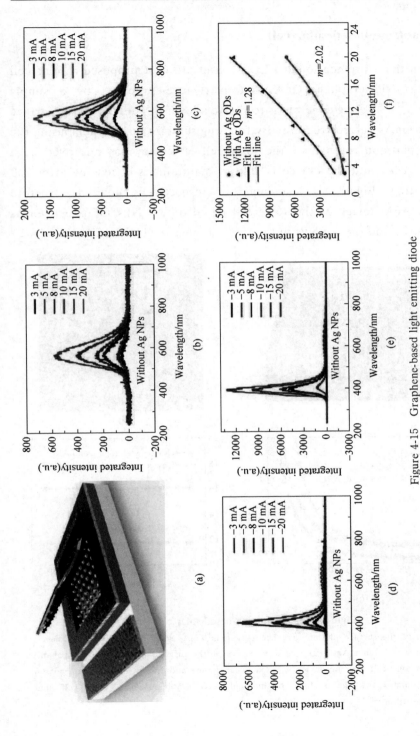

Figure 4-15 Graphene-based light emitting diode

(a) Schematic structure of the graphene/Ag NPs/p-GaN; (b) EL spectra of heterostructure at a forward bias graphene/p-GaN LED, (c) graphene/Ag NPs/p-GaN; (d) EL spectra of graphene/p-GaN LED at a reverse bias; (e) EL spectra of graphene/Ag NPs/p-GaN LED at reverse bias; (f) The variation of EL peak when increases the reverse current[120]

4.3.4　Graphene-based solar cell

Based on the aforementioned LED, Lin et al. first proposed a solar cell based on Au NPs/graphene/GaAs heterostructure prepared by a simple process[121-122,129-134]. The concrete is to rotate a layer of Au NPs on the surface of graphene/GaAs heterostructure. By focusing sunlight on graphene using the principle of plasmon resonance, the solar cell efficiency has improved by a whopping 16.2%, and the device is able to maintain its efficiency after 100 hours of lighting. Figures 4-16(a) and (b) respectively show the structure diagram and AFM image of the solar cell based on Au NPs/Graphene/GaAs heterostructure, and the Au NPs in the figure is clearly visible. Figures 4-16(c)

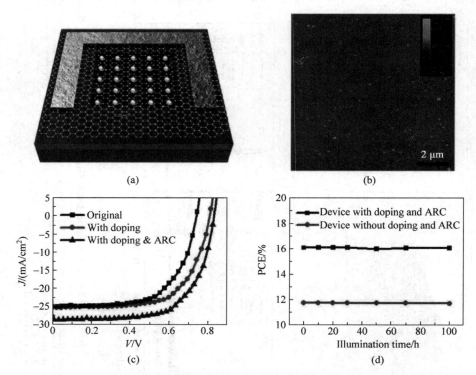

Figure 4-16　Graphene-based solar cell

(a) Illustration of the surface plasmon-enhanced graphene/GaAs solar cell; (b) AFM image of graphene coved with 80 nm Au NPs; (c) J-V curves of the original surface plasmon-enhanced graphene/GaAs solar cell and the device with chemical doping and ARC (antireflection coating) layer; (d) Stability of the surface plasmon-enhanced graphene/GaAs solar cell under AM1.5 G illumination[121]

and (d) respectively show the device performance curve and the stability result curve of solar cells under light conditions. It can be seen from the figure that, in addition to the improvement of the performance during the formation of heterostructure, the addition of Au NPs, which makes use of the characteristics of surface plasmon resonance focusing, improves the photoelectric characteristics of the device again.

Graphene/semiconductor heterogeneous devices have unique physical advantages: the adjustable Fermi level and the internal potential field of heterogeneous devices is located on the surface, which can efficiently absorb incident light and separate optical carriers. Researchers system analyzes the formed between graphene and semiconductor heterostructure devices physics principle, put forward on the basis of graphene/semiconductor heterogeneous feature of the different electronic energy band of the semiconductor device and the six to improve the performance of graphene/photoelectric semiconductor heterogeneous devices: graphene layer control, graphene semiconductor interface engineering, graphene Fermi level steady state regulation, 2D silicon carbon heterogeneous energy band engineering, semiconductor quantum dot light doping and surface plasmon enhancement technique.

4.3.5 Graphene-based ultrafast lasers

Ultrafast lasers are one of the most important applications of graphene[138-148]. The use of graphene as a saturated absorber has the following advantages: low saturation absorption threshold, due to the low electron density of graphene in the visible-near red region; fast saturation absorber, due to the relaxation time of ultrafast carrier of graphene; broadband absorption characteristics, from visible region to IR region; deep modulation depth, easy to achieve stable and efficient mode-locked ultrafast lasers; high optical damage threshold, applicable to solid-state high-power ultrafast lasers. As a new kind of nonlinear material, the electrons in graphene Dirac linear dispersion makes for any optical excitation wavelength, can create a pair of electron-hole (perfectly solved the problems in the CNT) before, plus the graphene ultra-fast dynamic process as well as the strong light absorption characteristics of carrier, the absorption of monolayer graphene 2.3% of vertical incident light, therefore considered to be an excellent wide spectrum saturation absorber.

2D graphene plasmon technology ensures rapid, compressed, and cheap development of active photonic elements because unlike other materials plasmon, graphene plasma can be modulated by doping. Such modulations use terahertz quantum cascade lasers to reversibly alter their emission[138]. These are two key steps: firstly, by stimulating the non-periodic lattice laser graphene plasma, and secondly, by engineering the photonic lifetime of the Fermi level and ring gain of the graphene. Modal gain and the resulting laser spectrum are highly sensitive to the doping of integrated and electrically controlled graphene layers. Using the unique properties of graphene plasma, the team developed a tunable terahertz laser. This tunable wavelength may change the state of terahertz lasers. To develop a new laser, the team used graphene instead of the metal in the laser because the wavelength of graphene can be changed in an electric field. They first placed them on a substrate through a series of GaAlAs quantum dots and different thicknesses of GaAs, and then covered the quantum dots with Au waveguides. Place another layer of graphene on top of the Au layer, and the researchers reduced cracks that forced electrons through the interwall tunnel. Finally, the sandwich structure is covered with polymer electrolyte and tuned by cantilever beam. Structure diagram of doped device is in Figure 4-17(a), the simulated electric field distribution of graphene when doped with different concentrations was shown at the frequency f = 2.8 THz in Figure 4-17(b). Figures 4-17(c) and (d) are the results for device. Laser emission spectra measured after electrolyte deposition for ungated (low n_s, low E_F) in Figure 4-17(e) and gated (high n_s, high E_F) graphene in Figure 4-17(f), collected just above laser threshold.

4.3.6 Graphene-based broadband image sensor array

Researchers have for the first time implemented single-chip integration of CMOS-integrated circuits with graphene, producing high-resolution image sensors based on hundreds of thousands of graphene-based and QDs photodetectors[149]. They incorporate them into digital cameras that are highly sensitive to both UV, visible and infrared light. This has never been done with existing imaging sensor technology. This technology allows graphene to be integrated with CMOS microcontrollers for a wide range of optoelectronic systems, such as low-power optical data communications and compact ultra-sensitive sensing systems.

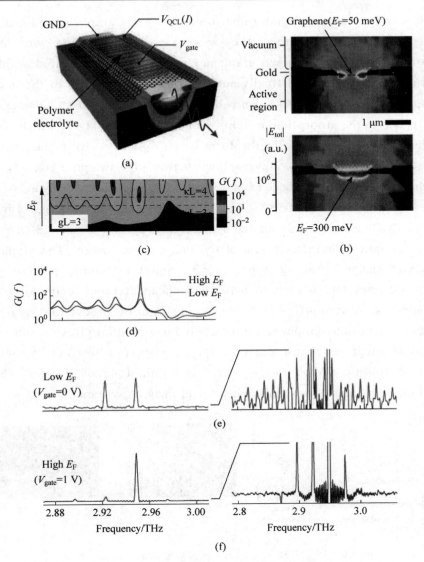

Figure 4-17 Sensitivity of laser emission to graphene doping
(a) Schematic of the polymer electrolytecovered device. (b) Simulated electric field intensity profiles within a single hologram pixel containing lowly doped (top) and highly doped (bottom) graphene, $f = 2.8$ THz. (c) and (d) Calculated $G(f)$ as E_F (and κ_L) is varied. Laser emission spectra measured after electrolyte deposition for (e) ungated (low n_s, low E_F) and (f) gated (high n_s, high E_F) graphene, collected just above laser threshold[138]

By using layering and patterning to create hybrid graphene and quantum dot systems on CMOS wafers, they solve a complex problem with simple solutions: firstly, they deposed graphene; secondly, they rendered graphene to define pixel shapes, and finally adding a layer of PbS QDs to the device. The optical response of this system is based on the grating effect. The grating effect is due to the absorption of light by the quantum dot layer, which is then transferred to the graphene in the form of an optical cavity or electron. It circulates due to the bias voltage applied between the two-pixel contacts. The high charge mobility of graphene makes the device highly sensitive, Figure 4-18(a) shows the device. Figures 4-18(b) and (c) show graphic and IR images of hybrid graphene-CQD-based image sensor and digital camera system. Figures 18 (d)~(f) are electro-optical characterization of the image sensor array. The graphene-QDs CMOS-image sensor does not require complex material processing or growth processes and is easy to make and cheap at room temperature and environmental conditions, which means the cost of production is greatly reduced. In addition, graphene can be easily integrated into flexible substrates and CMOS-integrated circuits due to its special properties. The development of such a monolithic CMOS image sensor is a milestone of low-cost, high-resolution, high-broadband and high-spectral imaging systems.

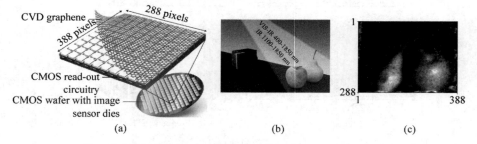

Figure 4-18   Graphene-based broadband image sensor array

(a) Computer-rendered impression of the CVD graphene transfer process on a single die (real dimensions 15.1 mm height, 14.3 mm width) containing an image sensor read-out circuit that consists of 388×288 pixels; (b) Digital camera set-up: the image sensor plus lens module captures the light reflected off objects that are illuminated by an external light source; (c) NIR and SWIR light photograph of an apple and pear; (d) Map of the conducting (blue) and non-conducting (gray) pixels; (e) Histogram of R_{pixel} before resistance compensation and after compensation (R_{pixel} + R_{comp}). Rcomp varies from 0 to 8 kΩ; (f) Histogram of the NEI for all pixels inside the dashed box in a, plotted per column (in total 255 pixels for each column). Light blue, pixels that are sensitive to moonlight; dark blue, pixels that are sensitive to twilight; black, pixels that are not sensitive to light. Photoresponse versus power at uniform illumination with λ = 633 nm and measured from twilight ($\sim 10^{-6}$ W·cm^{-2}) down to starlight ($10^{-10} \sim 10^{-9}$ W·cm^{-2}) conditions[149]

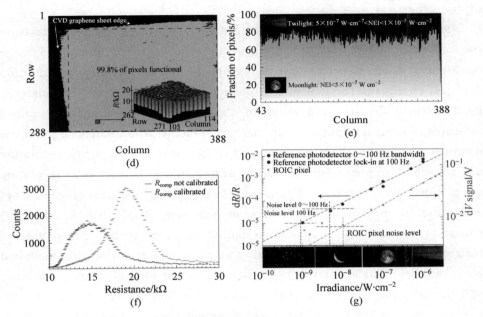

Figure 4-18(Continued)

4.4　Summary and outlook

Graphene is suitable for the development of high-performance optoelectronic devices with its high carrier mobility, high-efficiency adjustable Fermi level and wide-band photoelectric response. The superhigh carrier mobility makes it possible to realize ultra-high speed optoelectronic devices[150-156]. The high-efficiency and adjustable Fermi level makes it have novelty nonlinear characteristics and can realize the frequency transformation of photoelectric devices. The wide-band photoelectric response of graphene makes it possible to achieve a broad-spectrum response of optoelectronic devices, breaking through the limitations of traditional semiconductor optoelectronic devices that can only respond to specific wavelengths. The graphene-based optoelectronic devices make up for the deficiency of silicon in this field, display excellent properties different from traditional semiconductor materials, and truly give a play to the excellent properties of graphene, which can be used in such fields as terahertz, optical communication, IR imaging. It is the potential application field of graphene.

On the other hand, the zero-band-gap band structure of graphene restricts the practical application of graphene to some extent. More and more researchers using physical control (graphene nanomaterials configuration changes, changes in the structure of boundary, the topology of nano material itself defects, the size and shape of the set, etc.) and chemical control measures (chemical doping, functional group modification based on graphene nanomaterials, other 2D nanomaterials hybrid, etc.) to the specific regulation and change of graphene Fermi level, so as to achieve the purpose of regulating the photoelectric properties of graphene. In particular, the emergence of more and more heteromorphic structures in van der Waals has enabled 2D nanomaterials based on graphene to exhibit novel properties different from their own. When physical regulation and chemical regulation are combined with van der Waals heterogeneous structure technology, more and more optoelectronic devices are developed. The fabrication process of these devices is getting simpler and the performance of the devices is getting better. It shows great application prospect and value everywhere in various fields of science.

References

[1] AWSCHALOM D D, MCCORD M A, GRINSTEIN G. Observation of macroscopic spin phenomena in nanometer-scale magnets[J]. Physical Review Letters, 1990, 65(6): 783.

[2] FENDLER J H. Nanoparticles and nanostructured films: template synthesis of nanoparticles in nanoporous membranes[M]. Hoboken: Wiely, 2008.

[3] WILCOXON J P, THURSTON T R, MARTIN J E. Applications of metal and semiconductor nanoclusters as thermal and photo-catalysts[J]. Nanostructured Materials, 1999, 12(s 5-8): 993-997.

[4] NAKANISHI T, OHTANI B, UOSAKI K. Fabrication and characterization of cds-nanoparticle mono-and multilayers on a self-assembled monolayer of alkanedithiols on gold[J]. Journal of Physical Chemistry B, 1998, 102(9): 1571-1577.

[5] NOVOSELOV K S, GEIM A K, MOROZOV S V, et al. Electric field effect in atomically thin carbon films[J]. Science, 2004, 306(5696): 666-669.

[6] BONACCORSO F, SUN Z, HASAN T, et al. Graphene photonics and optoelectronics [J]. Nature Photonics, 2010, 4(9): 611-622.

[7] BLAKE P, HILL E W, CASTRO N A H, et al. Making graphene visible[J]. Applied

Physics Letters, 2007, 91(6): 063124.
- [8] WANG X, ZHI L, TSAO N, et al. Transparent carbon films as electrodes in organic solar cells[J]. Angewandte Chemie International Edition, 2008, 47(16): 2990-2992.
- [9] LOW T, AVOURIS P. Graphene plasmonics for terahertz to mid-infrared applications [J]. ACS Nano, 2014, 8(2): 1086.
- [10] SIERRA J F, NEUMANN I, CUPPENS J, et al. Thermoelectric spin voltage in graphene[J]. Nature Nanotechnology, 2018,13(2): 107-111.
- [11] NETO A H C, GUINEA F, PERES N M R, et al. The electronic properties of graphene[J]. Review of Modern Physics, 2009, 81(5934): 109.
- [12] ANDREY K, MACDONALD, ALLAN H. Graphene: exploring carbon flatland[J]. Phys Today, 2007, 60(8): 35-41.
- [13] DELIGEORGIS G, DRAGOMAN M, NECULOIU D, et al. Microwave propagation in graphene[J]. Applied Physics Letters, 2009, 95(7): 143111.
- [14] LEE C, WEI X, KYSAR J W, et al. Measurement of the elastic properties and intrinsic strength of monolayer graphene[J]. Science, 2008, 321(5887): 385-388.
- [15] BALANDIN A A, GHOSH S, BAO W, et al. Superior thermal conductivity of single-layer graphene.[J]. Nano Letters, 2008, 8(3): 902.
- [16] BOLOTIN K I, SIKES K J, JIANG Z, et al. Ultrahigh electron mobility in suspended graphene[J]. Solid State Communications, 2008, 146(9/10): 351-355.
- [17] NOVOSELOV K S, GEIM A K, MOROZOV S V, et al. Two-dimensional gas of massless dirac fermions in graphene[J]. Nature, 2005, 438: 197.
- [18] GEIM A K, NOVOSELOV K S. The rise of graphene[J]. Nature Materials, 2007, 6: 183-197.
- [19] ZHANG Y, TAN Y W, STORMER H L, et al. Experimental observation of the quantum Hall effect and Berry's phase in graphene[J]. Nature, 2005, 438(7065): 201-204.
- [20] KATSNELSON M I, NOVOSELOV K S, GEIM A K. Chiral tunnelling and the Klein paradox in graphene[J]. Nature Physics, 2006, 2(2): 620-625.
- [21] DU X, SKACHKO I, BARKER A, et al. Approaching ballistic transport in suspended graphene[J]. Nature Nanotechnology, 2008, 3(8): 491-495.
- [22] ANTONIO, CASTRO, NETO, et al. Drawing conclusions from graphene[J]. Physics World, 2006, 19(11): 33.
- [23] ANDO, TSUNEYA. The electronic properties of graphene and carbon nanotubes[J]. NPG Asia Materials, 2009, 1(1): 17-21.
- [24] FERNÁNDEZ-ROSSIER J, PALACIOS J J. Magnetism in graphene nanoislands[J]. Physical Review Letters, 2007, 99(17): 177204.
- [25] SON Y W, COHEN M L, LOUIE S G. Energy gaps in graphene nanoribbons[J]. Physical Review Letters, 2006.
- [26] YANG L, PARK C H, SON Y W, et al. Quasiparticle energies and band gaps in graphene nanoribbons.[J]. Physical Review Letters, 2007, 99(18): 186801.

[27] SON Y W, COHEN M L, LOUIE S G. Half-metallic graphene nanoribbons[J]. Nature, 2006, 444(7117): 347-349.

[28] DEAN C R, YOUNG A F, MERIC I, et al. Boron nitride substrates for high-quality graphene electronics[J]. Nature Nanotechnology, 2010, 5(10): 722-726.

[29] MOROZOV S V, NOVOSELOV K S, KATSNELSON M I, et al. Giant intrinsic carrier mobilities in graphene and its bilayer[J]. Physical Review Letters, 2008, 100: 1.

[30] CHEN J H, JANG C, XIAO S, et al. Intrinsic and extrinsic performance limits of graphene devices on SiO_2[J]. Nature Nanotechnology, 2008, 3(4): 206-209.

[31] HANSON G W. Dyadic Green's functions and guided surface waves for a surface conductivity model of graphene[J]. Journal of Applied Physics, 2008, 103(6): 19912.

[32] KIM J Y, LEE C, BAE S, et al. Far-infrared study of substrate-effect on large scale graphene[J]. Applied Physics Letters, 2011, 98(20): 267601.

[33] ZHANG Y, FENG Y, ZHU B, et al. Graphene based tunable metamaterial absorber and polarization modulation in terahertz frequency[J]. Optics Express, 2014, 22(19): 22743.

[34] GOSCINIAK J, TAN D T H. Graphene-based waveguide integrated dielectric-loaded plasmonic electro-absorption modulators[J]. Nanotechnology, 2013, 24(18): 185202.

[35] NAIR R R, BLAKE P, GRIGORENKO A N, et al. Fine structure constant defines visual transparency of graphene[J]. Science, 2008, 320(5881): 1308-1308.

[36] MAK K F, SFEIR M Y, WU Y, et al. Measurement of the optical conductivity of graphene[J]. Physical Review Letters, 2008, 101(19): 196-199.

[37] BLAKE P, HILL E W, CASTRO N A H, et al. Making graphene visible[J]. Applied Physics Letters, 2007, 91(6): 063124.

[38] WANG J, CAO S, DING Y, et al. Theoretical investigations of optical origins of fluorescent graphene quantum dots[J]. Entific Reports, 2016, 6: 24850.

[39] WANG J, CAO S, SUN P, et al. Optical advantages of graphene on the boron nitride in visible and SW-NIR regions[J]. RSC Advances, 2016, 6(112): 111345-111349.

[40] BAO Q, ZHANG H, WANG Y, et al. Atomic-layer graphene as a saturable absorber for ultrafast pulsed lasers[J]. Advanced Functional Materials, 2009, 19(19): 3077-3083.

[41] Nonlinear optics: theory, numerical modeling, and applications[M]. Boca Raton: CRC Press, 2003.

[42] MISHCHENKO E G. Dynamic conductivity in graphene beyond linear response[J]. Physical Review Letters, 2009, 103(24): 246802.

[43] LÓPEZ-RODRÍGUEZ F J, NAUMIS G G. Analytic solution for electrons and holes in graphene under electromagnetic waves: gap appearance and nonlinear effects[J]. Physical Review B, 2008, 78(20): 201406.

[44] MU X J, SUN M T. Interfacial charge transfer exciton enhanced by plasmon in 2D in-plane lateral and van der Waals heterostructures[J]. Appl. Phys. Lett., 2020,

117: 091601.
[45] BAO Q, LOH K P. Graphene photonics, plasmonics, and broadband optoelectronic devices[J]. ACS Nano, 2012, 6(5): 3677-3694.
[46] LUO S, WANG Y, TONG X, et al. Graphene-based optical modulators [J]. Nanoscale Research Letters, 2015, 10(1): 1-11.
[47] MIKHAILOV S A, ZIEGLER K. New electromagnetic mode in graphene[J]. Physical Review Letters, 2007, 99(1): 016803.
[48] JABLAN M, BULJAN H, SOLJA ČI Ć M. Plasmonics in graphene at infrared frequencies[J]. Physical Review B, 2009, 80(24): 245435.
[49] HANSON G W. Dyadic Green's functions and guided surface waves for a surface conductivity model of graphene[J]. Journal of Applied Physics, 2008, 103(6): 064302.
[50] JU L, GENG B, HORNG J, et al. Graphene plasmonics for tunable terahertz metamaterials[J]. Nature Nanotechnology, 2011, 6(10): 630-634.
[51] THONGRATTANASIRI S, KOPPENS F H L, DE ABAJO F J G. Complete optical absorption in periodically patterned graphene[J]. Physical Review Letters, 2012, 108(4): 047401.
[52] VAKIL A, ENGHETA N. Transformation optics using graphene[J]. Science, 2011, 332(6035): 1291-1294.
[53] KOPPENS F H L, CHANG D E, GARCIA DE A F J. Graphene plasmonics: a platform for strong light-matter interactions[J]. Nano Letters, 2011, 11(8): 3370-3377.
[54] CHEN J, BADIOLI M, ALONSO-GONZÁLEZ P, et al. Optical nano-imaging of gate-tunable graphene plasmons[J]. Nature, 2012, 487(7405): 77-81.
[55] HU S, LOZADA-HIDALGO M, WANG F C, et al. Proton transport through one-atom-thick crystals[J]. Nature, 2014, 516(7530): 227-230.
[56] LOZADA-HIDALGO M, HU S, MARSHALL O, et al. Sieving hydrogen isotopes through two-dimensional crystals[J]. Science, 2016, 351(6268): 68-70.
[57] ACHTYL J L, UNOCIC R R, XU L, et al. Aqueous proton transfer across single-layer graphene[J]. Nature Communications, 2015, 6(1): 1-7.
[58] LOZADA-HIDALGO M, ZHANG S, HU S, et al. Scalable and efficient separation of hydrogen isotopes using graphene-based electrochemical pumping [J]. Nature Communications, 2017, 8(1): 1-5.
[59] LOZADA-HIDALGO M, ZHANG S, HU S, et al. Giant photoeffect in proton transport through graphene membranes[J]. Nature Nanotechnology, 2018, 13(4): 300-303.
[60] NOVOSELOV K S, MISHCHENKO A, CARVALHO A, et al. 2D materials and van der Waals heterostructures[J]. Science, 2016, 353: 6298.
[61] LOW T, CHAVES A, CALDWELL J D, et al. Polaritons in layered two-dimensional materials[J]. Nature Materials, 2017, 16(2): 182-194.
[62] BASOV D N, FOGLER M M, DE ABAJO F J G. Polaritons in van der Waals materials[J]. Science, 2016, 354: 6309.

[63] ALONSO-GONZÁLEZ P, NIKITIN A Y, GAO Y, et al. Acoustic terahertz graphene plasmons revealed by photocurrent nanoscopy[J]. Nature Nanotechnology, 2017, 12(1): 31-35.

[64] LUNDEBERG M B, GAO Y, ASGARI R, et al. Tuning quantum nonlocal effects in graphene plasmonics[J]. Science, 2017, 357(6347): 187-191.

[65] IRANZO D A, NANOT S, DIAS E J C, et al. Probing the ultimate plasmon confinement limits with a van der Waals heterostructure[J]. Science, 2018, 360(6386): 291-295.

[66] GUO C C, ZHU Z H, YUAN X D, et al. Experimental demonstration of total absorption over 99% in the near infrared for monolayer-graphene-based subwavelength structures[J]. Advanced Optical Materials, 2016, 4(12): 1955-1960.

[67] MASSICOTTE M, SCHMIDT P, VIALLA F, et al. Photo-thermionic effect in vertical graphene heterostructures[J]. Nature Communications, 2016, 7(1): 1-7.

[68] WU S, WANG L, LAI Y, et al. Multiple hot-carrier collection in photo-excited graphene moiré superlattices[J]. Science Advances, 2016, 2(5): e1600002.

[69] YANKOWITZ M, XUE J, CORMODE D, et al. Emergence of superlattice Dirac points in graphene on hexagonal boron nitride[J]. Nature Physics, 2012, 8(5): 382-386.

[70] MOON P, KOSHINO M. Electronic properties of graphene/hexagonal-boron-nitride moiré superlattice[J]. Physical Review B, 2014, 90(15): 155406.

[71] WANG E, LU X, DING S, et al. Gaps induced by inversion symmetry breaking and second-generation Dirac cones in graphene/hexagonal boron nitride[J]. Nature Physics, 2016, 12(12): 1111-1115.

[72] PONOMARENKO L A, GORBACHEV R V, YU G L, et al. Cloning of Dirac fermions in graphene superlattices[J]. Nature, 2013, 497(7451): 594-597.

[73] TOMADIN A, HORNETT S M, WANG H I, et al. The ultrafast dynamics and conductivity of photoexcited graphene at different Fermi energies[J]. Science Advances, 2018, 4(5): eaar5313.

[74] LIU M, YIN X, ULIN-AVILA E, et al. A graphene-based broadband optical modulator[J]. Nature, 2011, 474(7349): 64-67.

[75] KIM K, CHOI J Y, KIM T, et al. A role for graphene in silicon-based semiconductor devices[J]. Nature, 2011, 479(7373): 338-344.

[76] LIU M, YIN X, ZHANG X. Double-layer graphene optical modulator[J]. Nano Letters, 2012, 12(3): 1482-1485.

[77] DING Y, ZHU X, XIAO S, et al. Effective electro-optical modulation with high extinction ratio by a graphene-silicon microring resonator[J]. Nano Letters, 2015, 15(7): 4393-4400.

[78] HAO R, DU W, CHEN H, et al. Ultra-compact optical modulator by graphene induced electro-refraction effect[J]. Applied Physics Letters, 2013, 103(6): 061116.

[79] POLAT E O, KOCABAS C. Broadband optical modulators based on graphene

supercapacitors[J]. Nano Letters, 2013, 13(12): 5851-5857.
[80] SORIANELLO V, MIDRIO M, CONTESTABILE G, et al. Graphene-silicon phase modulators with gigahertz bandwidth[J]. Nature Photonics, 2018, 12(1): 40-44.
[81] GOSCINIAK J, TAN D T H. Theoretical investigation of graphene-based photonic modulators[J]. Scientific Reports, 2013, 3(1): 1-6.
[82] MIDRIO M, BOSCOLO S, MORESCO M, et al. Graphene-assisted critically-coupled optical ring modulator[J]. Optics Express, 2012, 20(21): 23144-23155.
[83] QIU C, GAO W, VAJTAI R, et al. Efficient modulation of 1.55 μm radiation with gated graphene on a silicon microring resonator[J]. Nano Letters, 2014, 14(12): 6811-6815.
[84] PHARE C T, LEE Y H D, CARDENAS J, et al. Graphene electro-optic modulator with 30 GHz bandwidth[J]. Nature Photonics, 2015, 9(8): 511-514.
[85] YOUNGBLOOD N, ANUGRAH Y, MA R, et al. Multifunctional graphene optical modulator and photodetector integrated on silicon waveguides[J]. Nano Letters, 2014, 14(5): 2741-2746.
[86] LEE C C, SUZUKI S, XIE W, et al. Broadband graphene electro-optic modulators with sub-wavelength thickness[J]. Optics Express, 2012, 20(5): 5264-5269.
[87] GAO Y, SHIUE R J, GAN X, et al. High-speed electro-optic modulator integrated with graphene-boron nitride heterostructure and photonic crystal nanocavity[J]. Nano Letters, 2015, 15(3): 2001-2005.
[88] LI W, CHEN B, MENG C, et al. Ultrafast all-optical graphene modulator[J]. Nano Letters, 2014, 14(2): 955-959.
[89] ANSELL D, RADKO I P, HAN Z, et al. Hybrid graphene plasmonic waveguide modulators[J]. Nature Communications, 2015, 6(1): 1-6.
[90] DALIR H, XIA Y, WANG Y, et al. Athermal broadband graphene optical modulator with 35 GHz speed[J]. ACS Photonics, 2016, 3(9): 1564-1568.
[91] MITTENDORFF M, LI S, MURPHY T E. Graphene-based waveguide-integrated terahertz modulator[J]. ACS Photonics, 2017, 4(2): 316-321.
[92] KIM Y, KWON M S. Mid-infrared subwavelength modulator based on grating-assisted coupling of a hybrid plasmonic waveguide mode to a graphene plasmon[J]. Nanoscale, 2017, 9(44): 17429-17438.
[93] SHU H, SU Z, HUANG L, et al. Significantly high modulation efficiency of compact graphene modulator based on silicon waveguide[J]. Scientific Reports, 2018, 8(1): 1-8.
[94] HU X, WANG J. Design of graphene-based polarization-insensitive optical modulator [J]. Nanophotonics, 2018, 7(3): 651-658.
[95] KIM J T, CHOI H, CHOI Y, et al. Ion-gel-gated graphene optical modulator with hysteretic behavior[J]. ACS Applied Materials & Interfaces, 2018, 10(2): 1836-1845.
[96] MAIER M, NEMILENTSAU A, LOW T, et al. Ultracompact amplitude modulator by coupling hyperbolic polaritons over a graphene-covered gap[J]. ACS Photonics,

2018, 5(2): 544-551.

[97] WANG J, MA F, LIANG W, et al. Optical, photonic and optoelectronic properties of graphene, h-BN and their hybrid materials[J]. Nanophotonics, 2017, 6(5): 943-976.

[98] WANG J, MA F, LIANG W, et al. Electrical properties and applications of graphene, hexagonal boron nitride (h-BN), and graphene/h-BN heterostructures[J]. Materials Today Physics, 2017, 2: 6-34.

[99] WANG J, XU X, MU X, et al. Magnetics and spintronics on two-dimensional composite materials of graphene/hexagonal boron nitride [J]. Materials Today Physics, 2017, 3: 93-117.

[100] WANG J, MU X, WANG X, et al. The thermal and thermoelectric properties of in-plane C-BN hybrid structures and graphene/h-BN van der Waals heterostructures[J]. Materials Today Physics, 2018, 5: 29-57.

[101] WANG J, MA F, LIANG W, et al. Optical, photonic and optoelectronic properties of graphene, h-BN and their hybrid materials[J]. Nanophotonics, 2017, 6(5): 943-976.

[102] WANG J, LIN W, XU X, et al. Plasmon-exciton coupling interaction for surface catalytic reactions[J]. The Chemical Record, 2018, 18(5): 481-490.

[103] XIA F, MUELLER T, LIN Y M, et al. Ultrafast graphene photodetector [J]. Nature Nanotechnology, 2009, 4(12): 839-843.

[104] AN X, LIU F, JUNG Y J, et al. Tunable graphene-silicon heterojunctions for ultrasensitive photodetection[J]. Nano Letters, 2013, 13(3): 909-916.

[105] LV P, ZHANG X, ZHANG X, et al. High-sensitivity and fast-response graphene/crystalline silicon schottky junction-based near-IR photodetectors[J]. IEEE Electron Device Letters, 2013, 34(10): 1337-1339.

[106] ZHU M, LI X, GUO Y, et al. Vertical junction photodetectors based on reduced graphene oxide/silicon Schottky diodes[J]. Nanoscale, 2014, 6(9): 4909-4914.

[107] SHEN J, LIU X, SONG X, et al. High-performance Schottky heterojunction photodetector with directly grown graphene nanowalls as electrodes[J]. Nanoscale, 2017, 9(18): 6020.

[108] ZHU M, ZHANG L, LI X, et al. TiO_2 enhanced ultraviolet detection based on graphene/Si Schottky diode[J]. Journal of Materials Chemistry A, 2015, 3(15): 8133-8138.

[109] CHEN Z, LI X, WANG J, et al. Synergistic effects of plasmonics and electrons trapping in graphene short-wave infrared photodetectors with ultrahigh responsivity [J]. ACS Nano, 2016, 11(1): 430.

[110] LI X M, ZHU M, DU M, et al. High detectivity graphene-silicon heterojunction photodetector[J]. Small, 2016, 12(5): 595-601.

[111] KIM J F, JOO S S, LEE K W, et al. Near-ultraviolet-sensitive graphene/porous silicon photodetectors [J]. ACS Applied Materials & Interfaces, 2014, 6(23):

20880-20886.

[112] NI Z, MA L, DU S, et al. Plasmonic silicon quantum dots enabled high-sensitivity ultrabroadband photodetection of graphene-based hybrid phototransistors[J]. ACS Nano, 2017, 11(10): 9854-9862.

[113] SHEN J, LIU X, SONG X, et al. High-performance Schottky heterojunction photodetector with directly grown graphene nanowalls as electrodes[J]. Nanoscale, 2017, 9(18): 6020.

[114] ADOLFO DE S, GARETH F, DOMINIQUE J, et al. Extraordinary linear dynamic range in laser-defined functionalized graphene photodetectors[J]. Science Advances, 2017, 3(5): e1602617.

[115] SHEN J, LIU X, SONG X, et al. High-performance Schottky heterojunction photodetector with directly grown graphene nanowalls as electrodes[J]. Nanoscale, 2017, 9(18): 6020.

[116] LU G, WU T, YANG P, et al. Synthesis of high-quality graphene and hexagonal boron nitride monolayer in-plane heterostructure on Cu-Ni alloy[J]. Advanced Science Ence, 2017, 4(9): 1700076.

[117] HONG Q, JIAN Y, ZAIQUAN X, et al. Broadband photodetectors based on graphene-Bi2Te3 heterostructure.[J]. ACS Nano, 2015, 9(2): 1886-1894.

[118] XU H, HAN X, DAI X, et al. High detectivity and transparent few-layer MoS_2/glassy-graphene heterostructure photodetectors[J]. Advanced Materials, 2018, 30(13): e1706561.

[119] DU S, LU W, ALI A, et al. A broadband fluorographene photodetector[J]. Advanced Materials, 2017, 29(22): 1700463.

[120] WU Z, LU Y, XU W, et al. Surface plasmon enhanced graphene/p-GaN heterostructure light-emitting-diode by Ag nano-particles[J]. Nano Energy, 2016, 30: 362-367.

[121] Lin S S, Wu Z Q, Li X Q, et al. Stable 16.2% efficient surface plasmon-enhanced graphene/GaAs heterostructure solar cell[J]. Advanced Energy Materials, 2016, 6(21): 1600822.

[122] LIN S, LU Y, XU J, et al. High performance graphene/semiconductor van der Waals heterostructure optoelectronic devices[J]. Nano Energy, 2017, 40: 122-148.

[123] KIM B J, LEE C, JUNG Y, et al. Large-area transparent conductive few-layer graphene electrode in GaN-based ultra-violet light-emitting diodes[J]. Applied Physics Letters, 2011, 99 (14): 467.

[124] LEE J M, YI J, LEE W W, et al. ZnO nanorods-graphene hybrid structures for enhanced current spreading and light extraction in GaN-based light emitting diodes [J]. Applied Physics Letters, 2012, 100(6): 422.

[125] ZHANG S G, ZHANG X W, SI F T, et al. Ordered ZnO nanorods-based heterojunction light-emitting diodes with graphene current spreading layer[J]. Applied Physics Letters, 2012, 101(12): 121104-121104.

[126] LEE J, HAN T H, PARK M H, et al. Synergetic electrode architecture for efficient graphene-based flexible organic light-emitting diodes[J]. Nature Communications, 2016, 7: 11791.

[127] TCHERNYCHEVA M, LAVENUS P, ZHANG H, et al. InGaN/GaN core-shell single nanowire light emitting diodes with graphene-based P-contact[J]. Nano Letters, 2014, 14(5): 2456.

[128] WANG L, LIU W, ZHANG Y, et al. Graphene-based transparent conductive electrodes for GaN-based light emitting diodes: challenges and countermeasures[J]. Nano Energy, 2015, 12: 419-436.

[129] CASALUCI S, GEMMI M, PELLEGRINI V, et al. Graphene-based large area dye-sensitized solar cell modules[J]. Nanoscale, 2016, 8(9): 5368-5378.

[130] LEE C P, LAI K Y, LIN C A, et al. A paper-based electrode using a graphene dot/PEDOT: PSS composite for flexible solar cells[J]. Nano Energy, 2017, 36: 260-267.

[131] YIN Z, ZHU J, HE Q, et al. Graphene-based materials for solar cell applications [J]. Advanced Energy Materials, 2014, 4(1): 1-19.

[132] PARK H, CHANG S, JEAN J, et al. Graphene cathode-based ZnO nanowire hybrid solar cells[J]. Nano Letters, 2013, 13(1): 233-239.

[133] KAVAN L, YUM J H, GRAÄTZEL M. Optically transparent cathode for dye-sensitized solar cells based on graphene nanoplatelets[J]. ACS Nano, 2011, 5(1): 165-172.

[134] ZHANG D W, LI X D, LI H B, et al. Graphene-based counter electrode for dye-sensitized solar cells[J]. Carbon, 2011, 49(15): 5382-5388.

[135] AVOURIS, PHAEDON. Graphene: electronic and photonic properties and devices [J]. Nano Letters, 2010, 10(11): 4285-4294.

[136] BONACCORSO F, SUN Z, HASAN T, et al. Graphene photonics and optoelectronics [J]. Nature Photonics, 2010, 4(9): 611-622.

[137] SUN Z, HASAN T, TORRISI F, et al. Graphene mode-locked ultrafast laser[J]. ACS Nano, 2010, 4(2): 803-810.

[138] CHAKRABORTY S, MARSHALL O P, FOLLAND T G, et al. Gain modulation by graphene plasmons in aperiodic lattice lasers[J]. Science, 2016, 351(6270): 246-248.

[139] POLINI M. Tuning terahertz lasers via graphene plasmons[J]. Science, 2016, 351(6270): 229-231.

[140] DEGL'INNOCENTI R, KINDNESS S J, BEERE H E, et al. All-integrated terahertz modulators[J]. Nanophotonics, 2018, 7(1): 127-144.

[141] LEE E, FOLLAND T, NOVOSELOV K, et al. Graphene plasmon-modified THz laser waveguides[C]. CLEO: Science and Innovations. IEEE, 2016.

[142] WAN H, CAI W, WANG F, et al. High-quality monolayer graphene for bulk laser mode-locking near $2\mu m$[J]. Optical & Quantum Electronics, 2016, 48(1): 11.

[143] LI Y, GAO L, ZHU T, et al. Graphene-assisted all-fiber optical-controllable laser [J]. IEEE Journal of Selected Topics in Quantum Electronics, 2017, 24(3): 1-1.

[144] ZHU H, XU X, TIAN X, et al. A thresholdless tunable Raman nanolaser using a ZnO-graphene superlattice[J]. Advanced Materials, 2016, 29(2): 1604351.1-1604351.6.

[145] SOBON G, SOTOR J, PRZEWOLKA A, et al. Amplification of noise-like pulses generated from a graphene-based Tm-doped all-fiber laser.[J]. Optics Express, 2016, 24(18): 20359-20364.

[147] WENDLER F, MALIC E. Towards a tunable graphene-based Landau level laser in the terahertz regime[J]. Scientific Reports, 2015, 5: 12646.

[148] HAIDER G, RAVINDRANATH R, CHEN T P, et al. Dirac point induced ultralow-threshold laser and giant optoelectronic quantum oscillations in graphene-based heterojunctions[J]. Nature Communications, 2017, 8(1): 256.

[149] LU Z. Nanoscale electro-optic modulators based on graphene-slot[J]. Journal of the Optical Society of America B, 2012, 29(6): 1490-1496.

[150] XU J L, LI X L, HE J L, et al. Performance of large-area few-layer graphene saturable absorber in femtosecond bulk laser[J]. Applied Physics Letters, 2011, 99(26): 490.

[151] GUSYNIN V P, SHARAPOV S G, GUSYNIN V P, et al. Transport of Dirac quasiparticles in graphene: Hall and optical conductivities[J]. Phys. Rev. B, 2006, 73(24): 245411-245411.

[152] STAUBER T, PERES N M R, GEIM A K. The optical conductivity of graphene in the visible region of the spectrum[J]. Physical Review B, 2008, 78(8): 085432.

[153] GUSYNIN V P, SHARAPOV S G, CARBOTTE J P. Unusual microwave response of Dirac quasiparticles in graphene[J]. Physical Review Letters, 2006, 96(25): 256802.

[154] SUN Z, HASAN T, TORRISI F, et al. Graphene mode-locked ultrafast laser[J]. ACS Nano, 2010, 4(2): 803-810.

[155] POSPISCHIL A, HUMER M, FURCHI M M, et al. CMOS-compatible graphene photodetector covering all optical communication bands[J]. Nature Photonics, 2013, 7(11): 892-896.

[156] GAN X, SHIUE R J, GAO Y, et al. Chip-integrated ultrafast graphene photodetector with high responsivity[J]. Nature Photonics, 2013, 7(11): 883-887.

Chapter 5

Magnetics and Spintronics of 2D Graphene/h-BN Composite Materials

5.1 Graphene

The spin-orbital coupling interaction in graphene is very weak and there is almost no nuclear magnetic moment, and the electron spin transfer process is easier to control. Therefore, the unique magnetic characteristics of graphene make it ideal for the preparation of spintronic devices[1-5].

5.1.1 Lattice structure and electronic structure

The perfect graphene is a 2D crystal with a planar hexagonal lattice structure, which contains two unequal A and B carbon atoms[6-7]. In the four electrons outside each carbon atom, three electrons and three adjacent electrons of the adjacent carbon atoms are hybridized in the form of sp^2 to form the σ bond in the plane[8], so that the hexagonal honeycomb structure of the graphene is very stable; An electron perpendicular to the plane forms π bond in the p_z direction, and the π electrons are free to move, giving good conductivity to the graphene. The monolayer graphene has thickness of 0.350 nm, C-C bond length of 0.142 nm, lattice constant of 0.246 nm[9-11].

Figure 5-1(a) is a graphene model, in Figure 5-1(b), a_1 and a_2 are the original cell base vectors of graphene, Figure 5-1(c) is the Brillouin zone of graphene. Γ, M, K are the high-symmetry points of the Brillouin zone, b_1 and b_2 are the inverted vector.

Graphene is a zero-band-gap semiconductor, the conduction band and the valence band are in contact with the two unequal vertices K and K' in the hexagonal Brillouin zone[11-14]. From Figures 5-2 (a) and (b)[9,15], the

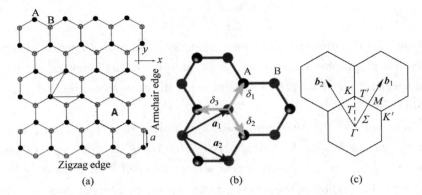

Figure 5-1 Lattice structure and electronic structure

(a) The model of graphene with zigzag edge and armchair edge[9]; (b) The original cell base vectors of graphene[10]; (c) The Brillouin zone of graphene original cell[11]

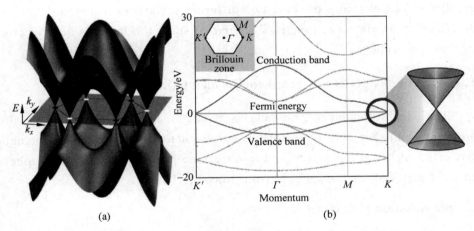

Figure 5-2 Graphene structure and electronic structure

(a) 3D band structure of graphene[15]; (b) 2D electronic band structure of graphene[9] (The blue conical illustration shows that the conduction band and valence band intersect at point K and point K' in Brillouin zone)

electron-conduction Fermi surface has a linear dispersion relation to the electrically conductive graphene monolayer just at the point of contact between the conduction band and the valence band (Dirac point) near the point of the conduction band and valence band structure, which leads to a fee the electron behavior near the rice surface can be quantitatively described by the massless relativistic Dirac equation[16]:

$$E = \frac{h\nu_F}{2\pi}\sqrt{k_x^2 + k_y^2}. \tag{5-1}$$

Equation (5-1) represents the linear dispersion relation between energy and momentum of graphene electrons, E is energy; h is Planck constant; ν_F is Fermi speed; k_x and k_y are the components in the x-axis direction and y-axis direction.

There will be the two non-equivalent points near the Dirac cone band structure, called K and K' valleys, respectively, K and K' valleys are two-component spins, which describe two sets of unequal A/B lattices[18-21]. This means that the graphene low-energy electrons or holes have two new degrees of freedom in addition to the electron spin and the degree of freedom of the coordinates, respectively, the pseudo-spin degrees of freedom and valley degrees of freedom. The former is derived from a graphene structure of graphene. The electrons of these two different sublattices in the equation are described by a spin wave function, similar to spin; the latter is derived from two unequal valleys[11-12,22].

5.1.2 The properties of graphene in magnetics and spintronics

The ideal monolayer graphene is bandless and exhibits semi-metallic properties and therefore does not possess magnetic properties. The magnetic properties of the graphene and the spin electron properties make the graphene an ideal material for the spin electron transport.

1. Spin quantum Hall effect

The energy of the electrons in the magnetic field can be written as: $E_l = \hbar\omega_c(l + 1/2)$[23], where l represents the number of quantum 0, 1, 2, ⋯, ω_c represents the frequency of electrons in the magnetic field. The Dirac fermions of graphene, at $l = 0$ with the zero-energy state, the Landau energy level movement $1/\hbar\ \omega_c$, degeneracy into other energy levels of 1/2. The presence of zero energy makes the graphene exhibit an anomalous quantum Hall effects (QHEs) and semi-integer quantum Hall conductivity σ_{xy}[24-27].

Graphene exhibits several QHEs: anomalous integer QHE[26], fractional QHE[27], and quantum spin Hall effect (QSHE)[28-30]. Figure 5-3 shows the spin Hall effect in graphene and nonlocal transport mediated by spin diffusion.

Figure 5-3(a) shows the relationship between Hall conductivity $I_e/V(e^2/h)$ and energy E_0. Figure 5-3(b) represents the relationship between the spin conductivity $I_s/V(e/4\pi)$ as the energy E_0 changes[28]. The illustrations in Figure 5-3(a) is the structural diagrams of the 4-and 6-terminal graphene Hall valves, respectively. The illustrations in Figure 5-3(b) shows the band structure of ferromagnetic graphene. M represents ferromagnetic exchange splitting.

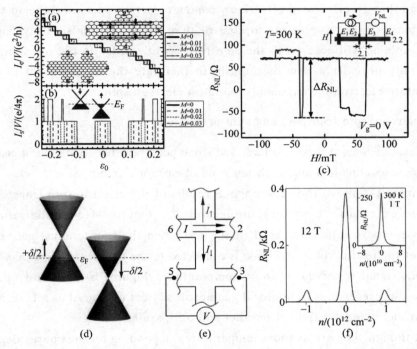

Figure 5-3 The spin Hall effect in graphene and nonlocal transport mediated by spin diffusion (a)~(b) Quantum spin Hall Effect[28]; (c) Nonlocal magnetoresistance scans of SLG spin valves measured at room temperature[32]; (d) Zeeman splitting at charge neutrality produces two pockets filled with electrons and holes having opposite spin[33]; (e) In the presence of the Lorentz force, I gives rise to transverse spin currents $I_\uparrow$ and $I_\downarrow$[33]; (f) R_{NL} predicted in our model for the QHE regime (main panel) and the quasiclassical regime (inset)[33]

2. Electron spin injection

In theory, the spin orbit coupling effect causes the lattice boundary of the graphene crystal[31-33]. In 2010, the researchers completed the tunneling spin implantation of graphene in order to solve the problem of low electron injection efficiency and spin life from ferromagnetic electrode to graphene. They inserted a few nanometer thick insulating layer *tunnel junction* between the ferromagnetic electrode and the graphene layer, which resulted in a 30-fold increase in the efficiency of the insulator's quantum tunneling of graphene. Figure 5-3(c) shows the result of the test[32].

Figures 5-3(d)~(f) show the spin quantum Hall effect of graphene[33], it is found that graphene can effectively conduct electron spin, and the *spin flow* is obviously larger and easier to control than other materials. They confirm that graphene will occur with the *spin Hall effect* similar to the phenomenon. This study provides a new mechanism to facilitate the development of next-generation electronic components based on electron spin.

3. Electronic spin transport and spin precession

Graphene unique band structure, and Dirac point near the separation of electrons and holes, resulting in a unique phenomenon of electronic transmission[34-37].

Figure 5-4 shows the experimental results of the electron spin transfer and spin precession after electron spin injection[34]. Figure 5-4(a) is the geometry of the *nonlocal* spin valve used in the test. And the four-terminal geometrical shape of the ferritic cobalt electrode is connected to the graphene. Figure 5-4(b) is the spin signal observed at room temperature. Figures 5-4(c) and (d) are Hanle spin precessions in nonlocal geometric devices measured as a function of the parallel magnetic field B parallel and antiparallel.

Although they are at room temperature, depending on the charge density, the extracted spin has a length of 1.5 mm to 2 mm. In contrast to the transmission distance of a few tenths of a micrometer material, graphene has farther spin information transmission, limited rotational loss. Based on this, making graphene known as a good choice for spin electronics.

On the other hand, Kamalakar et al. using CVD on the SiO_2/Si substrate to prepare graphene[37], tested, in the channel length shows pure spin transmission and rotation to 16 mm; spin lifetime reached 1.2 ns; at room temperature, spin diffusion length of 6 μm. The test results show a 6 times higher performance than ever before, opening up new horizons for the development of horizontal rotary storage and logic applications. Figures 5-4(e)

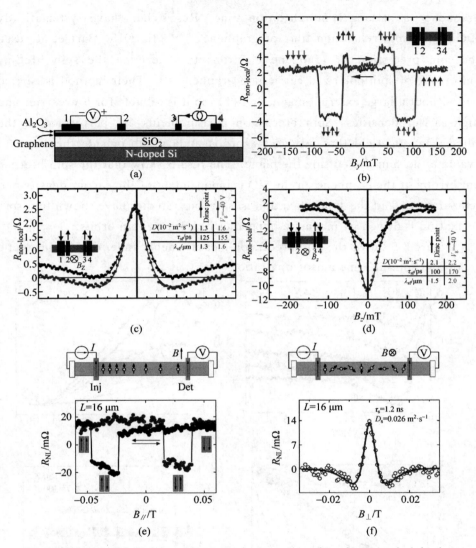

Figure 5-4 Electron spin transfer and spin precession after electron spin injection
(a) Test the structure of the device; (b) Under the action of magnetic field, the spin transfer diagram of electrons; (c)~(d) Hanle spin precessions[34]; (e)~(f) are the Long-distance spin transport in CVD graphene device at room temperature[37]

and (f) show the device and the test results of the device.

4. Determination of the spin-lifetime anisotropy

The spin-orbit coupling effect of the C atom is very small, so that the graphene has a long spin relaxation time. Researchers have systematically studied the spin relaxation time of graphene[35-44]. In 2015, Bart et al. used the spin precession of graphene electrons to determine the spin lifetime anisotropy of spin-polarized carriers in graphene[45]. Their method is carried out without a large external magnetic field, so it is reliable for low carrier and high carrier densities. Spin precession measurements are performed at the tilted magnetic field that produces a propaganda group outside the plane to evaluate the spin life outside the plane. The results show that the spin lifetime anisotropy of the graphene on the SiO is independent of the carrier density and temperature, and the spin relaxation is isotropic. Studies have shown that spin relaxation is driven by magnetic impurities or random spin orbits.

Figure 5-5 shows the spin precession measurements under oblique magnetic fields and spin-lifetime anisotropy ratio ζ [45].

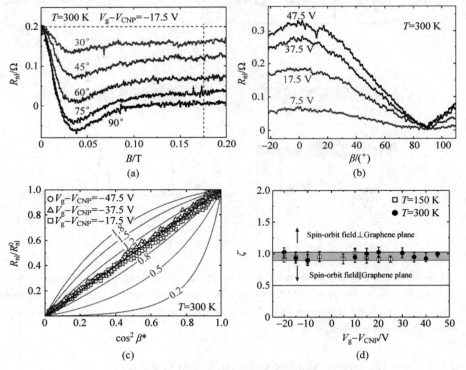

Figure 5-5 The spin procession measurement

Spin precession measurements under oblique magnetic fields in (a) and (b); Spin-lifetime anisotropy ratio ζ in (c) and (d)[45]

5. The magnetic induced by molecular adsorption

A large number of theories and experiments show that the adsorption of organic molecules such as graphene will be introduced based on the p-orbit magnetic, which fully proved the existence of carbon in the magnetic material[46-48]. Among them, the graphene through the adsorption of hydrogen atoms can achieve the electronic structure of the spin polarization, which for the carbon material of the magnetic research is of great significance[49-53].

In 2016, Héctor et al. demonstrated a magnetic moment caused by the adsorption of a single hydrogen atom on the surface of graphene by a characterization of a spin at about 60 meV at the Fermi level (Figure 5-6)[53].

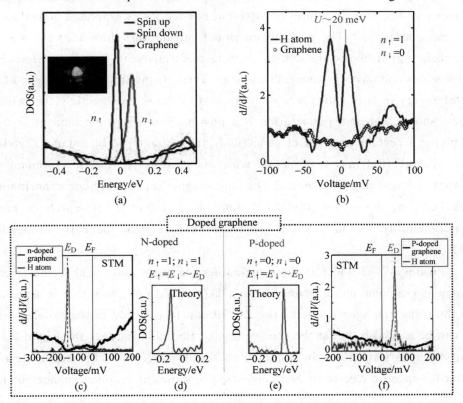

Figure 5-6 Spin-split state induced by atomic H on graphene
(a) The relevant data measured in the test. The insets for scanning electron microscopy observed the adsorption of hydrogen atoms on the surface of graphene; (b) Corresponding to A, the first principle of the calculation of the relevant data; (c) ~ (f) N-type doping and p-type doping conditions, the first principle of calculation and experimental data obtained

At the same time, they also used the first principles to calculate the completion of the scanning electron microscope to show that a self-polarized state is essentially located in the hydrogen atom of the chemical adsorption of the opposite carbon sub-lattice. This auto-modulated spin structure can extend a few nanometers away from the hydrogen atom, making the direct coupling between the long-distance magnetic moments. Using the tip of the scanning electron microscope to control the atomic precision of the hydrogen atom, the magnetization of the graphene region can be adjusted and selected.

6. The magnetic of graphene nanoribbons

The graphene nanometer has two basic types of the zigzag and the armchair. The nanostructures of different boundaries of graphene induce local magnetic moments. The density functional calculations show that the zigzag graphene nanoribbons are semiconducting materials with localized electronic edge states and have ferromagnetic and antiferromagnetic states[54-56]. And the total energy in both states is lower than that in the non-magnetic state, which again shows that spin polarization is a possible stabilizing mechanism. After applying a certain external electric field, the zigzag-type boundary (ZGNR) can be converted into semi-metal, which is possible for the development of graphene-based spin electronics. The edge magnetism of graphene nanoribbons has been proposed for the first time since 1996[57]. Researchers have experimentally demonstrated the presence of marginal magnetic properties in 2014[56-58].

Figures 5-7(a) and (b) show the antiferromagnetic states and ferromagnetic states of graphene nanoribbons. Figures 5-7(c) and (d) show that in addition to the transverse electric field, the ZGNR can be adjusted to semiconductors. It can be seen also that as the transverse electric field is added, the β spin band gap disappears and the spin band gap still exists. Figures 5-7(e) and (f) show the edge-specific electronic and magnetic properties of graphene nanoribbons.

7. Tunnel anisotropic magnetoresistance effect on the surface of graphene

Phillips et al., using a single layer of graphene and ferromagnetic $La_{0.67}Sr_{0.33}MnO_3$ (LSMO), electrodes bridged each other[59]. When the

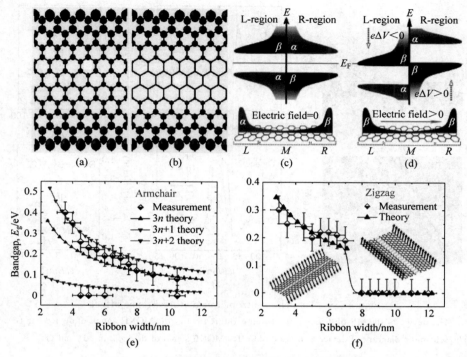

Figure 5-7 The magnetic of graphene nanoribbons

Antiferromagnetic states (a) and ferromagnetic states (b) of ZGNRs[58]; (c)~(d) Graphene nanoribbons present in the applied electric field show semi-metallic[54]; (e)~(f) The bandgap measured by tunneling spectroscopy as a function of ribbon width in armchair (e) and zigzag (f) ribbons[56]

temperature is at 5 K, the magnetoresistance (MR) change is observed to be 32~35 MΩ. At the same temperature, the light Kerr effect microscope shows that the MR is caused by the in-plane reorientation of the electrode magnetization, which further proves that the LSMO-graphene interface tunneling anisotropic reluctance. There is no spin-transfer large resistance connected through the non-magnetic channel, which in the graphene magnetic application is very attractive.

Figures 5-8(a), (b) and (e) show the device 1, the test results of MR and Magneto-optical Kerr effect (MOKE) data of device 1. Figure 5-8(c) shows the device 2, Figures 5-8(d) and (e) show the MR and MOKE data of device 2.

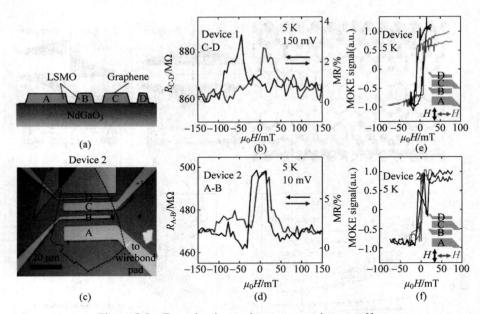

Figure 5-8 Tunnel anisotropic magnetoresistance effect
(a) The device 1, the MR of device 1 at 5 K temperature in (b), the MOKE signals of device 1 in (e). The schematic diagram of device 2 in (c), and MR, MOKE signal of device 2 in (d) and (f)[59]

5.1.3 The application of graphene in magnetic properties and spin electronics

Based on the weak coupling of the graphene spin-orbit, the electron spin transfer process is relatively easy to control, so the graphene is considered to be ideal for the manufacture of spin electrons (spin field effect transistors).

1. Graphene Nanoribbon Field-Effect Transistors

In 2008, a field-effect transistor (FET) based on graphene nanoribbons was fabricated. This is the Schottky barrier (SB) type FET, which is used to reduce the Schottky barrier of the P-type transistor through the carrier tunneling probability through the SB, which is used to reduce the P-type transistor's SB[60]. 10 nm thickness of the SiO_2 gate dielectric for the realization of higher ion ionization is very important. Figures 5-9(a) and (b) show the test results of the output characteristics of GNR-FETs. For the devices, Figures 5-9(c) and (d) show the metallic behavior because of vanishingly small band gaps.

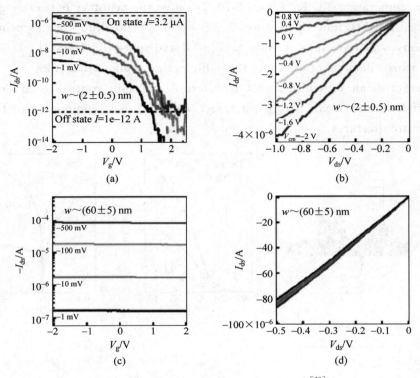

Figure 5-9 The test result of GNR-FETs[60]
(a)~(b) The output characteristics of GNRFETs; (c)~(d) Metallic behavior because of vanishingly small band gaps

2. Spintronic transistors

By combining graphene with another 2D material, researchers at Chalmers Polytechnic University have created the prototypes of transistor devices for future computers based on spin electronics[61]. The researchers chose to test the combination of graphene and another thin 2D material with contrasting spin properties. This 2D material is molybdenum disulfide (MoS_2), due to its high spin orbit coupling spin life is low. By performing nonlocal spin valves and Hanle measurements, they clearly demonstrate the gate-tunability of spin current and spin lifetime in graphene/MoS_2 van der Waals heterostructure at 300 K.

Figure 5-10(a) is the heterostructure's circuit diagram, Figures 5-10(b) and (c) show the transfer characteristic and output characteristic of the device at the temperature 300 K. Figure 5-10(d) shows the Schottky barrier height Φ obtained for different V_g. Figures 5-10(e) and (f) show the working principle diagram, which are spin-on state at $V_g < 0$ and spin-off state at $V_g > 30$ V. This work demonstrates that the all-electron spin electronics at room temperature can be constructed, transferred and spin controlled in the 2D material van der Waals heterostructures, which is a key component of future device architectures.

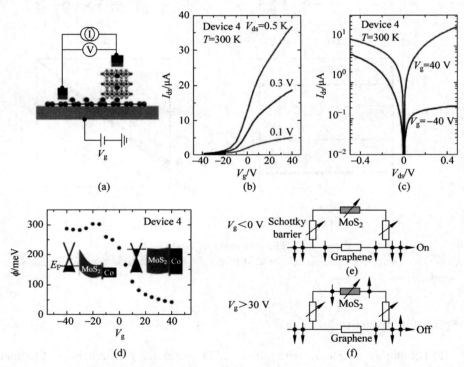

Figure 5-10 The working principle of the electron spin transistor constructed from graphene and MoS_2 heterostructures and the corresponding test results[61]

(a) The heterosucture's circuit diagram; (b)~(c) The transfer characteristic and output characteristic ofhe device with the temperature 300 K; (d) The Schottky barrier height Φ obtained for differen V_g; (e)~(f) Showed working principle diagram

3. Graphene spintronic devices with molecular nanomagnets

In view of the application of spin electronics in microelectronic devices, it is possible to manipulate the valley degree of graphene to realize the function of the electronic device. Based on the magnetic characteristics of the boundary of graphene nanoribbons, the valley-filter and valley-valve devices of *broad-narrow-wide* graphene nanoribbons are also constructed[62].

Figure 5-11(a) shows the device made of *wide-narrow-wide* graphene nanoribbons and $TbPc_2$ (Pc = phthalocyananine). This device was used to detect the magnetization reversal of molecules near graphene. The detection revealed that the spin reverses up to 20 % of the permeability signal, thus revealing the uniaxial magnetic anisotropy of the P quantum magnet. This kind of nanomagnets graphene spin-electronic devices have a single-molecule-level sensitivity of the long-effect nano-transfer. Figure 5-11(b) represents the relationship between the back-gate voltage and the magnetic permeability. Blue represents the magnetic permeability under increasing magnetic field, and red represents the change in permeability in the reverse magnetic field.

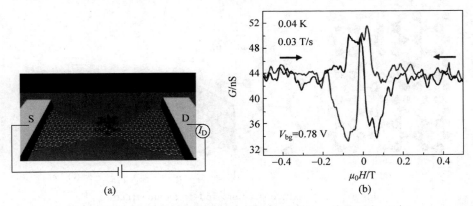

Figure 5-11　Graphene spintronic devices and experimental results

The device are made of graphene nanoribbons and $TbPc_2$ in (a); The test results of device in (b)[62]

5.2　Hexagonal boron nitride

As the 2D h-BN than the carbon material has better chemical stability and thermal stability, suitable for chemical complex environment and high-temperature environment, and very consistent with the needs of people, so the magnetic properties of boron nitride(BN) materials are also widely studied. It has been found that doping or generating defects in BN nanostructures without magnetic properties can lead to spontaneous magnetization.

5.2.1 Lattice structure and electronic structure

The h-BN and graphene have the same bonding properties and crystal structure, thus both are the same layered white crystal, so h-BN is also known as white graphene[63-65]. In the h-BN plane, B atoms and N atoms are sp^2 hybridized to form a honeycomb hexagonal ring structure with lattice constant: $a = b = 0.2504$ nm, and $c = 0.665$ nm (Figure 5-12). Due to the difference of electronegativity between the boron atom and the nitrogen atom, charge distribution in the hexagonal ring plane is biased toward the nitrogen atom, forming a strong polar covalent bond. The atoms between the layers are not directly bonded, and only weak interactions, such as electrostatic interaction, van der Waals interaction[66-68].

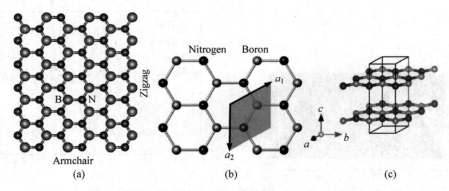

Figure 5-12 Lattice structure and dectronic structure

(a) 2D h-BN nanocrystals with zigzag edge and armchair edge[63]; (b) The lattice structure of N atom and B atom in h-BN[64]; (c) The h-BN layer is of *ABAB* type[65]

Because h-BN has good insulation (band gap of 3.0~7.5 eV), chemical stability and thermal stability (melting point greater than 3,000 K), so h-BN in the electrical insulation, high temperature, high pressure and complex chemical environment coufronts very broad application prospects.

In the 2D h-BN, each of the original cells contains B atom and N atom, both are sp^2 hybrid. In each B (N) atoms and the adjacent three N (B) atoms from the same formation of three σ bonds, there are six electrons filled to the σ orbit, the remaining two are in the $2p_z$ state with the formation of vertical plane π

bond[65,69-70]. 2D h-BN is a direct bandgap (Figure 5-13), wide band-gap semiconductor material with the band gap $E_g = 4.90$ eV, valence band top and conduction band bottom are at high-symmetry point K. The HOMO and LUMO of the system are determined by the localized π and π^* states on N atom and B atom, respectively.

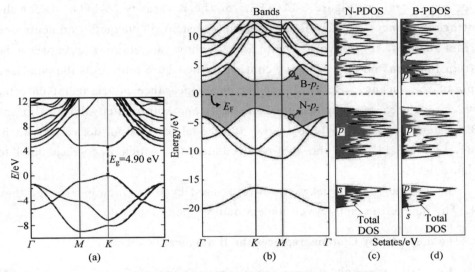

Figure 5-13 The band structure and DOS of h-BN
(a) The band structure of h-BN (Solid black bands correspond to the DFT, red dashed bands correspond to the downfolded Hamiltonian, E_g is a large band gap of h-BN in the Brillouin zone)[64];
(b) Calculated electronic structure of 2D BN honeycomb crystal together with total, TDOS, and partial density of states, PDOS, on N (c) and B (d) atoms. The orbital character of the states is also indicated[65]

5.2.2 Magnetic properties and spintronic of h-BN

Single-layer h-BN nitride has been successfully prepared in the experiment[71-72], with its unique electromagnetic properties and spin electronic properties used in electronic devices. Based on this, single-layer h-BN's magnetic research has also become a hot SPOC.

1. Defects induce magnetic properties

Recent researchers on focus on the single layer of h-BN magnetic, with the first principles of the study. The surface of the study, the surface of BN defects will lead to h-BN magnetic phenomenon[73-77]. Ouyang et al. calculated the electronic properties of biaxial strain versus void-modified monolayer h-BN using density functional theory in 2013[73]. They calculated from the theory of the N vacancy and B vacancy on the single layer of BN magnetic conversion. This study provides a theoretical basis for the preparation of low-latitude spin electronics devices. Figure 5-14(a) is for the B vacancy, 5-14(b) is for the nitrogen vacancy, 5-14(c) is the abnormal system of magnetic moments and strain changes, respectively. Figure 5-14(d) show the relationship between the strain function and d/r. The red contour in the figure represents the polarized spin of the vacancy. Du et al. studied the appearance of triangular defects, which would cause great magnetic moments and spin-related divisions in the BN system[74]. Figure 5-14(e) shows the model for B atom deficient with h-BN, Figure 5-14(f) is the spin-resolved density, which is corresponded to Figure 5-14(e)[74].

It is found that the magnetic force caused by B vacancy is stronger than N. The magnetism can be well understood by the Stoner rule.

2. The magnetic by C atoms replaces the B atoms or N atoms

Recently, researchers have shown that the use of C atoms in the h-BN nanoribbons to replace a single B atom or N atoms, can also induce h-BN system magnetic.

Wu et al. calculated the magnetic properties of carbon atoms doped with BN by GGA with a generalized gradient approximation method based on density functional theory[78]. When the carbon atom is substituted for the boron atom, the non-spin polarization is favored by 0.1 eV, and the calculated electron band structure shows the self-polarization near the Fermi energy. At the same time, the energy and curvature efficiency of spin polarization indicate that the BN system achieves the possibility of ferromagnetic ordering by carbon doping (Figure 5-15(a)~(c)).

Song et al. studied the system of carbon-doped h-BN systematically[79]. They found that when a carbon atom doping zigzag BN nanoribbons (ZBNNRs), the doping tends to replace the edge atoms, making the system a magnetic semiconductor. And when the carbon atoms are doped at the central position, the system is transformed from a magnetic semiconductor to a metal. Moreover, as the

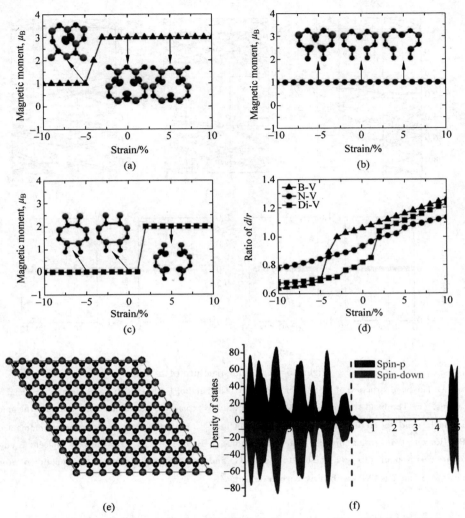

Figure 5-14 Electronic properties of biaxial strain versus void-modified monolayer hexagonal boron nitride

The strain dependence of magnetic moment for systems with a (a) boron vacancy, (b) nitrogen vacancy, and (c) divacancy. The corresponding evolutions of d/r as functions of the strain are shown in (d) for different systems. The red isosurfaces in (a)~(c) represent the polarized spin at vacancies[73]. (e) B atom deficient for h-BN nanosheets. (f) Spin-resolved density of states associated with (e)[74]

doping concentration of carbon atoms increases, the electrical conductivity of ZBNNRs increases, thereby realizing the conversion between the semi-metallic properties and the metallic properties of the ZBNNRs by the concentration of carbon doping (Figures 5-15(d)~(g)).

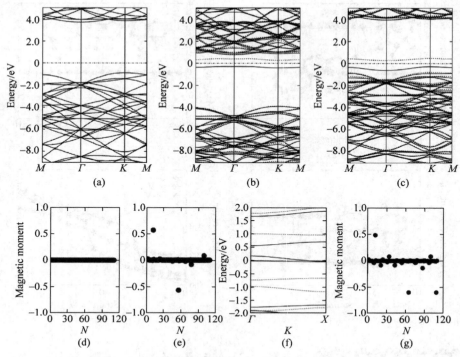

Figure 5-15 Band structure of h-BN

(a) The band structure of monolayer h-BN. (b) The band of h-BN, which the B atoms is replaced by C atoms. (c) The band structure of h-BN, which N atoms replaced by C atoms. Solid lines are the spin-up branches and dotted lines are the spin-down branches[78]. (d) and (e) show the magnetic moment of ZBNNRs with two carbon dopants, which is the adjacent B edge sites in (d), the non-adjacent B edge sites are corresponding to (e). The band structure (f) and atomic magnetic moment as a function of atom index (g) of the ZBNNRs with three carbon dopants[79]

3. The magnetic induced by hydrogenation or fluorination

Zhou et al. calculated the magnetic properties of BN nanosheets using hydrogen and fluorine functionalization based on the first principles of density functional theory[80]. It can be seen from the calculation that when the BN nanotubes are completely hydrogenated with hydrogen, the bandgap of BN changes from 4.7 eV to 0.6 eV and the BN nanosheets become semiconductors or semi-metals. Moreover, the hydrogenated and fluorinated BN nanosheets are rich in electrons and magnetic anisotropic structures. Different functions of the surface, BN nano-tablets can be made of ferromagnetic, antiferromagnetic and other materials. Figure 5-16(a) and (b) show the three magnetic coupling states of H-(h)BN with ferromagnetic (FM) states and antiferromagnetic (AF) states. Figure 5-16(c) shows the three magnetic coupling states of F-BN in FM.

On the other hand, experimental studies have also confirmed the theoretical calculations. Radhakrishnan et al. added fluorine to h-BN[81]. When about 5 % fluorine is added, the band gap of h-BN becomes smaller. Through experiments, they confirmed that the tension applied by the addition of fluorine atoms changed the spin of electrons in the nitrogen atom, thus affecting the magnetic moments of these electrons, resulting in a response to the magnetic field. This simple method allows h-BN to have the potential for the application of the underlying magnetic material and to again confirm the theoretical calculations. Figures 5-16(d) and (e) show the experimental results of the test.

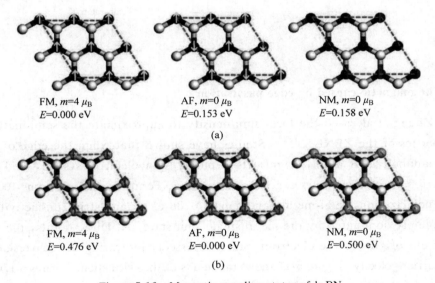

Figure 5-16 Magnetic coupling status of h-BN

Three magnetic coupling states of H-BN in MR states (a) and AF states (b). Three magnetic coupling states of F-BN states in FM states. (d) The room temperature hysteresis curve of F-BN with 8.1 % fluorine[80]. (e) The experimentally observed temperature-dependent susceptibility fitted with the Curie's law. The measurement was performed at an applied dc field of 500 Oe[81]

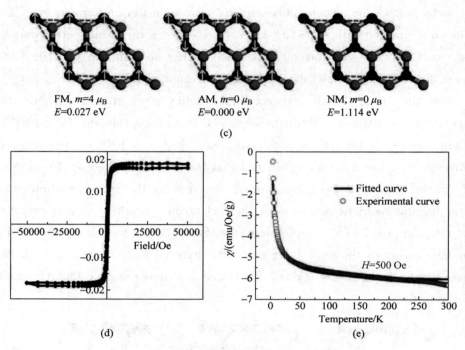

Figure 5-16(Continued)

4. The magnetic caused by edge passivation

Zheng et al. used the local spin density to approximate the semi-metallic properties of the ZBNNRs[82]. Studies have shown that when the edge of the BN nanobelt is B atom, it reflects a purely spin electron system, and the nanoribbons system shows a great spin splitting. Fermi level of electrons 100 % spin polarization, semi-metal band gap of 0.38 eV, nano-strip conductivity is completely dominated by the metal single spin state. At Dirac points, the two different origins of the electronic states intersect, indicating the conversion of the carrier velocity. Figure 5-17 shows the results of the calculation. Figures 5-17(a), (b) and (c) show the spin-up energy bands, spin-down energy bands and the total DOS. Figures 5-17 (d) and (e) show the partial charge density of electronic bands, α and β. Figure 5-17(f) shows the spatial distribution of the spin difference.

Chapter 5 Magnetics and Spintronics of 2D Graphene/h-BN Composite Materials

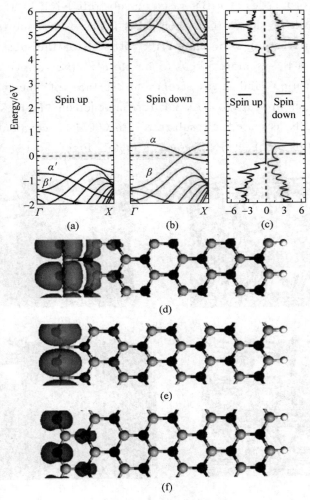

Figure 5-17 ZBNNRs properties
(a) Half metallic and (b) ferromagnetic behaviors of ZBNNRs[82]

5. The magnetic caused by substrate

In the study of BN preparation, the researchers introduced the substrate as a medium for the growth of BN. The principle of substrate-induced magnetic properties of BN is further revealed as BN and substrates are studied as the whole system.

Joshi et al.[83] used density functional theory to study the surface structure and electronic properties of h-BN based on substrate Ni(111) and substrate Co (0001). Studies have shown that when h-BN is based on a substrate Ni(111), the diffusion interaction is the main effect of h-BN on the substrate Ni(111). At this point, h-BN is more likely to be close to the surface of the strong chemical adsorption and away from the surface of the weak physical adsorption, which appears to be weak ferromagnetism (Figures 5-18(a) and (b)). When BN is based on substrate Co (0001), covalent interactions dominate. At this point, h-BN becomes a metal (Figures 5-18(c) and (d)).

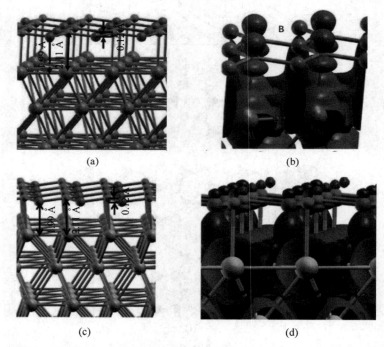

Figure 5-18 H-BN is based on substrate Ni(111) (a) and on substrate Co(0001) (c). Magnetic equivalent surface of h-BN on substrate Ni(111) in (b) and Co(0001) in (d)[83]

5.2.3 Application of h-BN in magnetics and spintronics

1. Magnetic tunnel junction

With the development of modern science and technology, the proliferation of data and the emergence of large data, spin electronics has always been the core of data storage. The key to achieving the data storage revolution is the magnetic tunnel junction (MTJ)[84-87].

Banci et al. used the CVD method to directly grow monolayer h-BN on Fe[84], and the magnetic tunnel junction based on h-BN was prepared. Specifically, they incorporated a 2D h-BN tunnel barrier with atomic thickness into the Co/h-BN/Fe tunnel junction. Studies have shown that the tunneling effect depends on the size of the h-BN layer (Figure 5-19). For a single-layer h-BN-based magnetic tunnel junction, the observed magnetic resistance of the tunnel is as high as 6 %. This is mainly due to the tunneling effect of spin polarization electrons of h-BN, with tunnel magnetoresistance (TMR) of 6 % and spin polarization P of 17 %. Then the previously reported magnetic tunnel junctions are two orders of magnitude higher barrier.

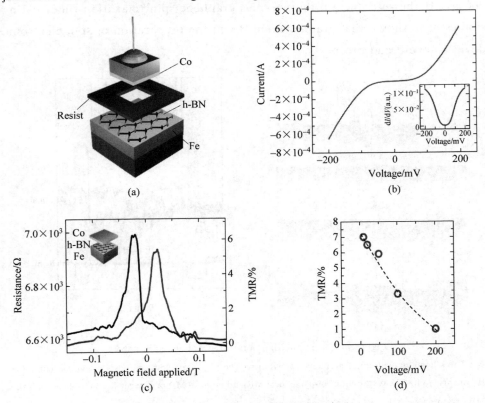

Figure 5-19 Magnetic tunnel junction
(a) The MTJ made of Co/h-BN/Fe heterostructure; (b) The V/I characteristic curve of device (a); (c) The relationship between the applied magnetic field and resistance; (d) The relationship of voltage and TMR of device (a)[84] (the red line is experimental results)

2. Tunnel barriers for graphene spintronics

The unique electrical and magnetic properties of h-BN make it play a unique role in the preparation of spin electronics. Fu et al. used CVD method[88] for the telescopic material preparation of graphene spin valve device. In this apparatus, h-BN acts as the tunnel barrier, which is placed on monolayer or bilayer graphene. Spin transport experiments were carried out using ferromagnetic out points deposited on the shield. It is found that the barrier of the single-layer tunnel is limited by the injection of the two materials due to the mismatch between the two materials (Figure 5-20). Using these devices, the spin relaxation time of the graphene material prepared by CVD is 260 ps. If the reference to the previous graphene spin relaxation time, eating studies have shown that graphene and BN in the preparation of spin electronic devices have broad prospects.

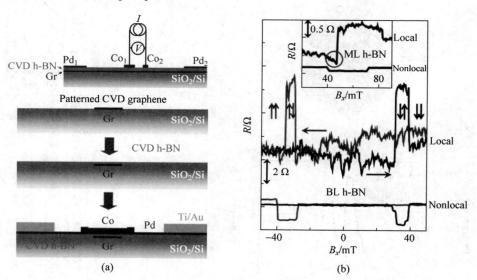

Figure 5-20 Tunnel barriers for graphene spintronics

(a) The schematic diagram of graphene spintronics (The diagram below (a) is spin device preparation flow chart); (b) The test result of in-plane local MR and no-local MR of device(The inset diagram is the MR of sample with $V_g = -32$ V)[88]

3. Ni(111)/graphene/h-BN junctions as ideal spin injectors

Spin electronics or spin electronics are designed to introduce conventional semiconductor-based electronics into metal-based magnetics to increase rotational degrees of freedom for metallic magnetic electrons. It is possible to solve the tunnel junction dependency or the Schottky barrier by spin injection rotation, although this spin polarization is not complete at room temperature.

Karpan et al. demonstrated how MTJS can be prepared by inserting layers of h-BN[89] made of Ni|h-BN|Gr_n/Ni heterostructure. As shown in Figure 5-21(a), when several layers h-BN are inserted into the middle of graphene and Ni, RA products (resistance-area) increase by more than three orders of magnitude, and there is no planned degradation. When comparing the performance of magnetic tunnel junctions with and without BN, they found that the use of five layers of graphene was the most desirable, essentially achieving a 100 % MR (magnetoresistance) ratio. Moreover, by adjusting the number of layers of graphene and BN, RA products can be arbitrarily exchanged. The results of adjusting the MR of the MTJ by adjusting the number of BN layers are given in Figure 5-21(b). It is clear from the figure that the MR disappears in the wide barrier limit. Studies have shown that the better BN and graphene lattice coincide with, the larger we can get MR and RA products.

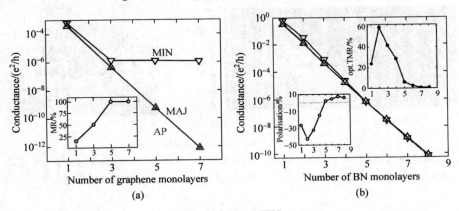

Figure 5-21 MTJS

(a) Test results of two sheets of h-BN are shown. Inset: MR as a function of n for junctions. (b) The relationship between conductance and the number of h-BN layers. Insets: optimistic MR as a function of m for junction (right) and polarization of the parallel conductance P[89]

5.3 Graphene/h-BN heterostructure

Graphene does not have d or f electrons by itself, but under certain conditions (vacancy, modification, boundary type change, etc.) will show paramagnetism, even ferromagnetism. The magnetism of graphene has a very important role in the application of graphene in spin electronics and spin-based graphene transistors. On the other hand, when the composite heterostructures of 2D graphene and 2D h-BN themselves are a composite of carbon (C) and h-BN (BN) systems, the C-BN system, whether theoretically or experimentally have predicted or observed some strange physical properties (as discussed earlier in this article) (Figure 5-22)[90-99]. In particular, researchers with the Geim's group at the University of Manchester in the UK have found that graphene on the surface of h-BN is a very good spintronics because of the magnetic field produced by graphene in the composite system can propagate to the macroscopic distance without attenuation. The graphene based on h-BN substrate exhibits a peculiar characteristic again in terms of magnetism and spintronics, and has a very broad prospect in the application of spin electronics.

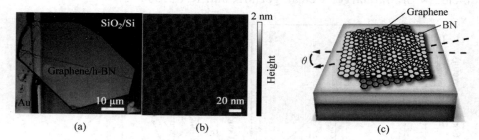

Figure 5-22 Graphene/h-BN heterostructure
(a) Optical micrograph image of graphene/h-BN heterostructure; (b) Raw image of graphene/h-BN, the moiré pattern of heterostructure[98]; (c) The model of graphene/h-BN heterostructure[99]

5.3.1 Lattice structure and electronic structure

1. Lattice structure of graphene/h-BN

After the rise of graphene research, the research of h-BN has quickly entered the field of view. The two have almost exactly the same hexagonal honeycomb lattice structure with a lattice mismatch of only 1.7 %. Nevertheless, the nature of them is very different. Graphene is a zero-band half-metal, while h-BN is a wide-bandgap insulating material. Therefore, the perfect lattice matching makes the graphene on the BN substrate show a strange nature. Theoretical studies have found the potential application of graphene/h-BN heterostructure in the semiconductor, and the electronic properties used to improve the mobility of graphene carriers are becoming more and more important for the heterostructure of graphene/h-BN.

The heterostructures are made of monolayer graphene and monolayer h-BN, which have been extensively studied over the past decade. Due to the difference in lattice between graphene and h-BN, there are moiré patterns in the heterostructure (Figure 5-23(a)). From the graphene and h-BN between the way of stacking, roughly divided into two main ways[90,100-107]: ①AA type (Figures 5-23(b)~(d)). The C atom is directly above the N atom and the B atom, where the lattice of the graphene completely coincides with the lattice of h-BN; ②AB type. Here, the three C atoms in a graphene lattice are located directly above the three N atoms, and the other three C atoms in the graphene are located at the center of the h-BN's lattice, which can be seen as AB type (Figures 5-23(e)~(g)). On the other hand, when the three C atoms in the

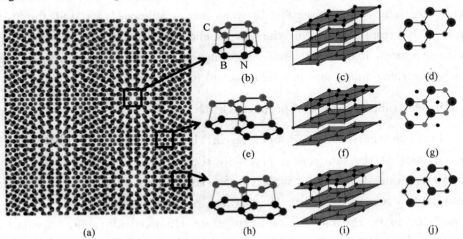

Figure 5-23 The moiré patterns of graphene/h-BN, and different configurations
(a) The moiré patterns of graphene/h-BN. The way of stacking: AA configuration in (b)~(d), AB configuration in (e)~(g), AB' configuration in (h)~(j)[107,90]

graphene lattice are directly above the B atom, the other three C atoms are in the center of the h-BN's lattice, and we consider the AB′ configuration (Figures 5-23(h)~(j)). In all the current research, theoretical research or experiments; by graphene and BN stacking methods people have studied a lot of species, but the most basic types of stacks are the above three types.

2. 3D structure of graphene/h-BN heterostructure

A large number of studies have shown that both the graphene and h-BN crystals are chemically inert and have almost the same lattice structure, although the lattice constant is only 1.7 %. Which makes them able to form a heterostructure, the lattice structure has a very good combination. Moreover, when a large area of h-BN substrate on the growth of graphene and the formation of large areas of heterostructure, the heterostructure formed a new periodic potential, in the highly de-oriented lattice weaekened, the nature and morphology is very similar to the area. At the same time, for the transmission electron microscopy (TEM) and scanning transmission electron microscopy (STEM)[108], it is clear that the heterostructures have obvious bending folds in the direction perpendicular to the surface of the sample, forming a period of wave-like structure consistent with the wavelength of the moiré patterns (Figure 5-24). Studies have shown that this benign structural formation originates from the interaction between the graphene layer and the h-BN layer. The results revealing the 3D atomic structure of the graphene/h-BN heterostructures, that is, in the self-supporting van der Waals heterostructures, there is a self-warping atomic structure that leads to interlayer lattice mismatch.

3. The electronic structure of graphene/h-BN van der Waals heterostructure

From the perspective of opening the band structure of graphene, the introduction of the substrate is gradually recognized as a means. The h-BN, due to its atomic level of flat surface, no extra suspension key, doping effect is weak and so on, to maximize the intrinsic physical properties of graphene is considered to be the most suitable substrate. In particular, the superlattice structure formed by graphene and BN can regulate the band structure of graphene, form an additional Dirac point, and further add a series of new physical phenomenas to graphene[90-120].

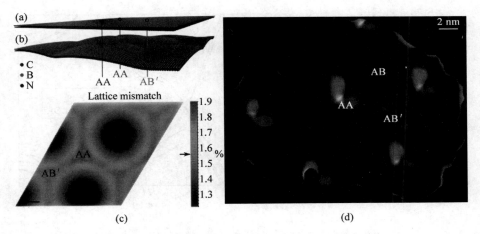

Figure 5-24   Graphene/h-BN and lattice mismatch

The model of graphene/h-BN before relaxation (a) and after fully relaxation (b); (c) The map of lattice mismatch in Van der Waals heterostructure; (d) The map of annular center of mass (ACOM) in heterostructure[108]

Based on the lattice mismatch between BN and graphene, the researchers found that heterostructures can regulate the electronic structure of graphene (Figure 5-25). In 2007, by using density functional theory, it was first predicted that h-BN substrate can open the band gap of graphene[90]. It is found that the AB configuration is the most stable configuration in the three-graphene lattice and h-BN lattice stacking schemes. This configuration can open a 53 meV band gap near the Dirac point of graphene, which can improve the pinch characteristics of the graphene FET at room temperature. As can be clearly seen in Figure 5-25(b)[109], a small bandgap in the band structure diagram based on the two computational methods is opened. Slawinska et al. calculated the electronic properties of the most stable graphene/h-BN configuration using a tight bond and a density functional method[109]. The results show that the electronic structure and the size of the band gap can be controlled by the applied electric field perpendicular to the plane. Then more studies focused on the electronic structure and bandgap of graphene/h-BN van der Waals heterostructure from 2007 to 2017[90-91,100,101,104,110-120].

The results of the theoretical study were quickly confirmed by experimental studies. Using the angular resolution photoelectron spectroscopy, the researchers first observed the bands of the original and the second-generation Dirac cones

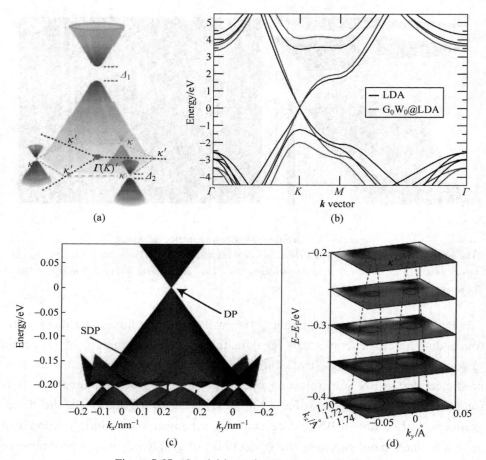

Figure 5-25 Spatial inversion symmetry breaking

(a) The band diagram of the original Dirac cone and the second-generation Dirac cone[98]; (b) The band gap, calculated by two method[109]; (c) The band structure of AA configuration[115]; (d) The second-generation Dirac cone, observed by the angular resolution photoelectron spectroscopy[98]

bands in the graphene/h-BN heterostructures. The band gap in the original Dirac cone band is as high as 160 meV, and the band gap in the second generation Dirac is as high as 100 meV. The experimental results show the importance of spatial inversion symmetry breaking in the band gap and band adjustment of heterostructures. Figure 5-25(a) shows the band diagram of the original Dirac cone and the second-generation Dirac cones[98]. It is worth noting that, Yang et al. successfully fabricated graphene/h-BN heterostructures with single crystal, high-quality, layer number control and zero-error lattice matching by epitaxial growth method. Figure 5-25(c) shows the band structure of a zero-turn, lattice-exact match graphene/h-BN heterostructures[115]. Figure 5-25(d) shows the second-generation Dirac cones observed directly by the angular resolution photoelectron spectroscopy[98].

Early theoretical studies have revealed the relationship between the changes and the distance between the graphene layer and the h-BN layer under different configurations of graphene/h-BN heterostructures. Figure 5-26(a) shows the relationship between the interlayer spacing and the energy in the heterostructure under three configurations[90]. As the spacing between the layers decreases, the bandgap becomes larger, but the degree of variation of the three configurations is different. At the same layer spacing (Figure 5-26(b)), the AA configuration structure has the largest bandgap and the AB configuration has the smallest bandgap. It can be seen from the figure that the AB configuration is the most stable. This is due to the gravitational interaction between the π electrons of graphene and BN monolayer and the mutual exclusion of the anions. The anions of N atom are more directly above the hexagonal lattice of C atom, so that the π electron density is very low here, and the cation of B atom is directly above the C atom to increase the mutual attraction. Although there are many similar calculations[101,105], the changes in energy and bandgap caused by the distance between the graphene layer and the BN layer are consistent, and the specific trends and conclusions are consistent, although the numerical size is different.

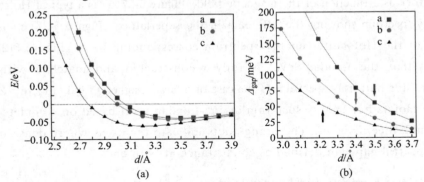

Figure 5-26 The relationship between the change in the distance between the graphene layer and the h-BN layer under different configurations of graphene/h-BN heterostructures

(a) The relationship of total energy and distance in graphene/h-BN heterostructure with three configurations. (b) The values of the gaps as a function of distance between monolayers. The calculated equilibrium separations are indicated by vertical arrows[90]

5.3.2 Magnetism and spintrons of graphene/h-BN van der Waals heterostructure

There has been a relative progress in the study of the magnetic properties and the spintrons properties of the graphene and h-BN complexes. That is, the composite system under different conditions (different edge effects, elemental deposition, vacancy defects, etc.), can appear ferromagnetic or antiferromagnetic. With the deepening of the study[121-125], the magnetic properties of graphene/h-BN heterostructures composed of the graphene on the h-BN substrate are gradually revealed.

1. Hofstadter butterfly

The addition of 2D crystal materials of different physical properties is highly likely to produce some new material structures and physical properties. New physical properties are observed in the graphene/h-BN heterostructures, confirming the theory of more than 40 years predict, for example: Hofstadter butterfly is a wonderful subtype pattern that describes the movement of electrons in the magnetic field[103,126-127].

The structure of the Hofstadter butterfly spectrum is complex and corresponds to the density of fractal small fractures following a simple linear trajectory as a function of the magnetic field. Figure 5-27(a) is a typical Hofstadter energy spectrum showing the MR data in the superlattice. Figure 5-27(b) is the band of the Hofstadter energy spectrum corresponding to A. Figure 5-27(c) shows that the Landau level is fully reconstructed and matched with the Hofstadter butterfly spectrum in the case of a large magnetic field. Figure 5-27(d) is the Hofstadter energy spectrum of the Landau level based on complete spin and sublattice division. The orange dots indicate the areas where dense bands and spectral gaps are defined as conductance at both ends.

2. Intrinsic band gap and Landau level renormalization

With the deepening of research, the intrinsic value and origin of graphene/h-BN heterostructures are also revealed. The intrinsic bandgap of the graphene/h-BN heterostructures that are perfectly aligned with the lattice

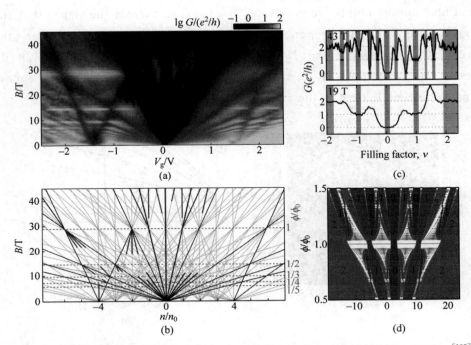

Figure 5-27 Hofstadter butterfly spectrum in superlittce of graphene/h-BN heterostructure[103] (a) A typical Hofstadter energy spectrum showing the magnetoresistance data in the superlattice; (b) The band of the Hofstadter energy spectrum corresponding to A; (c) Landau level; (d) The Hofstadter energy spectrum of the Landau level

is observed. At same time, people use the magnetic spectrum to detect the transitions of the heterostructure of the Landau level, thus revealing the band gap of 38 meV[107]. The effective Fermi velocity represents the transition of the Landau level transitions, which the magnetic field has a strong dependence.

Figure 5-28(a) shows the color rendition of the $-\ln[T(B)/T(B_0)]$ spectra, which corresponds to the relationship of energy and magnetic field. There are three dip features denoted as $T_1 \sim T_3$. As the magnetic field increases, the dip features systematically transit to a higher energy region. The effective mobility of the samples estimated from the resonance width of the spectrum is higher than 50,000 cm^2/(V·s). Figure 5-28(b) shows the Landau level (LL) transitions of T_1, T_2 and T_3. From the figure, the feature of T_3 is higher energies than the others. Figure 5-28(c) shows the change in energy versus magnetic field. Two samples are marked with red solid points and blue hollow points, respectively. It can be seen from the figure that the energy of T_1, T_2 and T_3 increases with the increase of magnetic field, and the trend of increasing T_3 and T_2 is obviously higher than T_1. Figure 5-28(d) shows the two samples in T_1 transition, the relationship between the energy and the magnetic field in the mode.

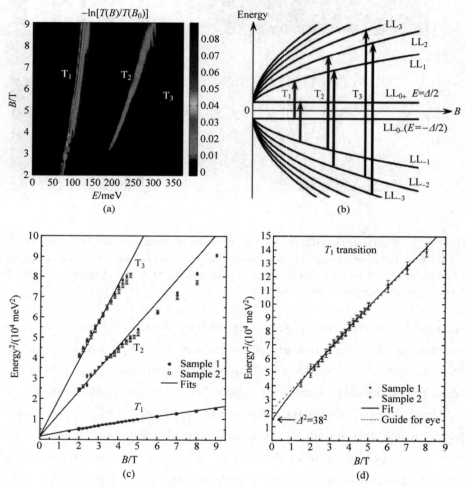

Figure 5-28 Magneto-transmission ratio spectra and LL transition energies
Magneto-transmission ratio spectra of graphene/h-BN in (a) and (b). LL transition energies of graphene/h-BN in (c) and (d)[107]

It is clear that for a heterogeneous structure that has a multibody system, it can be described by the same Dirac Hamiltonian operator, such as the valley-dependent magnetic moment. The observed recombination of the Landau level is of great significance for basic research, such as the magnetic control of the valley degree of freedom, and the spin-polarized magneto-optical response. It can be understood that the bandgap and multi-body interaction effects in the system result in such a result.

3. Anomalous quantum Hall effects

Quantum Hall effect and fractional quantum Hall effect are the focuses of the physical properties of 2D semiconductor materials. First, the semi-integer quantum Hall effect and the P Gabriel phase of Dirac Fermi were first discovered in graphene. Furthermore, in the suspended graphene samples, the fractional quantum Hall effect was also observed. Recently, researchers have found that this fractional state can be regulated by an electric field in high-quality graphene heterostructures with h-BN as the substrate. Through the in-plane magnetic field to change the spin orientation, the quantum Hall effect is also achieved in graphene.

The study found that the role of h-BN is not just a layer of high-quality dielectric materials. Experiments show that between graphene and h-BN due to the size and angle of the lattice does not match the reasons for the formation of periodic moiré crystals (Figure 5-23(a)). When the graphene and h-BN crystals are perfectly aligned, the period of the moiré lattice is the largest, about 14 nm, and the period becomes faster as the crystal orientation increases. And the electronic motion in the graphene is also modulated by the moiré lattice potential. In the high magnetic field, this moiré lattice also brought us a surprise. In the molybdenum crystal composed of graphene/h-BN, the cyclotron of electrons will be affected by the crystal field due to the magnetic length of the electron cyclotron (about 12.5 nm at 4 T magnetic field) and the size of the moiré lattice in an order of magnitude. The first observation of the Hofstadter butterfly predicted by the fractal Landau quantization(Figure 5-27(a)). The fractal energy spectrum of 2D electron gas can be observed in the extremely high-quality graphene/h-BN moiré crystal, that is, fractal quantum Hall effect (FQHE) and even fractal fraction quantum Hall effect[99,127-129].

Figure 5-29(a) shows FQHE of devices in Figure 5-29(b), which is made of graphene/h-BN van der Waals heterostructure. Figure 5-29(c) is fractional quantum Hall effect in the Hofstadter spectrum. The relationship of filling fraction and MR in Figure 5-29(d), conductivity and filling fraction in Figure 5-29(e). Figures 5-29(f) and (g) are the Landau fan diagram showing longitudinal resistance (R_{xx}) and Hall resistance (R_{xy}). Figures 5-29(h)~(j) show line cut view of the longitudinal Hall conductance and lateral Hall conductance at a constant magnetic field in the Landau fan diagram corresponding to Figures 5-29(f) and (g).

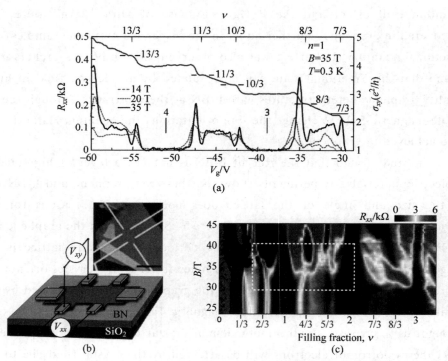

Figure 5-29 Magnetoresistance and Hall conductivity and FQHE of graphene/h-BN heterostructure

(a) Magnetoresistance (left axis) and Hall conductivity (right axis) in the $n = 0$ and $n = 1$ Landau levels at $B = 35$ T and $T \sim 0.3$ K of (b)[128]; (c)~(e) and (f)~(j) The FQHE of graphene/h-BN Van der Waals heterostructure[127,99]

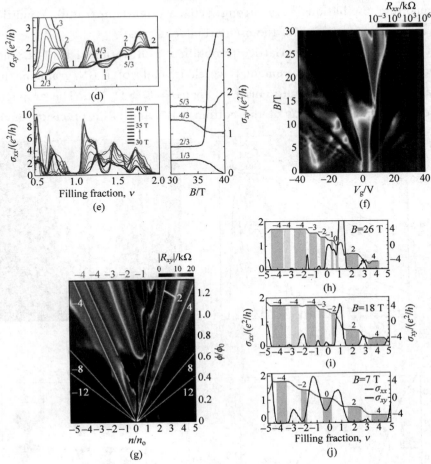

Figure 5-29(Continued)

4. High-temperature quantum oscillations caused by recurring Bloch states

The magnetic oscillation behavior and the Langdon quantization are usually caused by the cyclotron motion of the carriers in the metal or semiconductor. In the superlattice produced by the graphene based on the BN substrate, a quantum oscillation independent of the passages is supported, and this oscillation is distinguished from the previous quantum oscillations. Due to the complex changes in the electronic structure in the superlattices, the charge carriers in the superlattices effectively undergo a state where the magnetic flux in the unit superlattices is a fraction of the non-magnetic field, which makes the quantum oscillations very strong. Under high-temperature conditions, even in tiny magnetic fields, the oscillation continues.

Geim et al. used a superlattice to obtain a multi-terminal Hall bar device for carrier transport measurements, which is made of h-BN/graphene/h-BN van der Waals heterostructure (inset in Figure 5-30(a))[130]. The graphene is aligned with the crystal axis of the bottom h-BN and has a precision of more

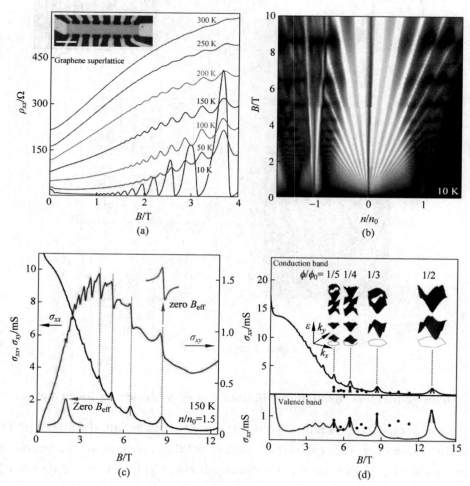

Figure 5-30 High-temperature quantum oscillations caused by recurring Bloch states
(a) The relationship between ρ_{xx} and magnetic field B at different temperatures. The upper left corner is illustrated as a test device. (b) The relationship between B and n/n_0 at 10. (c) The relationship between σ and B, σ_{xx} and σ_{xy} correspond to transverse magnetic transport and longitudinal magnetic transport, respectively. (d) Brown-Zak (BZ) oscillation in the Bloch state in the field of small and effective magnetic field[130]

than 2°, with a 15° deviation from the alignment of the upper h-BN crystal axis. So that the upper h-BN produced by the moiré periodicity is shorter, the upper part of the h-BN as an inert layer. In this way, in the high carrier concentration or ultra-high magnetic field to appear superlattice effect (Figure 5-30(c)). It can be seen from the measurement results, when ϕ_0 is a simple score, the charge carrier effectively experienced a zero magnetic field, the trajectory replaced by a straight line (Figures 5-30(a) and (b)). This results in a weaker Hall effect and a higher resistivity. The study also revealed that such quantum oscillations do not need to rely on monochromaticity and can continue to very high temperatures, even when the temperature reaches 1,000 K, quantum oscillations can be observed (Figure 5-30(d)).

5. Large magnetoresistance (MR)

At the atomic level, understanding the MR, especially the resistance changes with the external magnetic field, is fundamental and technically important. The emergence of van der Waals heterostructure provides an opportunity to explore the properties of MR. When the graphene is located on the h-BN substrate and forms the van der Waals heterostructures, the prominent localized MR is found[131].

The optical image of the MR test device is shown in Figure 5-31(a)[131]. The figure shows an irregular geometric device such as a Hall bar. In this experimental device, the deposition of multilayer graphene shows a large MR. At the same time, in the local geometry, that is, graphene/h-BN van der Waals heterostructure of the superlattice, the small size of the graphene samples at a temperature of 400 K, the MR reached 2,000 %, just like Figure 5-31(b). The results of the test show that this MR in the low and high magnetic fields can be used as the material of the magnetoresistive sensor. Even if the temperature is as high as 400 K, the material still has a very small resistance temperature coefficient. The test results clearly reflect the relationship between the heterostructure MR and the temperature and angle. From the test results of Figures 5-31(c) and (d), we can know that at the temperature 300 K, the magnetic field is 9 T, the tens of layers of the van der Waals heterostructures have a nonlocal MR of up to 9,000 % at room temperature. The local MR is understood to be generated by large differential transmission parameters on different layers of graphene on the normal magnetic field, whereas the local reluctance is due to the magnetic field induced Ettingshausen-Nernst effect. The nonlocal reluctance again expresses the possibility of a graphene-based gate-tunable thermal switch.

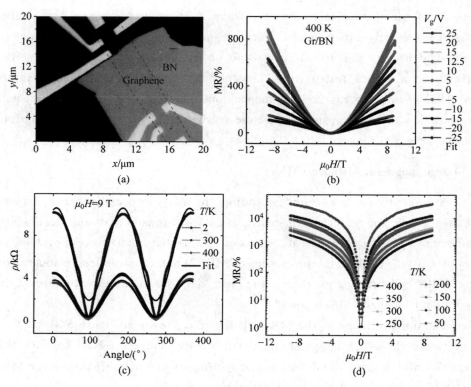

Figure 5-31 Large magnetoresistance

(a) Optical map of graphene/h-BN heterostructure; (b) The relationship between MR and $\mu_0 H$ with deferent V_G; (c) $\mu_0 H = 9$ T, MR based on different temperature and angle; (d) The relationship between MR and $\mu_0 H$, with deferent temperatures[131]

6. Gate-dependent pseudospin

The electronic pseudospin in graphene describes the contribution of two sublattices to the electron wave function, which is a two-component pseudospin relativistic Dirac-Will rotation description. The unique pseudospin structure of Dirac electrons in this graphene resulted in the emergence of many singular phenomena. The graphene forms a superlattice based on the BN substrate, and in the moiré period, the fast sublattice oscillations of the N atom and the B atom are superimposed. This rapid oscillation results in periodic spinor potential in graphene[132-139]. This unusual moiré superlattice leads to the anomalous hybridization of the spin potential electron pseudo-spin[140]. Since the optical transition is very sensitive to the excited state wave function, the infrared(IR) spectrum can be directly detected.

The top map in Figure 5-32(a) shows a schematic diagram of the apparatus for testing graphene/h-BN superlattices using IR spectroscopy, and the below one shows the optical micro-image of graphene/h-BN van der Waals heterostructure.

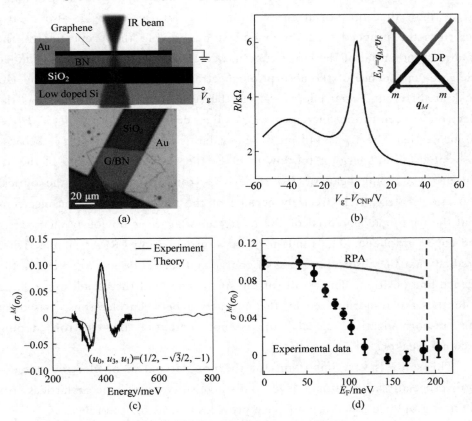

Figure 5-32 Gate-dependent pseudospin

(a) The schematic diagram of pseudospin based on graphene/h-BN heterostructure superlattice by using IR spectroscopy (top); The optical picture of Graphene/h-BN heterostructure(below); (b) The typical transport characteristics of graphene/h-BN superlattice; (c) The absorption spectrum of superlattice; (d) Gate-dependent moiré spinor potential[140]

Researchers tested Graphene/h-BN heterostructures, then typical room temperature transmission properties are shown in Figure 5-32(b). From the figure we can clearly see that there are two typical resistance peaks appearing at $V_g = 0$ and $V_g = -40$ V, respectively. This is mainly due to the original Dirac point and the superlattice inner Brillouin area at the M point of the hole side of the small Dirac point area. Figure 5-32(c) shows the absorption spectra observed during the experiment and the theoretical calculated absorption spectra, both of which are in good agreement. It is presumed that the coupling of C-N is weaker than that of C-B because the radius of p-orbit in B atom is larger than that of p-orbit in N atom. The absorption peak also clearly describes the properties of the pseudo-spin, proving the properties of the rotational potential of the moiré superlattice. Further, the researchers studied the gate dependence of the absorption spectra. Figure 5-32(d) shows the E_M peak at different Fermi energy. It can be seen from the figure that as the electron concentration increases, E_M value decreases sharply after $E_F = 50$ meV. Near the $E_F = 140$ meV, E_M value is almost zero. This is because associated Fermi energy is too low to block the electron transitions in the M-output. The rising peak of E_M is derived from the change of the optical transition matrix. It is precisely because of the interaction between electrons that the rotational potential of the moiré superlattice is obviously changed in the doped graphene. The random phase approximation (RPA) in the figure predicts that the absorption peaks of charge and carrier doping are weak in the orange line position. The results of the study show that the pseudospin-related potentials are re-normalized by the interaction between electrons. Based on this re-normalization effect, the spinor potential of electron doping becomes weaker.

This study is very important for the precise control of electron pseudo-spin, especially at the atomic level to distinguish two graphene sublattices, for the new graphene pseudo-spin device provides a basis for the release.

7. Magnetic characteristics of graphene/h-BN van der Waals heterostructure

The 2D material is made of graphene/h-BN heterostructure, as a composite system, the study of graphene magnetic is very important[141-142]. Especially in the magnetic devices and spintronics is particularly important.

Ding et al. prepared the graphene/h-BN heterostructure samples[143], and the samples are subjected to magnetic measurements at different temperatures. When the temperature from 5 K to 300 K, test the magnetic field for the 7T. The test results are shown in Figure 5-33(a) and (b), and the magnetization of graphene/h-BN heterostructures at different temperatures is related to the magnetic field. The test times for Figures 5-33(a) and (b) are 1.5 minutes and 5 minutes, respectively. It can be seen from the figure that the magnetization of pure h-BN at different temperatures is negative (antirustivity), as shown in Figure 5-33(c), and the magnitude of the magnetism signal does not vary with temperature.

However, the magnetization signal in the low temperature region ($T <$ 100K) becomes positive (paramagnetic) after the graphene is long on h-BN, and the size of the paramagnetic intensity is sensitive to the change in the growth time of the graphene. With the extension of the growth time of graphene, the paramagnetic intensity of the sample decreases. At the same time, according to the results of magnetization test, no ferromagnetic signal was observed in the measured temperature range of 5 ~ 300 K. When the measured temperature T is higher than 100 K, graphene/h-BN exhibits an antirust property, and the magnitude of the magnetization resistance is very close to the magnitude of the magnetic susceptibility of pure h-BN. When the temperature is below 20 K, the paramagnetic behavior satisfies the Curie paramagnetism law. When the temperature is 20 ~ 50 K, a magnetization platform appears in the magnetization curve, and the sample exhibits very strange paramagnetic characteristics.

Figure 5-33(d) shows the relationship between the mass susceptibility of the sample corresponding to Figures 5-33(a), (b) and the BN substrate as a function of temperature. The illustration shows that when $T < 20$ K, the linear relationship between the reciprocal of the susceptibility and the temperature satisfies Curie's law $X = C/T$. In the range of 20 ~ 50 K, a magnetization platform appears in the susceptibility curve, and the magnetization behavior does not satisfy Curie's law. When $T > 100$ K, graphene/h-BN heterostructures exhibit magnetic resistance, and the magnetostatic strength does not change with temperature. While the growth time of the two samples is different, resulting in the difference between the

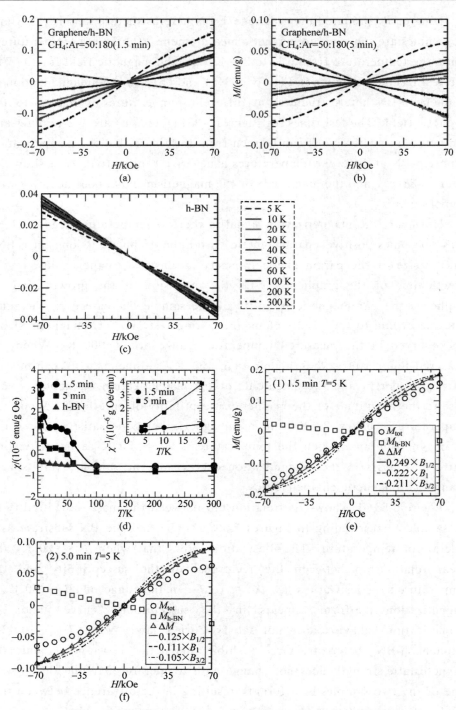

Figure 5-33　The test results of sample 1, sample 2 and h-BN

two curves. Figures 5-33(e) and (f) are two samples at $T = 5$ K when the measured magnetization curve is carried out by Brillouin formula fitting. The results show that both samples can only be fitted with $J = 1/2$ Brillouin curve.

8. Low B field magneto-phonon resonances

The rapid development of laser technology has brought unprecedented opportunities for the development and revival of Raman spectroscopy. Raman spectroscopy has a unique ability to identify the properties of graphene, doping, strain and lattice thermal information. When many of the Raman spectroscopy techniques and equipment are applied simultaneously with the magnetic field, we provide more efficient methods for studying the new peculiar physical phenomena of graphene[144-150].

When the graphene is in a magnetic field perpendicular to its plane, the electronic state in the graphene condenses into the Landau levels (LLs) that can interact with lattice vibrations. When the magnetic field changes, the Landau damping of the highest frequency optical mode at the point Γ can be tuned (G mode), and when the G mode is resonantly coupled to the energy-matched LLs conversion, the Raman shift occurs. This coupling mode, known as magnetic phonon resonance (MPR)[151].

The researchers performed magnetic Raman measurements on four systems: ①graphene/h-BN heterostructure; ② h-BN/graphene/h-BN heterostructure; ③electrically contacted h-BN/graphene/h-BN heterostructure; ④h-BN/graphene/graphene/h-BN heterostructure. Figure 5-34(a) shows a 2D pattern of color coding for the relationship of system ②, magnetic field strength and Raman shift. It can be seen from the figure that the Raman curve of the 2D material and the Raman curve of the h-BN are weakly dependent on the magnetic field. But the G-line shows a strong dependence on the magnetic field, which is due to the resonant coupling between the G mode and the single-phase LLs transition. When $B = 3.7$ T, the position indicated by the arrows of the G line is due to the coupling of the T_1 transition and the G mode. Figure 5-34(b) shows the graphene band structure. Different band, Landau level changes and transition are marked, Figures 5-34(c) and (d) it shows the relationship between the magnetic field (B) and ω_G, Γ_G. It can be seen from the figure that the evolution of ω_G and Γ_G with B changes. The position of the three arrows in the Figure 5-34(d) is $B = 2.1$ T, $B = 3.7$ T and $B = 5.8$ T, where are the three typical magnetic phonon resonance appears. Under these values, the energy change of the Landau levels (T_2, T_1 and L_1) matches the G-mode phonon energy ($\hbar \omega_{ph}$) at the zero-magnetic field. Figure 5-34(e) shows that the T_n mode and L_1 mode transition energy also increase as the magnetic field strength increases.

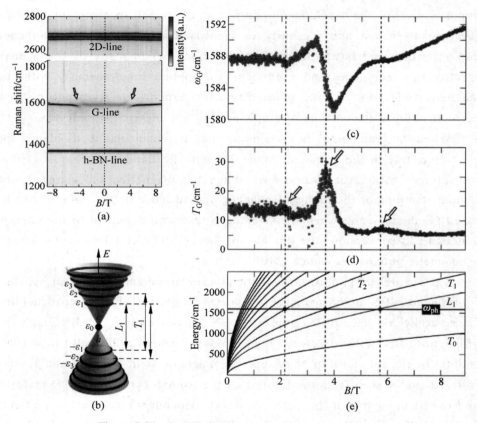

Figure 5-34 Low B field magneto-phonon resonances
(a) The relationship between Raman shift and B of graphene/h-BN heterostructure; (b) The typical graphene band structure; (c)~(e) The MPR of h-BN/graphene/h-BN heterostructure[151]

Same test method was also applied for the first time to test the system ④. Unlike monolayer graphene, a constant Fermi velocity is an effective way to explain the system ④. Compared with one-way graphene, the Fermi velocity decreases, indicating that the electron interaction in bilayer graphene plays an important role. Therefore, the G-line phonons of the monolayer and bilayer graphene for low magnetic fields depend on the magnetic field. Based on the monolayer graphene on the BN substrate, the Raman spectra under the action of the magnetic field show a significant magnetic phonon resonance state. These studies provide a better way to understand the multibody effects of electrons.

9. 100 % spin-injection and detection polarizations

Effective spin injection is one of the major challenges facing spintronics. How to inject a highly spin polarized stream from the ferromagnetic metal into the semiconductor effectively, and that there is no large loss in the process of spin polarization, which is one of the foci of spin injection.

It is found that h-BN as a barrier material for spin implanted graphene, inert h-BN as encapsulation material for graphene, can greatly improve the carrier mobility[152]. On the other hand, Britnell et al. studied the tunnel junctions of graphene/h-BN/graphene heterostructures[153], which revealed the pinhole characteristics of the tunnel barrier of h-BN. Yama et al. demonstrated the use of h-BN tunnel barriers and graphene assembled ferromagnetic materials for the first time[154]. They proved that the current spin injection and detection of non-local MR up to 300K.

Kamalakar et al. demonstrated the results of the study of spin-polarized electrons injected into the graphene through the atomic plane of h-BN[155]. The study of the use of ferromagnetic/h-BN contacts in the sisters of the tunnel spin injection and spin transport. They get the spin signal and spin life more than the previous study results by two orders of magnitude, spin life up to 1.46 nanoseconds. This suggests that the h-BN tunnel barrier provides a reliable, reproducible and alternative method that can incorporate the problem of the conductivity mismatch of spin implanted graphene. Even so, the interfacial resistance area of the single-layer h-BN barrier layer is relatively small, which requires the use of more layers of h-BN for non-invasive spin injection and detection. In the theoretical study, the spin-injection polarization has been predicted as a function of the deviation of the H-BN layer with the increase in the h-BN layer in the FM/h-BN/graphene system[156-159].

Gurram et al. studied the spin transport of heterostructures of graphene completely encapsulated by h-BN at room temperature[160]. They used a ferromagnetic cobalt electrode made of top h-BN as a tunnel barrier for spin implantation and detection in graphene. At bias voltage of +0.6 V, the large bias of the spin injection polarization is up to 50 %, and up to -70 % of the anti-polarization, the reverse bias is -0.4 V. The double-ended spin valve signal up to 800 Ω has a magneto-resistivity ratio of 2.7 % and a spin accumulation of up to 4.1 meV. In the case of a double-layer h-BN tunnel barrier, up to 100 %. All the test results are shown in Figure 5-35. As a function of the DC bias is applied at room temperature. These studies are closely related to the future development of graphene spin electrons.

10. Spin relaxation length and long-distance spin transport

The discovery of graphene has injected new vitality into the study of spin electronics. As the spin and the orbit coupling is weak, making the electron spin transmission process is relatively easy to control. The size of the spin transmission distance becomes a standard for measuring the spin electronics. The spin length can be expressed as: $\lambda = \sqrt{D_s \tau_s}$[34], λ is spin relaxation length, D_s is spin diffusion constant, and τ_s is spin relaxation time[161-171].

The emergence of h-BN has brought new hope for the study of the spin transport of graphene. In 2012, Zomer et al. prepared graphene/h-BN heterostructure[162], using four-terminal non-local technology to measure the characteristics of graphene spin transport. They inject a spin-polarized current from the end of the device into graphene, and the spin current passes through and diffuses into the detection area. And then exert an external magnetic field to control the spin polarization. Figure 5-36(a) shows the non-local resistance R_{nl} measured with the magnetic field of the spin valve. As can be seen from the results, the room temperature of this spin valve spacing is huge, the maximum length of the spin signal is to 20 μm. The inset on the right side is the assembly diagram of the test device. Figure 5-36(b) shows the measured Hanle spin precession value of the device under external magnetic field. The illustration is a precession for each magnetization geometry level. By calculating the corresponding values on the table has been given the spin relaxation distance, spin relaxation time and so on. From this study, we can see that monolayer graphene is measured and observed to achieve a maximum

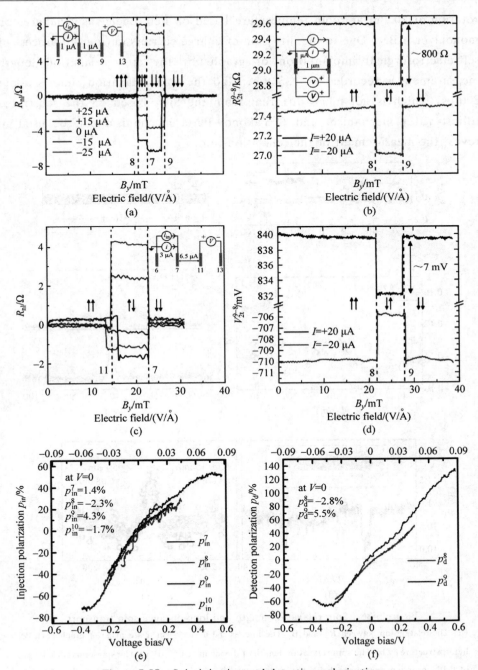

Figure 5-35 Spin-injection and detection polarizations

Nonlocal spin-valve and Hanle measurements at different DC bias across the injector in (a)～(d); (e), (f) Differential spin-injection (pin) and detection (pd) polarizations of the cobalt/bilayer-hBN/graphene contacts[160]

rotation of 20 μm at room temperature based on heterogeneous structures on monolayer h-BN. Due to the diffusion of charge carriers during the process of self-electrosurgical implantation of graphene, the spin relaxation length measurement is recorded as 4.5 μm. And the spin relaxation time is about 200 ps, which is not much different from the SiO_2 based on the substrate. Elliott-Yafet mechanism and D'Yakonov-Perel mechanism can be used to reveal the measurement of the relaxation rate.

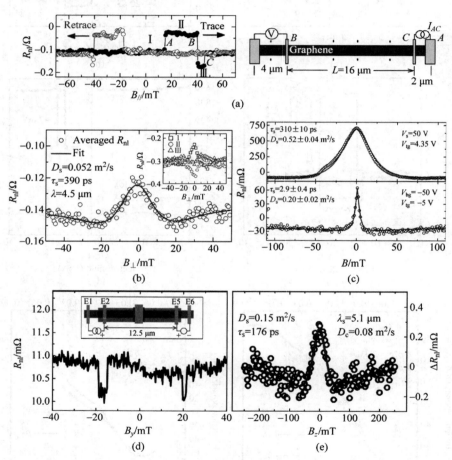

Figure 5-36　Spin relaxation length and long-distance spin transport

(a) The relationship between R_{nl} and B. The inset is the device, which are made of graphene/h-BN heterostructure; (b) The spin relaxation length of device in (a)[162]; (c) The spin relaxation length of h-BN/graphene/graphene/h-BN heterostructure[163]; (d) and (e) are the test results of h-BN/graphene/h-BN heterostructure[167].

Figure 5-36(c) shows the spin transport measurements for the h-BN/graphene/graphene/h-BN heterostructures of another group of researchers in 2015.[163] It can be seen from the figure that this double-hole high mobility double-layer graphene has a spin-relaxation length of up to 13 μm and a spin relaxation time of 2.5 ns. When the temperature is 4 K, the charge carrier diffusion coefficient increases, making the spin relaxation length up to 24 μm, spin relaxation time to 2.9 ns. Figures 5-36(d) and (e) show another group of researchers in 2016[167], tested the results of spin-transport of BN completely encapsulated graphene. As can be seen from the figure, the observed spin transport length was 12.5 μm.

11. Nanosecond spin relaxation times

Spin relaxation time is another factor in achieving spin implantation in BN as a spin barrier for spin implantation to study the spin transfer properties of graphene. A lot of experimental work and theoretical work are carried out, but the results of the experiment are often different from the theoretical calculations. BN as a good barrier material for spin injection, the access to high non-local spin signal and longer relaxation time has been affected by the numbers of layers[172-174].

Sinth et al. prepared a heterostructure of a single layer graphene (SLG) with several layers h-BN substrate (SLG/n-hBN heterostructure) in order to achieve a nanosecond spin transfer time. The results of the test are shown in Figure 5-37(a)[172]. The optical picture of the test apparatus is shown in the illustration of Figure 5-37(b). When the spin injection current is injected from the E_1 and E_2 contact points, the E_3 and E_4 electrodes measure the nonlocal voltage signal. The external magnetic field acts on this device, and the nonlocal voltage signal is used as a function of the applied magnetic field to obtain nonlocal resistance R_{nl}. Figure 5-37(b) shows the variation of the spin relaxation time as the gate voltage changes. It can be seen that the spin relaxation time is almost more than 1 ns. It can be seen that with the increase of the number of h-BN layers, the effect of tunnel barrier is more obvious, and the impurity and cleaning of heterogeneous structure interface are all affected by spin transport and spin relaxation time. In addition, they prepared a heterostructure of monolayer graphene based on a single-layer BN substrate

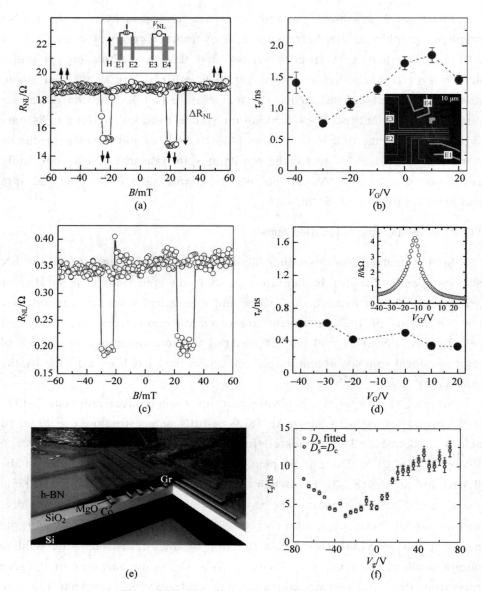

Figure 5-37　Nanosecond spin relaxation times

(a) The relationship between R_{nl} and B of the device. The inset is the structure of device 1. (b) The relationship between τ_s and V_G. The inset is the optical map of the device. (c) The relationship between R_{nl} and B of graphene/h-BN heterostructure. (d) The relationship between τ_s and V_G. The inset is the relationship between R_{nl} and V_G.[173] (e) The schematic diagram of the device. (f) The relationship between τ_s and V_G of the device (e)[174]

to study the spin-transport properties of the charge carriers (graphene/h-BN heterostructure). Similar to the previous test method, the test results are shown in Figures 5-37(c) and (d). Figure 5-37(d) extracts the τ_s as a function of V_G, which is the relationship between the gate voltage and the resistance. The results show that single-layer h-BN may not be the best way to effectively spin injection. When the number of h-BN layers is increased, the larger MR signal and longer spin relaxation time are achieved in the graphene spin valve. It is presumed that this is due to the fact that the multilayer h-BN has a more significant barrier, or that the multi-layer h-BN avoids the contamination of the monolayer h-BN during the migration process, resulting in a longer spin relaxation time.

An interesting heterostructure device was prepared to study the spin transport length and spin relaxation time of the graphene charge carriers in 2016. Figure 5-37(e) shows the schematic diagram of the device, made of graphene/h-BN/SiO$_2$/S$_i$ heterostructure. Co/MgO as the electrode, the spin current is injected from MgO. Figure 5-37(f) shows the test results of the device. At room temperature, single-layer graphene non-local spin transfer device, the spin life of up to 12.6 ns, spin transmission distance up to 30.5 μm.

In recent years, studies on spin length, spin relaxation time, etc. based on graphene/h-BN heterostructures are listed in Table 5-1. It is not difficult to find that, with the development of research technology and the improvement of research methods, it is possible to realize the longer transport length and the longer spin lifetime.

Table 5-1 Different configurations of heterostructures correspond to spin transport length and spin lifetime

Heterostructure	Year	τ_s	λ_s	Reference
Graphene/h-BN/SiO$_2$	2012	395 ps	20 μm	162
H-BN/bilayer graphene/h-BN	2013	55 ps	1.35 μm	154
H-BN/graphene/h-BN	2014	2 ns	12 μm	161
H-BN/graphene /MgO/Co/SiO$_2$/Si^{++}	2014	3.7 ns	10 μm	172
Co/h-BN(n)/graphene	2014	1.46 ns	1.4 μm	155
Co/Au/h-BN/graphene/ SiO$_2$/Si	2014	500 ps	1.6 μm	175
H-BN/graphene/h-BN	2015	2.9 ns	24 μm	163
H-BN/graphene/h-BN	2016	176 ps	12.5 μm	167

Continued

Heterostructure	Year	τ_s	λ_s	Reference
H-BN/graphene /MgO/Co/SiO$_2$/Si	2016	12.6 ns	30.5 μm	174
Graphene/n-hBN	2016	1.86 ns	5.78 μm	173
Graphene/2L-hBN	2016	490 ps	3.8 μm	173
2L-hBN/graphene /thick-hBN	2017	0.9 ns	5.8 μm	160

5.3.3 The recent application of graphene/h-BN van der Waals heterostructure in magnetic device and spintronics

The emergence of graphene has brought new opportunities for the development of spin electronics. The magnetization of graphene/h-BN heterostructures and the study of the properties of spintrons provide a theoretical basis for the preparation of magnetic devices and spintronic devices[175-179]. The unique spin-transport length and nanosecond spin relaxation time of graphene give new opportunities for the development of spin electronics. Combined with the different configurations of heterostructures, the prototype of new devices (resonators, tunneling transistors, FETs, etc.) are all prepared.

1. Field-effect transistors

Black phosphorus, as a new 2D crystal, due to its direct band gap and carrier mobility, in recent years, following the graphene, a 2D electronic device materials, a research focus. However, black phosphorus must achieve barrier-free contacts in the black phosphorus-related equipment to achieve its high performance[180-184].

Recently, the researchers used the graphene/h-BN heterostructure of the special performance[184], to achieve the black phosphorus necessary barrier. The specific method is that the use of graphene as the source-drain electrode, h-BN with its unique gate barrier characteristics as a packaging layer to characterize the inert gas conditions in the preparation of ultra-thin black phosphorus FET. It can be seen from the test results that the linear I_{SD}-V_{SD} behavior curve with negligible temperature can indicate the barrier-free contact caused by the graphene electrode. This solves the problem of Schottky barrier-limited transmission in technically related two-terminal FET geometries. On the other hand, the thicker conformal source-drain electrode also allows the black phosphorus to be completely sealed to avoid rapid degradation with the inert h-BN encapsulation layer. Such a structure can be extended to other sensitive 2D materials to achieve non-lagging transport characteristics.

Figure 5-38(a) shows the researchers' test results for the transistor[184]. At room temperature, when V_{TG} = -4 V, 0 V, 4 V, with the increase of V_{BG} ~ I_{SD} curve. It can be seen that when V_{TG} = 4 V, the graph shows the P-type behavior. The application of V_{TG} makes the threshold area significantly cheaper. I_{SD} shows a high negative V_{BG} saturation that is independent of V_{TG}. The magnitude of the saturation current in the graph increases as the V_{TG} decreases. It has been known from previous studies that this increase in saturation current observed with decreasing V_{TG} indicates an improvement in the quality of the base. Illustration of the diagram for the graphene, BN,

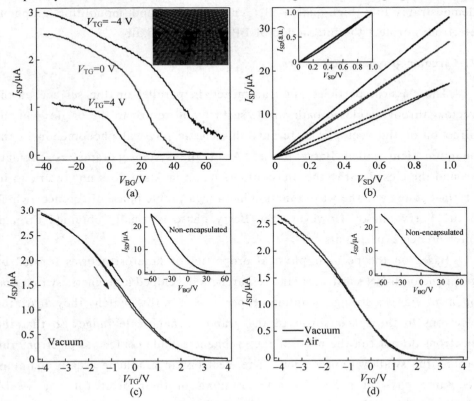

Figure 5-38 The test results of FETs, which is made of h-BN/graphene/BP heterostructure[184]

black phosphorus composition of the FET diagram. Figure 5-38(b) shows the transformation line of the I_{SD} as the V_{SD} increases. The values of the V_{TG} corresponding to the lines of various colors are the same as those of Figure 5-38(a). The illustration is the numerically re-normalized spectral line in Figure 5-38(b). This linear behavior indicates that the V_{TG} is tuned due to the gate modulation of the graphene red energy function. After that, the researchers further tested the stability of the equipment under vacuum and under environmental conditions. Figure 5-38(c) shows the relationship between I_{SD} and V_{TG}, when V_{BG} = 40 V and V_{SD} = 100 meV under vacuum conditions. Although the non-encapsulated device has a hysteresis of cotton thread under vacuum, the device has almost no hysteresis transmission characteristics. As can be seen from Figure 5-38(d), the packaged device has almost the same output characteristics in the external environment and the vacuum environment. Unpackaged devices in the external environment, a significant decline in conductivity. This fully demonstrates that high-quality BN encapsulation and monolithic graphene electrodes protect the interaction of BP surfaces with air.

2. Circular graphene resonators

The classical particles are transmitted in parallel on the surface of the vector, through a closed path on the surface and return to the origin, but the direction of the vector is different, this classic physical phenomenon in the quantum field called Berry phase. After the system parameters circulate around the closed path, the phase of the quantum state does not return to its original state, and the wave function has a measurable phase difference, which is the Berry phase. However, the Berry phase is usually obtained through interference experiments[185-188].

Based on the peculiar physical properties of heterostructures formed by graphene on h-BN substrates, the researchers developed a magnetic switch that opens or closes a strange quantum property[188]. In this switch, they drove the electrons to the special orbit of the graphene nanoscale range, so that the electrons do not run the orbit of the graphene sample center, as electrons run around the swallow center orbit. The electrons in graphene usually maintain the same physical properties after running in their orbit for one week. However, when the external magnetic field reaches a certain threshold, his role is like a switch, so that after a week of operation, not only the shape of the track changes, and will lead to electronics have different physical properties. The specific method is that in the case of relatively small critical magnetic field, the energy of the angular momentum of the ring-shaped graphene P-N junction resonator will suddenly increase significantly. It has been found that the change of π Berry phase and the topological properties of Dirac fermions in graphene have led to a sudden increase in the angular momentum energy. Interestingly, the Berry phase can be used in a very small magnetic field ($B = 10\ T$), regulating and switching the switch state. The result of this work is a special Berry phase related to the spectral display. This research work laid the theoretical foundation for the application of graphene optoelectronic devices.

Figure 5-39(a) shows a schematic diagram of the potential energy profile of a ring-shaped graphene resonator with a central region P-type doping, background region N-type doped. Figure 5-39(b) is a schematic diagram of the phase-space loop of the track in Figure 5-39(a). It is theoretically possible to know that the wave function of graphene is innate with a topological Berry phase, and that the transition between them is related to the sudden jump in the energy level of quantum mechanics. The upper part of Figure 5-39(c) is the potential distribution of the ring-shaped graphene resonator. The central region P-type doping, background area N-type-doped graphene ring resonator structure diagram. Figure 5-39(d) shows the simulation of the parametric calculation of the ring-shaped graphene resonator under the same parabolic potential. Figure 5-39(e) is the differential tunneling of the resonator. The horizontal axis represents the magnetic field measured at the center of the graphene resonator. Figure 5-39(f) is the LDOS for calculating the simulated graphene writing resonator. It can be seen that Figure 5-39(f) is in good agreement with Figure 5-39(e). Figure 5-39(g) shows the magnetic field distribution of the graphene resonator when $n = 4$, and the energy difference in the state of the magnetic field distribution $m = 1/2$ (Figure 5-39(h)). It can be clearly seen in the different external magnetic field range of the tuning switch effect.

3. Field-effect tunneling transistor

Graphene as a non-band-gap semiconductor material in practical applications

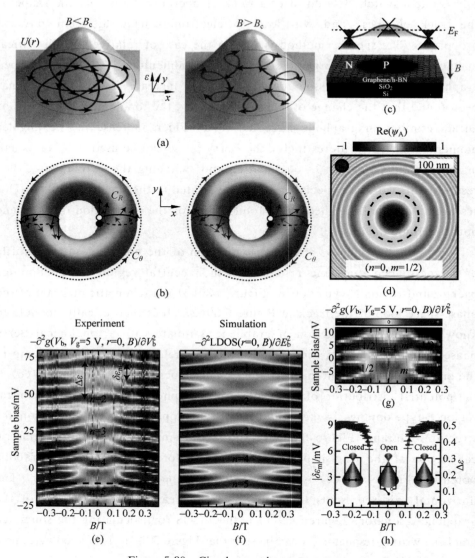

Figure 5-39 Circular graphene resonators

A schematic diagram of the potential energy profile of the ring-shaped graphene resonator in (a) and the phase-space ring of the orbit in (b); Schematic diagram of ring-shaped graphene resonator with potential distribution in (c); Simulation of parametric calculation of graphene resonator under different calibration conditions in (d); (e)～(h) Geometric phase switch state switching of ring-graphene resonator[188]

requires opening the bandgap to achieve low power consumption. The front and previous articles have been discussed, and the effect of the substrate is very common in opening the bandgap of graphene. The BN because of its unique lattice structure matching, inert electronic properties, high-dielectric properties, tunnel barriers and other properties, in the open gypsum band gap advantage can be described as unique. The heterostructures formed by the graphene based on h-BN substrates also have a wide range of applications in constructing field effect tunneling transistors[189-192].

A typical example of constructing a field effect tunneling transistor using a graphene/h-BN heterostructure is constructed in 2012 by Britnell[192]. After the corresponding test (Figure 5-40), the performance of the transistor also reflects the graphene/h-BN heterostructure in the construction of nano-device advantages. This field effect tunneling transistor is prepared on a Si wafer based on thicker BN as a gate electrode, a substrate, and a bottom encapsulation layer. The specific heterostructures are h-BN/graphene/n-hBN/graphene/h-BN (the inset in Figure 5-40(e)). Figure 5-40(a) shows the relationship between the voltage and resistivity of the two graphene layers GrB and GrT. The curve shows that the encapsulation of graphene is very small. Figures 5-40(b) and (c) show that as the amount of h-BN increases, the gate-induced electric field becomes larger and shielded by GrB. Thus, more electrons accumulate in the bottom of the graphene electrode, where only a few electrons reach the top. Figure 5-40(d) shows the graphs of the charges in the two layers with a linear change in V_g. Figure 5-40(e) shows the I-V characteristic of the transistor. Actually, the bias voltage V_b applied between GrB and GrT produces the tunnel current through the h-BN (intermediate position) barrier. The illustration shows the I-V characteristic curve described by the standard quantum tunneling formula, assuming that the momentum is conserved and that the momentum at the lattice mismatched graphene and BN interface is not conserved. V_g has a great influence on the tunneling current and has a stronger effect at lower bias voltages. And for up to ±50 V gate voltage, field effect is relatively slow. Figure 5-40(f) shows the low-bias tunneling conductivity of V_g as a function $\sigma^T = I/V_b$.

This study provides a theoretical basis for high-speed graphene-based electronic equipment. The time of tunneling electrons through the nanometer

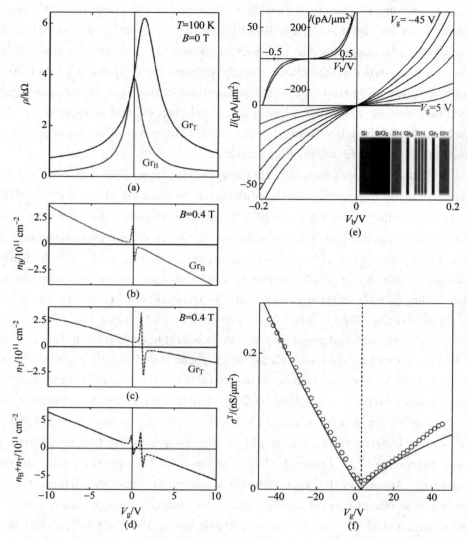

Figure 5-40 The test results of field-effect tunneling transistor, which is made of graphene/h-BN heterostructure[192]

thickness barrier is at the femtosecond scale, and the switching ratio exceeds 10 times the plane graphene FET at room temperature. If the tunnel transistor's lateral dimension is reduced to about 10 nanometers, the requirements for the integrated circuit can be achieved. The high-quality properties of BN provide high-quality gate dielectric and lower tunnel barriers.

4. Tunnel FETs

Since 2012, Britnell et al. have used graphene/h-BN heterostructures to fabricate field-effect tunneling transistors, more and more research works have been done in this area. Recently, the researchers have developed a method for manufacturing 2D heterostructures with rotational alignment. The working principle of the interfacial FET with different thickness of graphene electrode layer and different stacking order is studied in detail[193-196].

Researchers have studied the different layers and different stacking structure of graphene and BN heterostructure transistors. It is found that the reflection band structure of graphene with different thickness leads to the obvious change of interlayer I-V characteristics. Some band structures are more preferred than obtaining higher peak-to-valley current ratios (PVCRs). Moreover, with the opening of the graphene band gap, the formant can be split into smaller sub-peaks with a high bias due to the presence of h-BN and lead to an overall peak broadening.

Figure 5-41(a) shows a comparison of bilayer graphene in a Bernal stack configuration with a diamond-shaped stack configuration. After the analysis can be known, due to the accumulation of the results, making the formant more close. The interference effect of the former is obviously higher than that of the latter. The blue curve is lower than the strength of the valley of the red curve, which is the result of the first and second formants. Figure 5-41(b) shows the difference between dI and dV when two double graphene and different h-BN thicknesses are used to construct the transistor. Figure 5-41(c) shows the results of the double-layer double-film graphene versus the results of the double tetrahedron ITFET. The band structure of the even-layer graphene does not exhibit a Dirac cone, and there is no resonance condition. Figure 5-41(d) shows the normalized peak intensity of the first resonant normalized peak intensity treatment. In summary, the different combinations of different thicknesses and different layers of graphene and sandwich BN layers were studied, revealing the effect of electrode layer structure on ITFET. Although the increase in the number of layers of graphene increases the resonant enhancement, the device achieves a higher PVCR, increases the effect of interlayer capacitance on device characteristics, and reduces the effect of external gate capacitance.

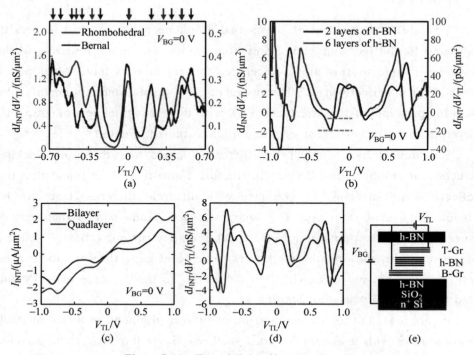

Figure 5-41 Tunnel field-effect transistors

The test results of ITFET in (a)~(d), the illustration of ITFET structure with graphene/h-BN heterostructure in (e)[196]

5.4 Summary and outlook

The band structure of the 2D material may be different from that of the parent material, so that the 2D material has the superior properties that the parent does not possess. On the other hand, 2D materials are more sensitive to external regulation than 3D materials. While the electrons in 2D material move in the plane, the movement of electrons in the magnetic field is also quantified to form the Landau energy level, to achieve quantum Hall effect. And when different 2D materials are stacked together to form a van der Waals heterogeneous structure, a strange nature of the new material was born.

The first representative of the van der Waals heterostructures is the graphene/h-BN heterostructures. When the same hexagonal lattice structures are stacked together, the unique moiré pattern is formed due to the difference in lattice or the presence of stacking angles. While the lattice is perfectly aligned, the moiré lattice is the largest, about 14 nm. This is also the most studied by people. Graphene electrons are also affected by the crystal grain potential. And in the original Dirac cone will appear under the degenerate second-order Dirac cone, and open the graphene band gap. The magnetic length of the electron cyclotron in the Mohr lattice is approximately equal to the size of the Mohr lattice, and the electron swing motion is also affected by the crystal field, thus observing the Hofstadter butterfly. The fractal energy spectrum of 2D electron gas is also observed in the heterostructure, that is, the fractal fraction quantum Hall effect. 2D h-BN, in addition to serving as a substrate, provides a uniform dielectric environment, especially in the configuration of encapsulated graphene. The research on optical properties, electrical properties, magnetism and spin electronics performance based on heterostructures will be recognized more[197-199].

With the improvement of the research level, the improvement of research methods and the input of more advanced laboratory equipment, more nano-level optoelectronic devices based on graphene/h-BN heterostructures, move and move spintronic devices will be developed and applied in real life.

References

[1] RAES B, SCHEERDER J E, COSTACHE M V, et al. Determination of the spin-lifetime anisotropy in graphene using oblique spin precession [J]. Nature Communications, 2016, 7(1): 1-7.

[2] KAMALAKAR M V, GROENVELD C, DANKERT A, et al. Long distance spin communication in chemical vapour deposited graphene[J]. Nature Communications, 2015, 6(1): 1-8.

[3] BARONE V, Peralta J E. Magnetic boron nitride nanoribbons with tunable electronic properties[J]. Nano Letters, 2008, 8 (8): 2210-2214.

[4] DANKERT A, DASH S P. Electrical gate control of spin current in van der Waals heterostructures at room temperature[J]. Nature Communications, 2017, 8(1): 1-6.

[5] AVSAR A, VERA-MARUN I J, TAN J Y, et al. Air-stable transport in graphene-contacted, fully encapsulated ultrathin black phosphorus-based field-effect transistors [J]. ACS Nano, 2015, 9(4): 4138-4145.

[6] IIJIMA S. Helical microtubules of graphitic carbon[J]. Nature, 1991, 354(6348): 56-58.

[7] NOVOSELOV K S, GEIM A K, MOROZOV S V, et al. Electric field effect in atomically thin carbon films[J]. Science, 2004, 306(5696): 666-669.

[8] LEE C, WEI X, KYSAR J W, et al. Measurement of the elastic properties and intrinsic strength of monolayer graphene[J]. Science, 2008, 321(5887): 385-388.

[9] ABBOTT'S I E. Graphene: exploring carbon flatland[J]. Phys. Today, 2007, 60(8): 35.

[10] NOVOSELOV K S, MOROZOV S V, MOHINDDIN T M G, et al. Electronic properties of graphene[J]. Physica Status Solidi (B), 2007, 244(11): 4106-4111.

[11] DRESSELHAUS M S, JORIO A, SAITO R. Characterizing graphene, graphite, and carbon nanotubes by Raman spectroscopy[J]. Annu. Rev. Condens. Matter Phys., 2010, 1(1): 89-108.

[12] NOVOSELOV K S, GEIM A K, MOROZOV S V, et al. Two-dimensional gas of massless Dirac fermions in graphene[J]. Nature, 2005, 438(7065): 197-200.

[13] ZHANG Y, TAN Y W, STORMER H L, et al. Experimental observation of the quantum Hall effect and Berry's phase in graphene[J]. Nature, 2005, 438(7065): 201-204.

[14] KATSNELSON M I, NOVOSELOV K S, GEIM A K. Chiral tunnelling and the Klein paradox in graphene[J]. Nature Physics, 2006, 2(9): 620-625.

[15] BEENAKKER C W J. Colloquium: andreev reflection and Klein tunneling in graphene [J]. Reviews of Modern Physics, 2008, 80(4): 1337.

[16] AVOURIS P, CHEN Z, PEREBEINOS V. Carbon-based electronics[M] Singapore: World Scintific, 2009.

[17] RYCERZ A, TWORZYD? O J, BEENAKKER C W J. Valley filter and valley valve in graphene[J]. Nature Physics, 2007, 3(3): 172-175.

[18] XIAO D, YAO W, NIU Q. Valley-contrasting physics in graphene: magnetic moment and topological transport[J]. Physical Review Letters, 2007, 99(23): 236809.

[19] ZHAI F, ZHAO X, CHANG K, et al. Magnetic barrier on strained graphene: a possible valley filter[J]. Physical Review B, 2010, 82(11): 115442.

[20] GUNLYCKE D, WHITE C T. Graphene valley filter using a line defect[J]. Physical Review Letters, 2011, 106(13): 136806.

[21] WU Z, ZHAI F, PEETERS F M, et al. Valley-dependent Brewster angles and Goos-Hänchen effect in strained graphene[J]. Physical Review Letters, 2011, 106(17): 176802.

[22] LI X, ZHANG F, NIU Q, et al. Spontaneous layer-pseudospin domain walls in bilayer graphene[J]. Physical Review Letters, 2014, 113(11): 116803.

[23] ASHCROFT N W, MERMIN N D. Solid state physics, holt, rinehart and winston[J]. Physik in Unserer Zt, 1978, 9(1): 33-33.

[24] JECKELMANN B, JEANNERET B. The quantum Hall effect as an electrical resistance standard[J]. Reports on Progress in Physics, 2001, 64(12): 1603.

[25] NOVOSELOV K S, JIANG Z, ZHANG Y, et al. Room-temperature quantum Hall effect in graphene[J]. Science, 2007, 315(5817): 1379-1379.

[26] NOVOSELOV K S, GEIM A K, MOROZOV S V, et al. Two-dimensional gas of massless Dirac fermions in graphene[J]. Nature, 2005, 438(7065): 197-200.

[27] DU X, SKACHKO I, DUERR F, et al. Fractional quantum Hall effect and insulating phase of Dirac electrons in graphene[J]. Nature, 2009, 462(7270): 192-195.

[28] SUN Q, XIE X C. C T-invariant quantum spin Hall effect in ferromagnetic graphene [J]. Physical Review Letters, 2010, 104(6): 066805.

[29] BERNEVIG B A, ZHANG S C. Quantum spin Hall effect[J]. Physical Review Letters, 2006, 96(10): 106802.

[30] KANE C L, MELE E J. Quantum spin Hall effect in graphene[J]. Physical Review Letters, 2005, 95(22): 226801.

[31] OHISHI M, SHIRAISHI M, NOUCHI R, et al. Spin injection into a graphene thin film at room temperature [J]. Japanese Journal of Applied Physics, 2007, 46(7L): L605.

[32] HAN W, PI K, MCCREARY K M, et al. Tunneling spin injection into single layer graphene[J]. Physical Review Letters, 2010, 105(16): 167202.

[33] ABANIN D A, MOROZOV S V, PONOMARENKO L A, et al. Giant nonlocality near the Dirac point in graphene[J]. Science, 2011, 332(6027): 328-330.

[34] TOMBROS N, JOZSA C, POPINCIUC M, et al. Electronic spin transport and spin precession in single graphene layers at room temperature [J]. Nature, 2007, 448(7153): 571-574.

[35] FRUCHART O, DIENY B. Magnetostatics of synthetic ferrimagnet elements[J]. Journal of Magnetism and Magnetic Materials, 2012, 324(4): 365-368.

[36] DANKERT A, KAMALAKAR M V, BERGSTEN J, et al. Spin transport and precession in graphene measured by nonlocal and three-terminal methods[J]. Applied Physics Letters, 2014, 104(19): 192403.

[37] KAMALAKAR M V, GROENVELD C, DANKERT A, et al. Long distance spin communication in chemical vapour deposited graphene[J]. Nature Communications, 2015, 6(1): 1-8.

[38] HAN W, KAWAKAMI R K, GMITRA M, et al. Graphene spintronics[J]. Nature Nanotechnology, 2014, 9(10): 794-807.

[39] HAN W, KAWAKAMI R K. Spin relaxation in single-layer and bilayer graphene[J]. Physical Review Letters, 2011, 107(4): 047207.

[40] JÓZSA C, MAASSEN T, POPINCIUC M, et al. Linear scaling between momentum and spin scattering in graphene[J]. Physical Review B, 2009, 80(24): 241403.

[41] VOLMER F, DRÖGELER M, MAYNICKE E, et al. Role of MgO barriers for spin and charge transport in Co/MgO/graphene nonlocal spin-valve devices[J]. Physical Review B, 2013, 88(16): 161405.

[42] ZOMER P J, GUIMARÃES M H D, TOMBROS N, et al. Long-distance spin transport in high-mobility graphene on hexagonal boron nitride[J]. Physical Review B, 2012, 86(16): 161416.

[43] PI K, HAN W, MCCREARY K M, et al. Manipulation of spin transport in graphene by surface chemical doping[J]. Physical Review Letters, 2010, 104(18): 187201.

[44] HAN W, CHEN J R, WANG D, et al. Spin relaxation in single-layer graphene with tunable mobility[J]. Nano Letters, 2012, 12(7): 3443-3447.

[45] RAES B, SCHEERDER J E, COSTACHE M V, et al. Determination of the spin-lifetime anisotropy in graphene using oblique spin precession[J]. Nature Communications, 2016, 7(1): 1-7.

[46] GARNICA M, STRADI D, BARJA S, et al. Long-range magnetic order in a purely organic 2D layer adsorbed on epitaxial graphene[J]. Nature Physics, 2013, 9(6): 368-374.

[47] WU M, LIU E Z, JIANG J Z. Magnetic behavior of graphene absorbed with N, O, and F atoms: A first-principles study[J]. Applied Physics Letters, 2008, 93(8): 082504.

[48] SEVINÇLI H, TOPSAKAL M, DURGUN E, et al. Electronic and magnetic properties of 3D transition-metal atom adsorbed graphene and graphene nanoribbons [J]. Physical Review B, 2008, 77(19): 195434.

[49] ZHOU J, WANG Q, SUN Q, et al. Ferromagnetism in semihydrogenated graphene sheet[J]. Nano Letters, 2009, 9(11): 3867-3870.

[50] XIE L, WANG X, LU J, et al. Room temperature ferromagnetism in partially hydrogenated epitaxial graphene[J]. Applied Physics Letters, 2011, 98(19): 193113.

[51] ELIAS D C, NAIR R R, MOHIUDDIN T M G, et al. Control of graphene's properties by reversible hydrogenation: evidence for graphane[J]. Science, 2009, 323(5914): 610-613.

[52] YI D, YANG L, XIE S, et al. Stability of hydrogenated graphene: a first-principles study[J]. RSC Advances, 2015, 5(26): 20617-20622.

[53] GONZÁLEZ-HERRERO H, GÓMEZ-RODRÓGUEZ J M, MALLET P, et al. Atomic-scale control of graphene magnetism by using hydrogen atoms[J]. Science, 2016, 352(6284): 437-441.

[54] SON Y W, COHEN M L, LOUIE S G. Half-metallic graphene nanoribbons[J]. Nature, 2006, 444(7117): 347-349.

[55] LEE H, SON Y W, PARK N, et al. Magnetic ordering at the edges of graphitic fragments: Magnetic tail interactions between the edge-localized states[J]. Physical Review B, 2005, 72(17): 174431.

[56] MAGDA G Z, JIN X, HAGYMÁSI I, et al. Room-temperature magnetic order on zigzag edges of narrow graphene nanoribbons[J]. Nature, 2014, 514(7524): 608-611.

[57] FUJITA M, WAKABAYASHI K, NAKADA K, et al. Peculiar localized state at zigzag graphite edge[J]. Journal of the Physical Society of Japan, 1996, 65(7): 1920-1923.

[58] PISANI L, CHAN J A, MONTANARI B, et al. Electronic structure and magnetic properties of graphitic ribbons[J]. Physical Review B, 2007, 75(6): 064418.

[59] PHILLIPS L C, LOMBARDO A, GHIDINI M, et al. Tunnelling anisotropic magnetoresistance at $La_{0.67}Sr_{0.33}MnO_3$-graphene interfaces[J]. Applied Physics

Letters, 2016, 108(11): 112405.
[60] WANG X, OUYANG Y, LI X, et al. Room-temperature all-semiconducting sub-10-nm graphene nanoribbon field-effect transistors[J]. Physical Review Letters, 2008, 100(20): 206803.
[61] DANKERT A, DASH S P. Electrical gate control of spin current in van der Waals heterostructures at room temperature[J]. Nature Communications, 2017, 8(1): 1-6.
[62] CANDINI A, KLYATSKAYA S, RUBEN M, et al. Graphene spintronic devices with molecular nanomagnets[J]. Nano Letters, 2011, 11(7): 2634-2639.
[63] AZEVEDO S, KASCHNY J R, DE CASTILHO C M C, et al. A theoretical investigation of defects in a boron nitride monolayer[J]. Nanotechnology, 2007, 18(49): 495707.
[64] EKUMA C E, DOBROSAVLJEVIÇ V, GUNLYCKE D. First-principles-based method for electron localization: application to monolayer hexagonal boron nitride[J]. Physical Review Letters, 2017, 118(10): 106404.
[65] TOPSAKAL M, AKTÜRK E, CIRACI S. First-principles study of two-and one-dimensional honeycomb structures of boron nitride[J]. Physical Review B, 2009, 79(11): 115442.
[66] PAKDEL A, ZHI C, BANDO Y, et al. Low-dimensional boron nitride nanomaterials [J]. Materials Today, 2012, 15(6): 256-265.
[67] YU W J, LAU W M, CHAN S P, et al. Ab initio study of phase transformations in boron nitride[J]. Physical Review B, 2003, 67(1): 014108.
[68] XU Y N, CHING W Y. Calculation of ground-state and optical properties of boron nitrides in the hexagonal, cubic, and wurtzite structures[J]. Physical Review B, 1991, 44(15): 7787.
[69] PARK C H, LOUIE S G. Energy gaps and stark effect in boron nitride nanoribbons [J]. Nano Letters, 2008, 8(8): 2200-2203.
[70] CHEN Z G, ZOU J, LIU G, et al. Novel boron nitride hollow nanoribbons[J]. ACS Nano, 2008, 2(10): 2183-2191.
[71] NOVOSELOV K S, JIANG D, SCHEDIN F, et al. Two-dimensional atomic crystals [J]. Proceedings of the National Academy of Sciences, 2005, 102(30): 10451-10453.
[72] HAN W Q, WU L, ZHU Y, et al. Structure of chemically derived mono-and few-atomic-layer boron nitride sheets[J]. Applied Physics Letters, 2008, 93(22): 223103.
[73] OUYANG B, SONG J. Strain engineering of magnetic states of vacancy-decorated hexagonal boron nitride[J]. Applied Physics Letters, 2013, 103(10): 102401.
[74] DU A, CHEN Y, ZHU Z, et al. Dots versus antidots: computational exploration of structure, magnetism, and half-metallicity in boron-nitride nanostructures[J]. Journal of the American Chemical Society, 2009, 131(47): 17354-17359.
[75] SI M S, XUE D S. Magnetic properties of vacancies in a graphitic boron nitride sheet by first-principles pseudopotential calculations [J]. Physical Review B, 2007, 75(19): 193409.

[76] MACHADO-CHARRY E, BOULANGER P, GENOVESE L, et al. Tunable magnetic states in hexagonal boron nitride sheets [J]. Applied Physics Letters, 2012, 101(13): 132405.

[77] YANG J H, KIM D, HONG J, et al. Magnetism in boron nitride monolayer: adatom and vacancy defect[J]. Surface Science, 2010, 604(19-20): 1603-1607.

[78] WU R Q, PENG G W, LIU L, et al. Possible graphitic-boron-nitride-based metal-free molecular magnets from first principles study[J]. Journal of Physics: Condensed Matter, 2005, 18(2): 569.

[79] SONG L L, ZHENG X H, HAO H, et al. Tuning the electronic and magnetic properties in zigzag boron nitride nanoribbons with carbon dopants[J]. Computational Materials Science, 2014, 81: 551-555.

[80] ZHOU J, WANG Q, SUN Q, et al. Electronic and magnetic properties of a BN sheet decorated with hydrogen and fluorine[J]. Physical Review B, 2010, 81(8): 085442.

[81] RADHAKRISHNAN S, DAS D, SAMANTA A, et al. Fluorinated h-BN as a magnetic semiconductor[J]. Science Advances, 2017, 3(7): e1700842.

[82] ZHENG F, ZHOU G, LIU Z, et al. Half metallicity along the edge of zigzag boron nitride nanoribbons[J]. Physical Review B, 2008, 78(20): 205415.

[83] JOSHI N, GHOSH P. Substrate-induced changes in the magnetic and electronic properties of hexagonal boron nitride[J]. Physical Review B, 2013, 87(23): 235440.

[84] PIQUEMAL-BANCI M, GALCERAN R, CANEVA S, et al. Magnetic tunnel junctions with monolayer hexagonal boron nitride tunnel barriers[J]. Applied Physics Letters, 2016, 108(10): 102404.

[85] DANKERT A, KAMALAKAR M V, WAJID A, et al. Tunnel magnetoresistance with atomically thin two-dimensional hexagonal boron nitride barriers [J]. Nano Research, 2015, 8(4): 1357-1364.

[86] YAZYEV O V, PASQUARELLO A. Magnetoresistive junctions based on epitaxial graphene and hexagonal boron nitride[J]. Physical Review B, 2009, 80(3): 035408.

[87] KARPAN V M, KHOMYAKOV P A, GIOVANNETTI G, et al. Ni (111) graphene h-BN junctions as ideal spin injectors[J]. Physical Review B, 2011, 84(15): 153406.

[88] FU W, MAKK P, MAURAND R, et al. Large-scale fabrication of BN tunnel barriers for graphene spintronics[J]. Journal of Applied Physics, 2014, 116(7): 074306.

[89] KARPAN V M, KHOMYAKOV P A, GIOVANNETTI G, et al. Ni (111) graphene h-BN junctions as ideal spin injectors[J]. Physical Review B, 2011, 84(15): 153406.

[90] GIOVANNETTI G, KHOMYAKOV P A, BROCKS G, et al. Substrate-induced band gap in graphene on hexagonal boron nitride: ab initio density functional calculations [J]. Physical Review B, 2007, 76(7): 073103.

[91] DEAN C R, YOUNG A F, MERIC I, et al. Boron nitride substrates for high-quality graphene electronics[J]. Nature Nanotechnology, 2010, 5(10): 722-726.

[92] XUE J, SANCHEZ-YAMAGISHI J, BULMASH D, et al. Scanning tunnelling microscopy and spectroscopy of ultra-flat graphene on hexagonal boron nitride[J].

Nature Materials, 2011, 10(4): 282-285.
[93] MAYOROV A S, GORBACHEV R V, MOROZOV S V, et al. Micrometer-scale ballistic transport in encapsulated graphene at room temperature[J]. Nano Letters, 2011, 11(6): 2396-2399.
[94] LIU Z, SONG L, ZHAO S, et al. Direct growth of graphene/hexagonal boron nitride stacked layers[J]. Nano Letters, 2011, 11(5): 2032-2037.
[95] TANG S, DING G, XIE X, et al. Nucleation and growth of single crystal graphene on hexagonal boron nitride[J]. Carbon, 2012, 50(1): 329-331.
[96] ZHANG A, TEOH H F, DAI Z, et al. Band gap engineering in graphene and hexagonal BN antidot lattices: a first principles study[J]. Applied Physics Letters, 2011, 98(2): 023105.
[97] MOON P, KOSHINO M. Electronic properties of graphene/hexagonal-boron-nitride moiré superlattice[J]. Physical Review B, 2014, 90(15): 155406.
[98] WANG E, LU X, DING S, et al. Gaps induced by inversion symmetry breaking and second-generation Dirac cones in graphene/hexagonal boron nitride[J]. Nature Physics, 2016, 12(12): 1111-1115.
[99] DEAN C R, WANG L, MAHER P, et al. Hofstadter's butterfly and the fractal quantum Hall effect in moiré superlattices[J]. Nature, 2013, 497(7451): 598-602.
[100] ZHONG X, YAP Y K, PANDEY R, et al. First-principles study of strain-induced modulation of energy gaps of graphene/BN and BN bilayers[J]. Physical Review B, 2011, 83(19): 193403.
[101] FAN Y, ZHAO M, WANG Z, et al. Tunable electronic structures of graphene/boron nitride heterobilayers[J]. Applied Physics Letters, 2011, 98(8): 083103.
[102] SAN-JOSE P, GUTIÉRREZ-RUBIO A, STURLA M, et al. Electronic structure of spontaneously strained graphene on hexagonal boron nitride[J]. Physical Review B, 2014, 90(11): 115152.
[103] HUNT B, SANCHEZ-YAMAGISHI J D, YOUNG A F, et al. Massive Dirac fermions and Hofstadter butterfly in a van der Waals heterostructure[J]. Science, 2013, 340(6139): 1427-1430.
[104] ZHOU S, HAN J, DAI S, et al. van der Waals bilayer energetics: generalized stacking-fault energy of graphene, boron nitride, and graphene/boron nitride bilayers [J]. Physical Review B, Condensed Matter and Materials Physics, 2015, 92(15): 155438.1-155438.13.
[105] SLOTMAN G J, DE WIJS G A, FASOLINO A, et al. Phonons and electron-phonon coupling in graphene-h-BN heterostructures[J]. Annalen Der Physik, 2014, 526(9-10): 381-386.
[106] ARGENTERO G, MITTELBERGER A, REZA A M M, et al. Unraveling the 3D atomic structure of a suspended graphene/hBN van der Waals heterostructure[J]. Nano Letters, 2017, 17(3): 1409-1416.
[107] CHEN Z G, SHI Z, YANG W, et al. Observation of an intrinsic bandgap and

Landau level renormalization in graphene/boron-nitride heterostructures[J]. Nature Communications, 2014, 5: 4461.

[108] ARGENTERO G, MITTELBERGER A, REZA A M M, et al. Unraveling the 3D atomic structure of a suspended graphene/hBN van der Waals heterostructure[J]. Nano Letters, 2017, 17(3): 1409-1416.

[109] SLAWIŃSKA J, ZASADA I, KLUSEK Z. Energy gap tuning in graphene on hexagonal boron nitride bilayer system[J]. Physical Review B, 2010, 81(15): 155433.

[110] BALU R, ZHONG X, PANDEY R, et al. Effect of electric field on the band structure of graphene/boron nitride and boron nitride/boron nitride bilayers[J]. Applied Physics Letters, 2012, 100(5): 052104.

[111] KAN E, REN H, WU F, et al. Why the band gap of graphene is tunable on hexagonal boron nitride[J]. The Journal of Physical Chemistry C, 2012, 116(4): 3142-3146.

[112] BRUGGER T, GÜNTHER S, WANG B, et al. Comparison of electronic structure and template function of single-layer graphene and a hexagonal boron nitride nanomesh on Ru (0001)[J]. Physical Review B, 2009, 79(4): 045407.

[113] BJELKEVIG C, MI Z, XIAO J, et al. Electronic structure of a graphene/hexagonal-BN heterostructure grown on Ru (0001) by chemical vapor deposition and atomic layer deposition: extrinsically doped graphene[J]. Journal of Physics: Condensed Matter, 2010, 22(30): 302002.

[114] TANG S, WANG H, ZHANG Y, et al. Precisely aligned graphene grown on hexagonal boron nitride by catalyst free chemical vapor deposition[J]. Scientific Reports, 2013, 3: 2666.

[115] YANG W, CHEN G, SHI Z, et al. Epitaxial growth of single-domain graphene on hexagonal boron nitride[J]. Nature Materials, 2013, 12(9): 792-797.

[116] SONG X, GAO T, NIE Y, et al. Seed-assisted growth of single-crystalline patterned graphene domains on hexagonal boron nitride by chemical vapor deposition[J]. Nano Letters, 2016, 16(10): 6109-6116.

[117] TANG S, WANG H, WANG H S, et al. Silane-catalysed fast growth of large single-crystalline graphene on hexagonal boron nitride[J]. Nature Communications, 2015, 6(1): 1-7.

[118] MENG J, ZHANG X, WANG Y, et al. Aligned growth of millimeter-size hexagonal boron nitride single-crystal domains on epitaxial nickel thin film[J]. Small, 2017, 13(18): 1604179.

[119] HÜSER F, OLSEN T, THYGESEN K S. Quasiparticle GW calculations for solids, molecules, and two-dimensional materials[J]. Physical Review B, 2013, 87(23): 235132.

[120] OSHIMA C, ITOH A, ROKUTA E, et al. A hetero-epitaxial-double-atomic-layer system of monolayer graphene/monolayer h-BN on Ni (111)[J]. Solid State Communications, 2000, 116(1): 37-40.

[121] OKADA S, OSHIYAMA A. Magnetic ordering in hexagonally bonded sheets with first-row elements[J]. Physical Review Letters, 2001, 87(14): 146803.

[122] CHOI J, KIM Y H, CHANG K J, et al. Itinerant ferromagnetism in heterostructured C/BN nanotubes[J]. Physical Review B, 2003, 67(12): 125421.

[123] RAMASUBRAMANIAM A, NAVEH D. Carrier-induced antiferromagnet of graphene islands embedded in hexagonal boron nitride[J]. Physical Review B, 2011, 84(7): 075405.

[124] BERSENEVA N, KRASHENINNIKOV A V, NIEMINEN R M. Mechanisms of postsynthesis doping of boron nitride nanostructures with carbon from first-principles simulations[J]. Physical Review Letters, 2011, 107(3): 035501.

[125] DING X, DING G, XIE X, et al. Direct growth of few layer graphene on hexagonal boron nitride by chemical vapor deposition[J]. Carbon, 2011, 49(7): 2522-2525.

[126] WANG P, CHENG B, MARTYNOV O, et al. Topological winding number change and broken inversion symmetry in a Hofstadter's butterfly[J]. Nano Letters, 2015, 15(10): 6395-6399.

[127] WANG L, GAO Y, WEN B, et al. Evidence for a fractional fractal quantum Hall effect in graphene superlattices[J]. Science, 2015, 350(6265): 1231-1234.

[128] DEAN C R, YOUNG A F, CADDEN-ZIMANSKY P, et al. Multicomponent fractional quantum Hall effect in graphene[J]. Nature Physics, 2011, 7(9): 693-696.

[129] NEUMANN C, REICHARDT S, DROÖGELER M, et al. Low B field magneto-phonon resonances in single-layer and bilayer graphene[J]. Nano Letters, 2015, 15(3): 1547-1552.

[130] KUMAR R K, CHEN X, AUTON G H, et al. High-temperature quantum oscillations caused by recurring Bloch states in graphene superlattices[J]. Science, 2017, 357(6347): 181-184.

[131] GOPINADHAN K, SHIN Y J, JALIL R, et al. Extremely large magnetoresistance in few-layer graphene/boron-nitride heterostructures[J]. Nature Communications, 2015, 6(1): 1-7.

[132] KATSNELSON M I, NOVOSELOV K S, GEIM A K. Chiral tunnelling and the Klein paradox in graphene[J]. Nature Physics, 2006, 2(9): 620-625.

[133] WANG F, ZHANG Y, TIAN C, et al. Gate-variable optical transitions in graphene[J]. Science, 2008, 320(5873): 206-209.

[134] LI Z Q, HENRIKSEN E A, JIANG Z, et al. Dirac charge dynamics in graphene by infrared spectroscopy[J]. Nature Physics, 2008, 4(7): 532-535.

[135] MIN H, BORGHI G, POLINI M, et al. Pseudospin magnetism in graphene[J]. Physical Review B, 2008, 77(4): 041407.

[136] JUNG J, ZHANG F, MACDONALD A H. Lattice theory of pseudospin ferromagnetism in bilayer graphene: competing interaction-induced quantum Hall states[J]. Physical Review B, 2011, 83(11): 115408.

[137] SAN-JOSE P, PRADA E, MCCANN E, et al. Pseudospin valve in bilayer graphene:

[138] towards graphene-based pseudospintronics [J]. Physical Review Letters, 2009, 102(24): 247204.

[138] PARK C H, YANG L, SON Y W, et al. New generation of massless Dirac fermions in graphene under external periodic potentials[J]. Physical Review Letters, 2008, 101(12): 126804.

[139] LUI C H, LI Z, MAK K F, et al. Observation of an electrically tunable band gap in trilayer graphene[J]. Nature Physics, 2011, 7(12): 944-947.

[140] SHI Z, JIN C, YANG W, et al. Gate-dependent pseudospin mixing in graphene/boron nitride moiré superlattices[J]. Nature Physics, 2014, 10(10): 743-747.

[141] LEHTINEN P O, FOSTER A S, MA Y, et al. Irradiation-induced magnetism in graphite: a density functional study[J]. Physical Review Letters, 2004, 93(18): 187202.

[142] FERNÁNDEZ-ROSSIER J, PALACIOS J J. Magnetism in graphene nanoislands[J]. Physical Review Letters, 2007, 99(17): 177204.

[143] DING X, SUN H, XIE X, et al. Anomalous paramagnetism in graphene on hexagonal boron nitride substrates[J]. Physical Review B, 2011, 84(17): 174417.

[144] ANDO T. Magnetic oscillation of optical phonon in graphene[J]. Journal of the Physical Society of Japan, 2007, 76(2): 024712-024712.

[145] CONG C, JUNG J, CAO B, et al. Magnetic oscillation of optical phonon in ABA- and ABC-stacked trilayer graphene[J]. Physical Review B, 2015, 91(23): 235403.

[146] GOERBIG M O, FUCHS J N, KECHEDZHI K, et al. Filling-factor-dependent magnetophonon resonance in graphene[J]. Physical Review Letters, 2007, 99(8): 087402.

[147] QIU C, SHEN X, CAO B, et al. Strong magnetophonon resonance induced triple G-mode splitting in graphene on graphite probed by micromagneto Raman spectroscopy [J]. Physical Review B, 2013, 88(16): 165407.

[148] BERCIAUD S, POTEMSKI M, FAUGERAS C. Probing electronic excitations in mono-to pentalayer graphene by micro magneto-Raman spectroscopy [J]. Nano Letters, 2014, 14(8): 4548-4553.

[149] FAUGERAS C, KOSSACKI P, BASKO D M, et al. Effect of a magnetic field on the two-phonon Raman scattering in graphene [J]. Physical Review B, 2010, 81(15):155436.

[150] JU L, VELASCO J, HUANG E, et al. Photoinduced doping in heterostructures of graphene and boron nitride[J]. Nature Nanotechnology, 2014, 9(5): 348-352.

[151] NEUMANN C, REICHARDT S, DROÖGELER M, et al. Low B field magneto-phonon resonances in single-layer and bilayer graphene[J]. Nano Letters, 2015, 15(3): 1547-1552.

[152] MAYOROV A S, GORBACHEV R V, MOROZOV S V, et al. Micrometer-scale ballistic transport in encapsulated graphene at room temperature[J]. Nano Letters, 2011, 11(6): 2396-2399.

[153] BRITNELL L, GORBACHEV R V, JALIL R, et al. Electron tunneling through ultrathin boron nitride crystalline barriers[J]. Nano Letters, 2012, 12(3): 1707-1710.

[154] YAMAGUCHI T, INOUE Y, MASUBUCHI S, et al. Electrical spin injection into graphene through monolayer hexagonal boron nitride[J]. Applied Physics Express, 2013, 6(7): 073001.

[155] KAMALAKAR M V, DANKERT A, BERGSTEN J, et al. Enhanced tunnel spin injection into graphene using chemical vapor deposited hexagonal boron nitride[J]. Scientific Reports, 2014, 4: 6146.

[156] WEN H, DERY H, AMAMOU W, et al. Experimental demonstration of XOR operation in graphene magnetologic gates at room temperature[J]. Physical Review Applied, 2016, 5(4): 044003.

[157] MAASSEN T, VERA-MARUN I J, GUIMARÃES M H D, et al. Contact-induced spin relaxation in Hanle spin precession measurements[J]. Physical Review B, 2012, 86(23): 235408.

[158] WU Q, SHEN L, BAI Z, et al. Efficient spin injection into graphene through a tunnel barrier: overcoming the spin-conductance mismatch[J]. Physical Review Applied, 2014, 2(4): 044008.

[159] LAZIĆ P, BELASHCHENKO K D, ŽUTIĆ I. Effective gating and tunable magnetic proximity effects in two-dimensional heterostructures[J]. Physical Review B, 2016, 93(24): 241401.

[160] GURRAM M, OMAR S, VAN WEES B J. Bias induced up to 100% spin-injection and detection polarizations in ferromagnet/bilayer-hBN/graphene/hBN heterostructures[J]. Nature Communications, 2017, 8(1): 1-7.

[161] GUIMARÃES M H D, ZOMER P J, INGLA-AYNÉS J, et al. Controlling spin relaxation in hexagonal BN-encapsulated graphene with a transverse electric field[J]. Physical Review Letters, 2014, 113(8): 086602.

[162] ZOMER P J, GUIMARÃES M H D, TOMBROS N, et al. Long-distance spin transport in high-mobility graphene on hexagonal boron nitride[J]. Physical Review B, 2012, 86(16): 161416.

[163] SAFEER C K, ONTOSO N, INGLA-AYNÉS J, et al. Large multidirectional spin-to-charge conversion in low-symmetry semimetal $MoTe_2$ at room temperature[J]. Nano Letters, 2019, 19(12): 8758-8766.

[164] KAMALAKAR M V, DANKERT A, KELLY P J, et al. Inversion of spin signal and spin filtering in ferromagnet | hexagonal boron nitride-graphene van der Waals heterostructures[J]. Scientific Reports, 2016, 6(1): 1-9.

[165] OCHOA H, NETO A H C, GUINEA F. Elliot-Yafet mechanism in graphene[J]. Physical Review Letters, 2012, 108(20): 206808.

[166] KOCHAN D, GMITRA M, FABIAN J. Spin relaxation mechanism in graphene: resonant scattering by magnetic impurities[J]. Physical Review Letters, 2014,

112(11): 116602.
[167] GURRAM M, OMAR S, ZIHLMANN S, et al. Spin transport in fully hexagonal boron nitride encapsulated graphene[J]. Physical Review B, 2016, 93(11): 115441.
[168] OCHOA H, NETO A H C, GUINEA F. Elliot-Yafet mechanism in graphene[J]. Physical Review Letters, 2012, 108(20): 206808.
[169] KOCHAN D, GMITRA M, FABIAN J. Spin relaxation mechanism in graphene: resonant scattering by magnetic impurities[J]. Physical Review Letters, 2014, 112(11): 116602.
[170] GUIMARÃES M H D, VELIGURA A, ZOMER P J, et al. Spin transport in high-quality suspended graphene devices[J]. Nano Letters, 2012, 12(7): 3512-3517.
[171] WANG L, MERIC I, HUANG P Y, et al. One-dimensional electrical contact to a two-dimensional material[J]. Science, 2013, 342(6158): 614-617.
[172] DROÖGELER M, VOLMER F, WOLTER M, et al. Nanosecond spin lifetimes in single-and few-layer graphene-hBN heterostructures at room temperature[J]. Nano Letters, 2014, 14(11): 6050-6055.
[173] SINGH S, KATOCH J, XU J, et al. Nanosecond spin relaxation times in single layer graphene spin valves with hexagonal boron nitride tunnel barriers[J]. Applied Physics Letters, 2016, 109(12): 122411.
[174] DROÖGELER M, FRANZEN C, VOLMER F, et al. Spin lifetimes exceeding 12 ns in graphene nonlocal spin valve devices[J]. Nano Letters, 2016, 16(6): 3533-3539.
[175] KAMALAKAR M V, DANKERT A, BERGSTEN J, et al. Spintronics with graphene-hexagonal boron nitride van der Waals heterostructures[J]. Applied Physics Letters, 2014, 105(21): 212405.
[176] LI L, YU Y, YE G J, et al. Black phosphorus field-effect transistors[J]. Nature Nanotechnology, 2014, 9(5): 372.
[177] KOENIG S P, DOGANOV R A, SCHMIDT H, et al. Electric field effect in ultrathin black phosphorus[J]. Applied Physics Letters, 2014, 104(10): 103106.
[178] XIA F, WANG H, JIA Y. Rediscovering black phosphorus as an anisotropic layered material for optoelectronics and electronics[J]. Nature Communications, 2014, 5(1): 1-6.
[179] LIU H, NEAL A T, ZHU Z, et al. Phosphorene: an unexplored 2D semiconductor with a high hole mobility[J]. ACS Nano, 2014, 8(4): 4033-4041.
[180] BUSCEMA M, GROENENDIJK D J, BLANTER S I, et al. Fast and broadband photoresponse of few-layer black phosphorus field-effect transistors[J]. Nano Letters, 2014, 14(6): 3347-3352.
[181] CHEN J R, ODENTHAL P M, SWARTZ A G, et al. Control of Schottky barriers in single layer MoS_2 transistors with ferromagnetic contacts[J]. Nano Letters, 2013, 13(7): 3106-3110.
[182] ENGLISH C D, SHINE G, DORGAN V E, et al. Improving contact resistance in MoS_2 field effect transistors[C]. 72nd Device Research Conference. IEEE, 2014.

[183] LATE D J, LIU B, MATTE H S S R, et al. Hysteresis in single-layer MoS_2 field effect transistors[J]. ACS Nano, 2012, 6(6): 5635-5641.

[184] AVSAR A, VERA-MARUN I J, TAN J Y, et al. Air-stable transport in graphene-contacted, fully encapsulated ultrathin black phosphorus-based field-effect transistors [J]. ACS Nano, 2015, 9(4): 4138-4145.

[185] ZWANZIGER J W, KOENIG M, PINES A. Berry's phase[J]. Annual Review of Physical Chemistry, 1990, 41(1): 601-646.

[186] XIAO D, CHANG M C, NIU Q. Berry phase effects on electronic properties[J]. Reviews of Modern Physics, 2010, 82(3): 1959.

[187] VELASCO JR J, JU L, WONG D, et al. Nanoscale control of rewriteable doping patterns in pristine graphene/boron nitride heterostructures[J]. Nano Letters, 2016, 16(3): 1620-1625.

[188] GHAHARI F, WALKUP D, GUTIÉRREZ C, et al. An on/off Berry phase switch in circular graphene resonators[J]. Science, 2017, 356(6340): 845-849.

[189] SCHWIERZ F. Graphene transistors[J]. Nature Nanotechnology, 2010, 5(7): 487.

[190] HAN S J, JENKINS K A, VALDES G A, et al. High-frequency graphene voltage amplifier[J]. Nano Letters, 2011, 11(9): 3690-3693.

[191] LIN Y M, VALDES-GARCIA A, HAN S J, et al. Wafer-scale graphene integrated circuit[J]. Science, 2011, 332(6035): 1294-1297.

[192] BRITNELL L, GORBACHEV R V, JALIL R, et al. Field-effect tunneling transistor based on vertical graphene heterostructures[J]. Science, 2012, 335(6071): 947-950.

[193] BRITNELL L, GORBACHEV R V, JALIL R, et al. Electron tunneling through ultrathin boron nitride crystalline barriers[J]. Nano Letters, 2012, 12(3): 1707-1710.

[194] BRITNELL L, GORBACHEV R V, GEIM A K, et al. Resonant tunnelling and negative differential conductance in graphene transistors[J]. Nature Communications, 2013, 4(1): 1-5.

[195] WANG L, MERIC I, HUANG P Y, et al. One-dimensional electrical contact to a two-dimensional material[J]. Science, 2013, 342(6158): 614-617.

[196] KANG S, PRASAD N, MOVVA H C P, et al. Effects of electrode layer band structure on the performance of multilayer graphene-hBN-graphene interlayer tunnel field effect transistors[J]. Nano Letters, 2016, 16(8): 4975-4981.

[197] WANG J, MA F, LIANG W, et al. Electrical properties and applications of graphene, hexagonal boron nitride (h-BN), and graphene/h-BN heterostructures[J]. Materials Today Physics, 2017, 2: 6-34.

[198] WANG J, MA F, LIANG W, et al. Optical, photonic and optoelectronic properties of graphene, h-BN and their hybrid materials[J]. Nanophotonics, 2017, 6(5): 943-976.

[199] WANG J, MA F, SUN M. Graphene, hexagonal boron nitride, and their heterostructures: properties and applications[J]. RSC Advances, 2017, 7(27): 16801-16822.

Chapter 6

The Thermal and Thermoelectric Properties of In-Plane C-BN Hybrid Structures and Graphene/h-BN van der Waals Heterostructures

In microelectronics, the thermophysical properties of nanomaterials have an important influence on the performance of microelectronic devices. The thermal conductivity (TC), thermal resistance (TR), thermal rectification, and thermoelectric effects of nanomaterials all affect the performance of microelectronic devices. Among the two dimensional (2D) nanomaterials, graphene and h-BN have broad application prospects in micro-nanoscale device preparation due to their excellent mechanical, electrical, and thermal properties, especially when they are combined to form a planar heterojunction and vertical stacking of van der Waals heterostructures. Thermal properties such as TC(k), TR, thermal rectification, and thermoelectric effect (ZT) have been shown to be significantly different from graphene and h-BN materials in these composite systems.

On the other hand, thermoelectric devices are distinguished from traditional refrigeration and power generation devices by virtue of their low pollution, stable operation, and low maintenance cost. The quality factor is an important parameter to measure the performance of thermoelectric devices. It is expressed as[1]:

$$ZT = \frac{S^2 \sigma}{k} T, \tag{6-1}$$

where, S is the Seebeck coefficient, σ is the electrical conductivity of the thermoelectric material, T is the absolute temperature, and k is the TC. Therefore, to improve the thermoelectric conversion efficiency of a thermoelectric device, a large Seebeck coefficient S, a high electrical conductivity σ, and a low TC k are required. This requires the reduction of the TC of the material as much as possible without affecting the electrical conductivity, thereby improving the quality of the thermoelectric device. In addition, heat capacity (TC) and thermal interface resistance are all factors that affect the thermoelectric material[1-2]. Both graphene and 2D hexagonal boron nitride (h-BN) have high TC and other physicochemical properties required as thermoelectric materials.

This article focuses on the TC, TR, and thermoelectric properties and applications of in-plane graphene/h-BN hybrid heterojunction and vertical graphene/h-BN van der Waals heterostructure, revealsing the nature of the thermal properties of 2D nanomaterials in the micro-nanoscale. In addition, we review their thermal properties, TR thermal resistance, thermal rectification, thermoelectric properties, and latest applications.

6.1 2D nanomaterials: graphene and h-BN

6.1.1 Structure and thermal properties of graphene

The ideal graphene is a 2D crystal with a honeycomb-like hexagonal configuration with a single-atomic-layer thickness. The 2s electrons of each carbon atom and the $2p_x$ and $2p_y$ electrons form σ bonds by sp^2 orbital hybridization. The $3p_z$ orbital electrons that do not participate in the hybrid form large π bonds delocalized perpendicular to this plane[3-9]. A strong interaction between the carbon atoms when subjected to external forces maintains the original hexagonal structure, and this stable structure also gives graphene good TC. The graphene electronic structure under tight binding depends mainly on the electrons of the π orbital[10-12]. A atoms and B atoms are two kinds of carbon atoms that are not equivalent to graphene (Figure 6-1(a)). The bond length was 1.420 Å and the lattice constant was 2.460 Å[13-15]. Based on the tight bond approximation, the band structure of graphene was obtained (Figure 6-1(b)). The band diagram directly reflects that the graphene is a zero-band-gap 2D crystal.

Such unique electronic structure gives graphene peculiar thermal characteristics. Studies have shown that the TC of graphene is as high as 3,500 ~ 5,300 W/m · K[16,17]. The average free path of phonons at room temperature is

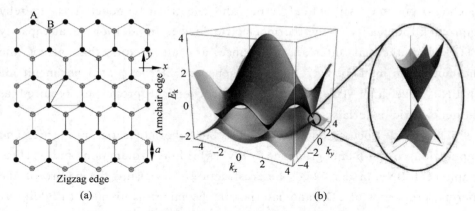

Figure 6-1　The graphene nanoribbons and band structure

(a) The graphene nanoribbons with zigzag edge and armchair edge[12]. (b) The band structure of graphene. The inset show the Dirac point, where is the intersection of the conduction band and the valence band[12]

775 nm, which is the main contribution to the thermal conduction of graphene. The interfacial scattering effect of the substrate in graphene leads to a decrease in the TC of the olefins[18-21]. Theoretical studies have shown that the high TC of graphene comes from the ballistic transport of its phonons, and its TC decreases with the increase in the number of graphene layers[17,22-25]. The thermal properties of graphene are affected by the influence of the boundary configuration and size[26-27]. In the micro-nanoscale range, the negative differential TC, TR, thermal rectification effect, and thermoelectric characteristics of graphene undergo great changes[28-30].

6.1.2　Structure and thermal properties of h-BN

The 2D h-BN belongs to the hexagonal system. B atoms and N atoms form an extremely strong covalent bond in the layer by sp^2 hybridization to form a regular honeycomb hexagonal structure[31-32]. The B atoms and N atoms in single-layer h-BN are alternately arranged. The B-N bond length is 0.144 nm, the lattice constant is 0.2504 nm, the interlayer spacing is 0.333 nm, and the layers are connected by weak van der Waals forces (Figure 6-2(a))[32-37]. The layers easily peel, are non-conductive, and have a wide band gap (5.1 eV, Figure 6-2(b)), Mohs hardness of 2, high melting point, and high-temperature reaching 2,000℃. Thermal performance along the c axis is better. As the number of layers increases, the TC decreases because of the scattering of phonons due to the layered structure.

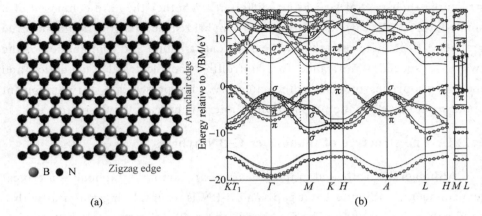

Figure 6-2 Structure electronic band of 2D h-BN
(a) The structure of 2D h-BN with zigzag edge and armchair edge[32]; (b) The electronic band structure along high-symmetry lines for bulk h-BN[37]

In 2D nanomaterials, the TC of h-BN is second only to that of graphene. Studies have shown that a single layer of h-BN at room temperature has a TC up to 600W/m · K[38]. When the number of layers increases to more than five, the TC of 2D h-BN is 250 W/m · K[39-42]. This value is close to the bulk TC of h-BN[43]. In addition, in the nanoscale range, the thermal properties of h-BN are greatly affected by the size effect and the boundary effect. The thermal rectification effect of the triangular vacancy defects in the h-BN nanoribbons (BNNRs) also affects the TC of h-BN[44]. The TC of the zigzag BNNRs is 20 % higher than that of the armchair-BNNRs[45]. In the thermoelectric effect, BNNRs are used for ballistic thermal power transmission. The transport properties are significantly higher than those of graphene nanoribbons (GNRs), and the lattice defects and boundary chirality all affect the thermoelectric properties of h-BN[46-50].

6.2 In-plane C-BN hybrid structure

The electronic properties of C atoms, N atoms, and B atoms determine the physical and chemical properties of both graphene and 2D h-BN. There is a strong ionic interaction between the B atoms and the N atoms, and the C atoms are covalently linked. Although the former two have the same lattice structure, their crystal lattices are different. Whether they are composed of a single-layer C-BN(carbon boron nitride) hybrid structure or a double-layered graphene/h-BN heterostructure, the different structure and combination of the two have special significance for controlling the electrical, magnetic, and thermal properties of composite materials. This possesses a very important application value for the preparation of micro-nanoscale functional devices.

6.2.1 The structure of monolayer C-BN hybrids

Although 2D h-BN and graphene lattice are similar, graphene is a typical semiconductor without a band gap, and h-BN is an insulating material with a band gap of 5.97 eV[51]. The C-C bond length (0.142 nm) is slightly smaller than the B-N bond length (0.144 nm), resulting in a lattice mismatch of only 1.7 %. Due to the lattice mismatch between the two, they display different behavior from h-BN and graphene, when both of them are complexed as a monolayer C-BN hybrid structure in physical characteristics[51-81].

1. In-plane monolayer C-BN hybrid structures

No matter how the system is combined, the appearance of such superlattices contains two types of interfaces, B/C or N/C. These two different interfaces play a key role in regulating the electronic properties of the system[51-81]. Theoretical work shows that when BNNRs are embedded in zigzag GNRs to form a monolayer structure of C/BN/C, the system has the properties of semimetals, which is related to the width of the GNR[52]. In turn, the formation of a BNC superlattice when zigzag GNRs are embedded in a monolayer of h-BN also displays semimetal properties[53]. Studies have shown that BNC superlattices exhibit half-semimetallic properties with BNNRs wide enough and GNRs narrow enough[54]. The electronic nature of the system determines the physical properties. In a 2005 experiment, researchers obtained the layered hybrid structure of graphene and h-BN[55]. In addition to the necessary electrical, magnetic, and spintronic properties, these systems are still to be expected in terms of thermal and thermoelectric properties.

Interface-type diagrams for the two calculated graphene/h-BN hybrid structures are provided in Figures 6-3(a)～(e)[56]. A schematic of the smooth interface between graphene and h-BN for an armchair edge is shown in Figure 6-3(a). Figures 6-3(b)～(e) show schematic diagrams of the smooth interfaces between graphene and h-BN at a zigzag edge with a C-N structure in 6-3(b), a C-B structure in 6-3(d), and an orthographic BNC hybrid structure in a zigzag boundary in Figure 6-3(c)～(d). Figure 6-3(f) shows the hybrid structure diagram of the single-layer h-BN and graphene that was actually prepared in the experiment[57]. The configuration of the two boundary interfaces can also be observed in the figure. The boundary configuration of an h-BN embedded in graphene is shown in Figure 6-3(g)[58]. The shape, size, width, and length of the embedded boron nitride(BN) or graphene can all vary.

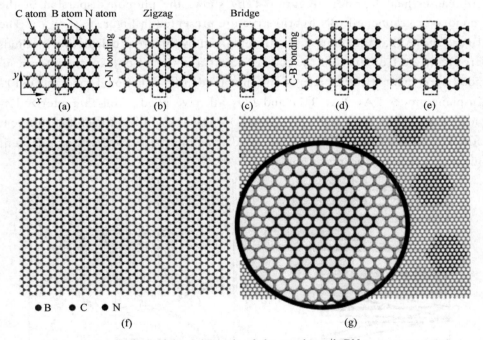

Figure 6-3 Schematic of the graphene/h-BN

(a) The armchair interface and the zigzag interface for (b) C-N zigzag, (c) C-N bridge, (d) C-B zigzag, and (e) C-B bridge bonding at the interface[56]; (f)～(g) Atomic model of the h-BNC film showing hybridized h-BN and graphene domains[57-58]

2. Phonon dispersions of monolayer C-BN hybrid structure

Differences between the phonon spectra of graphene and h-BN form different phonon modes in C-B-N superlattices[59]. Figure 6-4(a) shows that the phonon confine mode (CM) is confined within the periodic lattice of C or BN. In principle, these phonon modes vibrate in accordance with their respective characteristics without affecting each other. However, in the actual C-B-N superlattice, the regular hexagonal structure is destroyed. For example, the lattice of graphene changes from six C atoms to five C atoms and one B atom (N atom), as theoretically B atoms (N atoms) do not vibrate due to the restriction of the other five C atoms. However, in actual composites, B atoms (N atoms) vibrate. Although this vibration is smaller than that of C atoms, it still causes heat transfer. Figure 6-4(b) shows the phonon dispersion of the monolayer graphene-h-BN hybrid structure presented in Figure 6-3(g)[58]. The figure shows that the phonon spectra obtained by different calculation methods are slightly different, and the phonon spectra obtained by experiments are in good agreement with the results obtained by density functional theory (DFT). In-plane modes LA, TA, TO, and LO, all have good, well-characterized 4-nearest-neighbor force-constant (4NNFC). The phonon spectra of the different structures shown in Figure 6-4 are good examples of the effects of different phonon modes on thermal transfer.

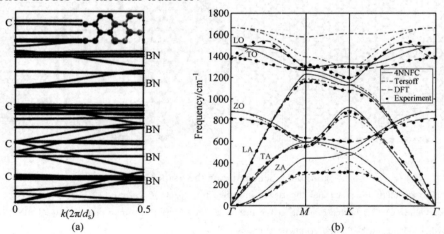

Figure 6-4 The phonon dispersion

(a) The phonon dispersions of C-B-N superlattice[59]; (b) The phonon dispersions of graphene/h-BN hybrid heterostructure, which is corresponding to Figure 6-3(g)[58]

6.2.2 The thermal properties of in-plane C-BN hybrid structures

1. TC of C-BN hybrid structures embedded with round h-BN in graphene

In 2011, Roche et al. studied the phonon-scattering characteristics and TC of the C-BN hybrid structure[58]. We have already discussed the phonon-scattering features, and the calculated model diagram is presented in Figure 6-3(g). The study revealed that BN is present in a composite structure in a circular geometry with a size variation in the range of 2~8 nm and a ratio of 0~100 % embedded in the graphene. The results show that the size effect and doping ratio have a significant effect on the TC of the C-BN hybrid structure. When the doping concentration is 50 %, the minimum mean free path is obtained. As the size decreases, the TC of the hybrid system also decreases. When the size of h-BN is 2 nm, the TC of the system decreases by 65 % (Figure 6-5).

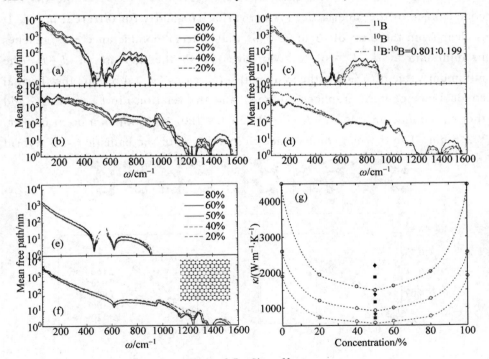

Figure 6-5 Size effects

(a) and (b) show the mean free paths (MFPs) for different concentrations at a fixed domain size of 2 nm are plotted for out-of-plane and in-plane modes. (c) and (d) show the MFPs for different B isotopic compositions at a 50 % concentration and 2 nm domain size for out-of-plane and in-plane modes. (e) and (f) show C atoms randomly distributed in the BN host. MFPs with different concentrations of C atoms for out-of-plane and in-plane modes. (g) shows room-temperature thermal conductivity for a 2 nm domain size versus the concentration of h-BN contributed by out-of-plane and in-plane modes[58]

2. Minimum thermal conductance in C-BN hybrid superlattices

Studies on the relationship between the minimum TC of a C-BN superlattice and the size of a supercell (d_s) and the thickness of a superlattice (L) have also recently been revealed[59]. Researchers used a non-equilibrium Green's function (NEGF) approach to study the relationship between the TC of C-BN superlattices and d_s. In this case, the phonon means the free path is much larger than the supercell size (d_s). Figure 6-6 shows the configuration of the C-BN hybrid superlattice. In a periodic superlattice of length $L = 10$, $d_s = 2$ represents the smallest lattice unit. Heat transport across the superlattice travels from left to right.

The variation of the thermal conductance of the C-BN superlattices is presented in Figures 6-6(a) and (b). At 300 K and 1,000 K, the thermal conductance of the superlattice system varies irregularly for different d_s/L. It is clear from the inset of Figure 6-6(a) that thermal conductance σ/σ_0 reaches its minimum at $d_s/L = 0.05$. Moreover, with the increase of d_s/L, the minimum value of σ/σ_0 of the system also increases (σ_0 represents the thermal conductance of pure graphene ribbons, and the relationship between TC and thermal conductance is $\kappa = \sigma L/S$). When $L = 120$, the minimum point near d_s is 2.6 nm. The thermal conductance characteristic of the ballistic C-BN hybrid

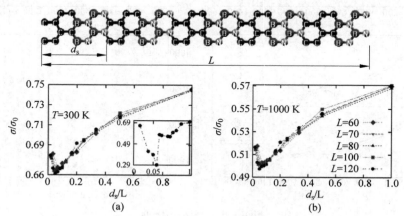

Figure 6-6 The configuration of C-BN with total length $L = 10$ and supercell length $d_s = 2$ unit cells above the figure

The thermal conductance versus supercell length d_s at (a) 300 K and (b) 1,000 K[59]

superlattice studied by NEGF, the minimum thermal conductance in the superlattice is obtained, where $d_s/L = 0.05$.

3. TC of in-plane C-BN hybrid heterostructures

Based on equilibrium molecular dynamics (MD), researchers calculated the thermal properties of different configurations of single-layer C-BN hybrid structures. They analyzed the influence of thermal properties, such as the energy, shape, and spacing of the contact interface between graphene and h-BN on different configurations. Under the C-BN superlattice configuration of the zigzag edge, the TC (κ) in the direction parallel to the interface is greater than that of the armchair edge. However, in the direction perpendicular to the interface, the boundary configuration has little effect on the TC of the superlattice but rather on the TC of h-BN. In addition, researchers found that embedded network structures with jagged-and armchair-type interfaces responded more strongly to heat transfer than superlattices[60].

Figures 6-7(a) and (b) show the TCs (of zigzag C-BN and armchair C-BN, $l_{\text{graphene}} = l_{\text{h-BN}}$) of the two superlattice interfaces in different directions. The different types of interfaces lead to different transmission coefficients. The zigzag superlattices have higher TC. This is due to the fact that armchair-type boundaries have a greater atomic density and affect heat transfer. Figures 6-7(c)~(d) show the TCs of graphene and h-BN at different lengths ($l_{\text{graphene}} \neq l_{\text{h-BN}}$). At this point $l_{\text{graphene}} + l_{\text{h-BN}} = 60$ nm remains constant. As the length of h-BN decreases, the parallel component of heat transfer increases toward the machine direction of graphene, while the vertical component is not higher than 700 W/m·K. When the period of h-BN is small, the interface phonon scattering makes the TC of the h-BN region close to the TC of pure h-BN. When $l_{\text{graphene}}/l_{\text{h-BN}} = 0.05$, the C-BN superlattice at the interface has the greatest influence on heat conduction. Whatever the configuration, the TC of zigzag C-BN in both directions is higher than that of armchair C-BN mainly due to the enhanced scattering at the zigzag boundary.

4. Negative TR in in-plane C-BN hybrid heterostructures

A schematic representation of a typical in-plane C-BN heterostructure is shown in the inset of Figure 6-8(a)[61]. Region A and B represent the

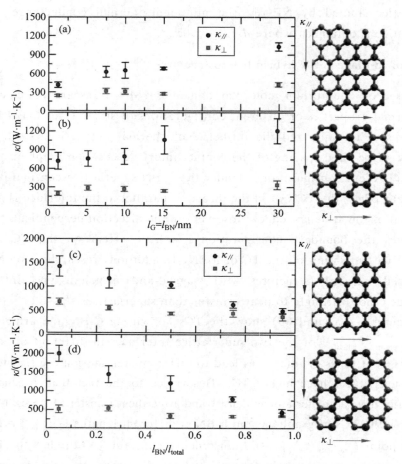

Figure 6-7　The TC of zigzag C-BN and armchair C-BN with different conditions[60]

graphene and h-BN, respectively. Researchers calculated the nonlinear heat transport characteristics of the zigzag edge C-BN hybrid structure using the non-equilibrium molecular dynamics (NEMD) method. It can be seen from the relationship between the heat flux (J) in different heat transfer directions ($J_{BN \to C}$ and $J_{C \to BN}$) and the normalized temperature difference (Δ) that the heat flux (J) flows from h-BN to graphene ($J_{BN \to C}$), Showing a clear thermal correction behavior, $J_{BN \to C}$ and temperature difference (Δ) displayed a linear relationship. $J_{C \to BN}$ direction increases with the increase of temperature difference Δ ($\Delta < 0.3$). When $\Delta > 0.3$, both have a nonlinear response. The two spectra of Figure 6-8(b) also well illustrates that both $J_{BN \to C}$ and $J_{C \to BN}$

are independent of in-plane phonon modes.

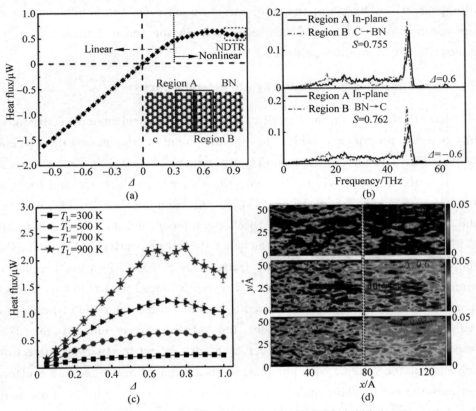

Figure 6-8 Negative thermal resistance in in-plane C-BN hybrid heterostructures
(a) Heat flux as a function of normalized temperature difference for C-BN hybrid heterostructure;
(b) The in-plane spectra of two groups of atoms near to the interface in graphene and h-BN domains for different heat flow dierctions[61]

Figure 6-8(c) shows the relationship between the heat flux J and temperature difference (Δ) at varying temperatures. When the temperature $T = 900$ K, the heat flux J changes greatly with Δ and the heat flux reaches the maximum near the temperature difference $\Delta = 0.85$. Figure 6-8(d) shows the local heat flux at varying temperature differences at $T = 700$ K. The results show that the increase of the temperature difference leads to the appearance of negative differential TR (NDTR), which is mainly due to the fact that the excited out-of-plane sound waves play an important role in the interface transmission. This is also the mismatch between the lattice vibration of graphene and h-BN hindered the heat transfer at the interface. This NDTR disappears as the length of the heterostructure increases and is independent of the width of the heterostructure. Research on the thermal management of nanoscale has theoretical guidance.

5. Thermal rectification in C-BN hybrid heterojunctions

Research on the thermal rectification of C-BN hybrid composite structures has become prominent, which has great potential for nanoscale thermal management and phonon information processing[61-62].

Researchers used NEMD to study the thermal conductance and thermal rectification behavior at the hybrid C-BN nanoribbon interface[62]. Figure 6-9(a) shows the relationship between heat flux and temperature difference for zigzag C-BN's hybrid nanoribbons and armchair C-BN's, respectively. Figure 6-9(b) shows the relationship between the frequencies of each vibrational mode of armchair C-BN hybrid nanoribbons. In this unbalanced heat transfer, a large number of optical phonons are excited to promote the conduction of heat. The red dots and black dots represent the different directions of the TC, respectively. The differences between them also reflect the thermal correction behavior of the system. This phenomenon is attributed to the resonance effect between the out-of-plane phonon modes of graphene and the h-BN phonons in the low-frequency region. Figures 6-9(c) and (d) show the thermal rectification ratios at different temperatures, different temperature differences, different sample lengths, and different interface densities, respectively. The results indicate that a low temperature, large temperature difference, shorter sample length, and small interface density are the best conditions for thermal rectification. Researchers also used phonon spectroscopy to analyze the large temperature difference of low-frequency transverse sound waves in the interface TC. This is the theoretical basis for the thermal rectification equipment based on C-BN nanomaterials.

6. Manipulation of heat current in interface of C-BN hybrid heterostructures

For the C-BN hybrid structure, the graphene nanomaterial is usually the main material. Different from this, when h-BN is the research object, it is

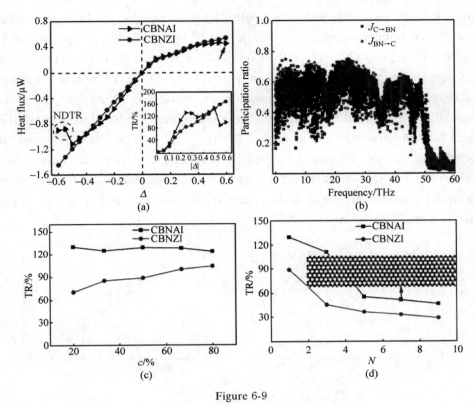

Figure 6-9
(a) Heat current versus that for the two kinds of heterostructure, the inset shows the temperature bias dependence of TR ratio; (b) Participation ratio for armchair C-BN nanoribbons under both forward and reverse biases, the dependence of the TR ratio on the (c) component; (d) number of interfaces for zigzag and armchair C-BN nanoribbons[62]

interesting that the heat current (J) and thermal resistance (TR) of the hybrid system are calculated by changing the number of C atoms in the C-BN hybrid structure[63-64].

Researchers have constructed three different types of graphene-BN interfaces in BNNRs: zigzag trigonal interface (CBN-TZ), armchair trigonal interface (CBN-TA), and square interface (CBN-C). They used MD simulations and NEGF methods to study the heat current and TR across these interfaces. The results indicate that heat current decreases linearly with the increase in C atoms, and the decrease of heat current caused by the zigzag trigonal interface is larger than that of the armchair trigonal interface. This is mainly due to the influence of the local phonon-scattering model on the thermal interface resistance. The results of MD simulation of the heat current at different temperatures ($T = 300$ K and $T = 1,000$ K) are presented in Figure 6-10(a) and (b). The NEGF transmission function of C-BN interfaces of different shapes and sizes is shown in Figure 6-10(c). The calculated thermal interface resistance for different configurations is provided in Figure 6-10(d). After calculation, the TR of the chamber zigzag interface is 7×10^{-10} m² · W/K and the TR of the armchair interface is 3.5×10^{-10} m² · W/K[63]. The same TR calculation was also measured in 2014 by another group of researchers. They measured the temperature difference between graphene and BN using electrically heated graphene and Raman spectroscopy to obtain a TR at the interface of 7.4×10^6 W/m² · K[64].

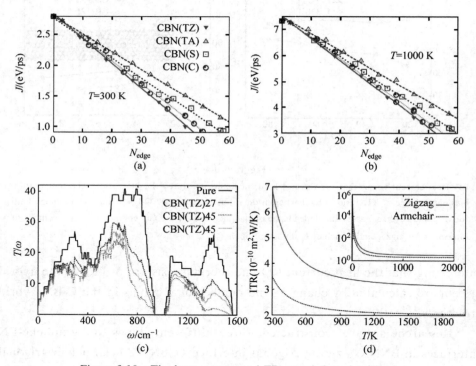

Figure 6-10 The heat current and TR across these interfaces

The heat current vs. N_{edge} for different C-BN interfaces in (a) and (b) from MD simulation[63]; The NEGF transmission function for C-BN interfaces of different shapes and sizes in (c); The interface TR vs. T for zigzag and armchair C-BN interfaces[64]

7. Topological defects enhance thermal conductance in C-BN hybrid structures

Eliminating or reducing the structural defects of the interface and increasing the TC of the material are the methods to manage heat in many advanced devices. However, when single-layer graphene and single-layer h-BN are combined to form the C-BN hybrid structure, the topological defects appearing at the interface can enhance the TC of the entire C-BN interface[65].

Figures 6-11 (a) and (b) show C-BN hybrid interfaces with zigzag edges: zigzag CB interface and zigzag CN interface, respectively, the heptagon topology on the h-BN side and the pentagonal topology on the graphene side. Figure 6-11(c) shows a typical temperature profile of the steady state in the C-BN hybrid structure with a clear view of the temperature distribution in the C-BN hybrid structure along the heat flux direction. Figures 6-11(d) and (e) show the phonon state densities for C atoms and N atoms at the interface of the zigzag C-N hybrid structure with and without topological defects, respectively. Figure 6-11 (f) shows the transverse heat flux density distribution in a C-BN heterostructure and a C-BN heterostructure with topological defects. The results indicate that for the topological structure with zigzag direction along the grain boundary and with a tilt angle of 0, there is a shift of the position of the five or seven in the grain boundary perpendicular to the grain boundary, resulting in an increase of 20% in the TC at the grain boundaries of the heterostructure and the phonon transport at the C-BN interface with topological defects has higher transmissibility. The phonon vibrational spectrum shows that the TC anomaly of this kind of topological interface is enhanced because of the localization of the stress field caused by the dislocation of the lattice and the structural deformation at the interface. This deformation of the structure causes the stress field to be confined to the center of the defect.

8. Coherence properties of phonon propagation and the relationship between phonon transport and thermal conductance

Acoustical phonon modes in C-BN hybrid superlattices have important research significance for the heat transfer of 2D superlattices. By studying the interaction between phonon population velocity and phonon relaxation time in different configurations of C-BN hybrid superlattices, researchers

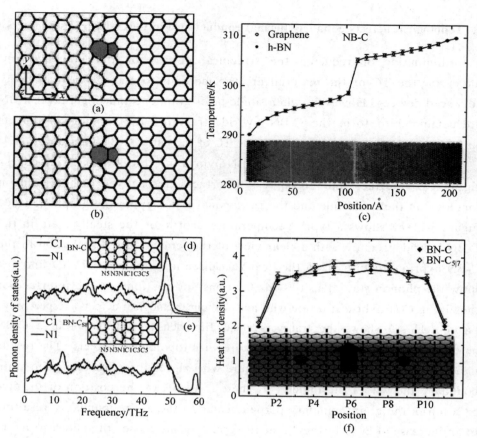

Figure 6-11 Topological defects enhance thermal conductance
in C-BN hybrid structures

The topological C-BN hybrid structure with different interfaces in (a) and (b). Temperature distribution along the heat flux direction in C-BN heterostructure in (c). Phonon density of states for C and N atoms at interfaces in C-BN and topological C-BN heterostructure in(d) and (e). Cross-sectional heat flux density distribution in C-BN and topological C-BN heterostructure in(f)[65]

revealed TC and phonon properties in different directions along the interface, as well as supercrystal coherent phonon transmission changes during the lattice period[66].

Figures 6-12(a)~(d) show 1×1 superlattice cells for the zigzag interface and the armchair interface, respectively. Using harmonic lattice dynamics (HLD) modeling, researchers calculated phonon group velocities and phonon frequencies for superlattice elements. They used density functional theory (DFT) to simulate the relaxation of the unit cells. Superlattice cell density perturbation theory (DFPT) validates the empirical phonon frequency curve. While an MD approach was used to calculate phonon relaxation time, the normal mode decomposition approach was used to discuss the atom's velocity. Finally, the researchers used the properties of a single-phonon mode to estimate the TC of two configurations of superlattices from the model-dependent equations used in kinetic theory, revealing the effects of acoustic phonon modes and superlattice sensitivity of period and interface configuration to TC. Figures 6-12(e) and (f) show the acoustic phonon spectra along the k-space for different sizes of superlattices (from 1×1 to 10×10). Figures 6-12(g)~(j) show the phonon lifetimes for different sizes of zigzag C-BN hybrid superlattices. Figures 6-12(k) and (l) show the phonon velocities for different sizes of zigzag C-BN hybrid superlattices. Figure 6-12(m) shows the TC of superlattices for different lattice periods and configurations (zigzag C-BN and armchair C-BN). From the first cycle length (0.44 nm) to the second cycle

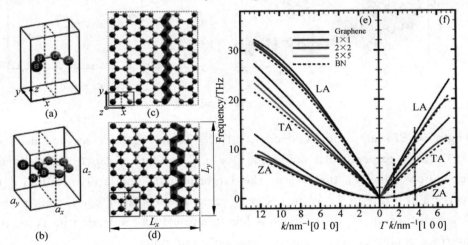

Figure 6-12 Coherence properties of phonon propagation and the relationship between phonon transport and thermal conductance

(a) The superlattice of C-BN hybrid heterostructure with zigzag edge and armchair edge in (a)~(d). The acoustic dispersion curves along k-space direction [0 1 0] and [1 0 0] in (e)~(f) for 1×1 superlattice. Phonon lifetimes for different size zigzag superlattices in (f)~(j). Squared phonon group velocities for zigzag superlattices in x direction and y direction in (k)~(l). Variation of the TC with the superlattice period and interface structure[66]

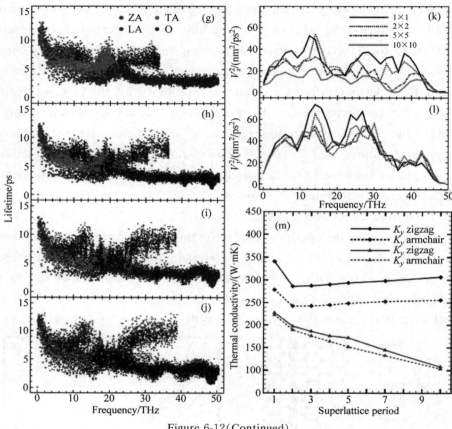

Figure 6-12(Continued)

length (0.87 nm), the TC decreases 16 % along the interface and declines 13 % at the perpendicular interface. Due to the coherent effect of the phonon population velocity, the TC in the larger-period superlattice continues to decrease at a rate of 11 W/m · K per cycle. The armchair boundaries have a 7 % reduction in TC over the cycle length and a 19 % reduction in TC at the interface compared to the zigzag edge configuration.

9. Poor thermoelectric performance and thermal transport in coplanar polycrystalline C-BN hybrid heterostructures

Studies of thermoelectric and heat transfer properties in polycrystalline heterostructures have recently been revealed, in particular for thermoelectric and heat in the cases of different sizes and distributions of graphene and h-BN in large-scale polycrystalline heterostructure transmission performance[67].

The left part of Figure 6-13(a) shows the structure of h-BN with different doping concentrations and the right is the C-BN hybrid structure of polycrystalline interface. Using quantum transport and MD simulations, researchers calculated the thermoelectric and thermal transport of the system with varying h-BN lattice sizes and distributions. Figures 6-13(b) and (c) show the TC of the system at different doping grain sizes and in different sizes. By increasing the grain density of h-BN from 0 to 100 %, the system transitions from the

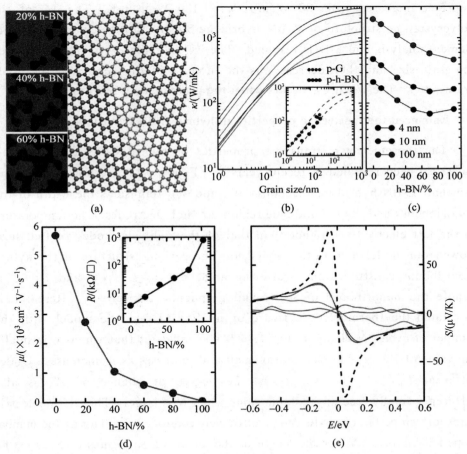

Figure 6-13 Poor thermoelectric performance and thermal transport in coplanar polycrystalline C-BN hybrid heterostructures

Magnification of polycrystalline structure showing a typical interface between graphene and h-BN grains in (a); TC of C-BN hybrid heterostructure with different size and grain density in (b)~(c); Mobility and Seebeck coefficient with different grain density in (d) and (e)[67]

conductor to the insulator and the TC of the system is minimized for the density of h-BN grains at 70 % density. Figure 6-13(d) shows the carrier mobility at different doping concentrations. Figure 6-13(e) presents the Seebeck coefficient variation of the system; the dashed line is the value of pure graphene. As shown in the figure, the Seebeck coefficient of the system is suppressed above 40 %. After estimating the thermoelectric conversion efficiency of the system, researchers found that the application of the system to energy collection is relatively low, while the TC of the polycrystalline interface increased from 30 W/m · K to 120 W/m · K. The TC of the polycrystalline structure of C-BN hybrid heterostructure calculated by finite element analysis and MD is revealed. The TC increases with the increase of the grain size and then decreases as the increase of the density of the h-BN grains rises. The worst TC occurs when the grain density reaches 70 %.

10. Enhanced thermoelectric properties in hybrid C-BN nanoribbons

The thermoelectric transport properties of hybrid C-BN nanomaterials have also been revealed in recent years[68]. Figure 6-14(a) shows zigzag and armchair C-BN hybrid nanoribbons. W_A and W_Z represents the width of the nanoribbons, and the red and blue regions are hot electrodes. The temperature of the left electrode is higher than that of the right electrode, so that heat flows from the left end to the right end. The periodic structure of the hybrid nanoribbons constitutes the scattering area. $L_{h\text{-BN}}$ ($L'_{h\text{-BN}}$) and L_C (L'_C) denote the nanoribbons of h-BN and graphene, respectively. Researchers calculated the thermoelectric figure of merit (ZT) of the model with the number of cycles N changes. Figure 6-14(b) indicates that when N is small, the value of ZT or ZT/ZT_0 is also small and increases as N increases. Under different $L_{h\text{-BN}}/L_C$ ($L'_{h\text{-BN}}/L'_C$), the degree of change of ZT is also different, finally reaching a stable value when N is larger. The thermoelectric transport properties of GNRs can be effectively controlled by adjusting the number of periodic lattices (N) or the length of the lattice (L). Figures 6-14(c)~(f) show the values of the electrical conductance ratio (σ/σ_0), thermal conductance ratio (κ/κ_0), Seebeck coefficient ratio (S/S_0), and transmission coefficient (T_e) of the system at different $L_{h\text{-BN}}$ and L_C, respectively. The results of the study indicate that the thermoelectric quality factor ZT of this periodic hybrid C-BN nanoribbons is significantly enhanced. For armchair C-BN nanoribbons, the ZT is 20 times greater than that of graphene and three times greater than that of zigzag C-BN nanoribbons. This study is instructive for the research on thermoelectric devices based on C-BN hybrid structure.

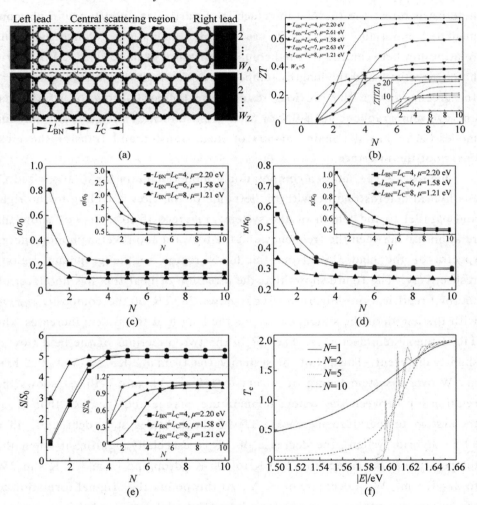

Figure 6-14 A-BCNNRs and Z-BCNNRs, and experimental results
(a) Schematics of A-BCNNRs and Z-BCNNRs. The results of test in (b)~(f)[68]

11. Effect of length, heat flux direction, and temperature on TR of C-BN hybrid heterostructures

Interface thermal conduction is the main research direction and factor of the thermal properties of nanostructured materials. For large C-BN hybrid heterostructures, the thermal interface is of particular importance in phonon-mediated materials. Researchers used NEMD simulations to investigate the relationship between the thermal contact resistance of large-sized C-BN hybrid heterostructures and the length, temperature, and heat flow of the system[69]. In agreement with the previous results, the temporal evolution of thermal energy from graphene to h-BN or vice versa indicates that acoustic phonon modes (ZA) are the main carriers of heat transfer and transfer the most energy at the interface.

Figure 6-15(a) shows a computational model diagram for a large-sized C-BN hybrid heterostructure with the left side of the flow traveling to the right and parallel to the length of the system. Figure 6-15(b) and (c) show the relationship between the frequency and the forward and reverse phonon energy spectra of the heat flow from the h-BN region to the graphene region, respectively. The figure shows that the direction of heat flux has no effect on thermal rectification. Figure 6-15(d) shows the TR of the composite system with the length of the system curve. As the length of the system increases, the TR of the system decreases. The TR in the two directions of the heat flow is slightly different. The TR of the system varies from $5.2 \text{ K} \cdot \text{m}^2/\text{W}$ to $2.2 \text{ K} \cdot \text{m}^2/\text{W}$ over a system length of 20 nm to 100 nm. Figure 6-15(e) shows the relationship between the system temperature change and the TR, that is, as the system temperature increases, TR of the hybrid system decreases. In a 40 nm hybrid system, the decrease of TR is especially significant when the temperature is increased from 200 K to 600 K, dropping from $4.1 \text{ K} \cdot \text{m}^2/\text{W}$ to $2.4 \text{ K} \cdot \text{m}^2/\text{W}$, a decrease of 42%. At this point, the original formation of the system due to the warming effect becomes undulating, and the appearance of this system is also a form of TC reflecting the phenomenon. Analogous to previous studies, this research has greatly improved the scale of the system, close to the practical application, and has practical significance for the thermal application of the composite system.

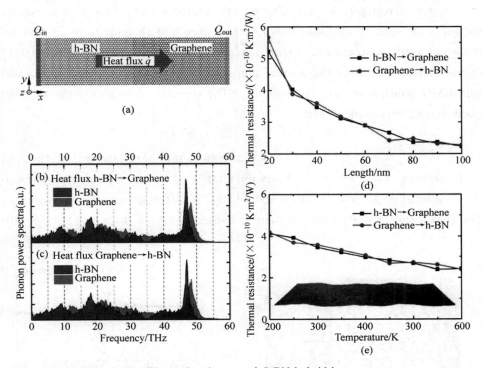

Figure 6-15 Thermal resistance of C-BN hybrid heterostructures
(a) Atomistic configurations of h-BN and graphene hybrid sheet; phonon power spectra of graphene-hBN hybrid sheet of two direction in (b) and (c); (d) Dependence of interfacial thermal resistance on the system length; (e) Thermal resistance variations with temperature from 200 K to 600 K[69]

12. Thermal giant magnetoresistance effects

GNRs and BNNRs exhibit excellent properties of giant magnetoresistance, spin polarization, and negative resistance at the nanometer scale. The monolayer GNRS and BNNRS complex into a hybrid structure, and the composite system showed excellent NDTR and thermal giant magnetoresistance effects (TCMR)[70].

Figure 6-16(a) shows a model of hybrid C-BN nanoribbons. Figure 6-16(b) presents the spin-charge distribution of the smallest unit ($H_2^C H_1^B (C2)_4 (BN)_4$). Researchers have studied the thermal spin-related magnetization and edge hydrogenation of zigzag GNR and zigzag BNNR hybrid structures in this system and considered the effects of system width and length on performance. Figure 6-16(c) shows the relationship between the thermal spin current of the unit under investigation and the system temperature. The figure clearly indicates the spin-up and spin-down current trends with temperature; that is, the greater the temperature difference, the greater the thermal spin current changes. After discussing the transmission spectra of parallel (P) and antiparallel spin (AP) configurations, they obtained the formula to calculate the thermal giant magnetoresistance (MR)[70]:

$$MR(\%) = (I^P - I^{AP})/I^{AP} \times 100, \qquad (6-2)$$

where $I^P = I^P_\uparrow + I^P_\downarrow$ and $I^P = I^P_\uparrow + I^P_\downarrow$ are the total charge current of the spin configuration. Figure 6-16(d) shows the change of reluctance corresponding to the structural unit shown in Figure 6-16(b) in different temperature ranges. The figure shows the perfect symmetry caused by the anti-parallel state with

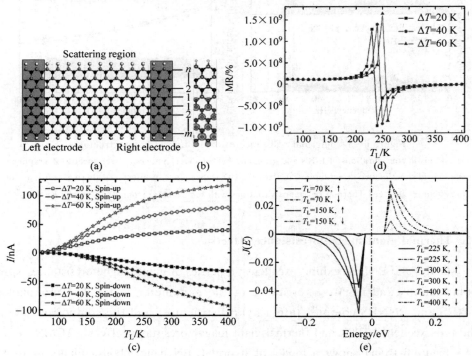

Figure 6-16 Thermal giant magnetoresistance effects

(a) The schematic for the thermal spin caloritronic device based on $H_2^C H_1^B (C2)_m (BN)_n$ heterojunctions; (b) The spin-charge distribution of $H_2^C H_1^B (C2)_4 (BN)_4$ heterojunctions; (c) The thermal spin-dependent current of a $H_2^C H_1^B (C2)_4 (BN)_4$ heterojunctions as a function of TL in the P spin configuration; (d) The magnetoresistance as a function of T_L for $H_2^C H_1^B (C2)_4 (BN)_4$ heterojunctions; (e) Spin-dependent current spectra for $H_2^C H_1^B (C2)_4 (BN)_4$ heterojunctions with different$_{TL}$ where $\Delta T = 60$ K[70]

reluctance greater than 106 %. By changing the direction of the current, researchers tested the change of the charging current with the temperature difference ΔT: the charging current of the parallel spin configuration increases with the increase of ΔT, the charging current of the antiparallel spin configuration is small, and negative heat magnetism resistance rises to 109 %. Concomitantly, the charge current of the parallel spin structure begins to drop at $T_L = 225$ K, resulting in the appearance of NDTR. Figure 6-16(e) shows the variation of the energy-related thermal current. Similarly, at $T < 225$ K, both the spin-up and spin-down thermal currents increase. When $T > 225$ K, the spin-down current drops further, resulting in the occurrence of NDTR.

13. Effect of adsorption on the thermoelectric properties of C-BN hybrid nanoribbons

Researchers have done a great deal of work in the design and study of nanoscale thermoelectric materials of the thermoelectric properties of C-BN hybrid heterostructures[68,71-73]. Using density functional theory combined with Green's function scattering method, researchers calculated the thermoelectric properties of C-BN hybrid heterostructures with foreign molecules adsorbed.[72]

Figure 6-17(a) shows the plane view and top view of tetracyanoethylene (TCNE) and tetrathiafulvalene (TTF) adsorbed on hybrid C-BN nanoribbons, respectively. Hybrid C-BN nanoribbons show enhanced thermoelectric properties due to the disruption of the symmetry of the Fermi level of the system. The hybrid system adsorbed other molecules, and thermoelectric power (S) and electron energy (E) are enhanced to form positive or negative thermoelectric materials. Figure 6-17(b) shows the thermal energy of the hybrid C-BN system as a function of the Fermi level difference after being TCNE and TTF adsorbed. The thermal quality factors of the three kinds of hybrid configurations with phonon thermal conductance of 2 nW/K and 0.2 nW/K are provided, respectively, as a function of Fermi level difference in Figures 6-17(c) and (d). The research shows that two different adsorption methods are available to adjust the thermal quality factor. The reduction of phonon TC will enhance the thermal figure of merit ZT. For TCNE-adsorbed systems, the room temperature thermopower of the adsorbed system increased to -284 μV/K and merit ZT increased by 0.9. For TTF-adsorbed systems, the room temperature thermopower increased to 210 μV/K and merit ZT increased by 0.9.

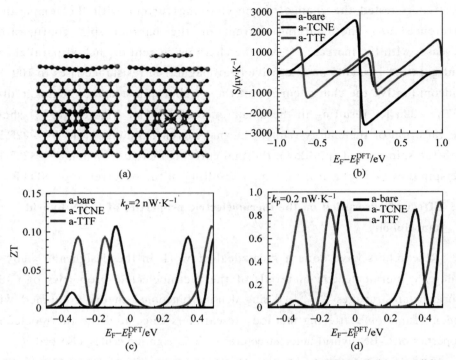

Figure 6-17 Adsorption on the thermoelectric properties of C-BN hybrid nanoribbons (a) Side and top views of optimized structure (1 BN-5G) for molecular complexs doped by electron acceptor-TCNE and donor-TTF; (b)~(d) The room-temperature thermopower S, the figure of merit ZT with different phonon thermal conductance[72]

14. Effect of vacancy on the thermoelectric properties and maximum thermoelectric values of merit (ZT) of C-BN hybrid heterostructures

Based on the excellent thermoelectric properties of the C-BN hybrid heterostructures, researchers have done a great deal of work in achieving high ZT[73-74]. An interesting monolayer C-BN hybrid heterostructure is simulated. Through interface engineering, the phonon scattering of armchair GNRs in hybrid systems is enhanced, which makes the electron conduction strongly reduce the phonon heat conduction under weak influence, resulting in the decrease of ZT. The introduction of corresponding vacancies in the system increased the system quality factor ZT by 1.48 at room temperature[73].

This peculiar hybrid configuration is presented in Figure 6-18(a). Its main body is the GNRs with armchair edges, and h-BN nanosheets are periodically attached to both sides of the GNRs. At $M_{CC} = 5$ and $M_{CC} = 6$, the maximum ZT of BN-C-BN in the active region were calculated. Compared to pure GNRs, the ZT of this hybrid heterostructure has been significantly enhanced. As the number of h-BNs increased, ZT increased from 0.6 to 0.81 with a lower chemical energy of 0.41 eV. The specific results are shown in Figure 6-18(b) and (c). Figures 6-18(d) and (e) present the effect of vacancies in graphene and h-BN on the thermoelectric properties of the system. The power factor P is provided in the illustration. This figure shows that the ZT of the system with vacancy defects is significantly higher than that of the system without vacancies. The ZT system also has a greater impact depending on the location of the vacancies. The presence of vacancies in h-BN always enhances the ZT of the system. This method of improving the thermoelectric performance by changing the configuration of the heterostructure and introducing vacancies has an important guiding significance in the preparation of a good thermoelectric material.

15. Giant seebeck coefficient in quasi-1D C-BN hybrid heterostructures

Compared to C-BN hybrid heterostructures, the thermoelectric properties of C-BN hybrids with quasi-1D or only one lattice width have also been investigated[74].

The inset in Figure 6-19(a) clearly shows the structure of this quasi-1D hybrid system. The width of the system is the width of one crystal lattice, and the length of the lattice is limited to the sum of the lattice of graphene and h-BN and is not more than 8. Using the ab initio method of density functional theory, the authors calculated miscellaneous the quasi-1D C-BN. The Seebeck coefficient of the system varies with the change of chemical potential under different zigzag interfaces and armchair interfaces and varying length configurations (m and n represent the number of lattices of graphene and h-BN, respectively). Compared to the pure GRNs, both boundary configurations of the hybrid system show a higher Seebeck coefficient. For the zigzag edge hybrid system, the Seebeck coefficient is the highest at $m = n = 2$, up to 1.78 mV/K, which is 6.3 times that of the ZGNR Seebeck coefficient and 22 times that of graphene. For the armchair edge hybrid configuration, when $m = n = 2$, the Seebeck coefficient is not the largest

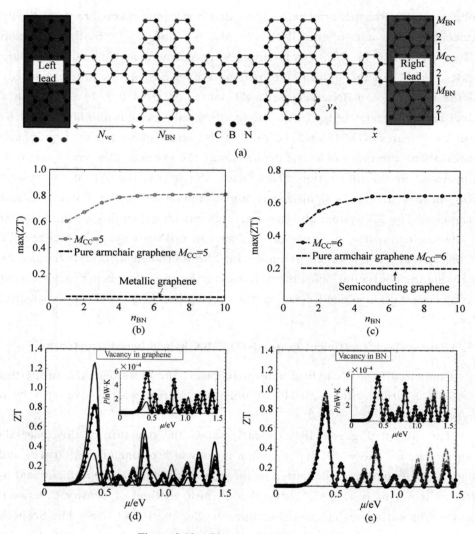

Figure 6-18　Giant Seebeck coefficient
(a) Schematic view of monolayer C-BN heterostructure. Maximum of ZT as a function of n_{BN} for (b) $M_{CC} = 5$ and (c) $M_{CC} = 6$. ZT and thermal power for different configurations of vacancy in (d) graphene and (e) h-BN at $T = 300$ K[73]

(Figure 6-19(b)). The Seebeck coefficient of the C-BN superlattice increases with the decrease of the ZGNR in the composition. This large Seebeck coefficient originates from the band structure with a preferential energy gap.

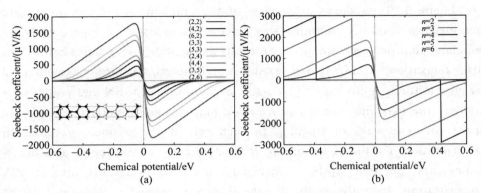

Figure 6-19 Seebeck coeffcients of (a) the ZGNR/BNNR superlattices and AGNR/BNNR superlattices at 300 K as a function of chemical potential felt by electrons[74]

6.3 Graphene/h-BN van der Waals heterostructures

Different 2D materials by van der Waals forces together can form different types of heterostructures and can exhibit a single 2D material that does not have the characteristics. Graphene/h-BN heterostructure is one of the most typical representatives[75-99].

6.3.1 Structures of van der Waals heterostructures

1. Graphene/h-BN van der Waals heterostructures

The initial introduction of h-BN as a substrate is due to h-BN and graphene having similar lattice structures[75-82], as the lattice mismatch between the two is only 1.7%. Unlike substrates such as Si, SiC, or SiO_2[83-87], h-BN has the atomic level of a flat surface, very low roughness, almost no dangling bonds on the surface, and weaker van der Waals forces than graphene. At this point, periodic moiré patterns appear on the surface of graphene. These moiré patterns are regarded as the regulation of the periodic potential of graphene by the BN substrate, resulting in the reconstruction of the graphene band. The results indicate that the BN substrate has a doping effect on the graphene meter, and both have good thermal properties. The research on the double-layer graphene/h-BN heterostructure has become popular.

Figure 6-20(a) shows a 3D view of the graphene/h-BN heterostructure obtained via scanning transmission electron microscopy[84]. Figure 6-20(b) simulates the configuration of the heterostructure in different states before and after relaxation[84]. Figure 6-20(c) shows a 2D heterostructure image obtained by atomic force microscopy[85]. As the lattice shape of h-BN and graphene is similar, the two pile together, forming a periodic moiré pattern. There are three different ways of stacking in each cell of the periodic pattern. In configuration I, the C atoms are located directly above the B and the N atoms, and the two layers of material are stacked vertically into an AA configuration. In configuration II, the B atom is located at the center of the regular hexagonal lattice of C atoms, while the N atom is just above the C atom, and both are vertically stacked in an AB configuration. In configuration III, the N atom is located in the center of the regular hexagonal lattice of C atoms and the B atom is just above the C atom, and both are vertically stacked in an AB′ configuration[84-85,88-90,92-97]. This periodic moiré pattern and the three stacked configurations are shown in Figure 6-20(d)[87].

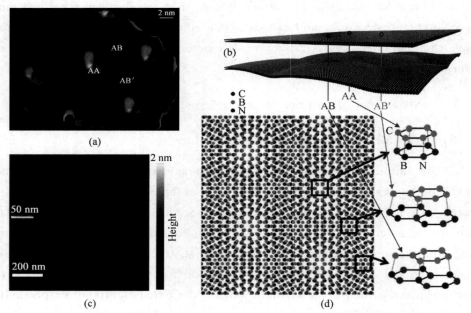

Figure 6-20 Graphene/h-BN heterostructure and periodic moiré patterns
The image of graphene/h-BN heterostructure in (a) ~ (c)[84-85]; Periodic moiré patterns and three configuration models in (d)[85]

2. Phonon dispersions of graphene/h-BN heterostructures

When graphene forms a heterostructure based on an h-BN substrate, the AA and the AB stacking modes also occasionally exist in the periodic moiré pattern, and the phonon spectra of the two configuration modes differ[98-99]. In general, the phonon branching of the heterostructure can be regarded as the common contribution of monolayer graphene phonon branching and monolayer h-BN phonon branching to the system, which is mainly because the difference between graphene and h-BN interaction is less affected. For the AA stacking mode, it is generally considered that the lattice of graphene (0.245 nm) relaxes for the length of the h-BN lattice (0.251 nm); the results of the calculation are presented in Figure 6-21(a)[99]. In this way, the phonon branching of graphene shifts to some extent: the K_6(ZA) phonon branch of h-BN moves to a higher frequency and has a certain contribution to P_3(47 mV); the Γ_6 or Γ_2 phonon of graphene (ZA) shifts to near P_2(36 meV); the

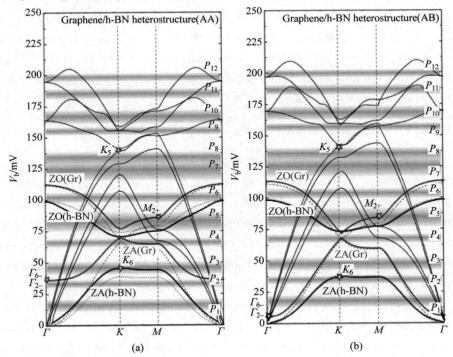

Figure 6-21 The phonon dispersions of graphene/h-BN hetetrostructure with AA configuration in (a) and AB configuration in (b), respectively[99]

M_2(ZO) phonon branch of graphene moves to near P_5 (86 meV); and the LA phonon branch of graphene softens slightly and the K_5 phonon approaches P_8.

The phonon dispersion relation for the heterostructure of the AB stacking mode is presented in Figure 6-21(b)[99]. Similar to the analysis of the AA stacking mode, we clearly observe that the phonon spectrum in the AB stacking mode is closer to the single-layer phonon spectrum of graphene, which proves once again that the AB stacking mode is more stable than the AA stacking mode. At this point, the planar motion of graphene and h-BN phonons outside the crystal lattice is relatively small. Unlike monolayers of graphene and H-BN, the contributions of the ZA, TA, and LA phonon modes to the total TC in the composite system are 10.4%, 33.2%, and 56.1%, respectively[97].

6.3.2 Thermal properties of graphene/h-BN van der Waals heterostructures

When graphene is combined with h-BN to form vertically stacked heterostructures, the interaction and coupling of the electrons between the layers of the two materials results in the different appearance of the heterostructures from the materials of the two in particular. The TC, TR, and other thermodynamic and thermoelectric properties of the research have aroused the concern of researchers[100-139].

1. Effect of length and stacking angles on TC

Suspended graphene requires the aid of a substrate during its use and migration. The h-BN is the best substrate for graphene demolition material because its lattice mismatch is small, it has fewer surface dangling bonds, there are smaller effects on the performance of graphene, and it provides outstanding performance, among other reasons[100-101]. Based on its influence on the properties of graphene, the authors used the MD simulation method to assess the TC of single-layer graphene based on a multilayer h-BN substrate[102].

Figure 6-22(a) shows a computational simulation based on a single layer of graphene on a multi-layer h-BN. The red and blue areas on both sides represent the heat source areas and the heat dissipation areas, respectively. Through simulation calculations, researchers studied three kinds of TC in graphene. Figure 6-22(b) shows the length dependency of the TC of the three

Chapter 6 The Thermal and Thermoelectric Properties of In-Plane C-BN Hybrid Structures and Graphene/h-BN van der Waals Heterostructures

kinds of stack-based substrate-based graphene on the h-BN substrate. Since the h-BN substrate suppresses the long mean free path (MFP) phonon, it leads to a decrease in the TC of graphene. The TC of graphene at 400 nm is approximately 70 % of the TC of graphene, up to (1,347.3 ± 20.5) W/m·K, much higher than the calculated value of the SiO_2 substrate and $\kappa_{AA} < \kappa_{AB} < \kappa_{AB'}$. Figure 6-22(c) compares the cumulative TC of phonon MFPs for different types of graphene. It is evident from the figure that the phonon of the MFP greater than 100 nm contributes 90 % of the TC of the suspended graphene, which is responsible for the extremely high TC and length dependence of the graphene. Due to the influence of the h-BN substrate, at this time, the main phonon MFP of graphene is compressed and becomes uniform still lies between 40~500 nm, precisely because the influence of the h-BN substrate on the dominant phonon MFP in graphene has a small effect, resulting in a high TC and length dependence of graphene. Figure 6-22(d) shows the relationship between the angle of rotation and the thermal conductance between graphene and

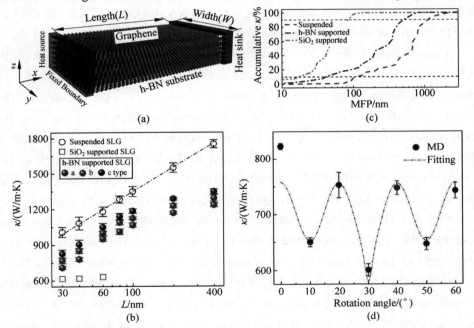

Figure 6-22 Thermal conductivity in graphene/h-BN heterostructure

Calculated simulation model in (a), (b)~(d) The results of thermal conductivity in graphene/h-BN heterostructure[102]

h-BN substrate. The bending of the graphene surface strengthens the warming effect of graphene, resulting in stronger phonon scattering, which leads to a decrease of the TC from (823.1 ± 5.6) W/m·K to (600.4 ± 11.2) W/m·K when the rotation angle changes from 0° to 30° and $\kappa_{0°} > \kappa_{\theta} > \kappa_{30°}$.

The TC and length change of graphene based on h-BN substrate have a strong dependence on the substrate angle. The effect of h-BN substrate on the phonon relaxation time and the phonon MFP of graphene is small, which is the main reason for higher TC of graphene than other substrates. This study provides a theoretical basis for the preparation of graphene thermal functional devices on the substrate.

2. Effect of stacking configuration and substrate number on TC

The effect of the substrate on the TC of graphene has been recently studied[103-107]. As early as 2013, Wang et al. used density functional theory (DFT) and the NEGF method to study the effects of layers (monolayer, bilayer, and trilayer) and different structures (AA stacked and AB stacked) on the TC of graphene/h-BN heterostructures[108].

Figure 6-23(a) shows the TC of the graphene/h-BN heterostructure in the AB stacking mode. The inset provides the AB stacking mode, AA stacking mode, and AB stacking mode of graphene on a bi-layer h-BN substrate. The results indicate that the coupling of h-BN substrate with graphene and the expansion of the graphene lattice lead to changes in the TC of graphene at different temperatures. The TC is reduced by 15 % at 100 K, which is the smallest decrease in the TC for all of the substrates. The figure also shows that the TC of graphene/h-BN heterostructures decreases by only 1 % under strain. Figure 6-23(b) shows the calculated phonon dispersion, transmittance, and spectral DOS for the graphene/h-BN heterostructure. Due to the substrate coupling and lattice strain changes, the high-frequency optical phonon mode has a blue shift of 40 cm^{-1}. The thermal transport of graphene is mainly determined by the ZA phonon at a temperature lower than 100 K. Strain decreases the TC. At high temperatures, the TA phonon and LA phonon determine the TC of graphene. Figure 6-23(c) compares the TC and phonon dispersion of the graphene/h-BN heterostructures for different stacking modes. The differences between the two are not significant, but varying stacking methods correspond to different interlayer distances with AA accumulation of 3.5 Å and AB stacking of 3.3 Å. The difference in interlayer distance results in a decrease in the coupling strength of the graphene from the h-BN substrate. The AA stacking method reduces the coupling strength, resulting in a further 3 % reduction in the TC of graphene at room temperature. Figure 6-23(d) compares the effects of different layers of h-BN substrates on the TC of graphene. The figure indicates that whether the AA or AB stacking structure is grown, the TC of graphene hardly changes when the number of layers of the substrate increases from 1 to 3, revealing that due to the h-BN lining in the bottom, only the contact layer with graphene will affect its TC.

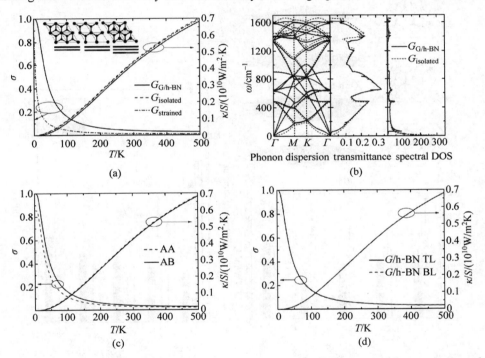

Figure 6-23   TC of graphene/h-BN heterostructure with different stacking methods and layers of h-BN substrates in (a)~(d)[108]

3. TC of different configurations of heterostructures

Research on the TC of different configurations of graphene/h-BN heterostructures has also recently been conducted by researchers. They used density functional theory (DFT) and classical MD methods to assess the TC of different configurations of heterostructures[109].

Figure 6-24(a) shows three different stacking patterns and four composite structures for graphene/h-BN heterostructures (from 1 to 3 h-BN substrates and h-BN-encapsulated graphene), respectively. Figure 6-24(b) indicates the dependence of TCs (κ) and thermal conductance (G) on the configurations of different heterostructures, respectively. The TC (κ) of graphene increases with the increase in the number of substrates and has the lowest TC under the h-BN encapsulation, which is smaller than that of single graphene, $\kappa_V < \kappa_{\text{II}} <$

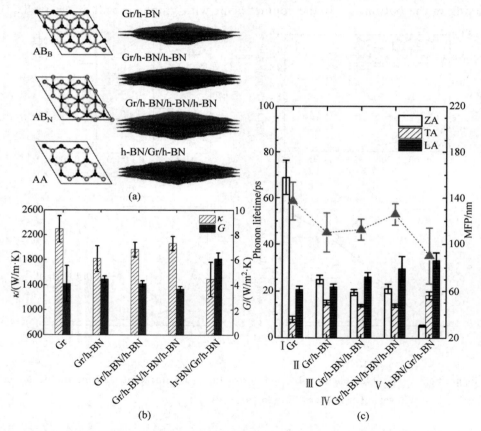

Figure 6-24 Thermal conductivity of different configurations of heterostructures
Three stacking configurations between graphene and h-BN, four types of heterostructures of graphene/h-BN in (a); The TC and thermal conductance of different heterostructure in (b); MFP and phonon lifetime in different heterostructure in (c)[109]

$\kappa_{\text{III}} < \kappa_{\text{IV}} < \kappa_{\text{I}}$. On the other hand, due to the influence of the coupling effect of the substrate, the TC G based on graphene on a single-layer substrate is larger than that of suspended graphene. Among them, h-BN-encapsulated graphene has the highest TC, $G_{\text{V}} > G_{\text{II}} > G_{\text{I}} \geqslant G_{\text{III}} > G_{\text{IV}}$. Figure 6-24(c) shows the phonon lifetime and the mean free path of LA, ZA, and TA for different configurations of heterostructures, respectively. As the figure indicates, as the lifetime of the acoustic phonon mode increases in plane, the heat transfer performance of graphene also increases under different configurations. This is due to the enhanced interaction between the layers because of the charge polarization of the h-BN substrate with graphene.

4. Thermal interface conductance across heterojunctions in experiments

Most studies on the TC of vertically stacked graphene/h-BN heterostructures have focused on theoretical modeling. Chen et al. designed a feasible experiment in 2014 in which the TC at interfaces between graphene and h-BN was measured[110].

First, they mechanically peeled the underlying h-BN from the bulk h-BN and deposited them on the Si/SiO$_2$ substrate. Graphene was then transferred to the layered h-BN using wet chemical etching. Figure 6-25(a) shows an optical microscopic image of this device. Figure 6-25(b) presents a schematic of the experimental setup and measurement. In the experiment, the current through GNRs increased from 0 to 1.5 mA. The laser wavelength of the Raman spectrometer at the center of the GGNRs was 532 nm. The Raman signal was collected for every 0.25 mA of current so that the temperature of graphene and h-BN could be observed with changes. The results of the measurements show that the G and 2D peaks in graphene and h-BN decrease linearly with increasing applied power. Studies have revealed that this change was not due to the thermal expansion or the strain of the two lattices, while the measurement also indicated that the electrostatic doping effect was negligible. Figures 6-25(c) and (d) show the temperature dependence of the electric heating power at several planes of graphene and h-BN. From the slope of the curve normalized to the interface area to define the thermal conductance (G) at the interface, an average thermal interface conductance of (7.41 ± 0.43) MW/m^2 · K was obtained.

The lattices of both are randomly stacked in graphene/h-BN heterostructure,

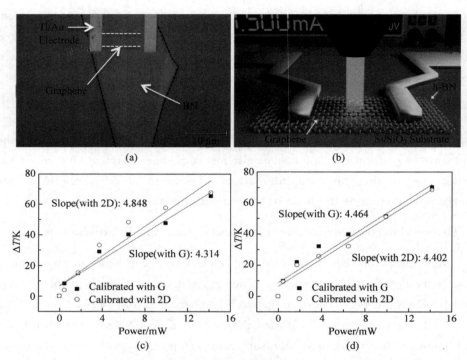

Figure 6-25 Experimentally measured optical microscopy images and schematic views of the graphene/h-BN heterostructure in (a) and (b). The results of test in (c) and (d)[11]

which was prepared using this mechanical stripping method. The conductance of the thermal interface is significantly lower than the thermal interface conductance of lattice-matched graphene grown on h-BN. Raman spectroscopy is used to measure the temperature change of the crystal lattice to obtain the interface TC method, which also provides a feasible approach for the TC measurement of nanomaterials.

5. Phonon thermal properties of graphene/h-BN heterostructures

In previous studies, researchers used the Boltzmann transport equation (BTE) to reveal that the TC of substrate-based graphene is lower than that of suspended graphene due to a decrease in the contribution of curved phonons. On the other hand, MD found that the lifetime of phonons based on graphene on the substrate will be shorter due to the substrate coupling effect[105,111-112]. Studying the phonon transport between graphene and the substrate plays a significant role in understanding the TC of the interface.

The researchers calculated the phonon properties of suspended graphene and substrate-based graphene using equilibrium MD simulations coupled with lattice dynamics[113]. Figure 6-26(a) shows that the coupling between the substrate and graphene has less effect on the phonon dispersion curve. Figure 6-26(b) shows the phonon lifetime of suspended graphene and substrate-based graphene. Due to the limitation of the substrate, the life of the graphene ZA mode is limited to within 3~40 ps, and the ZA mode causes the largest reduction in lifetime. The ZO phonon life expectancy decreased by 60 %, and the bending phonon life expectancy is also considered to be the cause of the above results. The TO mode and LO mode phonon lifetimes are not affected by h-BN. Figure 6-26(c) indicates that, for free graphene, the MFPs in the 90~800 nm range account for 80% of the total TC due to the

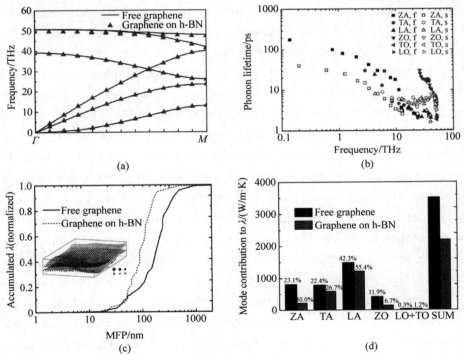

Figure 6-26 Phonon frequency, phonon lifetime, MFP and TC of graphene and graphene/h-BN in (a)~(d). The inset shows the graphene/h-BN heterostructure model[113]

feature length and the external phonon scattering mechanism dominates in thermal conduction. MFP reduction based on graphene on the h-BN substrate is 60~500 nm due to the reduction of MFP with the phonon of graphene affected by the substrate. The results of the TC measurements of suspended graphene and substrate-based graphene are shown in Figure 6-26(d). Based on the bending mode of graphene on the substrate, the TC decreased from 1,232 W/m·K to 367 W/m·K, the in-plane phonon TC decreased by 20.6%, and the total TC bending mode fell to 16.7%. The phonon thermal conductance rate of 2,200 W/m·K was mainly due to the bending of the phonon life expectancy changes.

6. Ultra-fast heat flux transfer

Researchers have long been concerned with the in-plane thermal properties of the graphene/h-BN heterostructure, ignoring the nature and characteristics of the out-of-plane heat flux. Recently, Spanish researchers observed the phenomenon of heat flux transfer between layers in the prepared van der Waals heterostructure formed by h-BN-encapsulated graphene. This out-of-plane heat flux transfer time is very fast, on the order of picoseconds, with heat flux from the graphene layer to the surrounding h-BN. This ultra-fast out-of-plane transmission is more advantageous for use in nanodevices[114-118].

Graphene has the ability to convert incident light into electrical heat (hot electrons), generating photocurrent. Researchers primarily determined the application of heterostructures by studying the length of the cooling time of the hot carriers in the graphene/h-BN heterostructure: longer cooling times favor photodetection sensitivity, cooling periods favor thermal management, and the high data switching rate of photodetectors in communications. They take advantage of the hyperbolic nature of graphene and h-BN formation, that is, the charge carriers in graphene couple with the hyperbola phonon polaritons in h-BN[115-117]. The high-momentum hyperbola phonon polarization exciton energy conversion in the near field, in the picosecond order of magnitude of the cooling time, the heat flow to the h-BN, making the hyperbolic cooling efficiency is particularly high. Generating a huge thermal energy density in high-density optical conditions, these hot carriers effectively couple the heat flow to hyperbolic phonon polaritons through near-field coupling with h-

BN[118]. Figure 6-27(a) shows a schematic of a graphene device with an h-BN package for ultrafast time-resolved photocurrent measurements. When light with an incident light of 800 nm is incident on the P-N junction, photovoltage is generated due to the photothermoelectric (PTE) effect. Changing the delay time Δt between pulses, the carrier dynamics are extracted from the optical voltage signal ΔV_{PTE}. The total hot carrier cooling and the carrier density in the high-mobility package have a linear decreasing trend. Changing the lattice temperature T_L, the cooling time decay is also fast, as shown in Figure 6-27(b). Figure 6-27(c) shows the change of near-equilibrium cooling time τ^*_{calc} and the measurement decay time τ_{exp}, from which it can be found that the hyperbola

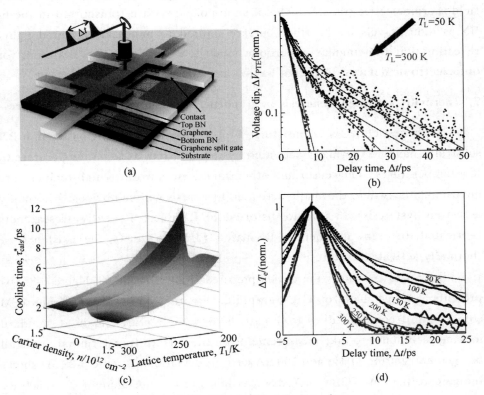

Figure 6-27 Ultrafast heat flux transfer

(a) Schematic drawing of the h-BN-encapsulated graphene device with graphene split gates; (b) The relationship of voltage dip and delay time at $T_L = 50$ K and $T_L = 300$ K; (c) The experiment and theory they vary the carrier density and lattice temperature; (d) Comparison of the complete cooling dynamics as measured (data points) and calculated (solid line) for $n = 0.06 \times 10^{12}$ cm^{-2} for varying lattice temperature[118]

h-BN cooling model shows high carrier density ($1.6 \times 10^{12}/cm^2$) and temperature lattice (200~300 K) throughout the experiment. At different initial electron temperatures corresponding to different laser powers, researchers obtained nearly balanced cooling times. They compared the measurement of photocurrent dynamics and time-domain cooling dynamics to determine the dependence of cooling time on temperature. Figure 6-27(d) shows this dependence of the carrier density at $0.06 \times 10^{12}/cm^2$.

The entire study revealed the heat transfer of graphene encapsulated by h-BN. It is through the incident light onto the graphene to stimulate the hot electrons and h-BN hyperbolic phonon-polarization coupled to achieve near-field thermoelectric transfer. The phonon polaritons are transmitted in the h-BN as light travels in the fiber but are limited to nanoscale infrared light, revealing that the singular hyperbolic mode is a very effective heat sink for optoelectronic devices. A design basis is provided for utilizing heat flow.

7. Thermally activated hysteresis in graphene/h-BN/graphene heterostructures

Graphene field effect transistors (FETs) made of graphene based on h-BN substrates have no thermally activated hysteresis effect at room temperature to low temperatures. The sensor and other nanodevices work in high-temperature environments, where thermally activated hysteresis will be obvious. Thermally activated hysteresis can lead to distorted FET values. Recently, researchers fabricated different graphene/h-BN-based FETs to verify the existence of thermally activated hysteresis[119].

Figure 6-28(a) shows the device prepared from GBN/SiO_2. Figure 6-28(b) presents a device made of GBN/graphite. Figure 6-28(c) features an AFM color picture corresponding to Figure 6-28(a). Researchers used standard locking techniques to take measurements on the current four-terminal and Hall bar geometry using 17 Hz and 100 nA current deviations. Figure 6-28(d) shows the gate voltage of a GBN/SiO_2 device scanned at a temperature of $T = 500$ K. The blue and red curves represent the forward and reverse sweep values, respectively. The figure shows that the shape of the two curves and the position of the charge-neutral point (CNP) are quite different, indicating a significant hysteresis in the resistance R. The P-type doped graphene has a CNP deviation of ± 4 V, and the N-type doped graphene has a CNP at $V_G =$

18 V. This device shows sub-lag. The scan results for the GBN/graphite device in Figure 6-28(e) show no hysteresis at the CNP. Figure 6-28(f) shows the gate bias as a function of temperature from − 40 V to + 40 V at a constant scan speed of 0.17 V/s vs the CNP position. The figure shows that there is no hysteresis effect on the resistance of the graphene in the GBN/SiO_2 device at a temperature lower than 4 K, mainly due to the surface charge of the graphene being frozen at a lower temperature. When the temperature rises to 375 K, hysteresis begins to occur, indicating that the thermal activation effect on the device has played a role. Figure 6-28(g) shows the CNP corresponding scan position for a GBN/SiO_2 device of h-BN with a thickness of 15 nm. The figure shows that, as the scan rate increases, the hysteresis effect becomes significantly larger and reaches saturation near 2 V/s. This temperature-dependent thermal hysteresis depends on the gate scan rate when the temperature is increased to 500 K.

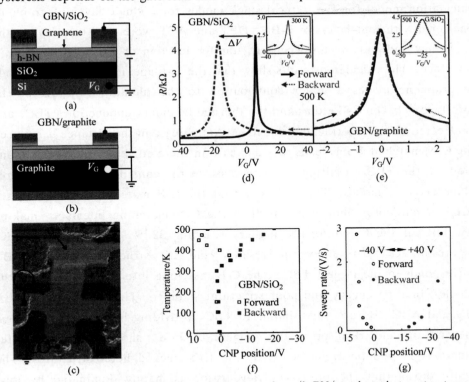

Figure 6-28 Thermally activated hysteresis in graphene/h-BN/graphene heterostructures Sketch of a graphene field effect device deposited on (a) GBN/SiO_2 and (b) GBN/graphite. (c) AFM false color image of GBN/SiO_2 device. The test of two kinds of devices in (d)~(g)[119]

8. Phonon transport at the interfaces of vertically stacked h-BN/graphene/2h-BN heterostructures

Regardless of how vertically graphene and BN are piled together to form a heterostructure, the performance of graphene is always taken into account over h-BN, which is present as a dielectric, substrate, or encapsulation material. Graphene encapsulated by h-BN has very high carrier mobility and ballistic transport[120-121]. In the nanoscale range, the packing material and graphene stacking mode also affects the phonon transport properties of the material.

Yan et al. used the atomistic Green's function (AGF) and density functional theory (DFT) to simulate the ballistic phonon transport at vertically stacked graphene/h-BN heterostructure interfaces[122]. Figure 6-29(a) shows a calculation model for five typical stack modes for a single layer of graphene packaged with four layers of h-BN. DFT and AGF were used to calculate the frequency and wave-vector-dependent transfer function of the heterostructure's interface. The calculated results show that the in-plane acoustic phonon mode of graphene makes a major contribution to the phonon transport at the interface between C-B atoms and the thermal boundary conduction (TBC) and is affected by the interfacial distance d. An increase in the distance d between the C-N matched atoms leads to a decrease in the contribution of the in-plane mode to the TBC, resulting in an increase in the contribution of the out-of-plane acoustic modes. This indicates that the C-B matching interface has a stronger phonon-phonon coupling than the C-N matching interface, making the TBC of the C-B interface higher, reaching $32.5 \sim 50.0$ MW/m^2 · K. Figures 6-29(b)~(c) show the results, and the structure of ABC (B) corresponds to the largest TBC. The C-B matching interface stacks with the smallest binding energy and poor structural stability. The structures of ABA (B) and ABC (B) are very similar to TBC, providing both types with good thermal and electrical properties. Figure 6-29(d) shows phonon transfer functions vs phonon frequencies for the five stacked heterostructures. The figure shows that the interface transmission is mainly dominated by low-frequency ($<$ 5 Hz) phonons.

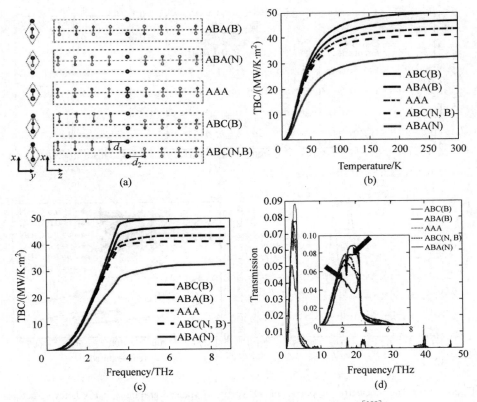

Figure 6-29 The results of simulation in (a)~(d)[122]

9. Thermoelectric properties of AB vertically stacked graphene/h-BN heterostructures

The thermal and thermoelectric properties of the most structurally stable AB configuration among the heterostructures formed by vertical stacking of graphene and h-BN have been revealed in recent studies[123-124]. In the most stable configuration, the distance between the graphene layer and the h-BN layer is 3.4 Å[95].

Wang et al. used the density functional theory (DFT) and NEGF to simulate the thermal transport and thermoelectric properties of the graphene/h-BN heterostructure in the AB stacking mode[123]. Figure 6-30(a)~(c) shows the conductance (G), electron TC (κ), and Seebeck coefficient (S) of the system, respectively. The figure indicates that the thermal transport characteristics of graphene in the heterostructure are lower than that of pure

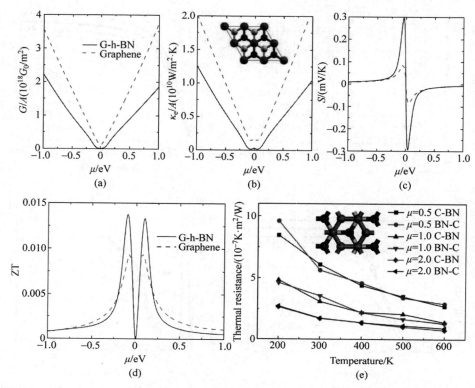

Figure 6-30 Thermoelectric properties of AB vertically stacked graphene/h-BN heterostructures (a) Conductance, (b) Electron TC, (c) Seebeck coefficient and (d) ZT in AB stacking graphene/h-BN heterostructure[123]; (e) The relationship between thermal flux direction and TR with deferent temperature[124]

graphene, but the Seebeck coefficient is enhanced due to the opening of the graphene band gap. Figure 6-30 (d) shows a ZT comparison of this heterostructure ZT with pure graphene. The figure indicates that the ZT of the graphene/h-BN superlattice is larger than that of pure graphene, with 44 % enhancement. Obviously, from the formula $ZT = GS^2 T/\kappa$, it can be seen that all of the factors that affect ZT are enhanced and lead to the enhancement of ZT itself. Li et al. calculated the interfacial TR (ITR) of the graphene/h-BN heterostructure in the AB stacking mode using MD[124]. The calculated ITR is on the order of $10^{-7} \sim 10^{-6}$ K · m^2/W and decreases with increasing temperature and interlayer coupling strength. The AB stacking heterostructure can reduce the ITR by increasing the coupling strength. As in previous studies, the direction of the heat flow has no effect on the ITR of the system. Figure 6-30(e) shows that there is almost no effect on the ITR of the heterostructure at different heat flow orientations in temperature ranges from 200 K to 600K.

6.3.3 Recent applications of thermal and thermoelectric in vertically stacked graphene/h-BN heterostructures

Whether an in-plane graphene/h-BN hybrid heterojunction or vertically stacking graphene/h-BN heterostructure, it is obviously different from pure graphene and pure h-BN in thermal conduction, thermal rectification, TR, and thermoelectric properties. Based on a large number of theoretical and experimental studies, more heterostructures have been used in nanodevices.

1. Lateral heat spreader based on vertically stacking graphene/h-BN heterostructures

Heat dissipation is a key factor affecting nanodevice performance[125-127]. The high TC of graphene and the enhancing effect of h-BN on the TC make it possible for graphene and h-BN to be fabricated into heat spreader. Graphene based on the h-BN substrate, the enhanced thermal flux density of 1,000 W/cm^2 under the hot spot temperature decreased by 8~10℃ [127].

Figure 6-31(a) provides a physical view of the transverse heat sink, and Figure 6-31(b) shows the TEM image of graphene/h-BN heterostructure. Figures 6-31(c) and (d) provide an infrared image of a bare wafer and a graphene/h-BN heterostructure chip at a thermal flux density of 1,200 W/cm^2. Figures 6-31(e) and (f) show the effect of the h-BN heat spreader on the hot spot temperature before and after graphene enhancement. Figure 6-31(e) indicates the hot spot temperature on the front of the chip, and Figure 6-31(f) shows the maximum temperature on the back of the chip. Testing revealed that the graphene/h-BN heterostructure reduces the hot spot temperature by 8~10℃ at a thermal flux density of 1,000 W/cm^2. Figure 6-31(g) shows the experimental results of the h-BN heat sink before and after graphene enhancement and the results of the finite element simulation. The calculated in-plane TC of the heterogeneous heat sink in the figure is about 10 W/m · K. The concentration of graphene in this lateral thermal spreader has little effect on the heat dissipation performance of the heat sink, but the combined effect of the composite structure has better heat dissipation.

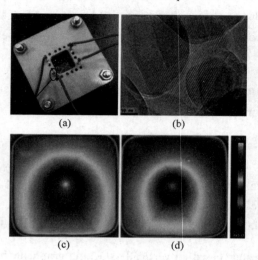

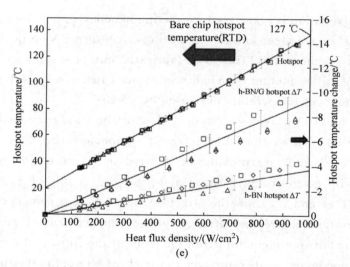

Figure 6-31 Lateral heat spreader based in vertically stacking graphene/h-BN heterostructures (a) Test structure mounted in a holder for four point measurements and infrared imaging; (b) TEM photomicrographs of enhanced graphene/h-BN. Infrared image of bare chip in (c) and graphene-enhanced h-BN chip in (d); The relationship between the heat flux density before and after the graphene enhancement and the chip front hot spot temperature and the chip back surface maximum temperature in (e)~(f); (g) Simulation results (lines) fitted to experimental data (symbols)[127]

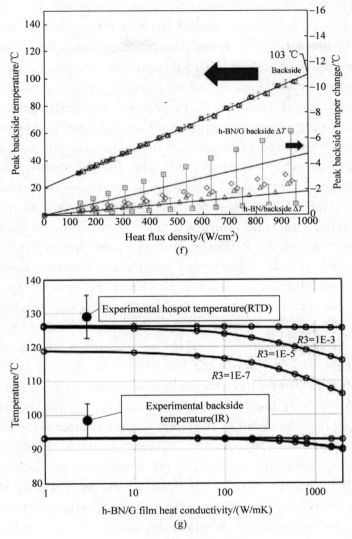

Figure 6-31(Continued)

2. Thermoelectric devices

Numerous studies have shown that vertical heat transfer dominates the vertically stacked graphene/h-BN heterostructures. This is mainly due to the atomic thickness of the device resulting in little heat conduction in the transverse plane[123,128].

Researchers used the AC-locked technique to separate the thermoelectric contribution from the vertically stacked graphene/h-BN/graphene heterostructure and the I-V properties[128]. Raman spectroscopy was used to measure the optical temperature gradient to determine the thermoelectric transport properties of the interface in the heterostructure. The researchers applied an alternating voltage of frequency ω to heat the top graphene layer, which produced a heating power/temperature of 2ω oscillations. The temperature gradient between the top graphene and the underlying graphene resulted in electron diffusion. The researchers used an AC lock-in amplifier to detect the thermal voltage ($\Delta V = 4$ mV). The temperature difference between the top and bottom graphene layers was determined by their 2D Raman frequency, and the difference in temperature was $\Delta T = 39$ K. Limited by the optical measurement range of the Raman spectrometer, the measured thermoelectric transmission was 1~2 nm. Researchers used the above measurements (ΔV and ΔT) to determine the Seebeck coefficient (S) of this device. Figure 6-32(a) shows the current vs voltage in the top and bottom graphene layers, respectively. Figure 6-32(b) presents plots of the second harmonic thermoelectric voltage at different frequencies plotted as a function of the applied AC voltage. Figure 6-32(c) shows the temperature of the top graphene layer obtained using the temperature coefficient acquired by DC calibration, with points 1 and 2 showing Joule heating. The Raman shift of the bottom graphene was zero, and the upper limit of the temperature was determined according to the Seebeck coefficient ($S = \Delta V/\Delta T$). Figure 6-32(d) shows the relationship between the thermoelectric voltage and the temperature gradient. The figure shows that the Seebeck coefficient at this time was 99.3 μV/K, higher than the Seebeck coefficient in the plane of the single-layer graphene at room temperature. The illustration is a schematic diagram of the device and measurement setup. At this point, h-BN inhibits quantum tunneling due to its lower conductivity. After calculation, the thermoelectric figure of merit of ZT = 1.05×10^{-6} for this heterostructure was obtained.

The results of this study showed the thermoelectric voltage generated at the material interface, with transmission lengths between 1 and 2 nm. Although the thermoelectric conversion energy of the device was low, the larger Seebeck coefficient and I-V characteristics show the thermoelectric performance of the device.

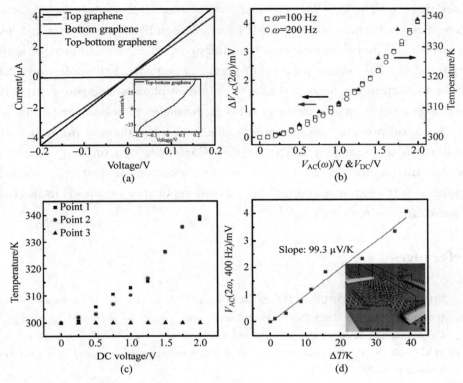

Figure 6-32 The results of the device in (a)~(d)[128]

6.4 Summary and outlook

A large number of theoretical and experimental studies have described the thermal properties of the graphene/h-BN heterostructure. These studies provide researchers with a theoretical and practical basis for the heating and cooling of devices. The hyperbolic material of h-BN has been proven to be the best substrate material for graphene, whether in electrical, magnetic, or thermal properties. The TC of graphene depends mainly on its acoustic phonon properties, and the thermal properties of h-BN as a typical hyperbolic material depend on the hyperbolic polarization. The coupling effect at the interface of the CB (CN) interface in hybrid-graphene/h-BN heterostructures with in-plane structures leads to changes in the thermal properties of the superlattice such as TC, TR, thermal rectification effects, and thermoelectric properties. In vertical graphene/h-BN heterostructures, heat transfer and heat dissipation are more advantageous than the planar structure and provide better applicability. The surface plasmon of graphene and hyperbolic-polaritons of h-BN in the interlayer behave similar to plasmon-phonon polaritons, and the long-range acoustic phonon modes of h-BN contribute more to heat conduction.

Based on previous studies and combined with the thermal properties and applications of graphene, h-BN and their composite structures, graphene/h-BN heterostructures in either in-plane hybridization or vertically stacking graphene/h-BN heterostructures, have broad application prospects in the field of nanoscale microelectronics[145-147].

References

[1] MINNICH A J, DRESSELHAUS M S, REN Z F, et al. Bulk nanostructured thermoelectric materials: current research and future prospects [J]. Energy & Environmental Science, 2009, 2(5): 466-479.

[2] MAJUMDAR A. Thermoelectricity in semiconductor nanostructures [J]. Science, 2004, 303(5659): 777-778.

[3] NOVOSELOV K S, GEIM A K, MOROZOV S V, et al. Electric field effect in atomically thin carbon films[J]. Science, 2004, 306(5696): 666-669.

[4] GEIM A K, NOVOSELOV K S. The rise of graphene[M]//Nanoscience and technology: a collection of reviews from nature journals, 2010: 11-19.

[5] LIU F, MING P, LI J. Ab initio calculation of ideal strength and phonon instability of graphene under tension[J]. Physical Review B, 2007, 76(6): 064120.

[6] ZAKHARCHENKO K V, KATSNELSON M I, FASOLINO A. Finite temperature lattice properties of graphene beyond the quasiharmonic approximation[J]. Physical Review Letters, 2009, 102(4): 046808.

[7] AHIN H, CAHANGIROV S, TOPSAKAL M, et al. Monolayer honeycomb structures of group-IV elements and III-V binary compounds: First-principles calculations[J]. Physical Review B, 2009, 80(15): 155453.

[8] BALANDIN A A, GHOSH S, BAO W, et al. Superior thermal conductivity of single-layer graphene[J]. Nano Letters, 2008, 8(3): 902-907.

[9] LEE C, WEI X, KYSAR J W, et al. Measurement of the elastic properties and intrinsic strength of monolayer graphene[J]. Science, 2008, 321(5887): 385-388.

[10] WALLACE P R, The band theory of graphite[J]. Physical Review, 1947, 71 (9): 622-634.

[11] REICH S, MAULTZSCH J, THOMSEN C, et al. Tight-binding description of graphene[J]. Physical Review B, 2002, 66 (3): 035412.

[12] NETO A H C, GUINEA F, PERES N M R, et al. The electronic properties of graphene[J]. Reviews of Modern Physics, 2009, 81(1): 109.

[13] AL-JISHI R, DRESSELHAUS G. Lattice-dynamical model for graphite[J]. Physical Review B, 1982, 26(8): 4514.

[14] LI C, CHOU T W. A structural mechanics approach for the analysis of carbon nanotubes[J]. International Journal of Solids and Structures, 2003, 40 (10): 2487-2499.

[15] ODEGARD G M, GATES T S, NICHOLSON L M, et al. Equivalent-continuum modeling of nano-structured materials[J]. Composites Science and Technology, 2002, 62(14): 1869-1880.

[16] BALANDIN A A, GHOSH S, BAO W, et al. Superior thermal conductivity of single-layer graphene[J]. Nano Lett., 2008, 8(3): 902-907.

[17] NIKA D L, GHOSH S, POKATILOV E P, et al. Lattice thermal conductivity of graphene flakes: comparison with bulk graphite[J]. Applied Physics Letters, 2009, 94(20): 203103.

[18] SEOL J H, JO I, MOORE A L, et al. Two-dimensional phonon transport in supported graphene[J]. Science, 2010, 328(5975): 213-216.

[19] PRASHER R. Graphene spreads the heat[J]. Science, 2010, 328(5975): 185-186.

[20] CAI W, MOORE A L, ZHU Y, et al. Thermal transport in suspended and supported monolayer graphene grown by chemical vapor deposition[J]. Nano Letters, 2010, 10(5): 1645-1651.

[21] WANG Z, XIE R, BUI C T, et al. Thermal transport in suspended and supported few-layer graphene[J]. Nano Letters, 2011, 11(1): 113-118.

[22] BALANDIN1 A A. Thermal properties of graphene and nanostructured carbon materials[J]. Nature Materials, 2011, 10(8): 569-581.

[23] GHOSH S, BAO W, NIKA D L, et al. Dimensional crossover of thermal transport in few-layer graphene[J]. Nature Materials, 2010, 9(7): 555-558.

[24] NIKA D L, POKATILOV E P, ASKEROV A S, et al. Phonon thermal conduction in graphene: role of umklapp and edge roughness scattering[J]. Physics Review B, 2009, 79(15): 155413-155424.

[25] KOH Y K, BAE M H, CAHILL D G, et al. Heat conduction across monolayer and few-layer graphenes[J]. Nano Letters, 2010, 10(11): 4363-4368.

[26] XU Y, CHEN X B, GU B L, et al. Intrinsic anisotropy of thermal conductance in graphene nanoribbons[J] Applied Physics Letters, 2009, 95(23): 233116-233118.

[27] XU Y, CHEN X B, GU B L, et al. Thermal transport in graphene junctions and quantum dots[J]. Physics Review B, 81(19): 195425-195431.

[28] PAN C N, XIE Z X, TANG L M, et al. Ballistic thermoelectric properties in graphene-nanoribbon-based heterojunctions [J]. Applied Physics Letters, 2012,

101(10): 103115.
[29] HU J, WANG Y, VALLABHANENI A, et al. Nonlinear thermal transport and negative differential thermal conductance in graphene nanoribbons[J]. Applied Physics Letters 2011, 99 (11): 113101.
[30] MA T, LIU Z, WEN J, et al. Tailoring the thermal and electrical transport properties of graphene films by grain size engineering[J]. Nature Communications, 2017, 8(1): 1-9.
[31] SICHEL E K, MILLER R E, ABRAHAMS M S, et al. Heat capacity and thermal conductivity of hexagonal pyrolytic boron nitride[J]. Physical Review B, 1976, 13(10): 4607.
[32] PAKDEL A, ZHI C, BANDO Y, et al. Low-dimensional boron nitride nanomaterials[J]. Materials Today, 2012, 15(6): 256-265.
[33] LIU L, FENG Y P, SHEN Z X. Structural and electronic properties of h-BN[J]. Physical Review B, 2003, 68(10): 104102.
[34] ALEM N, ERNI R, KISIELOWSKI C, et al. Atomically thin hexagonal boron nitride probed by ultrahigh-resolution transmission electron microscopy[J]. Physical Review B, 2009, 80(15): 155425.
[35] WANG C, GUO J, DONG L, et al. Superior thermal conductivity in suspended bilayer hexagonal boron nitride[J]. Scientific Reports, 2016, 6: 25334.
[36] ARNAUD B, LEBÈGUE S, RABILLER P, et al. Huge excitonic effects in layered hexagonal boron nitride[J]. Physical Review Letters, 2006, 96 (2): 026402.
[37] TABARRAEI A, WANG X. Anomalous thermal conductivity of monolayer boron nitride[J]. Applied Physics Letters, 2016, 108 (18): 181904.
[38] LINDSAY L, BROIDO D A. Enhanced thermal conductivity and isotope effect in single-layer hexagonal boron nitride[J]. Physical Review B, 2011, 84 (15): 155421.
[39] LINDSAY L, BROIDO D A. Theory of thermal transport in multilayer hexagonal boron nitride and nanotubes[J]. Physical Review B, 2012, 85 (3): 035436.
[40] JO I, PETTES M T, KIM J, et al. Thermal conductivity and phonon transport in suspended few-layer hexagonal boron nitride[J]. Nano Letters, 2013, 13 (2): 550-554.
[41] WANG C, GUO J, DONG L, et al. Superior thermal conductivity in suspended bilayer hexagonal boron nitride[J]. Scientific Reports, 2016, 6: 25334.
[42] ZHOU H, ZHU J, LIU Z, et al. High thermal conductivity of suspended few-layer hexagonal boron nitride sheets[J]. Nano Research, 2014, 7(8): 1232-1240.
[43] MORTAZAVI B, RÉMOND Y. Investigation of tensile response and thermal conductivity of boron-nitride nanosheets using molecular dynamics simulations[J]. Physica E: Low-dimensional Systems and Nanostructures, 2012, 44(9): 1846-1852.
[44] YANG K, CHEN Y, XIE Y, et al. Effect of triangle vacancy on thermal transport in boron nitride nanoribbons[J]. Solid State Communications, 2011, 151(6): 460-464.
[45] OUYANG T, CHEN Y, XIE Y, et al. Thermal transport in hexagonal boron nitride nanoribbons[J]. Nanotechnology, 2010, 21(24): 245701.

[46] ZEBARJADI M, ESFARJANI K, DRESSELHAUS M S, et al. Perspectives on thermoelectrics: from fundamentals to device applications[J]. Energy & Environmental Science, 2012, 5(1): 5147-5162.

[47] YAN X, POUDEL B, MA Y, et al. Experimental studies on anisotropic thermoelectric properties and structures of n-type Bi2Te2. 7Se0. 3[J]. Nano Letters, 2010, 10(9): 3373-3378.

[48] MUTO A, KRAEMER D, HAO Q, et al. Thermoelectric properties and efficiency measurements under large temperature differences[J]. Review of Scientific Instruments, 2009, 80(9): 093901.

[49] XIE Z X, TANG L M, PAN C N, et al. Ballistic thermoelectric properties in boron nitride nanoribbons[J]. Journal of Applied Physics, 2013, 114(14): 144311.

[50] SEVIK C, KINACI A, HASKINS J B, et al. Influence of disorder on thermal transport properties of boron nitride nanostructures[J]. Physical Review B, 2012, 86(7): 075403.

[51] WATANABE K, TANIGUCHI T, KANDA H. Direct-bandgap properties and evidence for ultraviolet lasing of hexagonal boron nitride single crystal[J]. Nature Materials, 2004, 3(6): 404-409.

[52] DUTTA S, MANNA A K, PATI S K. Intrinsic half-metallicity in modified graphene nanoribbons[J]. Physical Review Letters, 2009, 102(9): 096601.

[53] DING Y, WANG Y, NI J. Electronic properties of graphene nanoribbons embedded in boron nitride sheets[J]. Applied Physics Letters, 2009, 95(12): 123105.

[54] PRUNEDA J M. Origin of half-semimetallicity induced at interfaces of C-BN heterostructures[J]. Physical Review B, 2010, 81(16): 161409.

[55] NAKAMURA J, NITTA T, NATORI A. Electronic and magnetic properties of BNC ribbons[J]. Physical Review B, 2005, 72(20): 205429.

[56] ONG Z Y, ZHANG G, ZHANG Y W. Controlling the thermal conductance of graphene/h-BN lateral interface with strain and structure engineering[J]. Physical Review B, 2016, 93(7): 075406.

[57] CI L, SONG L, JIN C, et al. Atomic layers of hybridized boron nitride and graphene domains[J]. Nature Materials, 2010, 9(5): 430-435.

[58] SEVINÇLI H, LI W, MINGO N, et al. Effects of domains in phonon conduction through hybrid boron nitride and graphene sheets[J]. Physical Review B, 2011, 84(20): 205444.

[59] JIANG J W, WANG J S, WANG B S. Minimum thermal conductance in graphene and boron nitride superlattice[J]. Applied Physics Letters, 2011, 99(4): 043109.

[60] KINACI A, HASKINS J B, SEVIK C, et al. Thermal conductivity of BN-C nanostructures[J]. Physical Review B, 2012, 86(11): 115410.

[61] CHEN X K, LIU J, PENG Z H, et al. A wave-dominated heat transport mechanism for negative differential thermal resistance in graphene/hexagonal boron nitride heterostructures[J]. Applied Physics Letters, 2017, 110(9): 091907.

[62] CHEN X K, XIE Z X, ZHOU W X, et al. Thermal rectification and negative differential thermal resistance behaviors in graphene/hexagonal boron nitride heterojunction[J]. Carbon, 2016, 100: 492-500.

[63] JIANG J W, WANG J S. Manipulation of heat current by the interface between graphene and white graphene[J]. Europhysics Letters, 2011, 96(1): 16003.

[64] CHEN C C, LI Z, SHI L, et al. Thermal interface conductance across a graphene/hexagonal boron nitride heterojunction [J]. Applied Physics Letters, 2014, 104(8): 081908.

[65] LIU X, ZHANG G, ZHANG Y W. Topological defects at the graphene/h-BN interface abnormally enhance its thermal conductance[J]. Nano Letters, 2016, 16(8): 4954-4959.

[66] DA SILVA C, SAIZ F, ROMERO D A, et al. Coherent phonon transport in short-period two-dimensional superlattices of graphene and boron nitride [J]. Physical Review B, 2016, 93(12): 125427.

[67] BARRIOS-VARGAS J E, MORTAZAVI B, CUMMINGS A W, et al. Electrical and thermal transport in coplanar polycrystalline graphene-hbn heterostructures[J]. Nano Letters, 2017, 17(3): 1660-1664.

[68] YANG K, CHEN Y, D'AGOSTA R, et al. Enhanced thermoelectric properties in hybrid graphene/boron nitride nanoribbons[J]. Physical Review B, 2012, 86(4): 045425.

[69] HONG Y, ZHANG J, ZENG X C. Thermal contact resistance across a linear heterojunction within a hybrid graphene/hexagonal boron nitride sheet[J]. Physical Chemistry Chemical Physics, 2016, 18(35): 24164-24170.

[70] ZHU L, LI R, YAO K. Temperature-controlled colossal magnetoresistance and perfect spin Seebeck effect in hybrid graphene/boron nitride nanoribbons[J]. Physical Chemistry Chemical Physics, 2017, 19(5): 4085-4092.

[71] VISHKAYI S I, TAGANI M B, SOLEIMANI H R. Enhancement of thermoelectric efficiency by embedding hexagonal boron-nitride cells in zigzag graphene nanoribbons [J]. Journal of Physics D: Applied Physics, 2015, 48(23): 235304.

[72] ALGHARAGHOLY L A, AL-GALIBY Q, MARHOON H A, et al. Tuning thermoelectric properties of graphene/boron nitride heterostructures [J]. Nanotechnology, 2015, 26(47): 475401.

[73] TRAN V T, SAINT-MARTIN J, DOLLFUS P. High thermoelectric performance in graphene nanoribbons by graphene/BN interface engineering[J]. Nanotechnology, 2015, 26(49): 495202.

[74] YOKOMIZO Y, NAKAMURA J. Giant Seebeck coefficient of the graphene/h-BN superlattices[J]. Applied Physics Letters, 2013, 103(11): 113901.

[75] WANG M, ZHANG G, PENG H, et al. Energetic and thermal properties of tilt grain boundaries in graphene/hexagonal boron nitride heterostructures [J]. Functional Materials Letters, 2015, 8(03): 1550038.

[76] YANG K, CHEN Y, D'AGOSTA R, et al. Enhanced thermoelectric properties in

hybrid graphene/boron nitride nanoribbons[J]. Physical Review B, 2012, 86(4): 045425.

[77] DROST R, UPPSTU A, SCHULZ F, et al. Electronic states at the graphene-hexagonal boron nitride zigzag interface[J]. Nano Letters, 2014, 14(9): 5128-5132.

[78] KARAMANIS P, OTERO N, POUCHAN C. Electric property variations in nanosized hexagonal boron nitride/graphene hybrids[J]. The Journal of Physical Chemistry C, 2015, 119(21): 11872-11885.

[79] PARK J, LEE J, LIU L, et al. Spatially resolved one-dimensional boundary states in graphene-hexagonal boron nitride planar heterostructures[J]. Nature Communications, 2014, 5(1): 1-6.

[80] LIU Z, MA L, SHI G, et al. In-plane heterostructures of graphene and hexagonal boron nitride with controlled domain sizes[J]. Nature Nanotechnology, 2013, 8(2): 119-124.

[81] JUNG J, QIAO Z, NIU Q, et al. Transport properties of graphene nanoroads in boron nitride sheets[J]. Nano Letters, 2012, 12(6): 2936-2940.

[82] GAO G, GAO W, CANNUCCIA E, et al. Artificially stacked atomic layers: toward new van der Waals solids[J]. Nano Letters, 2012, 12(7): 3518-3525.

[83] DECKER R, WANG Y, BRAR V W, et al. Local electronic properties of graphene on a BN substrate via scanning tunneling microscopy[J]. Nano Letters, 2011, 11(6): 2291-2295.

[84] ARGENTERO G, MITTELBERGER A, REZA A M M, et al. Unraveling the 3D atomic structure of a suspended graphene/hBN van der Waals heterostructure[J]. Nano Letters, 2017, 17(3): 1409-1416.

[85] CHEN Z G, SHI Z, YANG W, et al. Observation of an intrinsic bandgap and Landau level renormalization in graphene/boron-nitride heterostructures [J]. Nature Communications, 2014, 5: 4461.

[86] HÜSER F, OLSEN T, THYGESEN K S. Quasiparticle GW calculations for solids, molecules, and two-dimensional materials[J]. Physical Review B, 2013, 87(23): 235132.

[87] XUE J, SANCHEZ-YAMAGISHI J, BULMASH D, et al. Scanning tunnelling microscopy and spectroscopy of ultra-flat graphene on hexagonal boron nitride[J]. Nature Materials, 2011, 10(4): 282-285.

[88] SACHS B, WEHLING T O, KATSNELSON M I, et al. Adhesion and electronic structure of graphene on hexagonal boron nitride substrates[J]. Physical Review B, 2011, 84(19): 195414.

[89] MOON P, KOSHINO M. Electronic properties of graphene/hexagonal-boron-nitride moiré superlattice[J]. Physical Review B, 2014, 90(15): 155406.

[90] ZHONG X, YAP Y K, PANDEY R, et al. First-principles study of strain-induced modulation of energy gaps of graphene/BN and BN bilayers[J]. Physical Review B, 2011, 83(19): 193403.

[91] WANG E, LU X, DING S, et al. Gaps induced by inversion symmetry breaking and second-generation Dirac cones in graphene/hexagonal boron nitride [J]. Nature

Physics, 2016, 12(12): 1111-1115.
[92] GIOVANNETTI G, KHOMYAKOV P A, BROCKS G, et al. Substrate-induced band gap in graphene on hexagonal boron nitride: ab initio density functional calculations [J]. Physical Review B, 2007, 76(7): 073103.
[93] SAN-JOSE P, GUTIÉRREZ-RUBIO A, STURLA M, et al. Electronic structure of spontaneously strained graphene on hexagonal boron nitride[J]. Physical Review B, 2014, 90(11): 115152.
[94] KIM S M, HSU A, ARAUJO P T, et al. Synthesis of patched or stacked graphene and h-BN flakes: a route to hybrid structure discovery[J]. Nano Letters, 2013, 13(3): 933-941.
[95] FAN Y, ZHAO M, WANG Z, et al. Tunable electronic structures of graphene/boron nitride heterobilayers[J]. Applied Physics Letters, 2011, 98(8): 083103.
[96] ZHOU S, HAN J, DAI S, et al. van der Waals bilayer energetics: generalized stacking-fault energy of graphene, boron nitride, and graphene/boron nitride bilayers [J]. Physical Review B, Condensed Matter and Materials Physics, 2015, 92(15): 155438.1-155438.13.
[97] KAN E, REN H, WU F, et al. Why the band gap of graphene is tunable on hexagonal boron nitride[J]. The Journal of Physical Chemistry C, 2012, 116(4): 3142-3146.
[98] GHOLIVAND H, DONMEZER N. Phonon mean free path in few layer graphene, hexagonal boron nitride, and composite bilayer h-BN/graphene[J]. IEEE Transactions on Nanotechnology, 2017, 16(5): 752-758.
[99] JUNG S, PARK M, PARK J, et al. Vibrational properties of h-BN and h-BN-graphene heterostructures probed by inelastic electron tunneling spectroscopy[J]. Scientific Reports, 2015, 5(1): 1-9.
[100] DEAN C R, YOUNG A F, MERIC I, et al. Boron nitride substrates for high-quality graphene electronics[J]. Nature Nanotechnology, 2010, 5(10): 722-726.
[101] MAYOROV A S, GORBACHEV R V, MOROZOV S V, et al. Micrometer-scale ballistic transport in encapsulated graphene at room temperature[J]. Nano Letters, 2011, 11(6): 2396-2399.
[102] ZHANG Z, HU S, CHEN J, et al. Hexagonal boron nitride: a promising substrate for graphene with high heat dissipation[J]. Nanotechnology, 2017, 28(22): 225704.
[103] CAI W, MOORE A L, ZHU Y, et al. Thermal transport in suspended and supported monolayer graphene grown by chemical vapor deposition[J]. Nano Letters, 2010, 10(5): 1645-1651.
[104] ONG Z Y, POP E. Effect of substrate modes on thermal transport in supported graphene[J]. Physical Review B, 2011, 84(7): 075471.
[105] CHEN L, KUMAR S. Thermal transport in graphene supported on copper[J]. Journal of Applied Physics, 2012, 112(4): 043502.
[106] GUO Z X, DING J W, GONG X G. Substrate effects on the thermal conductivity of epitaxial graphene nanoribbons[J]. Physical Review B, 2012, 85(23): 235429.

[107] CHEN J, ZHANG G, LI B. Substrate coupling suppresses size dependence of thermal conductivity in supported graphene[J]. Nanoscale, 2013, 5(2): 532-536.

[108] WANG X, HUANG T, LU S. High performance of the thermal transport in graphene supported on hexagonal boron nitride[J]. Applied Physics Express, 2013, 6(7): 075202.

[109] PAK A J, HWANG G S. Theoretical analysis of thermal transport in graphene supported on hexagonal boron nitride: the importance of strong adhesion due to π-bond polarization[J]. Physical Review Applied, 2016, 6(3): 034015.

[110] CHEN C C, LI Z, SHI L, et al. Thermal interface conductance across a graphene/hexagonal boron nitride heterojunction[J]. Applied Physics Letters, 2014, 104(8): 081908.

[111] QIU B, RUAN X. Reduction of spectral phonon relaxation times from suspended to supported graphene[J]. Applied Physics Letters, 2012, 100(19): 193101.

[112] WEI Z, YANG J, BI K, et al. Mode dependent lattice thermal conductivity of single layer graphene[J]. Journal of Applied Physics, 2014, 116(15): 153503.

[113] ZOU J H, CAO B Y. Phonon thermal properties of graphene on h-BN from molecular dynamics simulations[J]. Applied Physics Letters, 2017, 110(10): 103106.

[114] WANG D, CHEN G, LI C, et al. Thermally induced graphene rotation on hexagonal boron nitride[J]. Physical Review Letters, 2016, 116(12): 126101.

[115] CALDWELL J D, KRETININ A V, CHEN Y, et al. Sub-diffractional volume-confined polaritons in the natural hyperbolic material hexagonal boron nitride[J]. Nature Communications, 2014, 5(1): 1-9.

[116] BASOV D N, FOGLER M M, DE ABAJO F J G. Polaritons in van der Waals materials[J]. Science, 2016, 354(6309): 1992.

[117] DAI S, FEI Z, MA Q, et al. Tunable phonon polaritons in atomically thin van der Waals crystals of boron nitride[J]. Science, 2014, 343(6175): 1125-1129.

[118] TIELROOIJ K J, HESP N C H, PRINCIPI A, et al. Out-of-plane heat transfer in van der Waals stacks through electron-hyperbolic phonon coupling[J]. Nature Nanotechnology, 2018, 13(1): 41-46.

[119] CADORE A R, MANIA E, WATANABE K, et al. Thermally activated hysteresis in high quality graphene/h-BN devices[J]. Applied Physics Letters, 2016, 108(23): 233101.

[120] KRETININ A V, CAO Y, TU J S, et al. Electronic properties of graphene encapsulated with different two-dimensional atomic crystals[J]. Nano Letters, 2014, 14(6): 3270-3276.

[121] MAYOROV A S, GORBACHEV R V, MOROZOV S V, et al. Micrometer-scale ballistic transport in encapsulated graphene at room temperature[J]. Nano Letters, 2011, 11(6): 2396-2399.

[122] YAN Z, CHEN L, YOON M, et al. Phonon transport at the interfaces of vertically stacked graphene and hexagonal boron nitride heterostructures[J]. Nanoscale, 2016,

8(7): 4037-4046.
[123] WANG X M, LU S S. First-Principles study of the transport properties of graphene-hexagonal boron nitride superlattice[J]. Journal of nanoscience and nanotechnology, 2015, 15(4): 3025-3028.
[124] LI T, TANG Z, HUANG Z, et al. Interfacial thermal resistance of 2D and 1D carbon/hexagonal boron nitride van der Waals heterostructures[J]. Carbon, 2016, 105: 566-571.
[125] YAN Z, LIU G, KHAN J M, et al. Graphene quilts for thermal management of high-power GaN transistors[J]. Nature Communications, 2012, 3(1): 1-8.
[126] GAO Z, ZHANG Y, FU Y, et al. Thermal chemical vapor deposition grown graphene heat spreader for thermal management of hot spots[J]. Carbon, 2013, 61: 342-348.
[127] BAO J, EDWARDS M, HUANG S, et al. Two-dimensional hexagonal boron nitride as lateral heat spreader in electrically insulating packaging[J]. Journal of Physics D: Applied Physics, 2016, 49(26): 265501.
[128] CHEN C C, LI Z, SHI L, et al. Thermoelectric transport across graphene/hexagonal boron nitride/graphene heterostructures[J]. Nano Research, 2015, 8(2): 666-672.
[129] XU X G, GHAMSARI B G, JIANG J H, et al. One-dimensional surface phonon polaritons in boron nitride nanotubes[J]. Nature Communications, 2014, 5(1): 1-6.
[130] SHI Z, BECHTEL H A, BERWEGER S, et al. Amplitude-and phase-resolved nanospectral imaging of phonon polaritons in hexagonal boron nitride[J]. ACS Photonics, 2015, 2(7): 790-796.
[131] DAI S, MA Q, ANDERSEN T, et al. Subdiffractional focusing and guiding of polaritonic rays in a natural hyperbolic material[J]. Nature Communications, 2015, 6(1): 1-7.
[132] YOXALL E, SCHNELL M, NIKITIN A Y, et al. Direct observation of ultraslow hyperbolic polariton propagation with negative phase velocity[J]. Nature Photonics, 2015, 9(10): 674-678.
[133] WOESSNER A, LUNDEBERG M B, GAO Y, et al. Highly confined low-loss plasmons in graphene-boron nitride heterostructures[J]. Nature Materials, 2015, 14(4): 421-425.
[134] DAI S, MA Q, LIU M K, et al. Graphene on hexagonal boron nitride as a tunable hyperbolic metamaterial[J]. Nature Nanotechnology, 2015, 10(8): 682-686.
[135] SLOTMAN G J, DE WIJS G A, FASOLINO A, et al. Phonons and electron-phonon coupling in graphene-h-BN heterostructures[J]. Annalen Der Physik, 2014, 526(9-10): 381-386.
[136] BARCELOS I D, CADORE A R, CAMPOS L C, et al. Graphene/h-BN plasmon-phonon coupling and plasmon delocalization observed by infrared nano-spectroscopy [J]. Nanoscale, 2015, 7(27): 11620-11625.

[137]   LIN X, YANG Y, RIVERA N, et al. All-angle negative refraction of highly squeezed plasmon and phonon polaritons in graphene-boron nitride heterostructures [J]. Proceedings of the National Academy of Sciences, 2017, 114(26): 6717-6721.

[138]   WOESSNER A, PARRET R, DAVYDOVSKAYA D, et al. Electrical detection of hyperbolic phonon-polaritons in heterostructures of graphene and boron nitride[J]. npj 2D Materials and Applications, 2017, 1(1): 1-6.

[139]   ZHANG J, HONG Y, YUE Y. Thermal transport across graphene and single layer hexagonal boron nitride[J]. Journal of Applied Physics, 2015, 117(13): 134307.

[140]   VANDECASTEELE N, BARREIRO A, LAZZERI M, et al. Current-voltage characteristics of graphene devices: Interplay between Zener-Klein tunneling and defects[J]. Physical Review B, 2010, 82(4): 045416.

[141]   KANE G, LAZZERI M, MAURI F. High-field transport in graphene: the impact of Zener tunneling[J]. Journal of Physics: Condensed Matter, 2015, 27(16): 164205.

[142]   GUO Y, CORTES C L, MOLESKY S, et al. Broadband super-Planckian thermal emission from hyperbolic metamaterials[J]. Applied Physics Letters, 2012, 101(13): 131106.

[143]   BIEHS S A, TSCHIKIN M, BEN-ABDALLAH P. Hyperbolic metamaterials as an analog of a blackbody in the near field[J]. Physical Review Letters, 2012, 109(10): 104301.

[144]   YANG W, BERTHOU S, LU X, et al. A graphene Zener-Klein transistor cooled by a hyperbolic substrate[J]. Nature Nanotechnology, 2018, 13(1): 47-52.

[145]   WANG J, MA F, LIANG W, et al. Optical, photonic and optoelectronic properties of graphene, h-BN and their hybrid materials[J]. Nanophotonics, 2017, 6(5): 943-976.

[146]   WANG J, MA F, LIANG W, et al. Electrical properties and applications of graphene, hexagonal boron nitride (h-BN), and graphene/h-BN heterostructures[J]. Materials Today Physics, 2017, 2(2): 6-34.

[147]   WANG J, XU X, MU X, et al. Magnetics and spintronics on two-dimensional composite materials of graphene/hexagonal boron nitride [J]. Materials Today Physics, 2017, 3: 93-117.

Chapter 7

The Thermal, Electrical, and Thermoelectric Properties of Graphene Nanomaterials

7.1 Introduction of graphene

As an ideal two dimensional (2D) material, graphene has become a popular topic of scientific research due to its unique physical and chemical properties since it was first proposed in 2004[1-3]. Monolayer graphene has a hexagonal honeycomb structure comprised of the thinnest 2D crystal; graphene film is only one atom thick[2]. The sp^2 hybridization of C atoms also gives graphene its unique properties. As the basic unit of C-based materials, graphene can form zero-dimensional(0D) fullerenes, curl into one-dimensional (1D) C nanotubes, and form graphite by means of AA stacking or AB stacking, see Figure 7-1(a)[2,4].

Graphene is a transparent conductor that can be used to replace the current liquid crystal display materials[5]. It is one of the smallest known inorganic nanomaterials, very strong and hard, and is 100 times stronger than the best steel in the world[6]. The elastic modulus of the theoretical calculations and experimental measurements are 1.05~1.24 TPa and 1 TPa, respectively[7-9]. The TC of graphene is up to 5,300 W/m·K[10]. At room temperature, the carrier mobility of graphene is as high as 15,000 cm^2/V·s[2], and its carrier mobility is not affected by temperatures in the range of 10~100 K, which proves that defect scattering is the main scattering mechanism of the electrons in graphene[11]. The light absorption of single-layer graphene is only 2.3 %[12-14], and it has excellent nonlinear optical properties[15]. Graphene also has unique magnetic and spintronic properties[16].

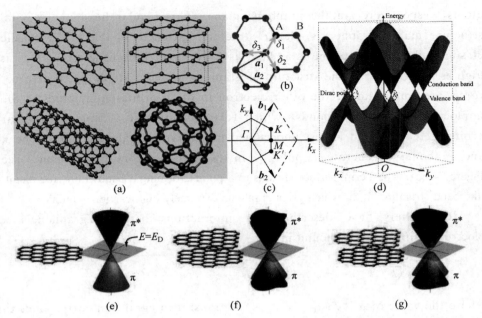

Figure 7-1 Current streamlines and potential map for viscous and ohmic flows. White lines show current streamlines, colors show electrical potential, arrows show the direction of current

(a) Carbon-based nanomaterials[2,4]. Copyright 2008, Springer. (b) The planar crystal structure of graphene, a_1 and a_2 are the lattice vectors of the unit cell, δ_1, δ_2 and δ_3 are mutually adjacent vectors each other[3]. (c) The Brillouin region corresponding to the graphene lattice[3]. Copyright 2009, 2009 American Physical Society. (d) Energy band diagram of graphene hexagonal lattice[18]. Copyright 2009, Springer. Electronic structure of a single (e), symmetric double layer (f), and asymmetric double layer (g) of graphene. The energy bands depend only on in-plane momentum because the electrons are restricted to motion in a 2D plane. The Dirac crossing points are at energy E_D[19]. Copyright 2006, Science

7.2 The crystal structure and electronic structure of graphene

The ideal graphene is the thinnest 2D crystal (0.35 nm). The bond length between C and C is 0.142 nm (a_{C-C}). In the crystal structure of graphene, each cell contains two C atoms, A and B. The A and B atoms are two types of carbon atoms that are not equivalent to graphene, see Figure 7-1(b). The lattice constant is 2.460 Å[17]. Each C atom has four valence electrons, three of which form three σ bonds in sp^2 orbital hybridization with neighboring C atoms, respectively, in the graphene plane. This regular hexagonal crystal structure makes graphene's planar structure extremely stable. The tensile elastic modulus of graphene is as high as 1 TPa and the tensile strength is as high as 130 GPa[7-9]. The structure of graphene remains stable when external forces are applied to it; the other electron in the p orbital contributes to the nonlocalized π and π^* bonds, which form the highest occupied molecular orbital (HOMO) and lowest unoccupied molecular orbital (LUMO). The π and π^* bonds degenerate at point K in the Brillouin zone of graphene, see Figure 7-1(c). The Fermi surface shrinks to a point, forming a bandwidth-free metal-like band structure[18-21], which gives graphene extremely high carrier mobility.

Researchers have described the interactions between graphene's π electrons using a tightly bound model[3,18,20]:

$$\varepsilon^{\pm}(k_x, k_y) = \pm \gamma_0 \sqrt{1 + 4\cos\frac{\sqrt{3}k_x a}{2}\cos\frac{k_y a}{2} + 4\cos^2\frac{k_y a}{2}}, \quad (7-1)$$

where the value of a is $\sqrt{3}\,a_{C-C}$, γ_0 is the transfer integral that corresponds to the matrix element between π orbitals of adjacent carbon atoms, with the magnitude ranging from 2.8 eV to 3.1 eV. The plus energy of Equation (7-1) corresponds to the π^* band, and the minus energy corresponds to the π bond. Figure 7-1(d) illustrates the electronic dispersion in the honeycomb lattice, which is the electronic structure of monolayer graphene. Since the electronic states near the Dirac point(DP) are composed of different sublattices, the dual component wave function corresponds to their relative contributions. The Schrödinger equation corresponding to these electrons is as follows[2]:

$$\hat{H} = \hbar v_F \begin{pmatrix} 0 & k_x - ik_y \\ k_x + ik_y & 0 \end{pmatrix} = \hbar v_F \boldsymbol{\sigma} \cdot \boldsymbol{k}. \quad (7-2)$$

The Fermi group velocity corresponds to $v_F = 1 \times 10^6$ m/s, $\boldsymbol{\sigma}$ corresponds to the Pauli matrix, and $\boldsymbol{k}$ corresponds to the momentum of graphene's quasiparticles. Figures 7-1(e)~(g) show the electronic structures of single-layer graphene and the electronic structure of double-layer graphene in the AA and AB stacking modes.

Graphene's unique physical and chemical properties are closely related to its electronic energy band structure. The energy level distribution of graphene can be calculated based on independent C atoms and the potential generated by the surrounding C atoms as perturbation. Spread out near the DP, the energy is linearly related to the wave vector (similar to the dispersion of photons), and a singularity occurs at the DP (DPs). This means that the effective mass of the electrons in graphene is zero near the Fermi energy level, which also explains the material's unique electrical properties.

7.3 Graphene's novel electronic properties

Graphene's unique crystal and electronic structures give it many novel and unique physical properties, including anomalous quantum Hall effects (QHEs)[20,22-23], ambipolar electric field effects (AEFEs)[24], Klein tunneling,[25-26] and ballistic transport[27-28]. With further research on graphene nanomaterials, more electrical characteristics have been discovered and recognized.

7.3.1 Current vortices, electron viscosity, and negative nonlocal resistance

Electron transport properties are similar to those of viscous fluid in strongly related electronics systems, known as the quantum critical effect, which is a typical collision-controlled mass transfer characteristic. However, the study of this phenomenon has been hindered by the lack of macroscopic features of electron viscosity. Levitov et al. determined the vortex characteristics associated with the easy verification of the macroscopic mass transfer performance[29-32]. The eddy current generated by viscous flow can be applied to the field of the opposite driving current, which results in a negative nonlocal voltage. This suggests that the latter plays an important role in the superconducting zero-resistance viscosity system. In addition to providing a diagnosis of viscous mass transfer from the ohmic flow, the electrical response of the signal changes, providing a powerful way to directly measure the viscosity and resistivity ratios. The intense electron hole plasma interaction in high-mobility graphene provides a unique connection between the critical quantum electron transport and many hydrodynamic phenomena.

It has been a prominent problem that the zero resistance of the viscous flow is related to superconductivity. The viscous flow caused by the vortex leads to the unique macroscopic transport behavior. Levitov et al. predicted that the vorticity of the shear flow with viscosity may cause the return current[33], which is opposite to the applied field. Figures 7-2(a) and (b) show a negative signature of the nonlocal voltage, which provides a clear signature

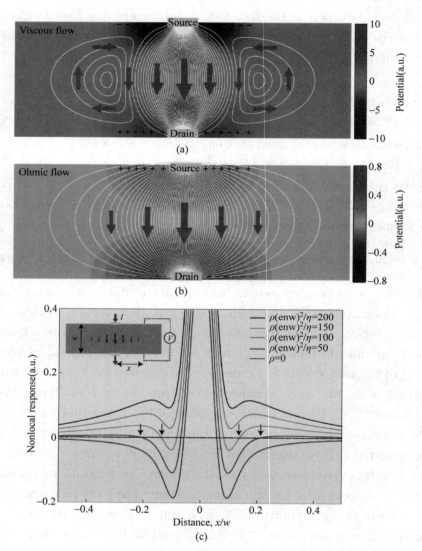

Figure 7-2　Current streamlines and potential map for viscous and ohmic flows. White lines show current streamlines, colors show electrical potential, arrows show the direction of current

(a) Mechanism of a negative electrical response: viscous shear flow generates vorticity and a backflow on the side of the main current path, which leads to charge buildup of the sign opposing the flow and results in a negative nonlocal voltage; (b) In contrast, ohmic current flow down the potential gradient, producing a nonlocal voltage in the flow direction; (c) Nonlocal response for different resistivity-to-viscosity ratios ρ/η[33]. Copyright 2016, Springer Nature

of collective viscosity. The characteristic symbols associated with this behavior change the spatial pattern of the potential, as shown in Figure 7-2, which can be observed directly using modern scanning microscopy techniques[34].

Figures 7-2(a) and (b) show the basic properties of the negative response from the shear flow. The collective behavior of the viscous system results from the momentum exchange of the fast exchange carrier while preserving the conservation of net momentum. The momentum is still referred to as a conserved quantity as it will produce a hydrodynamic momentum transfer mode. That is, the momentum flows in space, in a transverse diffusion to the source drain current and away from the nominal current path. As a result of this process, the shear flow is established to produce vorticity and the application field of the (non-compressible fluid) opposite reflux direction. The direct effect of such a complex and apparently non-current mode on the electrical response produces a reverse electric field acting on the opposite field driving the source drain current, see Figure 7-2(c). This results in the presence of a negative nonlocal resistance, even in the presence of quite significant ohmic flow, see Figure 7-2(c).

7.3.2 Transition between electrons and photos

Graphene's inimitable electronic properties make it a new infrared frequency domain plasmon waveguide and terahertz metamaterial[35-49]. By excitation of photons or electrons, the collective oscillations of electrons on the surface of a conductor is the surface plasmon. When photons and graphene's surface plasmon are coupled, they form surface plasmon polaritons (SPPs). As an alternative to conventional metal plasmon, graphene surface plasmons (GSP) can be used for optical metamaterials[40], optical absorption, and optical conversion[41-42].

Researchers have found that when light hits a single layer of graphene, it can slow down hundreds of times[48-49]. Researchers have studied the hot carriers located inside graphene, which form GSPs on graphene's surface, see Figure 7-3(a). Figures 7-3(b)~(g) show the test results and analysis of the experimental results. Slowing photons (particles of light) travel through a single layer of graphene many as electrons do through the same material. Electrons travel very fast in a single layer of graphene, at one million meters per second, about one-third of the speed of light in a vacuum. The speed of the two particles is close enough to strongly interact. When the material is adjusted to match the speed of the two particles, graphene slows the speed of the photons and the rapid movement of the electrons across a graphene surface. This suggests that graphene can be used to create a possible intrinsic effect, producing light instead of capturing it. Their theoretical research suggests this could generate light in a new way. This conversion is possible because electrons' drift velocity in graphene is close to light, breaking the *light barrier* just as shock waves produce sound when breaking the *sound barrier*. In graphene, this will result in a shock wave of light on a 2D level.

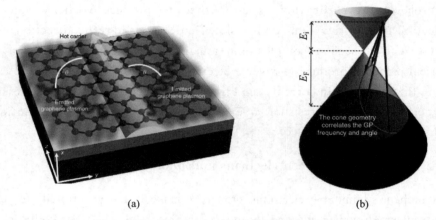

Figure 7-3 Hot carrier localized inside graphene

(a) Illustration of the plasmon emission from charge carriers in graphene via a 2D Čerenkov process. (b) Illustration of the possible transitions. The hot carrier (green dot) has a range of potential transitions (red arrows) with distinct final states (green curves and circles), emitting plasmons that satisfy conservation of momentum and energy (corresponding to the height and angle of the red arrows). This way the cone geometry correlates the GP frequency and angle. The projection of these arrows to a 2D plane predicts the in-plane angle y of the plasmonic emission, matching the (c) map of GP emission rate as a function of frequency and angle. (d) Spectrum of the ČE (Čerenkov effects) GP emission process. (e) Explaining the GP emission with the quantum ČE. The GP phase velocity is plotted as a red curve, with its thickness presenting the GP loss. (f) Illustration of the distribution of hot carriers, which is taken to be an exponential multiplied by the linear electron density of states in graphene. The exponential decay is (g) $e^{E_1/0.1\text{ eV}}$ with maximum hot carrier energy of $E_{i,\max} = 0.2$ eV corresponding to (d)[48]. Copyright 2016, Springer Nature

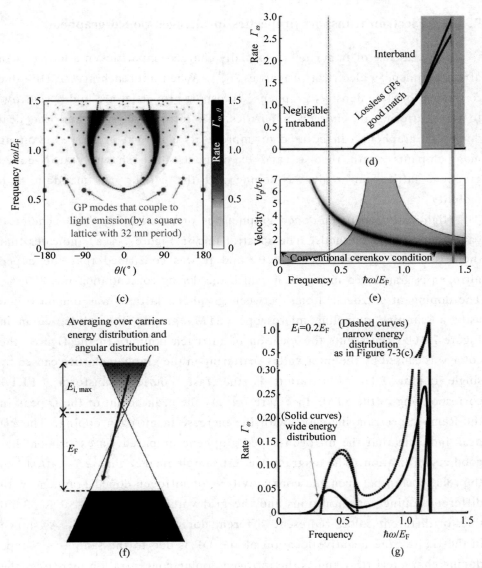

Figure 7-3(Continued)

This new approach may eventually become more efficient and adjustable. Highly efficient and controllable plasmon generation properties, compatible with current microchip technology, are a new method of creating core chips for optical circuits, which is a new research orientation in the evolution of computer technology toward smaller and more efficient devices.

7.3.3 Electron transport properties in nitrogen-doped graphene

The presence of impurities, especially charged impurities in a lattice, can affect graphene's electrical properties[50-54]. When approaching the DP, due to the loss of state density, graphene' transport properties are highly sensitive to the scattering of charged impurities. Therefore, physical and chemical doping of graphene can be used to improve its electrical properties. Among many dopants, both theories and experiments have shown that N-doped graphene lattice can realize N-doping and the carrier can maintain high mobility[55].

High-quality nitrogen-doped graphene is prepared using CVD (chemical vapor deposition) methods. The spectral line in Figure 7-4(a) indicates that there is a characteristic peak at 0, and it was confirmed that the doped nitrogen is graphitized nitrogen, in which the doping concentration was 2.0 %. The doping of nitrogen atoms between graphene lattices was demonstrated using a scanning tunneling microscope (STM) graph. The white region in Figure 7-4(b) represents the location of nitrogen doping. In the figure, the color was caused by the inter-valley scattering in the graphene lattice caused by single doping. The illustration is the fast Fourier transform (FFT) corresponding to the STM. In Figure 7-4(c), the peak height of the D peak in the Raman spectrum increased with the increase in nitrogen doping. The 2D peak indicates that the nitrogen-doped graphene prepared via experiment has good crystallization. The diagram is a test sample image. Figure 7-4(d) shows the relationship between the conductivity σ of nitrogen-doped graphene with different doping concentrations and the grid voltage V_g at $T = 9$ K. Solid lines of different colors represent different doping concentrations. As shown in the figure, the negative location of the DP is due to the sample's N-type doping characteristics, and as the nitrogen doping concentration increases, the DP moves toward negative values. Figure 7-4(e) shows the evolution of the charge impurity density with the nitrogen-doping concentration. Figure 7-4(f) shows that after the introduction of more scattering centers, carrier mobility in nitrogen-doped graphene decreases as the nitrogen-doping concentration increases.

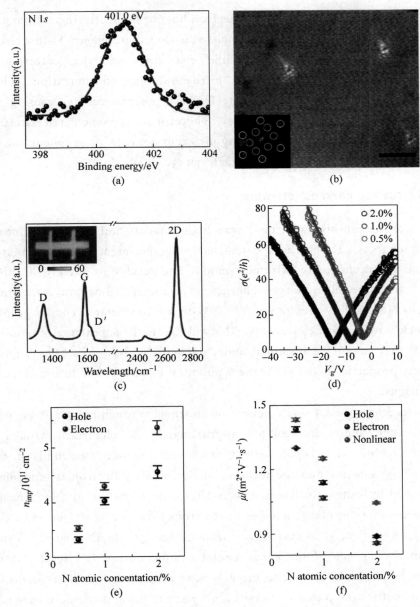

Figure 7-4 Electron transport properties in nitrogen-doped graphene
(a) N 1s core-level XPS spectra of an as-formed graphitic N-doped (2 % N atomic concentration) graphene film. (b) STM image of an as-formed graphitic N-doped graphene sample (V_{bias} = −10 mV and I_{set} = 100 pA). (c) Raman spectra of a graphitic N-doped (2 % N atomic concentration) graphene film. Transport properties of as-formed N-doped graphene films in (d) ~ (f)[55]. Copyright 2017, American Chemical Society

The test results confirmed the electron hole valley scattering asymmetry of the charge transport property of the nitrogen-doped graphene. Valley electron scattering is much stronger than hole scattering, and the scattering rate increases with the increase in the nitrogen-doping concentration. This is because the nitrogen-doped graphene forms a positive center, causing large angle scattering for electron transport. Therefore, graphene carrier scattering can be effectively regulated by adding different amounts of graphene nitrogen to control graphene's electrical characteristics.

7.3.4 Strong current tolerance

Graphene can also withstand very high currents and quickly balance out missing charges. Researchers conducted an experiment to prove that the electrons in graphene are extremely mobile and react very quickly. Collisions of Xe^+ with the supercharged charges on graphene films can detach many electrons from a very precise point[56]. Within a few femtoseconds, graphene can quickly resupply the electrons. This will lead to the appearance of ultrastrong currents that cannot be sustained under normal circumstances. These unusual electrical properties make graphene a potential candidate for future electronic applications.

Researchers used a single xenon ion to travel through the graphene layer, so that each xenon ion carried approximately 20 electrons through the graphene region. At this point, the C atom with a missing electron is squeezed out of the graphene, and the location of the missing electron is immediately replenished by other electrons, which lasts only a matter of femtoseconds. This means there is a large number of electrons flow in a short period of time, that is, a high density of current is transmitted on graphene. Figure 7-5 shows a schematic diagram of the experimental equipment and test results. Strong currents from areas adjacent to the graphene membrane quickly replenish the electrons before the positive charges repel each other and cause an explosion. To accomplish this, graphene must carry a current density approximately 1,000 times higher than any material would normally tolerate. The surrounding current density is 1,000 times that of a normally destructible material, but graphene can withstand these extreme currents without damage. The movement of extremely high charges on graphene is of great significance for a range of

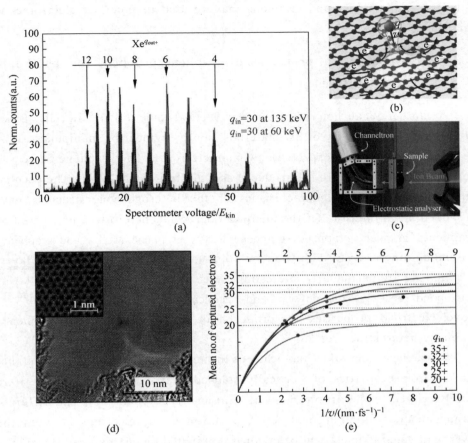

Figure 7-5 Experimental scheme and results

(a) Measured spectra of a $Xe^{[30+]}$ beam at kinetic energies of 135 and 60 keV (blue and red, respectively) transmitted through a freestanding SLG(single-layer graphene) sheet. Exit charge states q_{out} are calculated from the spectrometer voltage of the electrostatic analyzer. The exit charge state distribution shifts towards smaller average exit charge $\bar{q}_{out}$ for slower ions. (b) Schematic of the interaction process between freestanding SLG and an approaching highly charged ion (HCI). The HCI extracts a lot of charge from a very limited area on the femtosecond time scale leading to a temporary charge-up of the impact region. (c) Sketch of the experimental set-up with the target holder and electrostatic analyzer. (d) TEM image of a freestanding monolayer of graphene after irradiation with Xe^{40+} ions at 180 keV with an applied fluence of 10^{12} ions per cm^2 (about six impacts on the shown scale). No holes or nanosized topographic defects could be observed. (e) Average number of captured and stabilized electrons ($q_{in} - \bar{q}_{out}$) after transmission of $Xe^{q_{in}+}$ ions through a single layer of graphene as a function of the inverse projectile velocity for different incident charge states[56]. Copyright 2016, Springer Nature

potential applications, and graphene may be used in superfast electronics in the future.

7.3.5 Novel electrical properties of graphene/graphene van der Waals heterostructure

A van der Waals heterostructure is a vertical stack of 2D materials. Based on the rich functionality of 2D materials, more engineering manipulation can be realized to obtain unexpected new characteristics. The graphene-based van der Waals heterostructure is introduced as a substrate from h-BN. The peculiar physical and chemical characteristics of the heterogeneous structure have provoked much interest. Of the many van der Waals heterostructures based on graphene (graphene/graphene, graphene/h-BN, graphene/MoS_2, and graphene/Si, among others), van der Waals heterostructures produced by double layers of graphene are particularly interesting. Because the lattices of the two layers correspond precisely, the physical and chemical properties are equivalent[57-61]. Novel electrical properties are produced in the superlattices of graphene/graphene heterostructures[62-64].

Researchers from MIT (Massachusetts Institute of Technology) have found that the electron properties of vertically stacked graphene heterostructures are closely related to the arrangement of C atoms in graphene[62]. The different arrangement of atoms also affects the movement of electrons between the layers. In general, electrical behavior is dominated by energy, and the energy of electrons involved in the movement of electrons between atoms within a single layer of graphene is on an order of magnitude of electron volts, while the energy involved in the movement of electrons within a double layer of graphene is on an order of magnitude of several hundred millielectron volts. Figure 7-6 shows a diagram of experimental equipment and calculation results[62].

For graphene with a highly ordered structure, the electrical properties depend on symmetry. The MIT researchers created a rotating, twisted, two-layer graphene heterostructure that controls the electron state of the entire system through interactions between the electrons. The dislocation caused by rotation disaligns the electron band structure in the graphene layer and increases the single cell size. Twisted double layers of graphene produce two

Chapter 7 The Thermal, Electrical, and Thermoelectric Properties of Graphene Nanomaterials

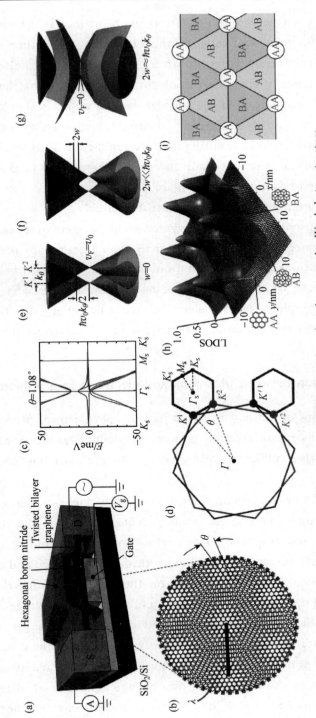

Figure 7-6 Novel electrical properties of graphene/graphene van der Waals heterostructure (a) Schematic diagram of twisted double graphene devices. (b) Moiré pattern in graphene/graphene heterostructure. (c) The band structure of the double graphene at an angle of $\theta = 1.08°$. (d) Mini Brillouin zones in graphene/graphene heterostructure. (e)~(g) Effect of hybridization between different layers. (h)~(i) Density of state of local heterostructure is calculated by first principles[62]. Copyright 2018, Springer Nature

new electron states[63-64]. One electronic state is the Mott insulated body that results from the strong repulsion between the electrons. The other is the superconducting state that results from the strong attraction between electrons to produce zero resistance. When the small rotation angle approaches the magic angle ($< 1.05°$), distortion of the vertical stack of atoms in the double-layer graphene area will form a narrow electronic band, enhancing the electronic interaction effect, thus resulting in a nonconductive Mott insulation state. In a Mott insulation state, a small charge carrier can be successfully converted into a superconducting state. Figure 7-7 shows the test results of superconductivity in a graphene/graphene heterostructure[63].

The MIT researchers created rotationally twisted double layers of graphene that control the electron states of the entire system through interactions between electrons. The dislocation caused by rotation misaligns the electron band structure in the graphene layer and increases the single cell size. The study found that in stacked graphene layers, electrical behavior is very sensitive to the arrangement of atoms, affecting the movement of electrons between layers.

7.3.6 The interaction between plasmons and electrons in graphene

The surface plasmons of 2D nanomaterials have unique electrical properties, and applications such as the interaction between plasmons and electrons and catalytic reactions are also well known, especially in surface plasmons based on graphene[65-70].

Cao et al. reported on a plasmon-exciton coupling device[69]. They first prepared a complementary metal-oxide semiconductor (CMOS)-like device composed of gold dots and graphene. The device can easily be added to the gate voltage or can inject current into the device through the source and drain electrodes. They included a series of studies on the electrical properties of plasma-exciton-coupled devices under different gate voltages and source-drain voltages. The source-drain bias voltages at different gate voltages were measured before and after the catalyzed molecules were placed in the center of the device. The source-drain bias increased at the same gate voltage after the molecules were placed and showed varying bias change behaviors at different gate voltages, see Figures 7-8(a) and (b). This occurred because the chemical

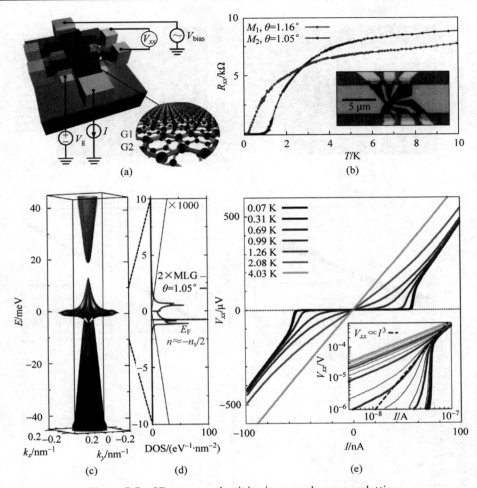

Figure 7-7 2D superconductivity in a graphene superlattice

(a) Schematic of a typical twisted bilayer graphene (TBG) device and four-probe measurement scheme. The stack consists of top h-BN, rotated graphene bilayers (G1, G2) and bottom h-BN; (b) Measured four-probe resistance ($R_{xx} = V_{xx}/I$) in two devices M1 and M2, with twist angles $\theta = 1.16°$ and $\theta = 1.05°$ respectively. (c) The band structure of TBG at $\theta = 1.05°$ in the first mini-Brillouin zone (MBZ) of the superlattice. (d) The density of states (DOSs) corresponding to the bands shown in (c), zoomed in to $-10 \sim 10$ meV. (e) I-V curves for device M2 measured at $n = -1.44 \times 1,012$ cm^{-2} and various temperatures. At the lowest temperature of 70 mK, the I-V curve shows a critical current of approximately 50 nA[63]. Copyright 2018, Springer Nature

potential of graphene is regulated by the gate voltage. Figures 7-8(c) and (d) show the operation of the device current as the bias voltage changes. First, the volt-ampere characteristics of the bias voltage and current showed a perfect

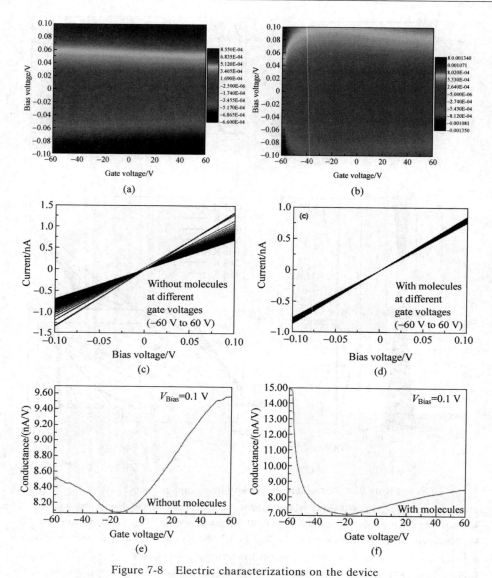

Figure 7-8 Electric characterizations on the device
(a)~(b) Gate-and bias-voltage-dependent electrical current for without and with molecules, (c)~(d) The current varies with bias voltage at different gate voltages, and (e)~(f) Gate-voltage-dependent conductance for devices without and with molecules, where $V_{Bias} = 0.1$ V[69]. Copyright 2017, Wiley

linear relationship. Second, the volt-ampere characteristic curve (slope) changed little at different gate voltages when no molecules were added to the device. When there was no molecule, the gate voltage had little ability to regulate the chemical potential of graphene. Third, after the device was added to the molecule, the slope of the volt-ampere characteristic curve was very different, indicating that the gate voltage had an increased regulation range for the device's chemical potential. These phenomena indicate that charge transfer occurs between the molecule and graphene, and the gate voltage also regulates the charge transfer. This can also be concluded by the relationship between the gate voltage and the conductivity, as shown in Figures 7-8(e) and (f). Focusing on the relationship between the volt-ampere characteristic curve and the gate voltage, the gate voltage can be regulated in both positive and negative directions for charge transfer. This is a very good property for photooxidative reduction catalysis. Since the oxidation reaction requires holes, the reduction reaction requires electrons. Such devices provide free holes or electrons for the reaction through different gate voltage controls.

7.4 The thermal and thermoelectric properties of graphene

7.4.1 The TC's measurement of graphene

In 2008, researchers used confocal micro-Raman spectroscopy to measure the heat transport of suspended graphene nanoribbons[71-72]. They found that the thermal conductivity (TC) (κ) of single-layer graphene (SLG) reached 3,500~5,300 W/m·K, and the mean free path of the phonons was 775 nm at room temperature[71]. The device is shown in Figure 7-9(a). The experimental results confirmed that the heat transfer of graphene mainly comes from the phonon contribution. The same study was conducted again in 2011. To be immune to air heat transfer, researchers measured suspended graphene at the pore size of 2.9 ~ 9.7 μm in a vacuum environment to eliminate the loss of ambient air heat. Figure 7-9(b) shows the device. The TC of graphene was measured up to $(2.6 \pm 0.9) \sim (3.1 \pm 1.0) \times 10^3$ W/m·K at a temperature of 350 K. The experimental device configuration is shown in Figure 7-9(b) near the TC of graphene up to $(2.6 \pm 0.9) \sim (3.1 \pm 1.0) \times 10^3$ W/m·K[72].

The phonon transmission mode and scattering mechanism of graphene's TC have a significant impact. Numerous studies have shown that the substrate of graphene has an unavoidable effect on its TC. After the 2D material contacts

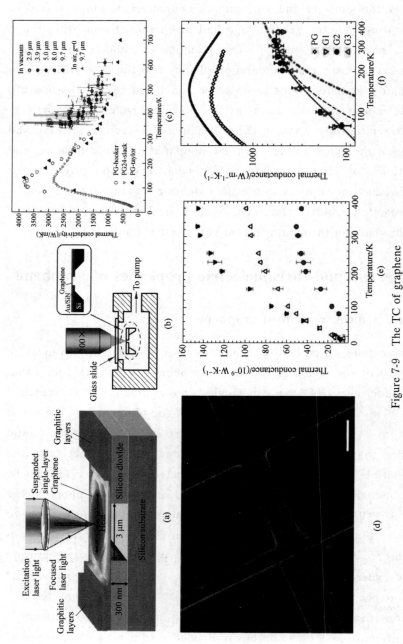

Figure 7-9 The TC of graphene

(a) Schematic of the experiment showing the excitation laser light focused on a graphene layer suspended across a trench[71]. Copyright 2008, American Chemical Society. (b) Schematic of the experimental setup for thermal transport measurement of suspended graphene[72]. (c) TC of the suspended CVD graphene as a function of the measured temperature of the graphene monolayer suspended in vacuum over holes of various diameters[72]. Copyright 2011, American Chemical Society. (d) SEM images of the suspended device, the central beam, and the folded edge of the SLG ribbon near the right electrode. (e) Measured thermal conductance of G2 before (solid downward triangles) and after (unfilled upward triangles) the SLG was etched, with the difference being the contribution from the SLG (circles). (f) The relation diagram of TC of different samples with temperature change[73]. Copyright 2010, Science

the substrate, the TC normally decreases significantly after the 2D material contacts the substrate because the thermal conduction is mainly determined by the phonon conduction. When the graphene comes into contact with the substrate[73-75], the surface or edge of the graphene is very sensitive. In 2010, researchers measured the TC of graphene deposited on silica substrates and found that it had been reduced due to scattering at the substrate interface, but it still reached 1,000 W/m · K[73]. Figure 7-9(c) shows the TC of the suspended CVD graphene as a function of the measured temperature of the graphene monolayer suspended in a vacuum over holes of various diameters. Figure 7-9(d) shows a STM representation of the graphene sample. The thermal properties of graphene nanostructures may play a very important role in future nanodevices. Figure 7-9(e) shows a comparison between the TC of graphene before and after etching and that of single-layer graphene. A solid inverted triangle represents the TC before etching, a hollow positive triangle represents the TC after etching, and a circle represents the TC of single-layer supported graphene. Figure 7-9(f) shows the relation diagram of thermal conductivity of different samples with temperature changes.

It is apparent from the measurement of the TC of graphene that all factors, such as the shape and thickness, affect graphene's TC. In general, the size of the graphene in the heat transfer direction is positively correlated with the TC. In general, the larger the linear size of graphene, the higher the TC; when graphene is on the substrate, the substrate has a great influence on the TC. This is because the interaction between the graphene and the substrate leads to changes in the graphene's lattice constant. The temperature also affects the magnitude of the TC.

7.4.2 Length-depended and temperature-depended TC of graphene

Based on the relationship between the TC and the length, researchers also specifically analyzed the relationship between the TC and the temperature[75]. Figure 7-10(a) shows the variation in the thermal conductance (σ) of a graphene nanoribbon with temperature (T) over a length of 9 μm to 300 nm. Figure 7-10(b) shows an image of the TC (κ) with length (L) at temperature of 120 K and 300 K. R_{total} is the thermal contact thermal resistance. The results indicate that all of the graphene nanoribbons exhibit the same properties as the temperature. Thermal conductance (σ) adds to the platform as the temperature increases. The TC (κ) increases with the length of the graphene nanoribbons over the entire temperature range.

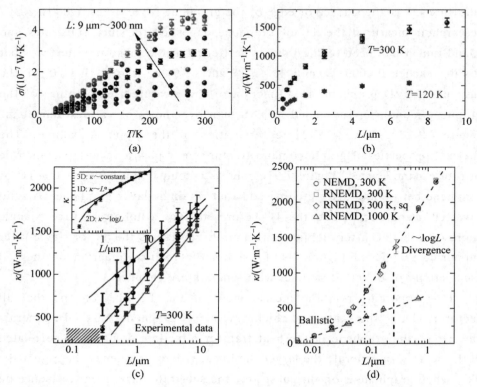

Figure 7-10   Thermal conduction versus temperature in (a) and (b). Experimental and simulation results on length-dependent TC in (c) and (d)[76]. Copyright 2014, Springer Nature

Earlier sections of this article cited examples of theoretical calculations for the TC of graphene[76]. However, the role of graphene in the ZA mode during heat transfer remains controversial. In particular, the solution to the Peierls-Boltzmann equation shows that the ZA mode is the main heat carrier for graphene transport and has a strong dimensional dependence[77-79]. Researchers through experimental data and NEMD simulation resulted in the perfect combination[80-81]. Figure 7-10(c) shows the relationship between the TC (κ) and the length (L)[75]. The value of κ increased with the increase in L at a temperature of 300 K. The temperature did not saturate in the longest $L = 9$ μm of graphene. k was proportional to log L at $L > 700$ nm. This is because the length of graphene is comparable to the phonon mean free path. The researchers also observed that the size far exceeded the ballistic length at 1,000 K and at 300 K (Figure 7-10(d)). After comparing the different aspect ratios of the graphene nanoribbons, their simulated results agree with Figure 7-10 (d). The relationship between TC and length of the single-layer graphene nanoribbons has been revealed experimentally.

The researchers used experimental measurements and non-equilibrium molecular dynamics to simulate the thermal conduction of a single layer of suspended graphene as a function of the temperature and sample length. Unlike bulk materials, the TC continues to increase at 300 K. Although the length of the sample is much larger than the mean free path of the phonons, the TC remains logarithmically discrete with the length of the sample. This is the result of the 2D properties of phonons in graphene, which provides a basis for understanding heat transfer in 2D materials.

7.4.3 Influence of boundary or configuration on thermal property and thermal rectification effect

Many theoretical studies and experiments have shown that the different boundaries and configurations of graphene nanoribbons have a certain impact on the TC[82-93].

The TC of zigzag-type graphene nanoribbons is 30 % higher than that of armchair-type graphene nanoribbons at room temperature, indicating obvious heat transport anisotropy[82,90]. Figures 7-11(a) and (b) demonstrate that the heat flux of Y-type graphene nanoribbons studied via molecular dynamic (MD) simulation shows a significant thermal rectification effect. Moreover, the bilayer Y-type nanoribbons can achieve a larger rectification ratio due to the interaction between the layers than the single-layered Y-type nanoribbons, thus providing a theoretical basis for the thermal management of nanoelectronics[84]. Figure 7-11 (c) shows the phonon properties of three-terminal graphene nanojunctions (GNJs)[85]. This study demonstrates that the transport direction of the heat flux moves along the narrow end to the wide end and produces an obvious ballistic thermal rectification effect. The thermal rectification ratio of 200% is dependent on the asymmetry of the nanojunction, indicating excellent ballistic heat transport properties, hence making the preparation of nanodevices possible.

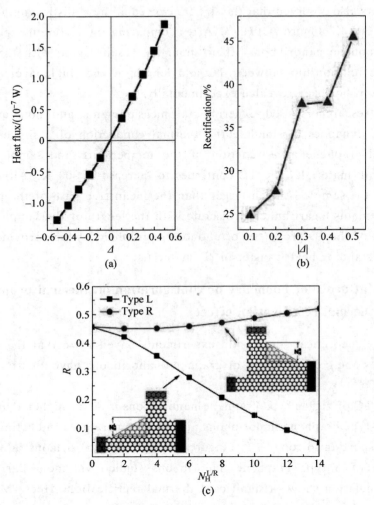

Figure 7-11 Thermal results
(a) Heat flux (J) versus Δ. (b) Rectifications versus Δ for the single-layer graphene Y junctions[84]. Copyright 2011, Royal Society of Chemistry. (c) The rectification ratio R versus the side height of corner $N_H^{L/R}$ for the two types of asymmetric TGNJs (three-terminal graphene nanojunctions) with armchair-edged corner[85]. Copyright 2011, American Physical Society

7.4.4 The effect of atomic edge variation and size change on TC

Using the non-equilibrium green function (NEGF) method combined with first-principles calculations, researchers revealed the influence of the edge atom position on the TC of graphene nanoribbons[94]. Figure 7-12(a) and (b) present the results of this study. When the width of the nanoribbons is $n >$ 12, the TC possesses obvious quantization characteristics and is independent of the nanoribbon width. When the nanoribbon width is $2 < n < 12$, due to the obvious boundary effect of the nanoribbons, the quantum heat transport is destroyed.

Another theoretical simulation revealed the dependence of the TC on the length of graphene[95]. Researchers simulated the relationship between TC and lateral dimension of graphene nanoribbons and again revealed the mean free path of acoustic phonons and the non-monotonic dependence of the nanoribbon length L. Moreover, past studies reported that the bulk 3D phonons in graphene greatly contribute to the TC in the graphene plane. Figure 7-12(c) shows the results.

Many theoretical calculations and experiments have revealed that graphene's TC can be changed through the hybridization of graphene[96], adsorption of metal atoms[96-97], gradient surface hydrogenation[76,98], grain size engineering[99], fluorination[100], carbon isotope doping[101], vacancies and defects engineering[102-103], and so on.

7.4.5 The thermoelectric properties of graphene

The performance of thermoelectric materials is generally measured by the thermoelectric optimal value ZT (the thermoelectric figure of merit)[58,104]:

$$ZT = \sigma S^2 T / (\kappa_e, \kappa_{ph}), \quad (7\text{-}3)$$

where σ is the electronic conductivity; S is the Seebeck coefficient, κ_e and κ_{ph} are the electronic thermal conductance and phonon thermal conductance, respectively; and T is the average temperature of the device. The formula indicates that the higher the ZT, the higher the conversion efficiency between the device's thermal energy and the electric energy. Outstanding thermoelectric materials must have higher Seebeck coefficients, good electrical conductivity, and very low TC. Therefore, it has become a focus of research to identify thermoelectric materials with higher Seebeck coefficient, improve their electronic conductivity, and reduce their TC[105-107].

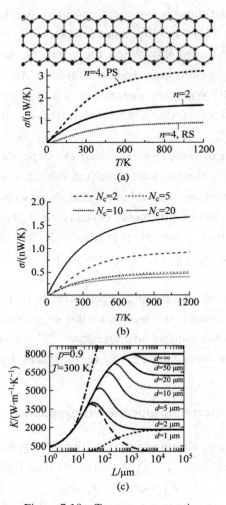

Figure 7-12 Transport properties

(a) Transport properties of GNRs with width $n = 2$ and $n = 4$. Dotted curve corresponds to the case for the relaxed structure (RS) of GNR-4 and dashed curve corresponds to the case for the perfect structure (PS) of GNR-4. Full circles represent the configuration of the perfect GNR-4. Circles with and without "+" represent two configurations after relaxation. (b) Thermal conductance of the sandwiched device versus temperature for different numbers of cells N_c in the central region. For comparison, the full line corresponds to the conductance of an infinite GNR-2[94]. Copyright 2009, American Physical Society. (c) Dependence of the TC of the rectangular graphene ribbon on the ribbon length L shown for different ribbon width d. The specular parameter is fixed at $p = 0.9$[95]. Copyright 2012, American Chemical Society

Chapter 7 The Thermal, Electrical, and Thermoelectric Properties of Graphene Nanomaterials

1. Thermoelectric properties in graphene nanoribbons (GNRs)

Thermoelectric conversion is currently a popular field of research. Researchers studied the ballistic thermoelectric properties of graphene-nanoribbon-based heterojunctions[106]. The results reveal that the binding structure affects the transport of electrons, whereas the fluctuations in electrons are strongly enhanced by the thermoelectric power. The first four panels in Figure 7-13 show the ZT at $T = 300$ K and the ZT at $T = 100$ K for two different graphene edge heterojunctions. The calculation of the ballistic thermoelectric properties based on graphene nanoribbons heterostructures

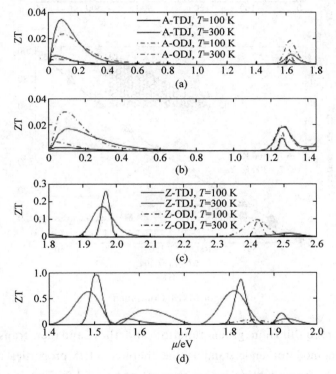

Figure 7-13 Thermoelectric properties of graphene

The figure of merit ZT at the two different temperatures at 100 K and 300 K in (a) and (b) For armchair edge junctions. (c) and (d) For zigzag-edge junctions. Thermal current (left vertical axis) and average temperature (right vertical axis) vs. temperature difference ΔT[106]. Copyright 2012, American Physical Society. The cashed boxes highlight NDTC in (e). The inset shows the structure of the GNR (1.5 nm × 6 nm). (f) Thermal current (left vertical axis) and average temperature (right vertical axis) vs. temperature difference ΔT in triangular GNRs shows in the inset. The dashed box highlights NDTC[107]. Copyright 2011, American Physical Society

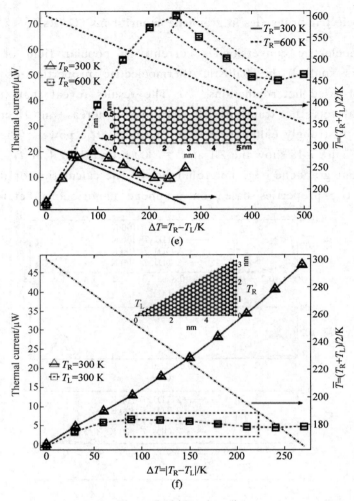

Figure 7-13(Continued)

using the non-equilibrium green function and the Landauer transport theory provides a method for understanding the thermoelectric properties of graphene. The phonon heat conduction under the influence of different heterostructures was basically similar, but the influence of the heterostructure geometry and geometric details on electron transport was substantial, and the change in electronic transport greatly enhanced the thermoelectric properties. These parameters further improve the thermoelectric properties of nanomaterials.

The nonlinear thermoelectric properties of triangular graphene nanoribbons at the armchair boundaries of AGNR and ZGNR were also modeled by MD[107]. Researchers observed that negative differential thermal conductance (NDTC) appeared in the system when GNRs had large temperature deviations and beyond the range of the linear influence, and the NDTC could be controlled by the temperature. For rectangular GNRs, as the GNR length increases, the NDTC gradually decreases and eventually disappears, see Figure 7-13(e). For triangular GNRs, the NDTC only exists when heat flows from the narrow end to the wide end, see Figure 7-13(f). These results provide the theoretical basis for thermal management and thermal signal processing of nanomaterials.

2. Thermoelectric spin voltage (TSV) in graphene

Spin-dependent thermal effects, the interaction between spin and thermal current, have been demonstrated in ferromagnetic materials, and the spin Seebeck effect is one of the most interesting phenomena[16,108-112].

The spin current caused by the thermal gradient has been detected by the inverse spin Hall effect[113-115]. Graphene, by virtue of its highly efficient spin transmission[116-118], energy-dependent carrier mobility, and unique density of states[3,119], has become the focus material in this direction.

$$S_{\text{Mott}} = \frac{\pi^2 k_B^2 T}{3e} \cdot \frac{\mathrm{d}\ln R}{\mathrm{d}\mu}\bigg|_{\mu=\mu_F}, \qquad (7\text{-}4)$$

where k_B is the Boltzmann constant and μ_F is Fermi energy.

Researchers prepared graphene-based detection devices and measured their properties and spin thermoelectric parameters. The results show that the thermal gradient of the carriers in a graphene-based transverse spin valve can lead to increased spin voltage in the areas near the graphene charge neutrality point (CNP). Similar to the thermal voltage in a thermocouple, the effect caused by the thermoelectric spin voltage can be enhanced by the thermal carrier generated by applying the current[120-123]. These results and research methods such as maintaining the purity of the spin signals through the thermal gradient and adjusting the remote spin accumulation by changing the spin injection bias voltage are very important for driving graphene-based spin electronic devices through thermal spin.

Figure 7-14(a) shows the experiment device, which consists of two graphene nanosheets (GNs)[124]. Different carrier concentrations on GNs lead

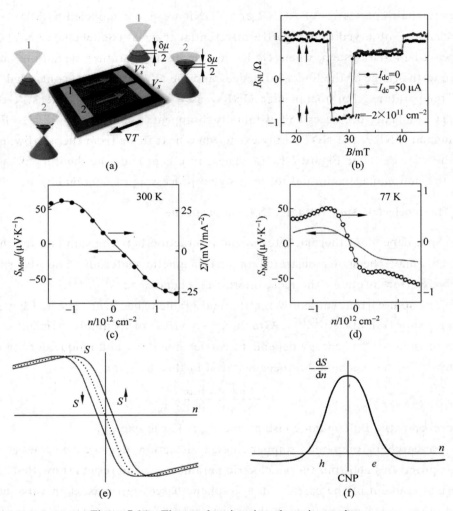

Figure 7-14 Thermoelectric spin voltage in graphene
(a) Schematic of the test device. (b) Non-local spin resistance R_{NL} versus magnetic field B along the magnetization of the electrodes for $I_{dc} = 0$ (black) and $I_{dc} = 50$ μA (red). (c) Comparison between the Mott Seebeck coefficient S_{Mott} and carrier density n (blue line and left axis) obtained from the graphene square resistance R at room temperature and the quadratic fitting coefficient σ versus n. Modelling and roles of Seebeck coefficient and the spin accumulation in (e) and (f), qualitative representation of S (a) and its derivative dS/dn (b) about the CNP[124]. Copyright 2018, Springer Nature

to varying Seebeck coefficients S_1 and S_2. The temperature difference ΔT on both sides of the nanosheets leads to thermoelectric voltage V_s ($V_s = V_s^+ - V_s^- = -(S_2 - S_1)\Delta T$), which is caused by the temperature gradient ∇T.

When the carrier concentrations $n_1 = -n_2$, $V_s = \delta\mu/e$. Figure 7-14(b) demonstrates nonlocal spin resistance R_{NL} changes slightly with the magnetic induction curve when the carrier concentration is $n = -2\times 10^{11}/\mathrm{cm}^{-1}$ and the electrode current is $I_{dc} = 0$ (black line) and $I_{dc} = 50$ μA (red line), respectively. Figure 7-14(b) clearly indicates that the ferromagnetic property of the system switches (the blue arrow is the relative direction of the ferromagnetic magnetization) when the magnetic induction intensity is at $B_1 = 30$ and $B_2 = 40$. Figures 7-14(c) and (d) demonstrate the characteristics of the device. Figures 7-14(e) and (f) show the qualitative representation of S and its derivative $\mathrm{d}S/\mathrm{d}n$ about the CNP.

7.5 The recent applications in electronic and thermal properties of graphene

The novel electrical and thermal properties of graphene have been gradually recognized, and more applications are being widely used in photoelectric and thermoelectric devices.

7.5.1 High-efficient TC composite film and flexible lateral heat spreaders

Polymer composite materials are ideal for horizontal heat dissipation in electronic equipment. Researchers prepared a highly efficient TC of polymer composites[125-129].

Ding et al. produced a graphene-nanocellulose composite film using vacuum-assisted self-assembly (VASA). The highly crystalline nanofibers driven by natural forces are conducive to the formation of thermal conduction paths, see Figure 7-15(a)[125]. Graphene's orientation was analyzed using effective medium approximation (EMA) to improve the TC of the composite. They changed the TC of the film by remedying the defects of graphene, see Figure 7-15(b). Through the qualitative and quantitative characterization of graphene's defects, the increase in the defects makes the TC decrease[126]. They fabricated composite films with high in-plane TC and thermal anisotropy using layer-by-layer assembly (LBL). The results show that when the content of graphene is reduced to 1.9 wt%, the in-plane TC of the composite film reaches 12.48 W/m · K and the anisotropy coefficient is 279 (Figure 7-15(c))[127]. Composite films have great research and application potential in various fields due to their excellent TC preparation and adjustment methods. Figure 7-15(d) shows the TC of a composite film prepared by another group of researchers[128].

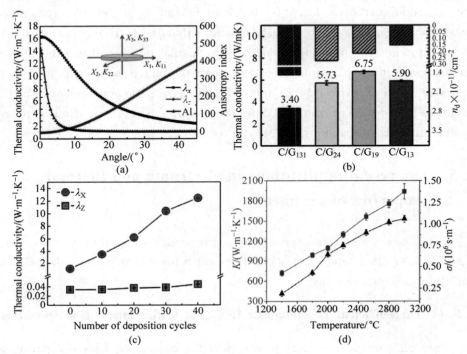

Figure 7-15　High-efficient thermal conductivity

(a) Effect of graphene sheet orientation on theoretical heat transfer calculated by EMA[125]. Copyright 2015, Royal Society of Chemistry. (b) The relationship between defect and TC of graphene.[126] Copyright 2017, Elsevier. (c) The test results of TC of composite film[127]. Copyright 2017, American Chemical Society. (d) The thermal and electrical conductivity of df-GFs (debris-free GO-based graphene film) annealed at different temperature[128]. Copyright 2018, Wiley

7.5.2　Thermal conductance modulator

Based on graphene's robust TC and strength, a graphene-nanoribbon-based TC modulator was developed in 2011[130]. By changing the geometry of the graphene nanoribbons, the researchers were able to control and reverse regulate the thermal conductance. The conductance of this regulation can range up to 40% of the unfolded graphene nanoribbons, as shown in Figures 7-16(a) and (b). At this point, the folding angle of the GNRs exhibits a linear relationship with the conductivity and changes with the distance between the graphene layers. This thermal device has potential for use in phonon circuits and nanoscale thermal management.

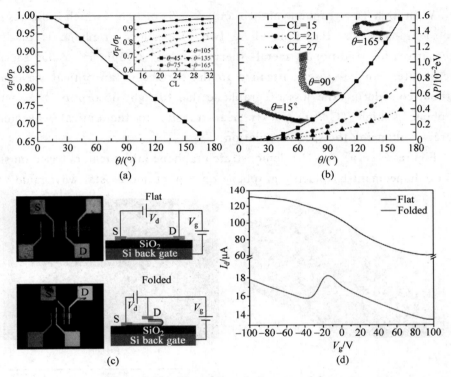

Figure 7-16　Thermal conductance modulator

(a) σ_F/σ_P versus θ for the A-FGNR-α with CL = 15 at T = 300 K. The inset shows the different θ value belongs to σ_F/σ_P versus CL[130]. Copyright 2011, American Physical Society. (b) The ΔP versus θ for the A-FGNR-α at different values of CL. (c) The optical images and circuit diagrams of graphene FETs, flat (top) and (bottom). (d) The transfer curves of the functionalized graphene FETs as a function of back-gate voltage in the flat (black line) and folded (red line) states[131]. Copyright 2017, Science

Another interesting study was based on the 3D graphene structure of a curved fire fold that was transformed by planar graphene[131]. The unique properties of self-folding 3D graphene using multiscale molecular dynamics models have been revealed, making it possible to encapsulate cells or construct folded transistors; Figures 7-16(c) and (d) show the results of this experiment.

7.5.3 Graphene microheaters based on slow-light-enhanced energy efficiency

With high TC, graphene absorbs only 2.3 % of light, which indicates it is almost transparent. Based on these two properties, graphene is the best alternative to traditional metal thermometers. Graphene as a thermal microheater can be closely attached to the surface of an optical waveguide without considering the loss of graphene due to light absorption[132], while graphene's high TC can quickly heat transfer to the optical waveguide, thereby enhancing the speed regulation.

Figures 7-17(a) and (b) demonstrate graphene microheaters based on slow light enhancement by placing graphene on a photonic crystal waveguide with

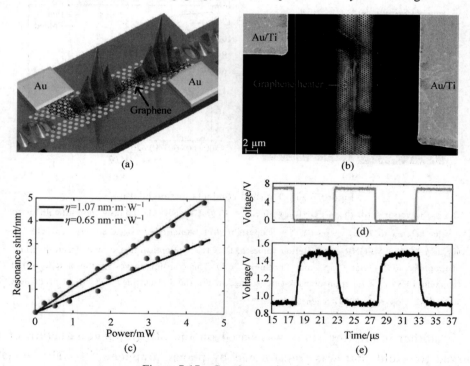

Figure 7-17 Graphene microheatrs
(a) Schematic of the slow-light-enhanced graphene heater. (b) False-color scanning electron microscope image of the slow-light-enhanced graphene heater. (c) Measured resonance shifts for the interference dips at 1,525.12 nm (blue) and 1,533.71 nm (red) as functions of the applied heating power. (d) Driving electrical signal and (e) corresponding temporal response signal[132]. Copyright 2017, Springer Nature

the light propagation speed reduced to 1/30 of the vacuum. The effective heating length of the optical signal is greatly increased, thus reducing the energy loss of the optical signal. Figure 7-17(c) shows the results of a graphene thermal microheater. The thermal regulation efficiency of the device is as high as 1.07 nm · m/W, which is nearly double that of traditional devices. The energy consumption of the optical signal reaching 2P phase shift is 3.99 mW, which is lower than that of traditional metal thermal heaters. The optical signal switching speed is 550 ns, which is 3 orders of magnitude faster than traditional metal thermal microheaters and is far from the fastest regulated nanothermal microheater, see Figures 7-17(d) and (e). The comprehensive evaluation index of the device is 2.5413 nWs, which is 30 times higher than the comprehensive evaluation index of the best nanothermal microheater. This study is expected to be widely used in integrated phased array radar systems and optical arbitrary waveform generators.

7.5.4 Hybrid graphene tunneling photoconductor

Composite photodetectors formed using high-efficient optical materials (such as quantum dots, carbon nanotubes, etc.) and graphene has attracted extensive attention. The photogenerated carriers in the absorbent materials can be effectively transferred to the graphene channel with high mobility to achieve super-high-light response gain[133-138]. However, due to a large number of trap states at the interface between the absorbent materials and the graphene, this type of photodetector is usually slow in response, which restricts its use in high-frequency applications.

Based on graphene/Si hybrid photodetectors, researchers inserted a single-layer MoS_2 between graphene and silicon to improve the performance of photoconductor[139]. The experimental results indicate that the photogenerated carriers flow out of the silicon and enter the graphene channel through the potential barrier of MoS_2 through the quantum tunneling effect under the condition of illumination. There are no suspension bonds on the surface of the molybdenum disulfide, which reduces the trap state and effectively passivates the interface. Comparing the detection performance of devices before and after MoS_2 insertion, the response speed of the latter is three orders of magnitude higher than that of the former, that is, the response time is 17 ns and the response degree is 3.4×10^4 A/W. Figure 7-18(a) presents a schematic diagram of the device. Figures 7-18(b), (c) and (d) show the device's

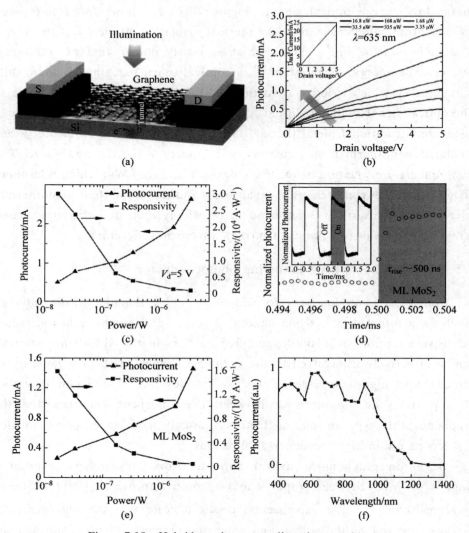

Figure 7-18 Hybrid graphene tunneling photoconductor

(a) Schematic diagram of hybrid graphene photoconductor. (b) Photocurrent vs. drain voltage under various light powers at 635 nm wavelength. The arrow indicates the direction of light power increase. The inset shows the dark current of the device. (c) Power-dependent photocurrent and photoresponsivity at 5 V drain voltage calculated from the data in (b). (d) Normalized photocurrent vs. illumination wavelength. (e) Transient characteristics of the hybrid graphene photoconductor with MoS_2 under 635 nm illumination, showing a rising time of ~500 ns. Inset is the switching performance over three periods of square-wave modulation. (f) Photocurrent and responsivity as functions of the illumination power of the device with MoS_2[139]. Copyright 2017, Springer Nature

photoelectric response characteristics. Figures 7-18(e) and (f) demonstrate the experimental results of the device. This type of nanodevice graphene-based heterostructure shows excellent performance, which provides new applications for graphene-based electrical characteristics.

7.5.5 Graphene electrodes

P-doped graphene-based electrodes improve the performance of the device by reducing its resistance. However, the resistance of the electrode will gradually increase in the environment, which will affect the actual use of the graphene electrode[140-146].

Researchers recently used perfluorinated polymeric sulfonic acid (PFSA) as a dopant molecule to conduct p-doped graphene and to build a PFSA-based p-doped graphene electrode[146]. The electric dipole of the sulfonate group proton in the PFSA molecule strongly attracts electrons, which leads to the high ionization potential of the perfluorocarbon skeleton. Doped with PFSA graphene electrodes, the device's surface resistance (R_{sh}) decreased by 56% and its surface potential increased by 0.8 V. Moreover, the graphene electrode of this configuration, although treated with a chemical agent, was stable under high temperatures and long-term exposure to air. This graphene-based electrode can be used to produce phosphorescent organic light-emitting diodes with high hole injection and substantial luminescence efficiency.

Figure 7-19(a) presents a structural diagram of the PFSA-doped graphene electrode. Figure 7-19(b) shows the calculated results of the device. Figures 7-19(c), (d) and (e) demonstrate the performance examination results of the electrode. Figure 7-19(f) shows the performance test results of OLED based on the graphene electrode. The construction of a doped graphene electrode has consistently been a popular topic of research, but many researchers have focus on the regulation of its electrical properties and ignore the problem of its stability. The study used organic macromolecules as dopants to improve the electrical properties of graphene and its stability. This work can promote the construction of a stable graphene electrode and its application.

7.5.6 Dirac-source field effect transistors (DS-FET)

The development trend of integrated circuits has changed from the pursuit

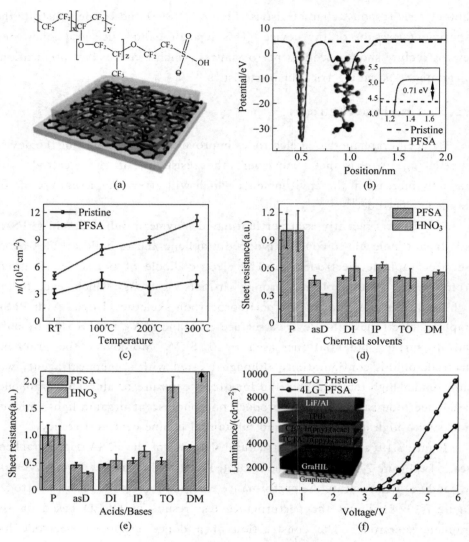

Figure 7-19 Graphene electrode
(a) Chemical structure of perfluorinated polymeric sulfonic acid (PFSA), and schematic drawings of graphene doped using PFSA (+: hole, −: electron). (b) Calculated electrostatic potential of the most stable configuration of PFSA-doped graphene (inset: difference in work function between pristine and PFSA-doped graphene). (c) Averaged n of thermally annealed pristine and PFSA-doped graphene with various Ta calculated from Raman spectroscopy results. (d) Various solvent treatments (e) Acid and base treatments as a function of exposure time. (f) Luminance vs. voltage of green phosphorescent OLEDs with pristine and PFSA-doped graphene anode (inset: Schematic device structure OLED)[146]. Copyright 2018, Springer Nature

of performance and integration to the most effective way to reduce power consumption, which is to reduce the working voltage. Currently, the working voltage of the integrated circuit (14/10 nm technical node) of a complementary metal-oxide semiconductor (CMOS) is reduced to 0.7 V, while the thermal excitation limit (60 mV/DEC) of the MOS transistor's subthreshold swing (SS) makes it impossible to reduce the working voltage of the integrated circuit to below 0.64 V. What existing can realize SS < 60 mV/DEC transistors are tunneling FET and negative capacitance FET, they have a low speed or important defects such as poor stability, unfavorable integration, and the lack of practical value. The ultralow power consumption transistor used in future integrated circuits not only must obtain SS < 60 mV/DEC, ensuring the open state current is sufficiently large, but also requires stable performance and simple preparation[147-157].

Researchers in Beijing recently reexamined the MOS transistor and the physical limits of its threshold swing[158]. They proposed a new type of ultralow power FET and adopted doped graphene as a *cold* electronic source with carbon nanotubes as the active channel. Semiconductor sources with high efficiency top grid structure have been built as DS-FET. The threshold value of swing experiments has been implemented at 40 mV/DEC at room temperature (Figure 7-20). The results of variable temperature measurement indicate that there is an obvious linear relationship between the DS-FET subthreshold amplitude and the temperature. This indicates that the carrier transport of transistors is a traditional thermal emission mechanism rather than a tunneling mechanism. DS-FET has excellent scalability. When the channel length of the device decreases to 15 nm, it can still achieve a subthreshold swing of 60 mV/DEC.

Most importantly, the DS-FET has a proposed driving current much higher than that of tunneling transistors compared to metal-oxide-semiconductor FET, and its SS < 60 mV/DEC spans a larger range of currents. As the key parameter of the comprehensive index of open and closed state characteristics of sub − 60 mV/DEC (that is, the current at SS = 60 mV/DEC), I_{60} = 40 $\mu A/\mu m$, which is 2,000 times the published best tunneling transistor and fully meets the standards of the international semiconductor development roadmap (ITRS) for the practical application of sub 60 mV/DEC devices. The open and closed current of a typical DS transistor at a working voltage of

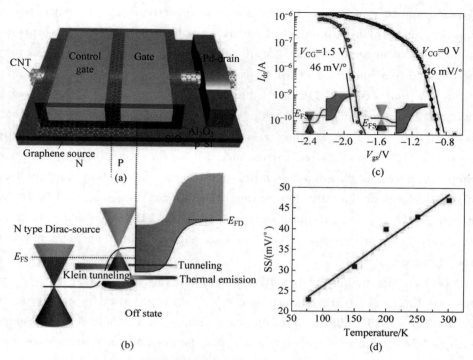

Figure 7-20 Dirac-source field-effect transistors

(a) Schematic diagram showing a DS-FET with a control gate in addition to the normal gate. (b) Schematic diagrams illustrating the off-state of the DS-FET. (c) Transfer characteristics of a typical DS-FET at different V_{CG}. Circles and lines represent respectively experimental and simulated results. Olive represents results obtained at $V_{CG} = 1.5$ V, and blue represents that at $V_{CG} = 0$ V. Inset figures are schematic band edge profiles for fitted data situations. (d) Temperature dependent SS of a typical DS-FET measured at temperatures between 77 K and 300 K, V_{CG} was set at 2 V to keep the device in Dirac-source mode. SS varied by more than 100 % from 77 K to 300 K. In all measurements, the substrate was biased with −20 V to keep the ungated region near the drain open[158]. Copyright 2018, Science

0.5 V is equivalent to that of a CMOS device at 14 nm (at a working voltage of 0.7 V). This indicates that the Dirac source transistor can meet the requirements of future ultralow power consumption ($V_{dd} < 0.5$ V) integrated circuits. Moreover, the device structure of the DS does not rely on semiconductor materials and may be used in conventional CMOS transistors and FETs in 2D materials.

7.6 Conclusion and prospect

Graphene, with its unique physical and chemical properties, has been increasingly applied to various fields of scientific research. When graphene's nanostructure changes (such as in the boundary configuration, shape, own defects, chemical doping, and the formation of heterogeneous structures), its physical and chemical properties show novel properties. With the improvements in the preparation of graphene nanomaterials and the enhancement of their measurement and regulation, more graphene nanomaterials and their hybrid structures have been applied to electronic, photothermal, thermoelectric, and photoelectric fields.

However, graphene's band gap characteristics limit its application. Graphene's physical and chemical properties are both closely related to its electronic properties. Graphene's electronic properties change its physical and chemical properties. There are two kinds of methods. One is a physical approach (for example, the change of the graphene nanostructures, the applied electric field or magnetic field, vertical configuration heterostructure or plane heterostructure, substrate). The other is a chemical method (chemical doping, other atoms or groups of adsorption, the use of chemical reagents, etc.). The fabrication level of graphene-based microscale or nanoscale devices also determines their application and development. Researchers are actively working in correlated fields, and more high-performance graphene-based devices will be prepared and used in the future.

References

[1] ABANIN D A, MOROZOV S V, PONOMARENKO L A, et al. Giant nonlocality near the Dirac point in graphene[J]. Science, 2011, 332(6027): 328-330.

[2] TOOHEY K S, SOTTOS N R, LEWIS J A, et al. Self-healing materials with microvascular networks[J]. Nature Materials, 2007, 6(8): 581-585.

[3] CASTRO N A H, GUINEA F, PERES N M R, et al. The electronic properties of graphene[J]. RvMP, 2009, 81(1): 109-162.

[4] GEIM A K, KIM P. Carbon wonderland[J]. Scientific American, 2008, 298(4):

90-97.

[5] NAIR R R, BLAKE P, GRIGORENKO A N, et al. Fine structure constant defines visual transparency of graphene[J]. Science, 2008, 320(5881): 1308-1308.

[6] LEE C, WEI X, KYSAR J W, et al. Measurement of the elastic properties and intrinsic strength of monolayer graphene[J]. Science, 2008, 321(5887): 385-388.

[7] LIU F, MING P, LI J. Ab initio calculation of ideal strength and phonon instability of graphene under tension[J]. Physical Review B, 2007, 76(6): 064120.

[8] ZAKHARCHENKO K V, KATSNELSON M I, FASOLINO A. Finite temperature lattice properties of graphene beyond the quasiharmonic approximation[J]. Physical Review Letters, 2009, 102(4): 046808.

[9] SAHIN H, CAHANGIROV S, TOPSAKAL M, et al. Monolayer honeycomb structures of group-IV elements and III-V binary compounds: first-principles calculations[J]. Physical Review B, 2009, 80(15): 155453.

[10] BALANDIN A A, GHOSH S, BAO W, et al. Superior thermal conductivity of single-layer graphene[J]. Nano Letters, 2008, 8(3): 902-907.

[11] CHEN J H, JANG C, XIAO S, et al. Intrinsic and extrinsic performance limits of graphene devices on SiO_2[J]. Nature Nanotechnology, 2008, 3(4): 206-209.

[12] AVOURIS P, CHEN Z, PEREBEINOS V. Carbon-based electronics [M]// Nanoscience and Technology: A Collection of Reviews from Nature Journals, 2010: 174-184.

[13] WANG J, MA F, SUN M. Graphene, hexagonal boron nitride, and their heterostructures: properties and applications[J]. RSC advances, 2017, 7(27): 16801-16822.

[14] WANG J, MA F, LIANG W, et al. Optical, photonic and optoelectronic properties of graphene, h-BN and their hybrid materials[J]. Nanophotonics, 2017, 6(5): 943-976.

[15] LI R, ZHANG Y, XU X, et al. Optical characterizations of two-dimensional materials using nonlinear optical microscopies of CARS, TPEF, and SHG[J]. Nanophotonics, 2018, 7(5): 873-881.

[16] KONG L, WANG J, MU X, et al. Porous size dependent g-C_3N_4 for efficient photocatalysts: regulation synthesizes and physical mechanism[J]. Materials Today Energy, 2019, 13: 11-21.

[17] AL-JISHI R, DRESSELHAUS G. Lattice-dynamical model for graphite[J]. Physical Review B, 1982, 26(8): 4514.

[18] ANDO T. The electronic properties of graphene and carbon nanotubes[J]. NPG Asia Materials, 2009, 1(1): 17-21.

[19] HAYASHI M, THOMAS L, MORIYA R, et al. Current-controlled magnetic domain-wall nanowire shift register[J]. Science, 2008, 320(5873): 209-211.

[20] NOVOSELOV K S, GEIM A K, MOROZOV S V, et al. Two-dimensional gas of

massless Dirac fermions in graphene[J]. Nature, 2005, 438(7065): 197-200.

[21] ELIAS D C, GORBACHEV R V, MAYOROV A S, et al. Dirac cones reshaped by interaction effects in suspended graphene[J]. Nat. Phys, 2012, 8: 172-172.

[22] ZHANG Y, TAN Y W, STORMER H L, et al. Experimental observation of the quantum Hall effect and Berry's phase in graphene[J]. Nature, 2005, 438: 201-204.

[23] NOVOSELOV K S, JIANG Z, ZHANG Y, et al. Room-temperature quantum Hall effect in graphene[J]. Science, 2007, 315(5817): 1379-1379.

[24] BERGER C, SONG Z, LI X, et al. Electronic confinement and coherence in patterned epitaxial graphene[J]. Science, 2006, 312(5777): 1191-1196.

[25] KATSNELSON M I, NOVOSELOV K S, GEIM A K. Chiral tunnelling and the Klein paradox in graphene[J]. Nature Physics, 2006, 2(9): 620-625.

[26] ALLAIN P E, FUCHS J N. Klein tunneling in graphene: optics with massless electrons[J]. The European Physical Journal B, 2011, 83(3): 301-317.

[27] DU X, SKACHKO I, BARKER A, et al. Approaching ballistic transport in suspended graphene[J]. Nature Nanotechnology, 2008, 3(8): 491-495.

[28] MIAO F, WIJERATNE S, ZHANG Y, et al. Phase-coherent transport in graphene quantum billiards[J]. Science, 2007, 317(5844): 1530-1533.

[29] JIAN S K, LIN C H, MACIEJKO J, et al. Emergence of supersymmetric quantum electrodynamics[J]. Physical Review Letters, 2017, 118(16): 166802.

[30] KOVTUN P K, SON D T, STARINETS A O. Viscosity in strongly interacting quantum field theories from black hole physics[J]. Physical Review Letters, 2005, 94(11): 111601.

[31] SON D T. Vanishing bulk viscosities and conformal invariance of the unitary Fermi gas[J]. Physical Review Letters, 2007, 98(2): 020604.

[32] KARSCH F, KHARZEEV D, TUCHIN K. Universal properties of bulk viscosity near the QCD phase transition[J]. Physics Letters B, 2008, 663(3): 217-221.

[33] LEVITOV L, FALKOVICH G. Electron viscosity, current vortices and negative nonlocal resistance in graphene[J]. Nature Physics, 2016, 12(7): 672-676.

[34] YOO M J, FULTON T A, HESS H F, et al. Scanning single-electron transistor microscopy: imaging individual charges[J]. Science, 1997, 276(5312): 579-582.

[35] MIKHAILOV S A, ZIEGLER K. Physical Review Letters 2007, 99: 016803.

[36] JABLAN M, BULJAN H, SOLJA ČI Ć M. Plasmonics in graphene at infrared frequencies[J]. Physical Review B, 2009, 80(24): 245435.

[37] HANSON G W. Dyadic Green's functions and guided surface waves for a surface conductivity model of graphene[J]. Journal of Applied Physics, 2008, 103(6): 064302.

[38] JU L, GENG B, HORNG J, et al. Graphene plasmonics for tunable terahertz metamaterials[J]. Nature Nanotechnology, 2011, 6(10): 630-634.

[39] THONGRATTANASIRI S, KOPPENS F H L, DE ABAJO F J G. Complete optical

absorption in periodically patterned graphene[J]. Physical Review Letters, 2012, 108(4): 047401.

[40] VAKIL A, ENGHETA N. Transformation optics using graphene[J]. Science, 2011, 332(6035): 1291-1294.

[41] KOPPENS F H L, CHANG D E, GARCIA DE A F J. Graphene plasmonics: a platform for strong light-matter interactions [J]. Nano Letters, 2011, 11 (8): 3370-3377.

[42] CHEN J, BADIOLI M, ALONSO-GONZÁLEZ P, et al. Optical nano-imaging of gate-tunable graphene plasmons[J]. Nature, 2012, 487(7405): 77-81.

[43] NOVOSELOV K S, MISHCHENKO A, CARVALHO A, et al. 2D materials and van der Waals heterostructures[J]. Science, 2016, 353(6298): 9439.

[44] LOW T, CHAVES A, CALDWELL J D, et al. Polaritons in layered two-dimensional materials[J]. Nature Materials, 2017, 16(2): 182-194.

[45] ZHANG Q, ZHEN Z, YANG Y, et al. Negative refraction inspired polariton lens in van der Waals lateral heterojunctions[J]. Applied Physics Letters, 2019, 114(22): 221101.

[46] ALONSO-GONZÁLEZ P, NIKITIN A Y, GAO Y, et al. Acoustic terahertz graphene plasmons revealed by photocurrent nanoscopy [J]. Nature Nanotechnology, 2017, 12(1): 31-35.

[47] LUNDEBERG M B, GAO Y, ASGARI R, et al. Tuning quantum nonlocal effects in graphene plasmonics[J]. Science, 2017, 357(6347): 187-191.

[48] TONGAY S, SAHIN H, KO C, et al. Monolayer behaviour in bulk ReS 2 due to electronic and vibrational decoupling[J]. Nature Communications, 2014, 5(1): 1-6.

[49] IRANZO D A, NANOT S, DIAS E J C, et al. Probing the ultimate plasmon confinement limits with a van der Waals heterostructure [J]. Science, 2018, 360(6386): 291-295.

[50] TAN Y W, ZHANG Y, BOLOTIN K, et al. Measurement of scattering rate and minimum conductivity in graphene[J]. Physical Review Letters, 2007, 99(24): 246803.

[51] CHEN J H, JANG C, ADAM S, et al. Charged-impurity scattering in graphene[J]. Nature Physics, 2008, 4(5): 377-381.

[52] CHEN J H, CULLEN W G, JANG C, et al. Defect scattering in graphene [J]. Physical Review Letters, 2009, 102(23): 236805.

[53] SCHEDIN F, GEIM A K, MOROZOV S V, et al. Detection of individual gas molecules adsorbed on graphene[J]. Nature Materials, 2007, 6(9): 652-655.

[54] JIA Z, YAN B, NIU J, et al. Transport study of graphene adsorbed with indium adatoms[J]. Physical Review B, 2015, 91(8): 085411.

[55] LI J, LIN L, RUI D, et al. Electron-hole symmetry breaking in charge transport in nitrogen-doped graphene[J]. ACS Nano, 2017, 11(5): 4641-4650.

[56] GRUBER E, WILHELM R A, PÉTUYA R, et al. Ultrafast electronic response of graphene to a strong and localized electric field[J]. Nature Communications, 2016, 7(1): 1-7.

[57] WANG J, MA F, LIANG W, et al. Electrical properties and applications of graphene, hexagonal boron nitride (h-BN), and graphene/h-BN heterostructures[J]. Materials Today Physics, 2017, 2: 6-34.

[58] WANG J, MU X, WANG X, et al. The thermal and thermoelectric properties of in-plane C-BN hybrid structures and graphene/h-BN van der Waals heterostructures[J]. Materials Today Physics, 2018, 5: 29-57.

[59] YU L, LEE Y H, LING X, et al. Graphene/MoS_2 hybrid technology for large-scale two-dimensional electronics[J]. Nano Letters, 2014, 14(6): 3055-3063.

[60] DE FAZIO D, GOYKHMAN I, YOON D, et al. High responsivity, large-area graphene/MoS_2 flexible photodetectors[J]. ACS Nano, 2016, 10(9): 8252-8262.

[61] GIUBILEO F, DI BARTOLOMEO A, IEMMO L, et al. Field emission from carbon nanostructures[J]. Applied Sciences, 2018, 8(4): 526.

[62] CAO Y, FATEMI V, DEMIR A, et al. Correlated insulator behaviour at half-filling in magic-angle graphene superlattices[J]. Nature, 2018, 556(7699): 80-84.

[63] CAO Y, FATEMI V, FANG S, et al. Unconventional superconductivity in magic-angle graphene superlattices[J]. Nature, 2018, 556(7699): 43-50.

[64] MELE E J. Novel electronic states seen in graphene[J]. Nature, 2018, 556: 37.

[65] LIN W, SHI Y, YANG X, et al. Physical mechanism on exciton-plasmon coupling revealed by femtosecond pump-probe transient absorption spectroscopy[J]. Materials Today Physics, 2017, 3: 33-40.

[66] YANG X, YU H, GUO X, et al. Plasmon-exciton coupling of monolayer MoS_2-Ag nanoparticles hybrids for surface catalytic reaction[J]. Materials Today Energy, 2017, 5: 72-78.

[67] LIN W, CAO Y, WANG P, et al. Unified treatment for plasmon-exciton co-driven reduction and oxidation reactions[J]. Langmuir, 2017, 33(43): 12102-12107.

[68] LIN W, CAO E, ZHANG L, et al. Electrically enhanced hot hole driven oxidation catalysis at the interface of a plasmon-exciton hybrid[J]. Nanoscale, 2018, 10(12): 5482-5488.

[69] CAO E, GUO X, ZHANG L, et al. Electrooptical synergy on plasmon-exciton-codriven surface reduction reactions[J]. Advanced Materials Interfaces, 2017, 4(24): 1700869.

[70] WANG J, LIN W, XU X, et al. Plasmon-exciton coupling interaction for surface catalytic reactions[J]. The Chemical Record, 2018, 18(5): 481-490.

[71] GRIGORENKO A N, POLINI M, NOVOSELOV K S. Graphene plasmonics[J]. Nature Photonics, 2012, 6(11): 749-758.

[72] CHEN S, MOORE A L, CAI W, et al. Raman measurements of thermal transport in suspended monolayer graphene of variable sizes in vacuum and gaseous environments [J]. ACS Nano, 2011, 5(1): 321-328.

[73] SEOL J H, JO I, MOORE A L, et al. Two-dimensional phonon transport in supported graphene[J]. Science, 2010, 328(5975): 213-216.

[74] NI Z, WANG Y, YU T, et al. Raman spectroscopy and imaging of graphene[J]. Nano Research, 2008, 1(4): 273-291.

[75] WANG Y Y, NI Z H, YU T, et al. Raman studies of monolayer graphene: the substrate effect[J]. The Journal of Physical Chemistry C, 2008, 112(29): 10637-10640.

[76] LUO F, WU K, SHI J, et al. Green reduction of graphene oxide by polydopamine to a construct flexible film: superior flame retardancy and high thermal conductivity[J]. Journal of Materials Chemistry A, 2017, 5(35): 18542-18550.

[77] NIKA D L, ASKEROV A S, BALANDIN A A. Anomalous size dependence of the thermal conductivity of graphene ribbons[J]. Nano Letters, 2012, 12(6): 3238-3244.

[78] BONINI N, GARG J, MARZARI N. Acoustic phonon lifetimes and thermal transport in free-standing and strained graphene[J]. Nano Letters, 2012, 12(6): 2673-2678.

[79] PEREIRA L F C, DONADIO D. Divergence of the thermal conductivity in uniaxially strained graphene[J]. Physical Review B, 2013, 87(12): 125424.

[80] LINDSAY L, BROIDO D A, MINGO N. Flexural phonons and thermal transport in graphene[J]. Physical Review B, 2010, 82(11): 115427.

[81] JUND P, JULLIEN R. Molecular-dynamics calculation of the thermal conductivity of vitreous silica[J]. Physical Review B, 1999, 59(21): 13707.

[82] XU Y, CHEN X, GU B L, et al. Intrinsic anisotropy of thermal conductance in graphene nanoribbons[J]. Applied Physics Letters, 2009, 95(23): 233116.

[83] NOSHIN M, KHAN A I, NAVID I A, et al. Thermal transport in defected armchair graphene nanoribbon: a molecular dynamics study[C]//TENCON 2017-2017 IEEE Region 10 Conference. IEEE, 2017: 2600-2603.

[84] ZHANG G, ZHANG H. Thermal conduction and rectification in few-layer graphene Y Junctions[J]. Nanoscale, 2011, 3(11): 4604-4607.

[85] OUYANG T, CHEN Y, XIE Y, et al. Ballistic thermal rectification in asymmetric three-terminal graphene nanojunctions[J]. Physical Review B, 2010, 82(24): 245403.

[86] NISSIMAGOUDAR A S, SANKESHWAR N S. Electronic thermal conductivity and thermopower of armchair graphene nanoribbons[J]. Carbon, 2013, 52: 201-208.

[87] BARNARD A S, SNOOK I K. Thermal stability of graphene edge structure and graphene nanoflakes[J]. The Journal of Chemical Physics, 2008, 128(9): 094707.

[88] OUYANG T, CHEN Y P, YANG K K, et al. Thermal transport of isotopic-superlattice graphene nanoribbons with zigzag edge[J]. EPL (Europhysics Letters), 2009, 88(2): 28002.

[89] YANG K, CHEN Y, XIE Y, et al. Resonant splitting of phonon transport in periodic T-shaped graphene nanoribbons[J]. Europhysics Letters, 2010, 91(4): 46006.

[90] HU J, RUAN X, CHEN Y P. Thermal conductivity and thermal rectification in graphene nanoribbons: a molecular dynamics study[J]. Nano Letters, 2009, 9(7): 2730-2735.

[91] YANG N, ZHANG G, LI B. Thermal rectification in asymmetric graphene ribbons [J]. Applied Physics Letters, 2009, 95(3): 033107.

[92] ZHONG W R, HUANG W H, DENG X R, et al. Thermal rectification in thickness-asymmetric graphene nanoribbons[J]. Applied Physics Letters, 2011, 99(19): 193104.

[93] PEI Q X, ZHANG Y W, SHA Z D, et al. Carbon isotope doping induced interfacial thermal resistance and thermal rectification in graphene[J]. Applied Physics Letters, 2012, 100(10): 101901.

[94] LAN J, WANG J S, GAN C K, et al. Edge effects on quantum thermal transport in graphene nanoribbons: Tight-binding calculations [J]. Physical Review B, 2009, 79(11): 115401.

[95] NIKA D L, ASKEROV A S, BALANDIN A A. Anomalous size dependence of the thermal conductivity of graphene ribbons[J]. Nano Letters, 2012, 12(6): 3238-3244.

[96] GOYAL V, BALANDIN A A. Thermal properties of the hybrid graphene-metal nano-micro-composites: applications in thermal interface materials [J]. Applied Physics Letters, 2012, 100(7): 073113.

[97] SI S, LI W, ZHAO X, et al. Significant radiation tolerance and moderate reduction in thermal transport of a tungsten nanofilm by inserting monolayer graphene [J]. Advanced Materials, 2017, 29(3): 1604623.

[98] LI Y, WEI A, DATTA D. Thermal characteristics of graphene nanoribbons endorsed by surface functionalization[J]. Carbon, 2017, 113: 274-282.

[99] NIKA D L, ASKEROV A S, BALANDIN A A. Anomalous size dependence of the thermal conductivity of graphene ribbons[J]. Nano Letters, 2012, 12(6): 3238-3244.

[100] HU J, RUAN X, CHEN Y P. Thermal conductivity and thermal rectification in graphene nanoribbons: a molecular dynamics study[J]. Nano Letters, 2009, 9(7): 2730-2735.

[101] SINGH S K, SRINIVASAN S G, NEEK-AMAL M, et al. Thermal properties of fluorinated graphene[J]. Physical Review B, 2013, 87(10): 104114.

[102] HAO F, FANG D, XU Z. Mechanical and thermal transport properties of graphene with defects[J]. Applied Physics Letters, 2011, 99(4): 041901.

[103] HASKINS J, KINACI A, SEVIK C, et al. Control of thermal and electronic transport in defect-engineered graphene nanoribbons[J]. ACS Nano, 2011, 5(5): 3779-3787.

[104] MINNICH A J, DRESSELHAUS M S, REN Z F, et al. Bulk nanostructured

thermoelectric materials: current research and future prospects [J]. Energy & Environmental Science, 2009, 2(5): 466-479.

[105] HASKINS J, KINACI A, SEVIK C, et al. Control of thermal and electronic transport in defect-engineered graphene nanoribbons[J]. ACS Nano, 2011, 5(5): 3779-3787.

[106] PAN C N, XIE Z X, TANG L M, et al. Ballistic thermoelectric properties in graphene-nanoribbon-based heterojunctions [J]. Applied Physics Letters, 2012, 101(10): 103115.

[107] HU J, WANG Y, VALLABHANENI A, et al. Nonlinear thermal transport and negative differential thermal conductance in graphene nanoribbons [J]. Applied Physics Letters, 2011, 99(11): 113101.

[108] JOHNSON M, SILSBEE R H. Thermodynamic analysis of interfacial transport and of the thermomagnetoelectric system[J]. Physical Review B, 1987, 35(10): 4959.

[109] BAUER G E W, SAITOH E, VAN WEES B J. Spin caloritronics [J]. Nature Materials, 2012, 11(5): 391-399.

[110] UCHIDA K, TAKAHASHI S, HARII K, et al. Observation of the spin Seebeck effect[J]. Nature, 2008, 455(7214): 778-781.

[111] UCHIDA K, XIAO J, ADACHI H, et al. Spin Seebeck insulator [J]. Nature Materials, 2010, 9(11): 894-897.

[112] JAWORSKI C M, YANG J, MACK S, et al. Observation of the spin-Seebeck effect in a ferromagnetic semiconductor[J]. Nature Materials, 2010, 9(11): 898-903.

[113] VALENZUELA S O, TINKHAM M. Direct electronic measurement of the spin Hall effect[J]. Nature, 2006, 442(7099): 176-179.

[114] SAITOH E, UEDA M, MIYAJIMA H, et al. Conversion of spin current into charge current at room temperature: Inverse spin-Hall effect[J]. Applied Physics Letters, 2006, 88(18): 182509.

[115] SINOVA J, VALENZUELA S O, WUNDERLICH J, et al. Spin hall effects[J]. Reviews of Modern Physics, 2015, 87(4): 1213.

[116] TOMBROS N, JOZSA C, POPINCIUC M, et al. Electronic spin transport and spin precession in single graphene layers at room temperature [J]. Nature, 2007, 448(7153): 571-574.

[117] AVSAR A, TAN J Y, KURPAS M, et al. Gate-tunable black phosphorus spin valve with nanosecond spin lifetimes[J]. Nature Physics, 2017, 13(9): 888-893.

[118] ROCHE S, VALENZUELA S O. Graphene spintronics: puzzling controversies and challenges for spin manipulation[J]. Journal of Physics D: Applied Physics, 2014, 47(9): 094011.

[119] WANG J, MU X, SUN M. The thermal, electrical and thermoelectric properties of graphene nanomaterials[J]. Nanomaterials, 2019, 9(2): 218.

[120] BERCIAUD S, HAN M Y, MAK K F, et al. Electron and optical phonon temperatures in electrically biased graphene[J]. Physical Review Letters, 2010, 104(22): 227401.

[121] BETZ A C, VIALLA F, BRUNEL D, et al. Hot electron cooling by acoustic phonons in graphene[J]. Physical Review Letters, 2012, 109(5): 056805.

[122] BETZ A C, JHANG S H, PALLECCHI E, et al. Supercollision cooling in undoped graphene[J]. Nature Physics, 2013, 9(2): 109-112.

[123] SIERRA J F, NEUMANN I, COSTACHE M V, et al. Hot-carrier Seebeck effect: diffusion and remote detection of hot carriers in graphene[J]. Nano Letters, 2015, 15(6): 4000-4005.

[124] SIERRA J F, NEUMANN I, CUPPENS J, et al. Thermoelectric spin voltage in graphene[J]. Nature Nanotechnology, 2018, 13(2): 107-111.

[125] SONG N, JIAO D, CUI S, et al. Highly anisotropic thermal conductivity of layer-by-layer assembled nanofibrillated cellulose/graphene nanosheets hybrid films for thermal management[J]. ACS Applied Materials & Interfaces, 2017, 9(3): 2924-2932.

[126] SONG N, CUI S, JIAO D, et al. Layered nanofibrillated cellulose hybrid films as flexible lateral heat spreaders: the effect of graphene defect[J]. Carbon, 2017, 115: 338-346.

[127] SONG N, JIAO D, CUI S, et al. Highly anisotropic thermal conductivity of layer-by-layer assembled nanofibrillated cellulose/graphene nanosheets hybrid films for thermal management[J]. ACS Applied Materials & Interfaces, 2017, 9(3): 2924-2932.

[128] PENG L, XU Z, LIU Z, et al. Ultrahigh thermal conductive yet superflexible graphene films[J]. Advanced Materials, 2017, 29(27): 1700589.

[129] SONG N, HOU X, CHEN L, et al. A green plastic constructed from cellulose and functionalized graphene with high thermal conductivity[J]. ACS Applied Materials & Interfaces, 2017, 9(21): 17914-17922.

[130] OUYANG T, CHEN Y, XIE Y, et al. Thermal conductance modulator based on folded graphene nanoribbons[J]. Applied Physics Letters, 2011, 99(23): 233101.

[131] XU W, QIN Z, CHEN C T, et al. Ultrathin thermoresponsive self-folding 3D graphene[J]. Science Advances, 2017, 3(10): e1701084.

[132] YAN S, ZHU X, FRANDSEN L H, et al. Slow-light-enhanced energy efficiency for graphene microheaters on silicon photonic crystal waveguides[J]. Nature Communications, 2017, 8(1): 1-8.

[133] XIA F, MUELLER T, LIN Y, et al. Ultrafast graphene photodetector[J]. Nature Nanotechnology, 2009, 4(12): 839-843.

[134] RASHEED A, WEN W, GAO F, et al. Development and validation of KASP assays for genes underpinning key economic traits in bread wheat[J]. Theoretical and

Applied Genetics, 2016, 129(10): 1843-1860.

[135] URICH A, UNTERRAINER K, MUELLER T. Intrinsic response time of graphene photodetectors[J]. Nano Letters, 2011, 11(7): 2804-2808.

[136] LIMMER T, FELDMANN J, DA COMO E. Carrier lifetime in exfoliated few-layer graphene determined from intersubband optical transitions[J]. Physical Review Letters, 2013, 110(21): 217406.

[137] CHEN Z, LI X, WANG J, et al. Synergistic effects of plasmonics and electron trapping in graphene short-wave infrared photodetectors with ultrahigh responsivity [J]. ACS Nano, 2017, 11(1): 430-437.

[138] AN X, LIU F, JUNG Y J, et al. Tunable graphene-silicon heterojunctions for ultrasensitive photodetection[J]. Nano Letters, 2013, 13(3): 909-916.

[139] TAO L, CHEN Z, LI X, et al. Hybrid graphene tunneling photoconductor with interface engineering towards fast photoresponse and high responsivity[J]. npj 2D Materials and Applications, 2017, 1(1): 1-8.

[140] BAE S, KIM H, LEE Y, et al. Roll-to-roll production of 30-inch graphene films for transparent electrodes[J]. Nature Nanotechnology, 2010, 5(8): 574.

[141] HAN T H, LEE Y, CHOI M R, et al. Extremely efficient flexible organic light-emitting diodes with modified graphene anode[J]. Nature Photonics, 2012, 6(2): 105-110.

[142] HWANG J O, PARK J S, CHOI D S, et al. Workfunction-tunable, N-doped reduced graphene transparent electrodes for high-performance polymer light-emitting diodes [J]. ACS Nano, 2012, 6(1): 159-167.

[143] XU W, WANG L, LIU Y, et al. Controllable n-type doping on CVD-grown single- and double-layer graphene mixture[J]. Advanced Materials, 2015, 27(9): 1619-1623.

[144] KIM Y, RYU J, PARK M, et al. Vapor-phase molecular doping of graphene for high-performance transparent electrodes[J]. ACS Nano, 2014, 8(1): 868-874.

[145] HAN T H, KWON S J, LI N, et al. Versatile p-type chemical doping to achieve ideal flexible graphene electrodes[J]. Angewandte Chemie, 2016, 128(21): 6305-6309.

[146] KWON S J, HAN T H, KO T Y, et al. Extremely stable graphene electrodes doped with macromolecular acid[J]. Nature Communications, 2018, 9(1): 1-9.

[147] CHANG L, FRANK D J, MONTOYE R K, et al. Practical strategies for power-efficient computing technologies[J]. Proceedings of the IEEE, 2010, 98(2): 215-236.

[148] IONESCU A M, RIEL H. Tunnel field-effect transistors as energy-efficient electronic switches[J]. Nature, 2011, 479(7373): 329-337.

[149] LI X, ZHU M, DU M, et al. High detectivity graphene-silicon heterojunction photodetector[J]. Small, 2016, 12(5): 595-601.

[150] CHOI Y S, KANG H, KIM D G, et al. Mussel-inspired dopamine-and plant-based cardanol-containing polymer coatings for multifunctional filtration membranes[J].

ACS Applied Materials & Interfaces, 2014, 6(23): 21297-21307.

[151] NI Z, MA L, DU S, et al. Plasmonic silicon quantum dots enabled high-sensitivity ultrabroadband photodetection of graphene-based hybrid phototransistors[J]. ACS Nano, 2017, 11(10): 9854-9862.

[152] CHEN Z, LI X, WANG J, et al. Synergistic effects of plasmonics and electron trapping in graphene short-wave infrared photodetectors with ultrahigh responsivity [J]. ACS Nano, 2017, 11(1): 430-437.

[153] SEABAUGH A C, ZHANG Q. Low-voltage tunnel transistors for beyond CMOS logic[J]. Proceedings of the IEEE, 2010, 98(12): 2095-2110.

[154] GOPALAKRISHNAN K, GRIFFIN P B, PLUMMER J D. Impact ionization MOS (I-MOS)-part I: device and circuit simulations[J]. IEEE Transactions on electron devices, 2004, 52(1): 69-76.

[155] SALAHUDDIN S, DATTA S. Use of negative capacitance to provide voltage amplification for low power nanoscale devices[J]. Nano Letters, 2008, 8(2): 405-410.

[156] KHAN A I, YEUNG C W, HU C, et al. Ferroelectric negative capacitance MOSFET: Capacitance tuning & antiferroelectric operation[C]. 2011 International Electron Devices Meeting, IEEE, 2011.

[157] GNANI E, REGGIANI S, GNUDI A, et al. Steep-slope nanowire FET with a superlattice in the source extension[J]. Solid-State Electronics, 2011, 65: 108-113.

[158] QIU C G, LIU F, XU L, et al. Dirac-source field-effect transistors as energy-efficient, high-performance electronic switches[J]. Science, 2018, 361(6400): 387-392.

Chapter 8

Properties and Applications of New Superlattices: Twisted Bilayer Graphene

8.1 Twisted bilayer graphene (TwBLG)

The atoms of single-layer Two dimensional (2D) materials are on the surface, which makes the properties of 2D materials significantly different from those of their bulk phase materials[1-5]. The physical and chemical properties of 2D materials can be manipulated by changing the electrons of the atoms[6-16]. Bilayer graphene (BLG), Lego-like heterogeneous structure not only regulates the physical properties of the 2D material but also enables more novel properties and applications[17-21]. TwBLG formed by stacking double layers of graphene is the simplest heterostructure[22-23]. By adjusting the twisted angle between the layers of graphene[24-27], the electronic structure and energy band of single-layer graphene (SLG) can be regulated to form novel electrical properties[28-30], optical properties, and other physical properties[31-34], such as superconducting states and Mott insulators[13,35-41]. With the development of preparation technology and the improvement in observation and measurement levels[42-45], more properties of TwBLG have been discovered and applied.

8.1.1 Graphene and BLG

Graphene is a 2D crystal with a regular hexagonal honeycomb structure[1]. Its unit cell is formed by two C atoms A and B, which are defined by the lattice vectors a_1 and a_2 as shown in Figure 8-1(a)[4]. The three electrons in the outer shell of the C atom are hybridized with the electrons of the neighboring C atom to form a σ bond with sp^2, and the fourth electron forms a π bond.

The length of a_{C-C} is 0.246 nm, and the thickness of SLG is 0.335 nm. There are three σ bonds in the lattice, and the p orbitals of all of the C atoms are perpendicular to the sp^2 hybridization plane. According to theoretical calculations, in the Brillouin zone (BZ) corresponding to the hexagonal lattice of graphene (Figure 8-1(b)), the conduction band (CB) and valence band (VB) intersect at the K point (Figures 8-1(c) and (d)), which is called the Dirac point (DP), and the electrons in graphene are massless Dirac fermions[46].

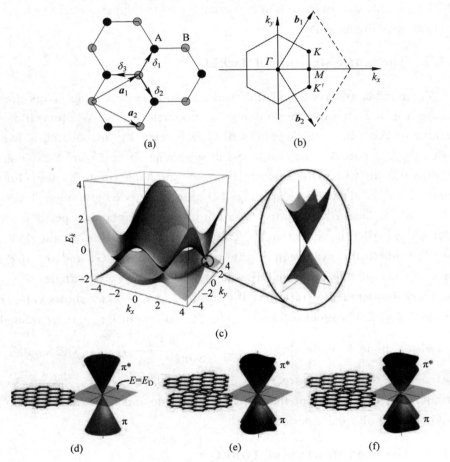

Figure 8-1 Graphene and bilayer graphene
(a) Lattice structure of graphene (a_1 and a_2 are the lattice unit vectors, and δ_i, $i = 1$, 2,3 are the nearest-neighbor vectors); (b) BZ of graphene; (c) Electronic dispersion in the honeycomb lattice[4]; Electronic structure of a single (d) symmetric double layer (e); and asymmetric double layer (f) of graphene[6]

BLG is formed by the vertical stacking of single graphene[4,6,47-50]. According to the stacking mode of C atoms in the upper and lower layers of graphene, it can be divided into AA stacking and AB stacking. BLG stacked by AA stacking has the same electronic energy spectrum and band structure as SLG, mainly because the band structure of BLG stacked by AA is approximately the overlapping of the two SLG energy bands (Figure 8-1(e))[6]. The electronic band structure of BLG stacked by AB has an obvious band gap at the DP, which is mainly determined by the asymmetry of the upper and lower layers of the C atoms (Figure 8-1(f))[6].

8.1.2 The lattice structure of TwBLG

The number of layers and the stacking mode between the layers are the reasons for the changes in graphene's properties[51-55]. Whether it is AA stacking or AB (BA) stacking, TwBLG is formed by introducing a twisted angle ($\theta_{gra\text{-}gra}$) between the two layers of graphene[56-73]. Figure 8-2(a) shows the direction in which the double layers of graphene rotate as they form a superlattice[54]. The layer spacing between the two graphene layers is 0.335 nm[55]. Due to the relative rotation of the BLG plane, a periodic moiré superlattice (MS) is generated in TwBLG. The formation of the MS also directly leads to the expansion of the unit cells of TwBLG, and the size and range of the unit cells decrease with the increase in the twisted angle ($\theta_{gra\text{-}gra}$). The black bordered illustration at the top of Figure 8-2(a) shows cells from SLG and TwBLG's superlattice[56]. The period length ($\lambda_{gra\text{-}gra}$) is related to the twisted angle θ in this MS, $\lambda_{gra\text{-}gra} = \dfrac{a_{C\text{-}C}}{2\sin(\theta_{gra\text{-}gra}/2)}$ [59]. The smaller the twisted angle $\theta_{gra\text{-}gra}$, the larger $\lambda_{gra\text{-}gra}$ (Figures 8-2(b) and (c))[59,72]. In the periodic MS, the C atom of the AA stacking mode and the C atom of the AB stacking mode also exist periodically[51-73].

8.1.3 The band structure of TwBLG

The electronic band structure at the DP K is linear whether it is SLG or AA stacking BLG[6,74]. The electronic band structure of BLG stacked by AB at point K is a quadratic curve[6,74]. The structural complexity of TwBLG determines the singularity of its electronic band structure. The BZs of the

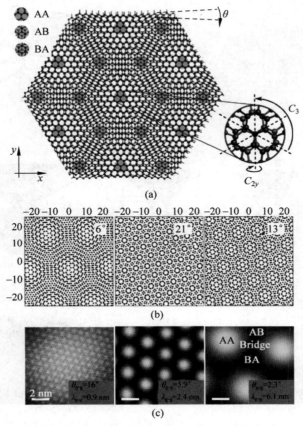

Figure 8-2 Structural diagram of TwBLG
(a) TwBLG formed by rotation in the C_3 direction. The above illustration shows the unit cell of SLG and the enlarged unit cell in TwBLG superlattice. The following illustration shows the rotation direction of TwBLG formed by the rotation of two layers of graphene[54]. (b) Different MS of TwBLG with twist angles θ = 6°, 21°, 13°[59]. (c) STM (scanning tunneling microscopy) topographic images of TwBLG with different twist angles θ = 16°, 5.9°, 2.3°[72]

upper graphene and lower graphene rotate relative to each other[66,75-78], resulting in a certain angle ($\theta_{\text{gra-gra}}$) of rotation in the BZ of the two layers (Figure 8-3(a))[77]. Due to the interaction of the electrons in the upper-lower BLG and the coupling of their respective Dirac cones in the BZ, the electronic band structure of TwBLG is redistributed, and the electronic band structure near the Fermi surface is similar to that of SLG, with a linear distribution. The electronic band structure of TwBLG is formed along the anti-crossing Dirac

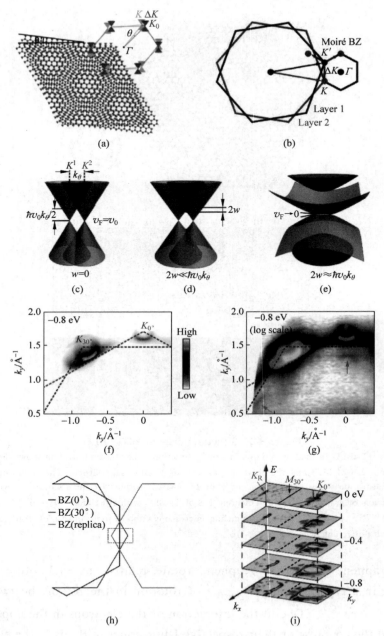

Figure 8-3 The EBS of TwBLG

(a) Schematic structure of two misoriented graphene sheets with a twist angle and schematic two Dirac cones[77]; (b) The first BZ of each graphene layer, and of the moiré BZ[69]; (c)~(e) The EBS of TwBLG is formed by the interaction of the Dirac cone of the upper graphene and lower graphene[79]; (f)~(i) Observation of mirrored Dirac Cone in 30° TwBLG[80]

cone formed by K_1 and K_2 (Figure 8-3(b))[79]. There are obvious saddle points in the electronic band structure and van Hove singularities (VHSs) in the density of states (DOS). These saddle points are caused by the overlap of the electronic band structure between the upper-lower layers of graphene (Figures 8-3(c)~(e))[79]. A previous study reported the splitting of the electronic band structure near point M (Figures 8-3(f)~(i)) that is caused by the saddle point near point M[80].

The dispersion relationship of the low-energy electronic band structure near the K valley of TwBLG for quasiparticles is[81]:

$$E(k,\bar{k}) = \pm \frac{\hbar^2}{m(\theta)} \left| k - \frac{\Delta K}{2} \right| \left| k + \frac{\Delta K}{2} \right|, \quad (8\text{-}1)$$

where $\Delta K = \frac{1}{\sqrt{2}}(\Delta K_x + i\Delta K_y)$, $k = \frac{1}{\sqrt{2}}(k_x + ik_y)$, and $m(\theta)$ is the θ-dependent effective mass. If $\kappa = k - \frac{\Delta K}{2}$, the massless behavior $E \approx \pm \sqrt{2}\hbar \tilde{v}_F |\kappa|$ near $\kappa = 0$ and $\tilde{v}_F = \frac{\hbar |\Delta K|}{\sqrt{2} m(\theta)}$ is the renormalized Fermi speed. As shown in Figures 8-2(c)~(e), as the angle decreases ($\theta_{\text{gra-gra}}$), an approximately flat electronic band structure appears between the two saddle points. The electronic band structure of TwBLG at the twist angle is determined by the twist angle, and the electronic band structure of TwBLG at different twist angles also varies.

8.1.4 Superlattices with different symmetric structures

Strictly speaking, there are two commensurate structures (CSs) in the superlattice of TwBLG: the sublattice exchange odd (SE-odd) structure and the sublattice exchange even (SE-even) structure[82-84]. The SE-even and SE-odd moiré patterns result in quite different structures of superlattice DP because of their distinguishing real and reciprocal structures.

In TwBLG, researchers used m and n to describe the twisted angle of CS[84], the twisted angle is:

$$\theta_{\text{gra-gra}} = \arccos\left(\frac{3m^2 - n^2}{3m^2 + n^2}\right). \quad (8\text{-}2)$$

To distinguish the different structures of TwBLG, $\delta = 3/\gcd(m,3)$ was defined as a parameter. When $\delta = 3$, the vectors of primitive commensuration cell of SE-odd structure are

$$\begin{cases} L_1 = \dfrac{1}{\varepsilon}(m+n)a_1 - \dfrac{1}{\varepsilon}(2n)a_2, \\ L_2 = -\dfrac{1}{\varepsilon}(m-n)a_1 + \dfrac{1}{\varepsilon}(m+n)a_2. \end{cases} \quad (8\text{-}3)$$

When $\delta = 1$, the vectors of primitive commensuration cell of SE-even structure are

$$\begin{cases} L_1 = \dfrac{1}{\varepsilon}(m+3n)a_1 + \dfrac{1}{\varepsilon}(m-3n)a_2, \\ L_2 = -\dfrac{1}{\varepsilon}(2m)a_1 + \dfrac{1}{\varepsilon}(m+3n)a_2; \end{cases} \quad (8\text{-}4)$$

where $\varepsilon = \gcd(m+3n, m-3n)$, a_1 and a_2 are direct-lattice vectors. Figures 8-4(a) and (d) show the superlattice of TwBLG with different CSs, corresponding to SE-odd structure and SE-even structure, respectively. Figures 8-4(b) and (e) are schematic diagrams of the BZ and the mini-BZ corresponding to SE-odd and SE-even. Figures 8-4(g) and (h) show the band structures of SE-odd and SE-even. As can be clearly seen from the figure, in the SE-odd structure, six DPs of the superlattice are generated, while in the SE-even structure, the DPs of the superlattice in the center of K and K_θ are degenerated due to the interlayer coupling, so a total of ten DPs of the superlattice are formed. At this point, the DPs of the superlattice appear at the position $\pm E_{SD}$. Figure 8-4(i) shows the relationship between ΔE_{SD} of DPs with CS and the twisted angle $\theta_{\text{gra-gra}}$.

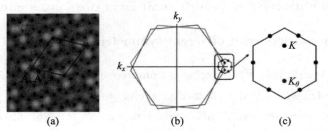

(a)　　　　　　　(b)　　　　　　　(c)

Figure 8-4 Superlattices with different symmetric structures

Schematic structural model of two misoriented graphene layers with SE-odd (a) and SE-even (d) structures. The red parallelogram represents the superlattice. BZ and enlarged minizone of TwBLG with SE-odd (b) ~ (c) and SE-even (e) ~ (f) structures. Energy of charge carriers in a twisted graphene bilayer with (g) the SE-odd structure and (h) the SE-even structure. The energy separation of the superlattice DPs ΔE_{SD} as a function of the rotation angle θ[84]

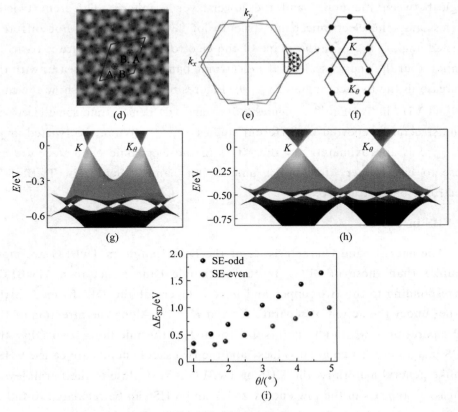

Figure 8-4(Continued)

8.2 The properties of TwBLG

For a 2D material, the electronic band structure determines the physical properties of the material. TwBLG has a novel electronic band structure with obvious physical properties different from graphene. Due to the twisted angle ($\theta_{\text{gra-gra}}$), the electronic band structure of TwBLG becomes more adjustable, which also results in richer electrical properties.

8.2.1 Electronic properties of TwBLG

1. The relationship between the band structure and twisted angle ($\theta_{\text{gra-gra}}$)

The electronic band structure of TwBLG is determined by the twisted angle between the upper and the lower-layer graphene. Different twisted angles cause the electronic band structure of TwBLG to vary. For different twisted angles ($\theta_{\text{gra-gra}}$), the shape of the electronic band structure is basically similar, but the energy scale of the electronic band structure decreases with the decrease in the twisted angle ($\theta_{\text{gra-gra}}$). Figures 8-5(a) and (b) show the mini BZ and VHS in TwBLG[66]. Figures 8-5(c) and (d) demonstrate some electronic band structures reported by different studies[67,79,85]. When the twisted angle ($\theta_{\text{gra-gra}}$) is approximately 1°, the VHSs in the electronic band structure are close to each other, forming a unique flat band structure in TwBLG's electronic band structure.

2. VHS in TwBLG

The energy band characteristics of the Dirac points in TwBLG are more complex than those of SLG. In the electronic band structure of TwBLG, corresponding to the two (upper and lower) layers of the DPs formed in the corner under the action of separate K_1 and K_2[66]. Along the direction of the DP separation, two saddle points are formed in the middle of two DPs, the DOS images in the two great values, and the two saddle points are called VHS. Unlike general materials, the VHS of TwBLG is very close to the Fermi level, that is, it appears in the low energy zone, and VHS can be regulated to fall at the Fermi energy level. This feature provides TwBLG with many new properties.

Both in terms of experimental observations and theoretical calculations, many studies have shown that the position of the VHS is closely dependent on the twisted angle ($\theta_{\text{gra-gra}}$)[26-27,66,86]. Figure 8-6(a) shows the periodic MS of TwBLG observed in an experiment[25]. Point 1 is AA stacking, point 2 is AB stacking, and point 3 is BA stacking. In Figure 8-6(b)[25], the black line corresponds to the DOS in AA stacking. The two raised peaks in the black line in Figure 8-5(b) relate to the concave points in the electronic band structure in Figure 8-6(c) by a dotted line[25], namely VHS. Figures 8-6(d) and (e) show the energy difference of the VHS's ΔE_{VHS} as a function of the twisted angles as reported in different studies[25-26]. The specific relationship between ΔE_{VHS} and the twisted angle ($\theta_{\text{gra-gra}}$) satisfies the following formula[25,27,66]:

$$\Delta E_{\text{VHS}} = \hbar \nu_F \Delta K, \qquad (8-5)$$

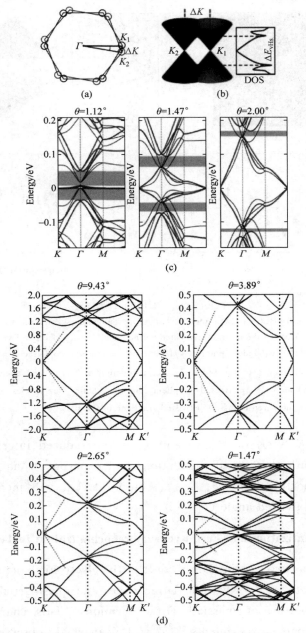

Figure 8-5 Electronic properties of TwBLG

(a) The mini BZ of TwBLG[66]. (b) Emergence of VHS as a consequence of the rotation in reciprocal space[66]. (c) Band structure of TwBLG with rotation angles 1.12°, 1.47°, 2.0°. The black lines are bands for the system[82]. (d) Band structure of TwBLG with rotation angles 9.43°, 3.89°, 2.65°, 1.47°. Dashed slopes around the K point indicate the monolayer's band dispersion[67]

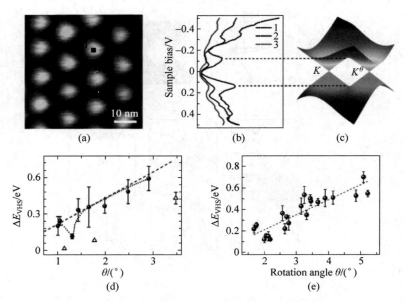

Figure 8-6　VHS in TwBLG

(a) STM image of TwBLG, twisted angle is 1.1°. (b) Tunneling spectra recorded on bright and dark regions of the MS at the positions indicated on the panel (a). (c) EBS of TwBLG with a finite interlayer coupling. Two saddle points (VHS) in the middle of two Dirac cones, K and K_θ; (d) The energy difference of the two ΔE_{VHS} as a function of the twisted angles ($\theta_{\text{gra-gra}}$)[25]. (e) Angle-dependent VHS in twisted graphene bilayers. Solid blue circles are the average experimental data measured in different twisted graphene bilayers[26]

where $\Delta K = 2K\sin(\theta/2)$, and t_θ is the reduced interlayer coupling coefficient. Both theoretical calculations and experiments have proved that ΔE_{VHS} is proportional to ($\theta_{\text{gra-gra}}$), that is, ΔE_{VHS} will increase with the increase in the twisted angle ($\theta_{\text{gra-gra}}$).

3. Landau quantization and Fermi velocity renormalization in TwBLG

TwBLG has two existing Dirac cones at the electronic band structure of its twisted angle ($\theta_{\text{gra-gra}}$). Near the Dirac cone, the quasiparticles in TwBLG have the properties of massless Dirac fermions. This characteristic was reported in previous experiments[87,88-90]. Luican et al. observed the Landau quantization of quasiparticles in TwBLG for the first time via scanning tunneling microscopy (STM) under a magnetic field[76]. The Landau level (LL) characteristics of the massless Dirac fermions satisfy the following formula[76]:

$$E_n = E_D + \text{sgn}(n)\sqrt{2e\,\hbar v_F^2 |n|B}, \quad n = 0, \pm 1, \pm 2, \cdots, \tag{8-6}$$

where E_D is the energy at the DPs, n is an integer, and v_F is the Fermi velocity. Wang et al. observed TwBLG based on an SiC surface with STM and described TwBLG's LL using the tight binding method[91]. This type of TwBLG system, applicable to different twisted angles, also reveals that interlayer coupling is one reason for the emergence of novel TwBLG characteristics.

Theoretical calculations show that the excitation spectrum of TwBLG can be described by massless Dirac fermions, and the Fermi velocity (v_F) is renormalized at this point. The formula is as follows[76,92-95]:

$$\frac{\tilde{v}_F(\theta)}{v_F} = 1 - 9\left(\frac{t_\theta}{\hbar v_F \Delta K}\right)^2, \tag{8-7}$$

where t_θ is the interlayer coupling for AB stacking.

Figures 8-7(a) and (b) show the relationship between the Fermi velocity and the rotation angle of TwBLG on a graphite substrate and an SiC substrate, respectively[76,94]. Figure 8-7(c) demonstrates that the Fermi velocity normalized to that of SLG v_F as a function of θ[95]. Figures 8-7(d) and (e) show the LL spectra of different twisted angles ($\theta_{\text{gra-gra}}$) as reported in previous studies[91,93]. The results of these reports indicate that the renormalization of the Fermi velocity of TwBLG is the result of the combined action of the twisted angle ($\theta_{\text{gra-gra}}$) and the interlayer coupling t_θ[93]. Figures 8-7(f) and (g) are the scanning tunneling spectroscopy (STS) spectra taken in various magnetic fields in TwBLG with different $\theta_{\text{gra-gra}}$[94]. It can be seen from the figure that, under the high magnetic field, the experiment finds that there are very obvious Landau quantization energy levels in TwBLG with different angles. The properties of LLs satisfy the massless Dirac fermion. The researchers measured the Fermi velocity of TwBLG at different angles. They found that the Fermi velocity renaming in TwBLG is not only dependent on the twisted angle $\theta_{\text{gra-gra}}$ but also on the strength of the interlaminar coupling t_θ. Table 8-1 lists the relevant research results.

Table 8-1 The relationship between $\theta_{\text{gra-gra}}$ and v_F, t_θ

Twisted angle ($\theta_{\text{gra-gra}}$)	2.3°	2.8°	3.5°	3.6°	6.0°	21.8°
Fermi velocity (10^6 m/s)	1.08	1.03	0.87	0.81	1.1	1.1
t_θ/meV	—	50	110	130	—	—
Reference	94	93	76	93	93	76

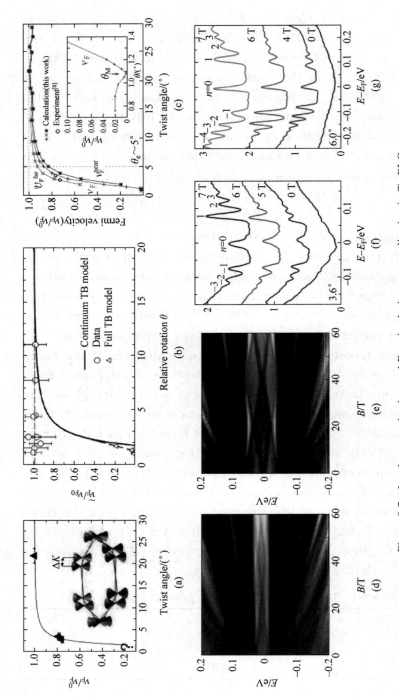

Figure 8-7 Landau quantization and Fermi velocity renormalization in TwBLG
(a) Angle dependence of the Fermi velocity renormalization. Line is theoretical prediction. Triangles are experimental data. (Inset picture is Dirac cones of TwBLG)[76]. (b) Comparison of the measured v_F versus the predicted renormalized velocity $\tilde{v}_F$ as a function of relative rotation angle[94]. (c) Fermi velocity normalized to that of SLG v_F as a function of θ[95]. LL spectra for TwBLG at commensurate angle $\theta = 1.067°$ in (d) and incommensurate angle $\theta = 1.3°$ in (e)[91]. STS spectra taken in various magnetic fields in TwBLG with $\theta = (3.6 \pm 0.1)°$ in (f), $\theta = (6.0 \pm 0.1)°$ in (g)[93]

4. Quantum Hall effect (QHE) of TwBLG

Many studies on the QHE of TwBLG were conducted from 2011 to 2018[96-102]. The unique carrier linear dispersion property of SLG makes its QHE significantly different from the traditional 2D system[103-104]. BLG with AB stacking exhibits different Hall conductivity (HC): $\sigma_{xy} = (e^2/h)\nu_{tot}$, $\nu_{tot} = \pm 4, \pm 8, \pm 12, \cdots$[105]. Another group of researchers prepared the layers of graphene on the surface of 4H-SiC-0001, with TwBLG areas in the sample. Through experimental observation and first-principles calculation, they confirmed that the band structure of TwBLG is significantly different from that of AB stacked BLG or SLG[106].

TwBLG was prepared on a SiC surface and its twisted angle ($\theta_{gra\text{-}gra}$) was distributed in a range of $(0 \pm 2.2)°$ and $(30 \pm 2.2)°$ as the center. Figure 8-8(a) shows that the density function of HC (σ_{xy}) due to the backgate voltage (V_{bg}), $n_{app} = \alpha(V_{bg} - V_N)$ is the total density relation ($\alpha = 7.2 \times 10^{10}/\text{cm}^2 \cdot \text{V}$, $V_N = 18.5$ V). At this point, the HC is characterized by quantization, expressed as $\sigma_{xy} = i(e^2/h)$, $i = \pm 4, \pm 8, \cdots$. This sequence is identical to commensurate Bernal-stacked bilayer graphene (BsBLG). Figure 8-8(b) shows the LL spectrum of TwBLG, which is obviously different from the LL spectrum of BsBLG. Figure 8-8(c) demonstrates the HC as a function of the charge carrier density (CCD) rescaled by $4eB/h$.

The test results clearly indicate that the QHE of TwBLG is affected by the strength of the interlayer coupling, or if the interlayer spacing is too large, the QHE of SLG is mainly shown. In addition, TwBLG's Fermi velocity is renormalized at a small twisted angle ($\theta_{gra\text{-}gra}$) below 10°, while at a large twisted angle ($\theta_{gra\text{-}gra}$), the Fermi velocity remains the same as that of SLG.

5. Magic angles and superconductivity in TwBLG

The size of the interlayer coupling (t_θ) depends on the stack configuration of the C-C atoms, the spacing between the layers, and the size of the twisted angle ($\theta_{gra\text{-}gra}$). The electronic band structure near the charge neutral point is flattened by a small twisted angle ($\theta_{gra\text{-}gra}$). The flat electronic band structure has an insulating state when the carrier is semi-filled. This new electronic state is called the Mott insulating state, which is mainly due to the strong repulsion

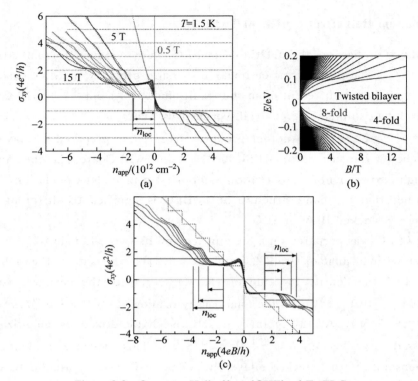

Figure 8-8 Quantum Hall effect (QHE) of TwBLG

(a) HC as a function of n_{app}. (b) LL spectra of TwBLG. The degeneracy of the energy levels is color coded: red corresponds to eightfold degenerate and black to fourfold degenerate. (c) The HC as a function of charge carrier density rescaled by $4eB/h$. Black dashed line shows the HC expected for BsBLG for a comparison[96]

between electrons. Cao et al. prepared TwBLG of the twisted angle ($\theta_{gra\text{-}gra}$) at 1.08°, and the conductive performance was tested and discussed[40-41].

The relative rotation of the two layers of graphene opens an energy gap on the Dirac cone, where the Fermi velocity at the DP is renormalized at certain angles (Figure 8-3(e))[40], the Fermi velocity becomes zero. These angles became known as *magic angles* and the first of these appeared around 1.1°. When the two layers of graphene are twisted close to the *magic angle* predicted by the theory, the superlattice is formed (Figures 8-9(a) and (b))[36,40], the mini BZ is formed (Figure 8-9(c)), too. The energy band structure near the neutral charge generated becomes flat due to the strong interlayer coupling (Figure 8-9(d)). When small rotation angle to magic angle ($\theta < 1.05°$), distortion of vertical stack of atoms in the BLG area will form the narrow electronic energy band, the electronic interaction effect increases, resulting in a nonconductive Mott insulation state. The superconducting state can be successfully converted by adding a small number of charge carriers in the Mott insulated state.

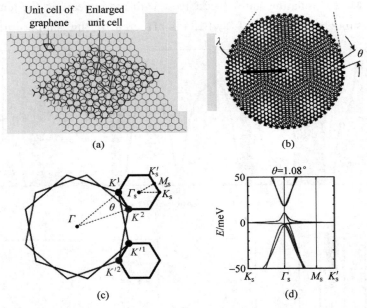

Figure 8-9 Magic angles and superconductivity in TwBLG
(a) Enlarged unit cell of TwBLG (red line)[36]. (b) Moiré superlattice in TwBLG with a twisted angle θ, $\lambda = a/[2\sin(\theta/2)]$ is moire wavelength. (c) Moire BZ of superlattice. (d) The energy band of TwBLG with twisted angle $\theta = 1.08°$. The blue lines show the flat bands[40]

When the electronic band structure is full, the carrier concentration is ns and the half-full carrier concentration is ns/2. When the value of n continuously increases or decreases, in the ns region, the conductance is zero, that is, insulation occurs, which is consistent with the energy band theory. However, in the ns/2 region, when the energy band is fractional occupied state, there is an insulating platform, which does not conform to the energy band theory, and the insulator that does not conform to the electronic band structure theory is called the Mott insulator (Figure 8-10(a)). This is the first similarity observed between TwBLG and high-temperature superconductors (Figures 8-10(b), (c) and (d)). Therefore, the phase diagrams of the electron states of parabolic superconducting TwBLG and high-temperature superconducting TwBLG at different carrier concentrations are basically consistent (Figures 8-10(e) and (f)). When the magnetic field applied on the test device is 0, the part corresponding to the red line experiences a yellow SC (superconducting state) and a blue Mott (insulating state) successively with the increase in the temperature. Thus, the superconductivity of TwBLG is affected by the temperature. When

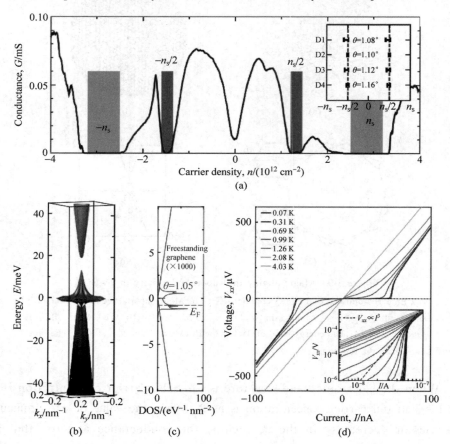

Figure 8-10 Magic angles and superconductivity in TwBLG

(a) Measured conductance G of TwBLG with $\theta = 1.08°$ and $T = 0.3$ K[79]. (b) The EBS of TwBLG at $\theta = 1.05°$ in the first mini BZ of the superlattice. (c) The DOS corresponds to the bands of (b). (d) The I-V curve of the device at different temperatures when the carrier concentration is E. (e) and (f) are the electron phase diagram of TwBLG with twisted angle ($\theta_{gra\text{-}gra}$) are 1.16° and 1.05°. SC is super-conductivity said superconducting area, it is parabolic. Schematic phase diagrams of TwBLG for different magnetic fields in (g), (h) and (i)[40]

Chapter 8 Properties and Applications of New Superlattices: Twisted Bilayer Graphene

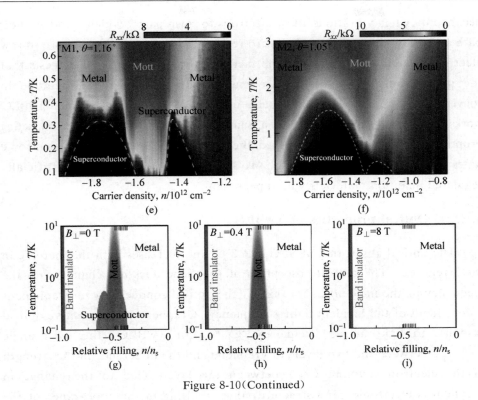

Figure 8-10(Continued)

$B = 0.4$ T, the corresponding part of the red line only has a Mott state, and the SC state disappears. When $B = 0.8$ T, both the Mott state and SC state disappear, and the TwBLG demonstrates metal conductivity. This indicates that the superconductivity of TwBLG is affected by a magnetic field (Figures 8-10(g), (h) and (i)).

Although studies indicate that TwBLG has superconductivity, which is affected by the temperature (T), carrier concentration (n), and magnetic field (B), researchers do not believe that TwBLG is a Mott insulator[56]. Previous studies have elucidated the superconducting mechanism of TwBLG via theoretical calculations[38,107-110]. Cao's work provides an ideal research platform for the study of high-temperature superconductor (HTS) materials.

Although ordinary materials exhibit superconductivity at ultralow temperatures. However, at the same temperature, the electron density of graphene can be a superconducting material as long as it reaches one thousandth of that of a conventional superconductor. In conventional superconductors, this phenomenon occurs only when vibrations allow electrons to form pairs, which stabilize their path and allow them to flow with zero resistance. But because there are so few electrons available in graphene, the fact that they can be paired suggests that the interactions in the system are much stronger than they would be in conventional superconductors. Although physicists are not sure whether TwBLG works in the same way as conventional superconductors, TwBLG's superconducting properties have undoubtedly opened the door to the regulation of electronic states in the low-dimensional world. We believe that more and more artificially regulated quantum materials will appear soon.

8.2.2 Optical properties of TwBLG

The optical absorption of SLG is 2.3 % and increases with the increase in the layers[111]. The optical absorption of graphene does not change with the frequency of the incident light, that is, the optical conductivity of graphene is independent of the incident light's frequency. Graphene has a sensitive optical response to light at any frequency[111-113]. For TwBLG, the twist angle ($\theta_{gra\text{-}gra}$) between the two layers of graphene and the difference in the strength of the electron coupling (t_θ) between the layers lead to the change in graphene's electronic band structure, thus resulting in the appearance of the VHS on the DOS. The DOS is a factor affecting the optical conductivity (σ) of graphene.

Figures 8-11(a) and (b) show the position of the optical conductivity peak of TwBLG with different rotation angles[114]. The electronic band structures of TwBLG with varying twist angles differ, leading to the position change in the optical conductivity peak. During the change from 0° to 30°, peak (i) changes in the direction of high-energy region, peak (ii) changes in the direction of low energy, and peak (iii) remains unchanged. For the optical conductivity (σ) peak (i), when the twist angle ($\theta_{gra\text{-}gra}$) is large, the absorption peak is located in the visible region. As the rotation angle ($\theta_{gra\text{-}gra}$) decreases, the absorption peak gradually moves toward the direction of the middle infrared region[114]. Figure 8-11(c) shows the result of another study on the optical absorption characteristics of TwBLG at different twist angles ($\theta_{gra\text{-}gra}$)[115]. In the short wavelength and terahertz regions, studies demonstrated that the optical conductivity (σ) of TwBLG is strictly affected by the VHS[116-117].

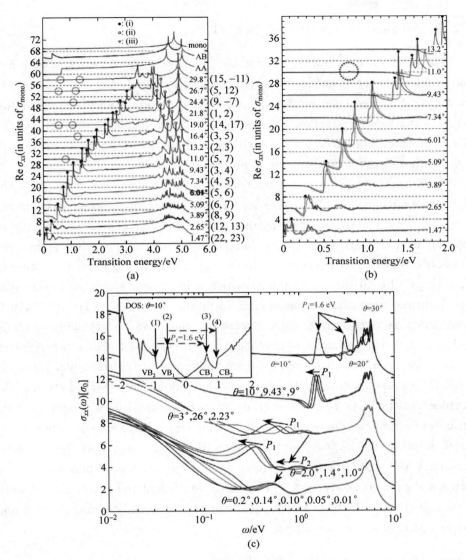

Figure 8-11 Optical properties of TwBLG

(Color online) Dynamical conductivities of TwBLG with various twist angles in (a) wide and (b) narrow frequency ranges, calculated by the tight-binding model [solid (black) lines] and the effective continuum model [dashed (red) lines, only for (b)][114]; Optical conductivity of a series of TwBLG configurations with the commensurate and incommensurate twist angles ranging from large to extremely tiny values in (c)[115]; Inset shows the dominant transition processes contributing to the formation of the optical conductivity peaks

8.2.3 Magnetic properties of TwBLG

The electron distribution outside the nucleus of the C atom determines that graphene is not magnetic. However, the electrical and magnetic properties of graphene can be regulated by doping, atomic adsorption, boundary configuration changes, and the introduction of substrate to make graphene magnetic[14,21]. The electronic band structure of TwBLG is obviously different from that of SLG, which gives TwBLG magnetic properties.

A study on the magnetic properties of angular graphene found that the electrons near the VHS of TwBLG are unstable due to the interactions between them[32]. This electron-pair instability causes the Fermi energy level to be adjusted to the VHS in graphene's conducting band. At this point, the symmetry of the dispersion decreases to the C_{3v} group, leading to the divergence of the susceptibility at the saddle point, which is integrable along the Fermi line. This indicates that the ferromagnetic instability near the VHS of TwBLG dominates. As the rotation angle decreases, the VHS energy band tends to flatten and ferromagnetism is amplified[32]. Figure 8-12 shows the research results. Alex et al. studied the anti-ferromagnetism of the triangular lattice in TwBLG[54], demonstrating that the local Coulomb repulsion between the electrons in TwBLG resulted in the anti-ferromagnetism of the triangular lattice and some magnetic properties. Gonzalez-Arraga et al. studied the magnetic properties of AA stacking in TwBLG and found that the magnetic properties of TwBLG can be regulated by the application of biasing pressure between the TwBLG layers[62]. They also confirmed the instability of anti-ferromagnetism in TwBLG and the transformation of anti-ferromagnetism and ferromagnetism under biasing pressure.

8.2.4 Thermal properties of TwBLG

SLG has excellent thermal and thermoelectric properties that mainly depend on the transmission of long wavelength phonons[118-122]. The thermal properties of graphene, such as the thermal conductivity(TC), are affected by the number of layers, size, and configuration.

The TC of suspended TwBLG was studied using the optothermal Raman technique (OTRT). Li et al. prepared SLG, BLG, and TwBLG on the surface

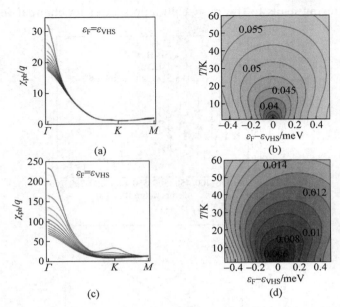

Figure 8-12 Plot of the electron-hole susceptibility (in units of the inverse of electron volt times L_n^2) for temperatures corresponding (from top to bottom) to $k_B T = 0.1$ meV, 0.5 meV, 1 meV, 1.5 meV, 2 meV, 2.5 meV, 3 meV, 3.5 meV, 4 meV, 4.5 meV, 5 meV (a) and (c), (b) and (d) contour lines of the critical coupling for the interaction strength $v_\perp/L_n^2$ (in eV) as a function of temperature and doping with respect to VHS, for TwBLG with $n = 10$ ($\theta \approx 3.15°$) and $n = 22$ ($\theta \approx 1.47°$)[32]

of Cu using chemical vapor deposition[123]. They employed OTRT technology to obtain non-contact measurements using optoelectric transmission or absorption. At this point, the formula of thermal conductivity is $k = \dfrac{\ln(R/r_0)}{2\pi t R_g}\alpha$, where R is the radius of the holes, t is graphene's thickness, r_0 is the Gaussian laser beam radius, and α is the factor accounting for the Gaussian beam profile. Figure 8-13(a) shows the curves of TC in SLG, BLG, and TwBLG changing with the temperature. At this point, the TC of TwBLG at 314 K is $k = (1,896 \pm 410)$ W/m·K. Through molecular dynamics (MD) simulation and experimental testing, the TC of suspended TwBLG at room temperature (RT) is $k = (1,412 \pm 390)$ W/m·K. Research shows that phonon transmission in TwBLG is different from electron transmission[123-124]. Phonon dispersion leads to a lower TC in TwBLG than SLG and BLG, and phonon scattering leads to the mini-Umklapp scattering channel for phonons.

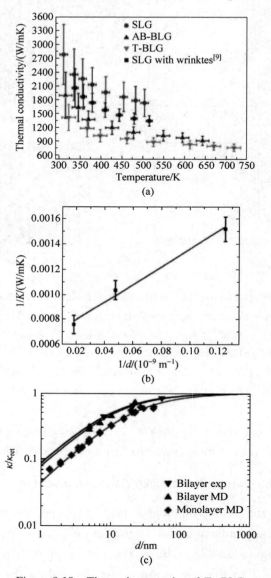

Figure 8-13 Thermal properties of TwBLG

(a) The TC of suspended SLG, Bernal stacked BLG and TwBLG as a function of the temperature[123]; (b) Inverse of TC versus the inverse of grain size (d); (c) Normalized TC k/k_{ref} as a function of grain size[125]

Limbu et al. prepared TwBLG with different grain sizes and studied the characteristics of TC in TwBLG changing with grain size at room temperature[125]. Figure 8-13(b) shows the change curve of TwBLG's TC with the grain size, and the obtained TC was $(1,510 \pm 10^3)$ W/m·K. Figure 8-13(c) illustrates the relationship between the normalized TCs of SLG and TwBLG obtained by experimental measurements and MD simulation with the grain size. The results showed that the TC of TwBLG decreases with the decrease in the grain size and decreases more rapidly in the small grain state.

8.3 TwBLG preparation methods

TwBLG superlattice structure was experimentally discovered more than 20 years ago. Due to the influence of experimental technology, measuring methods, instrument conditions and other factors, the preparation of TwBLG is relatively slow. However, with the rapid development of graphene's preparation technology, TwBLG's preparation technology has also been developed. Several main methods for preparing TwBLG are listed below, and their advantages and disadvantages are compared in Table 8-2.

8.3.1 SiC-based epitaxial growth

SiC-based epitaxial growth is one of the first methods used to prepare graphene. The number of layers of graphene prepared via this method is generally large, and the size of the resulting graphene nano-sheet is relatively small, that is, the number of layers and size are not easy to control. In the prepared graphene, when the researchers observed the experimental samples through STM, they found the typical TwBLG superlattice. However, this preparation method cannot fulfill the needs of scientific research and practical applications, because neither the number of layers nor the size and rotation angle can be controlled. Figure 8-14 shows the images of TwBLG from different studies, demonstrating that there are many superlattices in TwBLG[126-128].

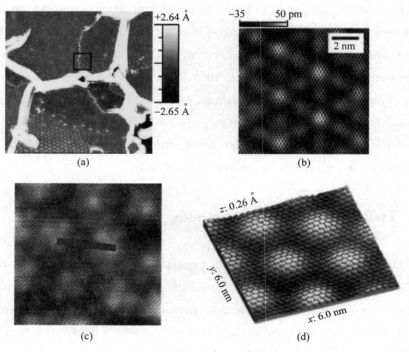

Figure 8-14 TwBLG preparation methods

(a) The image of TwBLG graphitized SiC surface (150×150 nm^2)[126]; (b) STM topographs of TwBLG[127]; (c) STM image of a moiré pattern (MP) in TwBLG on SiC (10.6×10.6 nm^2)[128]; (d) STM image at sample bias $V = +0.5$ V of a MP with $P = 2.32$ nm ($\theta = 6.08°$)[128]

8.3.2 Chemical vapor deposition

The first report on hexagonal CVD graphene domain was given by Yu et al. This work opened the door to CVD BLG[129]. From then on, the preparation of graphene by CVD has been widely used[130-136]. TwBLG has been successfully prepared due to the constant improvements in the CVD technique[132-136]. As shown in Figure 8-15, the dimensions of the double-layer area are commonly a few microns to dozens of micrometers. This method has the advantage of substantial production and can basically satisfy the requirement of experiments for the twist angle. Moreover, the quality of graphene grown by CVD is superior. However, the current technology is only suitable for the growth of TwBLG, and this method is not applicable for multi-layer angular graphene. Figure 8-15 shows images of TwBLG grown using the CVD method.

Chapter 8 Properties and Applications of New Superlattices: Twisted Bilayer Graphene

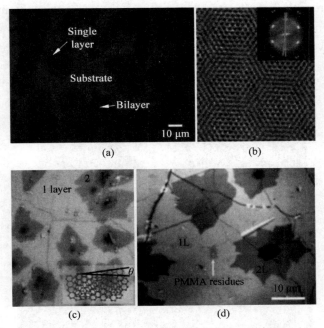

Figure 8-15 Images of TwBLG grown by CVD

(a) Optical image of TwBLG domains transferred onto Si substrate with 300 nm thermal oxide[131-132]; (b) (high-resolution) HR-TEM image of TwBLG with twist angle 6.0°[133]; (c) Optical reflection image of CVD graphene transferred to Si/285 nm SiO$_2$ and a large area wide field G band Raman image of the same region. (Inset: Structure of TwBLG with a twist angle θ)[134]; (d) Optical micrographs of TwBLG[135]

8.3.3 Folding SLG

Mechanical exfoliation is the first method used to prepare graphene. Graphene prepared via this approach has very high quality and has always been the first choice for scientists studying the properties of graphene. In this process, there is a certain probability that the graphene samples will be folded, as shown in Figure 8-16. These folded areas form multilayer twist angle graphene structures. The size of TwBLG prepared by this method is generally within a few microns to more than a dozen microns, which can basically fulfill the needs of Raman and other measurements[137-141]. The disadvantage of this method of folding graphene prepared via mechanical exfoliation is that the sample yield is very low and there is generally only one twist angle on each sample, which is not conducive to the systematic study of the changes in the performance of TwBLG with twist angle.

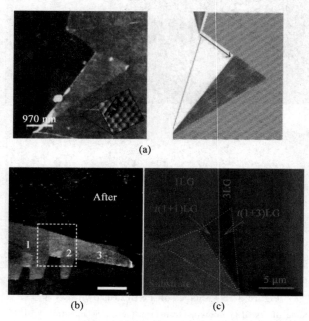

Figure 8-16　Folding SLG

(a) AFM image of a folded graphene. The inset shows a high-resolution AFM image which determines the zigzag crystallographic orientation of the bottom layer. The rotational angle $\theta = 6.2°$ was determined as twice the angle between the zigzag orientation and the folding line. The right schematics shows the folding procedure[136]. (b) Three pieces of TwBLG obtained by AFM folding[137]. (c) Optical image of a flake comprising a t(1 + 1)LG and a t(1 + 3)LG[138]

8.3.4　Vertically stacking SLG

In this method, two pieces of SLG, one fixed on the substrate and the other stacked on the SLG of the substrate at different angles, are used to obtain TwBLG samples with twist angles of $\theta_{\text{gra-gra}}$ [134,142-144], as shown in Figure 8-17. Figure 8-17(a) demonstrates the process of preparing TwBLG. The area of the angular graphene samples prepared using this method is related to the size of the SLG crystals. By stacking two large polycrystalline graphene monolayers on top of each other, many double layers of graphene with different rotation angles can be produced at the same time, which can fulfill the requirements of experiments for different angles (Figures 8-17(b)～(d)).

Chapter 8 Properties and Applications of New Superlattices: Twisted Bilayer Graphene

This method can stack more layers of angular graphene but cannot control the relative rotation angle between the layers.

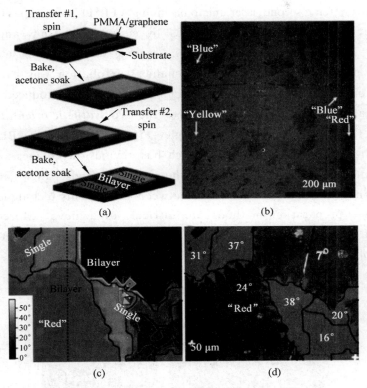

Figure 8-17 Vertically stacking SLG

(a) Schematic of the process flow used in forming BLG from single-layer CVD graphene films; (b) Optical microscope image of a TwBLG film; (c) False color map of the LEED (low-energy electron diffraction) angular orientations of a TwBLG film; (d) Optical microscope images of the TwBLG film measured in (c)[144]

Cao et al. studied TwBLG's superconducting properties at the magic-angle $\theta_{\text{gra-gra}}$, which is to stack SLG vertically to form TwBLG. They divided the SLG into two parts and fixed the first half as the bottom layer. The other part can be placed on the bottom layer of the graphene at a specific angle via the mechanical transfer method to obtain TwBLG. This method is simple and easy to obtain any twist angle of TwBLG but is not suitable for the preparation of a large number of materials.

8.3.5 Cutting-rotation-stacking (CRS)

Based on femtosecond laser micromachining (FLM) and a specific transfer technique, Liu et al. prepared TwBLG by cutting, rotating, and stacking graphene. They call this method the *cutting-rotating-stacking* (CRS) method[145]. Graphene prepared via high-quality mechanical exfoliation was patterned by FLM technology. Two parallel *synthetic orientations* are introduced to control the twist angle $\theta_{gra\text{-}gra}$. The angle between the two *synthetic orientations* after graphene rotates is equal to its twist angle. Figure 8-18 shows TwBLG via the CRS method. This preparation approach has obvious advantages and precise angle control. Using this method, the graphene twist angle and design deviation have a less than 0.1° angle between them. This technique is useful when other preparation methods are difficult to achieve. Moreover, the quality of angular BLG prepared by this method is comparable to that of TwBLG obtained using other preparation methods, see Table 8-2.

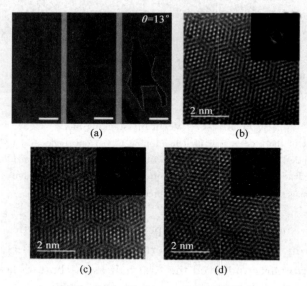

Figure 8-18 Cutting-rotationstacking

(a) Fabrication process of TwBLG with a twist angle θ = 13° by the CRS method; HR-TEM (high-resolution transmission electron microscopy) images of TwBLG with twist angles of 7°, 10° and 13° in (b), (c) and (d)[145]

Chapter 8   Properties and Applications of New Superlattices: Twisted Bilayer Graphene

Table 8-2   Comparison of different preparation methods

Methods	Complexity	Quality and Size	Yield	Advantage	Disadvantage
SiC-based epitaxial growth	simple	low $(10 \sim 100) \times 10^{-10}$ m	low	Technical maturity	The twisted angle is not controllable; below needs; low quality
CVD	simple	low $(1 \sim 100) \times 10^{-6}$ m	high	High yield; meet the experimental needs	Low quality
Folding SLG	complex	high $(1 \sim 20) \times 10^{-6}$ m	low	Twisted angle can be controlled	Low yield; single rotation angle
Vertically stacking SLG	simple	high (depends on the size of SLG)	high	Meet the experimental needs	Twisted angle can't be controlled
Cutting-rotation-stacking (CRS)	complex	high $(1 \sim 50) \times 10^{-6}$ m	low	Twisted angle can be controlled	Low yield

8.4 TwBLG's latest research results

Research on TwBLG has attracted considerable attention, especially the recent results of TwBLG superconductivity, which prompted an upsurge in this field of study. As a type of VdWH, the energy band structure of TwBLG is significantly adjusted due to the electronic coupling between the layers, which is highly dependent on the twist angle $\theta_{\text{gra-gra}}$. The novel electronic properties directly determine the physical properties of TwBLG.

8.4.1 Optoelectronic device of TwBLG

Liu et al. focused on the physical and chemical properties of 2D heterostructures and TwBLG optoelectronic devices[146-147].

They synthesized TwBLG on a Cu substrate using CVD (Figures 8-19(a) and (b))[147], demonstrating that the coupling of electronic states between layers of TwBLG resulted in the overlap of two Dirac cone energy bands of graphene. The positions of the VHS singularity differed with the changes in the twist angle (Figures 8-19(a) and (d)), which also resulted in changes in the optoelectronic properties of TwBLG. When the incident photon energy is different from that of two VHS (Figure 8-19(c)), when they coincide, TwBLG's optoelectronic interaction improves, which is specifically reflected in the enhancement of the optoelectronic reactivity, optoelectronic response, and selectivity of the wavelength. Under the excitation of a 532 nm laser, the

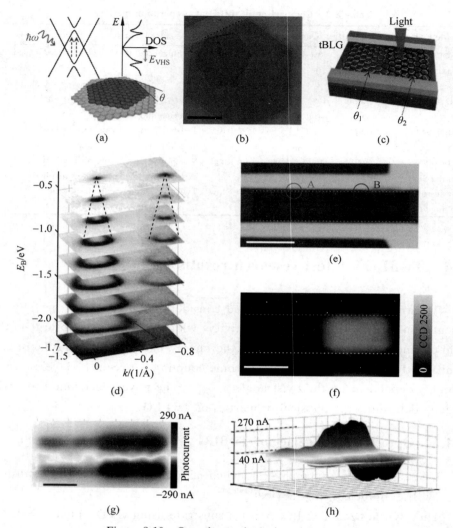

Figure 8-19 Optoelectronic device of TwBLG

(a) Schematics for band structure with minigaps (top left) and the corresponding DOS with VHSs (top right) in TwBLG (bottom). Blue arrows describe the photoexcitation process as the energy interval of two VHSs ($2E_{VHS}$) matches the energy of incident photon. (b) Scanning electron microscopy (SEM) images of TwBLG domains with a twist angle (13°) on SiO_2/Si. (c) Schematic illustration of a TwBLG photodetection device. The channel comprises of two adjacent TwBLG domains with different twist angles of θ_1 and θ_2. (d) Stacking plot of constant-energy contours at different binding energies (EBs) of TwBLG. (e) Optical image of the TwBLG photodetection device. The θ_1 and θ_2 are 7° and 13°, respectively. (f) Raman G-band intensity mapping image under 532 nm (2.33 eV) laser. 13° TwBLG domain exhibits an enhanced G-band intensity. (g) Scanning photocurrent images of the same TwBLG device. (h) 3D view of the scanning photocurrent image of the same TwBLG device[147]

photocurrent of TwBLG develops a twist angle of 13°, significantly better than 7° (Figures 8-19(e)～(h)). When the wavelength of the incident light is 633 nm, the device displays an obvious photoelectric response. On this basis, Liu et al. coupled TwBLG with the nanostructure of the plasmon, which further enhanced the photocurrent by nearly 80 times. This work provides a new opportunity for the application of TwBLG for ultrafast, highly sensitive, and highly selective photoelectric detection.

8.4.2 Photonic crystals for nano-light

Photonic crystals use their own periodic structures to effectively regulate electromagnetic waves. Similarly, photonic crystals can isolate the transmission of electromagnetic waves through their own band gaps and can regulate the surface ionic excitons with local wave fields and extremely short wave fields. However, traditional methods of preparing photonic crystals require top-down nano-fabrication, which is expensive and low in fabrication accuracy.

A previous study used the moiré superlattice (MS) in TwBLG with a twist angle to construct a periodic barrier[148]. This type of superlattice is a plasmon with a short wavelength, which is a photonic crystal supporting the transmission of nano-photons. The researchers achieved the regulation of the Mohr graphene superlattice periodic potential by adjusting the TwBLG torsion angle. When the twist angle was small, atomic reconstruction occurred in the BLG, transforming the superlattice into photonic crystals with periodic potential field distribution, thus avoiding the use of top-down nano-micromachining. Due to the small size of the photonic crystal, it can support the transmission of surface ionic excitons with extremely short wavelengths that can be regulated by changing the TwBLG twist angle, interlayer coupling, and applied voltage. The researchers also introduced the Moiré superlattice topological theory, which posits that the potential field of periodic arrangement is formed by topologically protected boundary states. Figure 8-20 shows the results of the experiment. This work provides a new mechanism and technology for the preparation and characterization of nano-photonic size photonic crystals and further ideas for the application and use of topological materials.

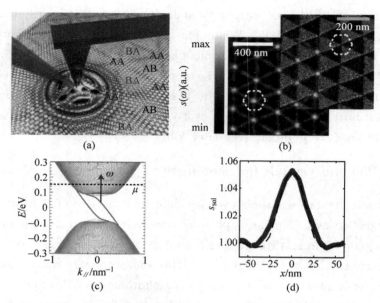

Figure 8-20 Nano-light photonic crystal formed by a network of solitons in TwBLG[148] (a) Schematic of the IR nano-imaging experimental setup. AB, BA, and AA labels periodically occurring stacking types of graphene layers. (b) (Left) Visualizing the nano-light photonic crystal formed by the soliton lattice. (Right) Dark-field TEM image of a TwBLG sample showing contrast between AB and BA triangular domains. (c) Electronic band structure of a single infinitely long soliton (only the K valley is shown). (d) Experimental (solid) and calculated (dashed) near-field signal $S_{sol}(x)$ across a single soliton line.

8.4.3 Tuning superconductivity of TwBLG

Researchers from MIT found that when the twist angle between the two layers of graphene was 1.1°, the entire system was superconducting. However, it is not easy to prepare TwBLG with a twist angle of 1.1°. Moreover, the electron states of TwBLG vary greatly with the twist angle.

Another study used an external force applied to the TwBLG test apparatus to convert TwBLG with different twist angles from a metallic state to an insulating state or a superconductor[149], depending on the number of electrons in the TwBLG (Figure 8-21(a)). h-BN was used as a dielectric to encapsulate TwBLG to form a h-BN/TwBLG/h-BN heterostructure. When the h-BN/TwBLG/h-BN was subjected to 10,000 standard atmospheres, insulating and superconducting states were both observed in the TwBLG when the ambient

Chapter 8   Properties and Applications of New Superlattices: Twisted Bilayer Graphene

temperature was slightly higher than 3°C above absolute zero (Figure 8-21(b)). The pressure on the system and the critical temperature of the superconducting state increased under low temperature, high magnetic field, and high pressure (Figures 8-21(c)~(f)).

In this study, applying pressure on the TwBLG had the same effect on the electronic state of the system as adjusting the twist angle. Adjusting the electronic coupling between the TwBLG layers led to a new electronic state. This technology should be applied and tested in other electronic materials (twist angle h-BN/graphene VdWH, $MoSe_2/WSe_2$ VdWH, and graphene/$MoSe_2$ VdWH). Thus, there are two different ways in which one can regulate the electronic state of a superlattice: one is magic-angle, the other is pressure.

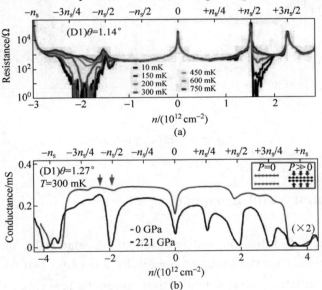

Figure 8-21 Tuning superconductivity of TwBLG

(a) Temperature dependence of the resistance of device D1 over the density range necessary to fill the moiré unit cell, $n \in [-n_s, n_s]$ at $D = 0$. (b) Conductance of device D2 (1.27°) measured over the entire density range necessary to fill the moiré unit cell at two values of pressure: 0 GPa (gray) and 2.21 GPa (blue) at $T = 300$ mK. (c) Resistance of device D2 over a small range of carrier density near $-n_s/2$ versus. An insulating phase at $-n_s/2$ neighbors a superconducting pocket at slightly larger hole doping. (d) Similar map as a function of B. (e) Resistance as a function of T at 2.21 GPa at $-n_s/2$ (red) and at optimal doping of the superconductor (blue). (f) Map of dV/dI versus I_{dc} and B at $n = -2.1 \times 10^{12}/cm^2$, $T = 300$ mT, and 1.33 GPa[149]

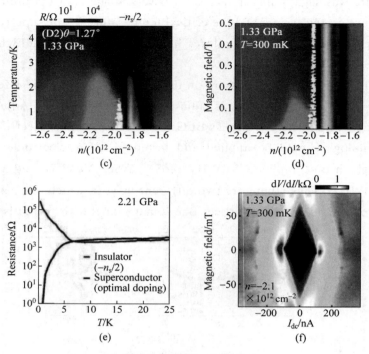

Figure 8-21(Continued)

8.5 Summary and prospect

With the wide attention and research of 2D materials, more and more novel materials are synthesized and studied by scientists, just like VdWH and TwBLG superlattice. These metamaterials are only a few atoms thick, and the vertical stacking of 2D materials rotates between layers, giving new superlattices completely different structures, properties, and functions than ever before. The weak interlayer interaction makes the TwBLG interlayer coupling, thus forming a completely new electronic state which is obviously different from SLG. This new electronic state is linked to the TwBLG's twisted angle. A growing number of researchers have found that because the study of TwBLG is limited by the level of preparation, the preparation of TwBLG at specific angles is not very accurate. The new electronic state of TwBLG with different twisted angles can completely achieve the ideal electronic state under any twisted angle by applying external electric field or external force to the superlattice of TwBLG.

Due to the mutual coupling of the electrons between the layers, the electronic states of the entire system improved, with many novel physical properties. The relative rotation angle between the 2D single-layer materials is an important factor affecting the van der Waals heterostructure characteristics, that is, varying angles create different periodic superlattices. *Electronic quantum metamaterials* involves the twisting and stacking of single-layer 2D materials to form new superstructures[150]. More studies have focused on such metamaterials, and their physical properties have been thoroughly analyzed. These metamaterials are widely used in the field of optoelectronics.

The superconducting properties found in TwBLG have provided new insight into its characteristics. This indicates that electron interactions and subwavelength characteristics can interact together in quantum metamaterials to generate novel physical properties. With different torsional angles, the formation of a new superlattice leads to a new electronic state. This new electronic state produces novel physical properties. We believe that with the emergence and change of the new optical and magnetic properties, the application of TwBLG superlattices in the field of optoelectronics will be more extensive. As for the application of thermology, photothermoelectric property should also become a new research direction. Although the preparation method, the level of observation technology, the extreme experimental conditions limit its application, TwBLG may lead to new applications for 2D materials.

References

[1] NOVOSELOV K S, GEIM A K, MOROZOV S V, et al. Electric field effect in atomically thin carbon films[J]. Science, 2004, 306 (5696): 666-669.

[2] PAKDEL A, ZHI C, BANDO Y, et al. Low-dimensional boron nitride nanomaterials [J]. Materials Today, 2012, 15 (6): 256-265.

[3] RADISAVLJEVIC B, RADENOVIC A, BRIVIO J, et al. Single-layer MoS_2 transistors[J]. Nature Nanotechnology, 2011, 6 (3): 147-150.

[4] KANGPENG W, JUN W, JINTAI F, et al. Ultrafast saturable absorption of two-dimensional MoS_2 nanosheets[J]. ACS Nano, 2013, 7 (10): 9260.

[5] LI L, KIM J, JIN C, et al. Direct observation of the layer-dependent electronic

structure in phosphorene[J]. Nature Nanotechnology, 2016, 12 (1): 21.

[6] TAISUKE O, AARON B, THOMAS S, et al. Controlling the electronic structure of bilayer graphene[J]. Science, 2006, 313 (5789): 951-954.

[7] HE B, ZHANG W J, YAO Z Q, et al. p-type conduction in beryllium-implanted hexagonal boron nitride films[J]. Applied Physics Letters, 2009, 95 (25): 252106.

[8] WANG J, MU X, WANG X, et al. The thermal and thermoelectric properties of in-plane C-BN hybrid structures and graphene/h-BN van der Waals heterostructures[J]. Materials Today Physics, 2018, 5: 29-57.

[9] WU M, ZENG X C, JENA P. Unusual magnetic properties of functionalized graphene nanoribbons[J]. The Journal of Physical Chemistry Letters, 2013, 4 (15): 2482-2488.

[10] OUYANG B, SONG J. Strain engineering of magnetic states of vacancy-decorated hexagonal boron nitride[J]. Applied Physics Letters, 2013, 103 (10): 102401.

[11] DU B, CHEN Y, ZHU Z, et al. Dots versus antidots: computational exploration of structure, magnetism, and half-metallicity in Boron-Nitride nanostructures[J]. Journal of the American Chemical Society, 2009, 131 (47): 17354-17359.

[12] DEAN C R, YOUNG A F, MERIC I, et al. Boron nitride substrates for high-quality graphene electronics[J]. Nature Nanotechnology, 2010, 5(10): 722-726.

[13] PONOMARENKO L A, GEIM A K, ZHUKOV A A, et al. Tunable metal-insulator transition in double-layer graphene heterostructures[J]. Nature Physics, 2011, 7 (12): 958-961.

[14] WANG J, XU X, MU X, et al. Magnetics and spintronics on two-dimensional composite materials of graphene/hexagonal boron nitride [J]. Materials Today Physics, 2017, 3: 93-117.

[15] GEIM A K, GRIGORIEVA I V. Van der Waals heterostructures[J]. Nature, 2013, 499 (7459): 419-425.

[16] CHEN Y X, SHI T T, LIU P Y, et al. Insights into the mechanism of the enhanced visible-light photocatalytic activity of black phosphorus/$BiVO_4$ heterostructure: a first-principles study[J]. Journal of Materials Chemistry A, 2018, 6 (39): 19167-19175.

[17] GONG Y, LIN J, WANG X, et al. Vertical and in-plane heterostructures from WS_2/MoS_2 monolayers[J]. Nature Materials, 2014, 13 (12): 1135-1142.

[18] BERTOLAZZI S, KRASNOZHON D, KIS A. Nonvolatile memory cells based on MoS_2/graphene heterostructures[J]. ACS Nano, 2013, 7 (4): 3246-3252.

[19] WANG J G, MA F C, LIANG W J, et al. Optical, photonic and optoelectronic properties of graphene, h-NB and their hybrid materials[J]. Nanophotonics, 2017, 6(5): 943-976.

[20] GONG Y, LEI S, YE G, et al. Two-step growth of two-dimensional WSe_2/$MoSe_2$ heterostructures[J]. Nano Letters, 2015, 15 (9): 6135-6141.

[21] WANG J, MA F, LIANG W, et al. Electrical properties and applications of graphene, hexagonal boron nitride (h-BN), and graphene/h-BN heterostructures[J]. Materials Today Physics, 2017, 2: 6-34.

[22] PHILIP H. Bilayer graphene: a little twist with big consequences[J]. Nature Materials, 2013, 12 (10): 874-875.

[23] HE W Y, CHU Z D, HE D. Chiral tunneling in a twisted graphene bilayer[J]. Physical Review Letters, 2013, 111 (6): 066803.

[24] CHOI M Y, HYUN Y H, KIM Y. Angle dependence of landau level spectrum in twisted bilayer graphene[J]. Physical Review B, 2011, 84 (19): 195437.

[25] WEI Y, LAN M, LIU M, et al. Angle-dependent van Hove singularities and their breakdown in twisted graphene bilayers[J]. Physical Review B, 2014, 90 (11): 5758-5785.

[26] YAN W, LIU M, DOU R F, et al. Angle-dependent van Hove singularities in a slightly twisted graphene bilayer[J]. Physical Review Letters, 2012, 109 (12): 126801.

[27] HAVENER R W, ZHUANG H, BROWN L, et al. Angle-resolved Raman imaging of interlayer rotations and interactions in twisted bilayer graphene[J]. Nano Letters, 2012, 12 (6): 3162-3167.

[28] UCHIDA K, FURUYA S, IWATA J I, et al. Atomic corrugation and electron localization due to Moiré patterns in twisted bilayer graphenes[J]. Physical Review B, 2014, 90 (15): 155451.

[29] MELE E J. Band symmetries and singularities in twisted multilayer graphene[J]. Physical Review B, 2011, 84 (23): 235439.

[30] KIM Y, HERLINGER P, MOON P, et al. Charge inversion and topological phase transition at a twist angle induced van hove singularity of bilayer graphene[J]. Nano Letters, 2016, 16 (8): 5053-5059.

[31] HAVENER R W, LIANG Y, BROWN L, et al. Van Hove singularities and excitonic effects in the optical conductivity of twisted bilayer graphene[J]. Nano Letters, 2014, 14 (6): 3353-3357.

[32] GONZALEZ J. Magnetic and Kohn-Luttinger instabilities near a van Hove singularity: monolayer versus twisted bilayer graphene[J]. Physical Review B, 2013, 88 (88): 125434.

[33] YIN J, WANG H, PENG H, et al. Corrigendum: Selectively enhanced photocurrent generation in twisted bilayer graphene with van Hove singularity[J]. Nature Communications, 2016, 7: 10699.

[34] MOON P, SON Y W, KOSHINO M. Optical absorption of twisted bilayer graphene with interlayer potential asymmetry[J]. Physical Review B, 2014, 90 (15): 155427.

[35] YUAN N F Q, LIANG F. A model for metal-insulator transition in graphene superlattices and beyond[J]. Physical Review B, 2018, 98 (4): 045103.

[36] WANG J, LIN W, XU X, et al. Plasmon-exciton coupling interaction for surface catalytic reactions[J]. Chemical Record, 2018, 18 (5): 481-490.

[37] PO H C, ZOU L, VISHWANATH A, et al. Origin of Mott insulating behavior and superconductivity in twisted bilayer graphene[J]. Physical Review X, 2018, 8(3): 031089.

[38] OCHI M, KOSHINO M, KUROKI K. Possible correlated insulating states in magic-angle twisted bilayer graphene under strongly competing interactions[J]. Physical

Review B, 2018, 98 (8): 081102.
[39] SANCHEZ-YAMAGISHI J D, TAYCHATANAPAT T, WATANABE K, et al. Quantum Hall effect, screening, and layer-polarized insulating states in twisted bilayer graphene[J]. Physical Review Letters, 2012, 108 (7): 076601.
[40] CAO Y, FATEMI V, FANG S, et al. Unconventional superconductivity in magic-angle graphene superlattices[J]. Nature, 2018, 556 (7699): 43-50.
[41] GONZÁLEZ J, STAUBER T. Kohn-Luttinger superconductivity in twisted bilayer graphene[J]. Physical Review B, 2019, 122: 026801.
[42] WANG W X, JIANG H, ZHANG Y, et al. Scanning tunneling microscopy and spectroscopy of finite-size twisted bilayer graphene[J]. Physical Review B, 2017, 96: 115434.
[43] YIN L J, JIANG H, QIAO J B, et al. Direct imaging of topological edge states at a bilayer graphene domain wall[J]. Nature Communications, 2016, 7: 11760.
[44] MARCHENKO D, EVTUSHINSKY D V, GOLIAS E, et al. Extremely flat band in bilayer graphene[J]. Science Advances, 2018, 4 (11): eaau0059.
[45] CASTRO NETO A H, GUINEA F, PERES N M R, et al. The electronic properties of graphene[J]. Reviews of Modern Physics, 2009, 81: 109-162.
[46] NOVOSELOV K S, GEIM A K, MOROZOV S V, et al. Two-dimensional gas of massless Dirac fermions in graphene[J]. Nature 2005, 438 (7065): 197-200.
[47] OOSTINGA J B, HEERSCHE H B, LIU X, et al. Gate-induced insulating state in bilayer graphene devices[J]. Nature Materials, 2008, 7 (2): 151-157.
[48] HSU Y F, GUO G Y. Anomalous integer quantum Hall effect in AA-stacked bilayer graphene[J]. Physical Review B, 2010, 82 (16): 165404.
[49] SANJAY K, SANJAY A. Quasi-particle spectrum and density of electronic states in AA-and AB-stacked bilayer graphene[J]. European Physical Journal B, 2013, 86 (3): 1-9.
[50] LIU L, ZHOU H, CHENG R, et al. High-yield chemical vapor deposition growth of high-quality large-area AB-stacked bilayer graphene[J]. ACS Nano, 2012, 6 (9): 8241-8249.
[51] ZHANG L, ZHANG Y, CAMACHO J, et al. The experimental observation of quantum Hall effect of I = 3 chiral quasiparticles in trilayer graphene[J]. Nature Physics, 2011, 7 (12): 953-957.
[52] LUI C H, LI Z, MAK K F, et al. Observation of an electrically tunable band gap in trilayer graphene[J]. Nature Physics, 2011, 7 (12): 944-947
[53] JU L, SHI Z, NAIR N, et al. Topological valley transport at bilayer graphene domain walls[J]. Nature, 2015, 520 (7549): 650-655.
[54] ALEX T, SHUBHAYU C, SUBIR S, et al. Triangular antiferromagnetism on the honeycomb lattice of twisted bilayer graphene[J]. Physical Review B, 2018, 98(7): 075109.
[55] ANDRES P L, RAMIREZ R, VERGES J A. Strong covalent bonding between two

graphene layers[J]. Physical Review B, 2008, 77: 045403.

[56] MELE E J. Novel electronic36 states seen in graphene[J]. Nature, 2018, 556 (7699): 37-38.

[57] UCHIDA K, FURUYA S, IWATA J I, et al. Atomic corrugation and electron localization due to Moiré patterns in twisted bilayer graphenes[J]. Physical Review B, 2014, 90 (15): 155451.

[58] MELE E J. Band symmetries and singularities in twisted multilayer graphene [J]. Physical Review B, 2011, 84 (23): 235439.

[59] SYMALLA F SHALLCROSS S, BELJAKOV I, et al. Band-gap engineering with a twist: formation of intercalant superlattices in twisted graphene bilayers[J]. Physical Review B, 2015, 91 (20): 205412.

[60] NISHI H, MATSUSHITA Y, OSHIYAMA A. Band-unfolding approach to moiré-induced band-gap opening and Fermi level velocity reduction in twisted bilayer graphene[J]. Physical Review B, 2017, 95 (8): 085420.

[61] LOPES D S J M B, PERES N M R, CASTRO N A H. Continuum model of the twisted graphene bilayer[J]. Physical Review B, 2012, 86 (15): 155449.

[62] GONZALEZ-ARRAGA L A, LADO J L, GUINEA F, et al. Electrically controllable magnetism in twisted bilayer graphene[J]. Physical Review Letters, 2017, 119(10): 107201.

[63] RAMIRES A, LADO J L. Electrically tunable gauge fields in tiny-angle twisted bilayer graphene[J]. Physical Review Letters, 2018, 121 (14): 146801.

[64] XU Y, JIN G. Electron retroreflection and spin beam splitting in a twisted graphene bilayer[J]. Solid State Communications, 2016, 247: 72-77.

[65] HUANG S, KIM K, EFIMKIN D K, et al. Topologically protected helical states in minimally twisted bilayer graphene [J]. Physical Review Letters, 2018, 121(3): 037702.

[66] BRIHUEGA I, MALLET P, GONZÁLEZ-HERRERO H, et al. Unraveling the Intrinsic and robust nature of van Hove singularities in twisted bilayer graphene by scanning tunneling microscopy and theoretical analysis[J]. Physical Review Letters, 2012, 109 (19): 196802.

[67] MOON P, KOSHINO M. Energy spectrum and quantum Hall effect in twisted bilayer graphene[J]. Physical Review B, 2012, 85 (19): 195458.

[68] STAUBER T, KOHLER H. Quasi-flat plasmonic bands in twisted bilayer graphene [J]. Nano Letters, 2016, 16 (11): 6844-6849.

[69] KIM K, DASILVA A, HUANG S, et al. Tunable moiré bands and strong correlations in small-twist-angle bilayer graphene[J]. Proceedings of the National Academy of Sciences, 2017, 114 (13): 3364-3369.

[70] NAM N N T, KOSHINO M. Lattice relaxation and energy band modulation in twisted bilayer graphenes[J]. Physical Review B, 2017, 96 (7): 075311.

[71] KIM K S, WALTER A L, MORESCHINI L, et al. Coexisting massive and massless

Dirac fermions in symmetry-broken bilayer graphene[J]. Nature Materials, 2013, 12(10): 887-892.
[72] WONG D, YANG W, JUNG J, et al. Local spectroscopy of moiré-induced electronic structure in gate-tunable twisted bilayer graphene[J]. Physical Review B, 2015, 92(15): 155409.
[73] ZHANG F, MACDONALD A H, MELE E J. Valley Chern numbers and boundary modes in gapped bilayer graphene[J]. Proceedings of the National Academy of Sciences, 2013, 110 (26): 10546-10551.
[74] MELE E J. Commensuration and interlayer coherence in twisted bilayer graphene[J]. Physical Review B, 2010, 81 (16): 161405.
[75] XU Y, JIN G. Chirality crossover of Andreev reflection in a twisted graphene bilayer [J]. Europhysics Letters, 2015, 111 (6): 67006.
[76] LUICAN A, LI G, REINA A, et al. Single-layer behavior and its breakdown in twisted graphene layers[J]. Physical Review Letters, 2011, 106 (12): 126802.
[77] LI S, LIU K, YIN L, et al. Splitting of van Hove singularities in slightly twisted bilayer graphene[J]. Physical Review B, 2017, 96 (15): 155416.
[78] VAEZI A, LIANG Y, NGAI D H, et al. Topological edge states at a tilt boundary in gated multilayer graphene[J]. Physical Review X, 2013, 3 (2): 1-6.
[79] YUAN C, FATEMI V, DEMIR A, et al. Correlated insulator behaviour at half-filling in magic-angle graphene superlattices[J]. Nature, 2018, 556 (7699): 80-84.
[80] WEI Y, WANG E, BAO C, et al. Quasicrystalline 30° twisted bilayer graphene as an incommensurate superlattice with strong interlayer coupling[J]. Proceedings of the National Academy of Sciences, 2018, 184 (27): 50.
[81] CHOI M, HYUN Y, KIM Y. Angle dependence of the Landau level spectrum in twisted bilayer graphene[J]. Physical Review B, 2011, 84 (19): 195437.
[82] SHALLCROSS S, SHARMA S, PANKRATOV O A. Quantum interference at the twist boundary in graphene[J]. Physical Review Letters, 2008, 101 (5): 056803.
[83] MELE E J. Commensuration and interlayer coherence in twisted bilayer graphene[J]. Physical Review B, 2010, 81(16): 161405(R).
[84] CHU Z D, HE W Y, HE L. Coexistence of van Hove singularities and superlattice Dirac points in a slightly twisted graphene bilayer[J]. Physical Review B, 2013, 87(15): 155419.
[85] CARR S, FANG S, JARILLO-HERRERO P, et al. Pressure dependence of the magic twist angle in graphene superlattices[J]. Physical Review B, 2018, 98 (8): 085144.
[86] LI G, LUICAN A, LOPES D S J M B, et al. Observation of van Hove singularities in twisted graphene layers[J]. Nature Physics, 2009, 6 (2): 119-113.
[87] MILLER D L, KUBISTA K D, RUTTER G M, et al. Observing the quantization of zero mass carriers in graphene[J]. Science, 2009, 324 (5929): 924-927.
[88] LEE D S, RIEDL C, BERINGER T, et al. Quantum Hall effect in twisted bilayer graphene[J]. Physical Review Letters, 2011, 107 (21): 216602.

[89] GOERBIG M O. Electronic properties of graphene in a strong magnetic field[J]. Review of Modern Physics, 2011, 83 (4): 1193-1243.

[90] SONG Y J, OTTE A F, KUK Y, et al. High-resolution tunnelling spectroscopy of a graphene quartet[J]. Nature, 2010, 467 (7312): 185-189.

[91] WANG Z F, LIU F, CHOU M Y. Fractal Landau-level spectra in twisted bilayer graphene[J]. Nano Letters, 2012, 12 (7): 3833-3838.

[92] HU F, DAS S R, LUAN Y, et al. Real-space imaging of the tailored plasmons in twisted bilayer graphene[J]. Physical Review Letters, 2017, 119 (21): 247402.

[93] YIN L J, QIAO J B, WANG W X, et al. Landau quantization and Fermi velocity renormalization in twisted graphene bilayers[J]. Physical Review B, 2015, 92(20): 201408(R).

[94] HICKS J, SPRINKLE M, SHEPPERD K, et al. Symmetry breaking in commensurate graphene rotational stacking: comparison of theory and experiment [J]. Physical Review B, 2011, 82 (20): 205403.

[95] UCHIDA K, FURUYA S, IWATA J I, et al. Atomic corrugation and electron localization due to Moiré patterns in twisted bilayer graphenes[J]. Physical Review B, 2014, 90 (15): 155451.

[96] LEE D S, RIEDL C, BERINGER T, et al. Quantum Hall effect in twisted bilayer graphene[J]. Physical Review Letters, 2011, 107 (21): 216602.

[97] MOON P, KOSHINO M. Energy spectrum and quantum Hall effect in twisted bilayer graphene[J]. Physical Review B, 2012, 85(19): 195458.

[98] SANCHEZ-YAMAGISHI J D, TAYCHATANAPAT T, WATANABE K, et al. Quantum Hall effect, screening, and layer-polarized insulating states in twisted bilayer graphene[J]. Physical Review Letters, 2012, 108 (7): 076601.

[99] FINOCCHIARO F, GUINEA F, SAN-JOSE P. Quantum spin Hall effect in twisted bilayer graphene[J]. 2D Materials, 2017, 4 (2): 025027.

[100] LÖFWANDER T, SAN-JOSE P, PRADA E. Quantum Hall effect in graphene with twisted bilayer stripe defects[J]. Physical Review B, 2013, 87(20): 205429.

[101] FALLAHAZAD B, HAO Y, LEE K, et al. Quantum Hall effect in Bernal stacked and twisted bilayer graphene grown on Cu by chemical vapor deposition[J]. Physical Review B, 2012, 85(20): 201408.

[102] KIM Y, PARK J, SONG I, et al. Broken-symmetry quantum hall states in twisted bilayer graphene[J]. Scientific Report, 2016, 6: 38068.

[103] NOVOSELOV K S, GEIM A K, MOROZOV S V, et al. Two-dimensional gas of massless Dirac fermions in graphene[J]. Nature, 2005, 438 (7065): 197-200.

[104] ZHANG Y, TAN Y W, STORMER H L, et al. Experimental observation of the quantum Hall effect and Berry's phase in graphene[J]. Nature, 2005, 438: 201-204.

[105] NOVOSELOV K S, MCCANN E, MOROZOV S V, et al. Unconventional quantum Hall effect and Berry's phase of 2p in bilayer graphene[J]. Nature Physics, 2006, 2(2): 177-180.

[106] HASS J, VARCHON F, MILLÁN-OTOYA J E, et al. Why multilayer graphene on 4H-SiC (0001) behaves like a single sheet of graphene[J]. Physical Review Letters, 2008, 100 (12): 125504.

[107] LIU C C, ZHANG L D, CHEN W Q, et al. Chiral spin density wave and d + id superconductivity in the magic-angle-twisted bilayer graphene[J]. Physical Review Letters, 2018, 121: 217001.

[108] KENNES D M, LISCHNER J, KARRASCH C. Strong correlations and d + id superconductivity in twisted bilayer graphene[J]. Physical Review B, 2018, 98: 241407(R).

[109] PO H C, ZOU L, VISHWANATH A, et al. Origin of Mott insulating behavior and superconductivity in twisted bilayer graphene[J]. Physical Review X, 2018, 8(3): 031089.

[110] YUAN N F Q, LIANG F. Model for the metal-insulator transition in graphene superlattices and beyond[J]. Physical Review B, 2018, 98(4): 045103.

[111] NAIR R R, BLAKE P, GRIGORENKO A N, et al. Fine structure constant defines visual transparency of graphene[J]. Science, 2008, 320(5881): 1308-1308.

[112] WANG J G, MA F C, LIANG W J, et al. Optical, photonic and optoelectronic properties of graphene, h-NB and their hybrid materials[J]. Nanophotonics, 2017, 6(5): 943-976.

[113] MAK K F, SFEIR M Y, WU Y, et al. Measurement of the optical conductivity of graphene[J]. Physical Review Letters, 2008, 101(19): 196405.

[114] MOON P, KOSHINO M. Optical absorption in twisted bilayer graphene[J]. Physical Review B, 2013, 87(20): 205404.

[115] ANH LE H. Electronic structure and optical properties of twisted bilayer graphene calculated via time evolution of states in real space[J]. Physical Review B, 2018, 97(12): 125136.

[116] HAVENER R W, LIANG Y, BROWN L, et al. Van Hove singularities and excitonic effects in the optical conductivity of twisted bilayer graphene[J]. Nano Letters, 2014, 14(6): 3353-3357.

[117] ZOU X, SHANG J, LEAW J, et al. Terahertz conductivity of twisted bilayer graphene[J]. Physical Review Letters, 2013, 110(6): 067401.

[118] WANG J, MA F, LIANG W, et al. Electrical properties and applications of graphene, hexagonal boron nitride (h-BN), and graphene/h-BN heterostructures[J]. Materials Today Physics, 2017, 2: 6-34.

[119] WANG J, XU X, MU X, et al. Magnetics and spintronics on two-dimensional composite materials of graphene/hexagonal boron nitride [J]. Materials Today Physics, 2017, 3: 93-117.

[120] WANG J, MU X, WANG X, et al. The thermal and thermoelectric properties of in-plane C-BN hybrid structures and graphene/h-BN van der Waals heterostructures[J]. Materials Today Physics, 2018, 5: 29-57.

[121] WANG J, MA F, SUN M. Graphene, hexagonal boron nitride, and their heterostructures: properties and applications[J]. RSC Advances, 2017, 7 (27): 16801-16822.

[122] WANG J, MU X, SUN M. The thermal, electrical and thermoelectric properties of graphene nanomaterials[J]. Nanomaterials, 2019, 9: 218.

[123] LI H, YING H, CHEN X, et al. Thermal conductivity of twisted bilayer graphene [J]. Nanoscale, 2014, 6(22): 13402-13408.

[124] COCEMASOV A I, NIKA D L, BALANDIN A A. Phonons in twisted bilayer graphene[J]. Physical Review B, 2013, 88(3): 035428.

[125] LIMBU T B, HAHN K R, MENDOZA F, et al. Grain size-dependent thermal conductivity of polycrystalline twisted bilayer graphene[J]. Carbon, 2017, 117: 367-375.

[126] VARCHON F, MALLET P, MAGAUD L, et al. Rotational disorder in few-layer graphene films on 6H-SiC (000-1): a scanning tunneling microscopy study [J]. Physical Review B, 2008, 77(16): 165415.

[127] MILLER D L, KUBISTA K D, RUTTER G M, et al. Structural analysis of multilayer graphene via atomic moiré interferometry[J]. Physical Review B, 2010, 81(12): 125427.

[128] BRIHUEGA I, MALLET P, GONZÁLEZ-HERRERO H, et al. Unraveling the intrinsic and robust nature of van hove singularities in twisted bilayer graphene by scanning tunneling microscopy and theoretical analysis[J]. Physical Review Letters, 2012, 109(19): 196802.

[129] YU Q Y, JAUREGUI L A, WU W, et al. Control and characterization of individual grainsand grain boundaries in graphene grown bychemical vapour deposition[J]. Nature Materials, 2011, 10(6): 443-449.

[130] WANG Y, SU Z, WU W, et al. Resonance Raman spectroscopy of G-line and folded phonons in twistedbilayer graphene with large rotation angles[J]. Applied Physics Letters, 2013, 103(12): 123101.

[131] WANG Y, SU Z, WU W, et al. Four-fold Raman enhancement of 2D band intwisted bilayer graphene: evidence for adoubly degenerate Dirac band and quantuminterference [J]. Nanotechnology, 2014, 25(33): 335201.

[132] HE R, CHUNG T F, DELANEY C, et al. Observation of low energy raman modes in twisted bilayer graphene[J]. Nano Letters, 2013, 13(8): 3594-3601.

[133] LU C C, LIN Y C, LIU Z, et al. Twisting bilayer graphene superlattices[J]. ACS Nano, 2013, 7(3): 2587-2594.

[134] HAVENER R W, ZHUANG H, BROWN L, et al. Angle-resolved Raman imaging of interlayer rotations and interactions in twisted bilayer graphene[J]. Nano Letters, 2012, 12(6): 3162-3167.

[135] CAMPOS-DELGADO J, CANÇADO L G, ACHETE C A, et al. Raman scattering study of the phonon dispersion in twisted bilayer graphene[J]. Nano Research, 2013, 6(4): 269-274.

[136] TA H Q, PERELLO D J, DUONG D L, et al. Stranski-Krastanov and Volmer-Weber CVD growth regimes to control the stacking order in bilayer graphene[J]. Nano Letters, 2016, 16(10): 6403-6410.

[137] CAROZO V, ALMEIDA C M, FERREIRA E H M, et al. Raman signature of graphene superlattices[J]. Nano Letters, 2011, 11(11): 4527-4534.

[138] CAROZO V, ALMEIDA C M, FRAGNEAUD B, et al. Resonance effects on the Raman spectra of graphene superlattices[J]. Physical Review B, 2013, 88(8): 085401.

[139] WU J B, ZHANG X, IJÄS M, et al. Resonant Raman spectroscopy of twisted multilayer graphene[J]. Nature Communications, 2014, 5: 5309.

[140] NI Z, WANG Y, YU T, et al. Reduction of Fermi velocity in folded graphene observed by resonance Raman spectroscopy[J]. Physical Review B, 2008, 77(23): 235403.

[141] NI Z, LIU L, WANG Y, et al. G-band Raman double resonance in twisted bilayer graphene: evidence of band splitting and folding[J]. Physical Review B, 2009, 80(12): 125404.

[142] PONCHARAL P, AYARI A, MICHEL T, et al. Raman spectra of misoriented bilayer graphene[J]. Physical Review B, 2008, 78(11): 113407.

[143] KIM K, COH S, TAN L Z, et al. Raman spectroscopy study of rotated double-layer graphene: misorientation-angle dependence of electronic structure [J]. Physical Review Letters, 2012, 108(24): 246103.

[144] ROBINSON J T, SCHMUCKER S W, BOGDAN DIACONESCU C, et al. Electronic hybridization of large-areastacked graphene films[J]. ACS Nano, 2013, 7(1): 637-644.

[145] CHEN X D, XIN W, JIANG W S, et al. High-precision twist-controlled bilayer and trilayer graphene[J]. Advanced Materials, 2016, 28(13): 2563-2570.

[146] LIAO L, WANG H, PENG H, et al. van Hove singularity enhanced photochemical reactivity of twisted bilayer graphene[J]. Nano Letters, 2015, 15(8): 5585-5589.

[147] YIN J, WANG H, PENG H, et al. Selectively enhanced photocurrent generation in twisted bilayer graphene with van Hove singularity[J]. Nature Communications, 2016, 7: 10699.

[148] SUNKU S S, NI G X, JIANG B Y, et al. Photonic crystals for nano-light in moiré graphene superlattices[J]. Science, 2018, 362(6419): 1153-1156.

[149] YANKOWITZ M, CHEN S, POLSHYN H, et al. Tuning superconductivity in twisted bilayer graphene[J]. Science, 2019, 363(6431): 1059-1064.

[150] SONG J C W, GABOR N M. Electron quantum metamaterials in van der Waals heterostructures[J]. Nature Nanotechnology, 2018, 13: 986-993.

Chapter 9

Two Dimensional Black Phosphorus: Physical Properties and Applications

9.1 Introduction

Two dimensional (2D) nanomaterials, with their excellent mechanical, thermal, electrical, and optical properties, have become popular topics of research. Among them, the massless Dirac fermion characteristics of graphene significantly improve its carrier mobility, but the characteristics of the zero band gap limit its application in semiconductor fields[1]. However, for 2D transition metal dichalcogenides (TMDCs)[2-6], the vacancy of some elements reduces the carrier mobility and limits its application in transistors. 2D black phosphorus (BP) has extremely high carrier mobility at room temperature and a certain band gap that ensures its large current switching ratio[7-9]. As a result, BP has become a popular topic of research for the preparation of nanometer devices[10-12]. In recent years, research on 2D materials, especially two-dimensional materials, has become increasingly popular in the field of materials. There are many theoretical and experimental studies of TMDCs, graphene, and hexagonal boron nitride (h-BN). These 2D materials have extraordinary mechanical, thermal, optical, electrical, and magnetic properties that differ from bulk materials. 2D BP, also known as phosphorene, is an emerging 2D material. Due to its great application potential in the field of field-effect transistors (FETs)[13], it has gradually become a popular topic for many researchers. Research into BP began in the 1920s. It was not until 2000 that BP was taken seriously by researchers, and related research papers increased (see Figure 9-1). In 2014, Zhang et al. were inspired by 2D materials such as graphene to separate a small amount of BP and apply it to

FETs[13]. Due to the extraordinary performance of this FET, 2D BP attracted the attention of many researchers.

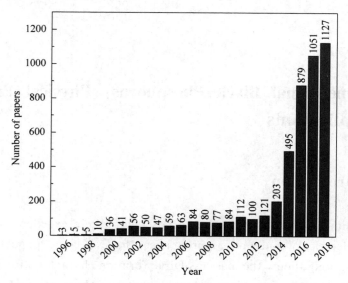

Figure 9-1　Number of academic papers based on BP published in more than a decade (Statistics on citation reports for BP in web of science)

Combining different 2D materials can present properties that differ from individual materials. A structure that combines different materials is called a heterostructure. BP can also be combined with other 2D materials to form a heterostructure to obtain excellent properties that are not possessed by monomers. In recent years, research on heterogeneous structures based on BP has shown an explosive growth trend, and researchers have discovered many applications[14-23]. In addition to forming a heterostructure with graphene[24-29], BP forms a heterojunction with TMDCs, which have become popular in recent years[30-33]. Researchers have begun to pay attention to the electrical properties[22], regulatory properties[30-32], and surface plasmon properties[33] of heterostructures using first principles and experiments.

9.1.1　2D crystal structure of BP

Similar to graphite, BP is a type of van der Waals force between layers of materials[34-37], but the difference is it has fold layers. Each layer of the phosphorus atoms combines through sp^3 hybridization and the adjacent three phosphorus atoms[38], so that it occupies three valence electrons of the phosphorus atoms to form a stable loop structure. The atomic arrangement within each layer plane produces two inequitable directions: namely along the zigzag type structure of atoms, and the vertical and armchair type structures of the atomic arrangement direction[39]. This leads to the anisotropy of the crystal structure of BP, and the anisotropy of the crystal structure leads to the anisotropy of its electrical and optical properties along three different crystal axes. The spacing between layers is 5 Å as shown in Figure 9-2(a)[10]. The layers have strong covalent bonds and a single electron pair, so each atom is saturated. The atoms between the layers are joined by van der Waals forces[40]. (Some studies showed that the overlapping of the wave functions is very important for the interactions between the layers).

The 2D BP in a single layer is not a perfect plane structure, as it presents a folded state in the plane. There are two types of p-p bonds inside the crystal lattice: one is a p-p bond in the same plane whose bond length is L_{p-p} = 0.2224 nm and whose bond angle is 96.3° and the other is a p-p bond in a different plane whose bond length is L_{p-p} = 0.2244 nm and whose bond angle is 102.10°[41-44]. Similar to the structure of graphene nanoribbons, the boundary of 2D BP is divided into an armchair type and a zigzag type according to the configuration, see Figure 9-2(b)[39]. Figure 9-2(c) shows the unit lattice structure of massive BP and its corresponding Brillouin zone (BZ)[43]. Figure 9-2(d) shows the unit lattice structure of 2D BP; the corresponding BZ is on the right[43]. It can be seen from the illustration on the right side in Figure 9-2(c) that the crystal structure of the block BP belongs to the orthorhombic system. Since the monolayer of BP is stripped from the bulk BP, it still belongs to the orthorhombic system (because theoretical analysis requires a vacuum layer to approximate a single layer of electronic structure, the 2D material crystal structure is not a strict 2D unit cell). However, the method embodying the 2D electronic structure is a new class of quasi-momentum quantum numbers in k-space, k-point. Using symmetry to classify K points, take a series of highly symmetric points into a path that is the point and line in the right illustration in Figures 9-2(c) and (d). These points and lines are closely related to the band structure and are important factors in determining the electronic structure of BP.

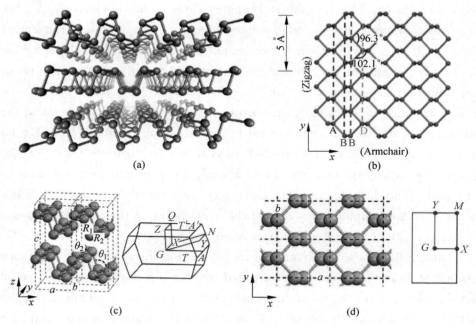

Figure 9-2　2D crystal structure of BP

(a) Atomic structure of BP[10]. (b) Top view of the lattice of single-layer BP. The bond angles are shown. The corresponding x, y, and z directions are indicated in both a and b. The x and y directions correspond to the armchair and zigzag directions of BP, respectively[39]. (c) Crystal structure of bulk BP marked with coordinate axes (x, y, and z), lattice vectors (a, b, and c), and structural parameters (R_1, R_2, y_1, and y_2). Right inset: Brillouin zone path of a primitive BP cell[43]. (d) Top view of the atomic structure of the monolayer BP and the associated Brillouin zone[43]

9.1.2　Electronic structure of BP

　　Three covalent bonds occupy all three valence electrons in a phosphorus atom, and single-layer BP (phosphorene) is predicted to have a direct band gap of approximately 2 eV. Both theoretical calculations and angular resolution photoelectron energy spectrum experiments show that when the thickness is greater than 4 nm or the number of atomic layers is greater than 8, the BP crystal has a direct band gap semiconductor of approximately 0.3 eV[43], which is between the graphene with zero band gap and the transition metal sulfide with a larger band gap. The electronic structure of BP has been calculated using different models. The distance between the top and bottom of the guide band at the Z point of BP in the block was 0.3 eV[43], and the curvature of the energy band was the greatest at this point. The curvature along the Z-Q direction is greater than along Z-G, that is to say, the effective carrier mass in the Z-Q direction is the smallest. As the number of layers decreased to approximately 6 layers[45], the electronic structure of thin BP crystals showed significant layer-dependent characteristics, and the band gap shifted blue. The theoretical calculation of the band-gap value of monolayer BP crystal is between 2.0 eV and 1.4 eV, but most results indicate that monolayer BP has a direct band gap structure of 1.51 eV at the reciprocal G point[46]. Liang et al. obtained clean single-layer BP samples at the original position using a scanning tunneling microscope (STM) in a high-vacuum environment and measured the band gap of single-layer BP as 2.0 eV through the difference curve[46].

The electronic band structures of BP shown in Figures 9-3(a) and (b)[10,44] are obtained using theoretical calculations and angle-resolved photo-emission spectroscopy (ARPES)[10]. Figure 9-3(c) shows the monolayer BP band structure by theoretical calculation[45]. As shown in this diagram, as the number of layers increases, the BP band gap decreases gradually.

9.1.3 Electronic structure of BP-based heterostructures with TMDCs

The theoretical calculation of the band structure of the BP-molybdenum disulfide heterostructure is carried out. To use this heterogeneous structure more flexibly in practical applications, Huang et al. conducted a theoretical study on the regulation performance of this heterostructure under an electric field[31]. Figures 9-4(a)~(c) show the energy band structure of the heterostructure under different electric fields. First, the applied electric field can significantly change the band gap of the heterostructure. As previously demonstrated, the regulation of the energy band regulates the light absorption of the material. Second, the applied electric field has no effect on the energy-band properties and curvature of the low-conduction band and the top of the valence band. This means that the effective mass of the material does not change; in other words, the carrier mobility of the material does not change. Huang et al. also calculated the charge density map under different electric fields as demonstrated in Figures 9-4(d)~(f). The forward electric field shifts the

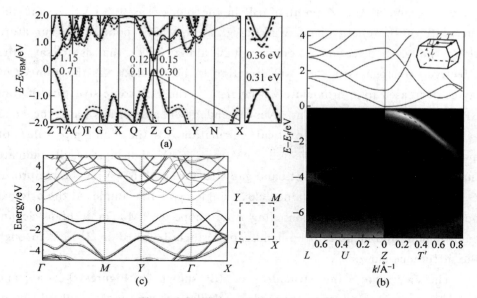

Figure 9-3   Electronic structure of BP

(a) Electronic band structures for bulk BP calculated with the HSE06 functional (red solid line) and the mBJ potential (blue dashed line). At the right of the image, a zoomed-in plot shows the direct band gap at Z. E_{VBM} is the energy of the valence-band maximum[10]. (b) The band structure of bulk BP mapped uses ARPES measurements. A band gap is clearly observed. Superimposed on top are the calculated bands of the bulk crystal. The blue solid and red dashed lines denote empty and filled bands, respectively[43]. (c) The density functional theory-calculated (dashed lines) and GW-calculated (solid lines) band structures of monolayer BP[45]

charge of the phosphorus atom near the molybdenum disulfide side of the BP to the molybdenum disulfide side. Static interactions occur between the heterostructures.

9.1.4   Electronic structure of BP and blue phosphorus heterostructures

As previously mentioned, Huang et al. conducted a theoretical study on the electronic structure of BP and molybdenum disulfide heterojunction under different applied electric fields[18]. Before this work, they also conducted a theoretical study on the heterojunction electronic structure formed between BP and blue phosphorus, which is also an allotrope of phosphorus as shown in Figure 9-5 (a)[18]. The effects of different applied electric fields on the electronic structure were also studied in detail. First, Huang et al. calculated the energy band of a heterogeneous structure composed of pure BP, pure blue

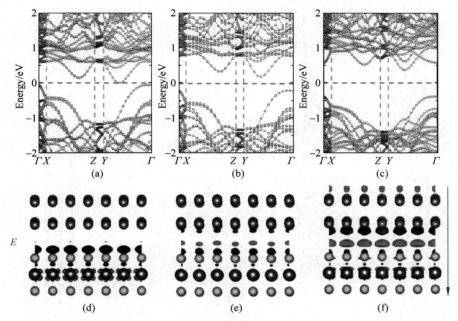

Figure 9-4 Electronic structure of BP-based heterostructures with TMDCs
(a)～(c) Projected electronic energy-band structures and (d)～(f) isosurface of the charge accumulation (blue) and depletion (violet) of a BP-molybdenum disulfide heterostructure under an applied external field of -0.4, 0, and 0.4 V/Å. The green arrows indicate the direction of the electric field[31]

phosphorus, and both. The band-gap edges were contributed by different compositions when forming a heterostructure. The valence band top of the heterostructure energy band was contributed by the BP, while the low-conduction band was contributed by the blue phosphorus. This conclusion contended that the light absorption caused by the inter-band transition of this heterostructure is the charge transfer absorption and the charge transfer between heterostructures. Second, Huang et al. calculated the energy-band structure of heterostructures under different electric field disturbances as shown in Figure 9-5(e). The band gap varies with the electric field, but the band curvature and properties do not change substantially. Third, the band gap of the heterostructure has a certain range depending on the magnitude of the external electric field. When the external electric field changes linearly with the electric field in the range of ± 0.5 V/Å, when the external electric field reaches ± 1 V/Å, the band gap abruptly changes to zero. The main research

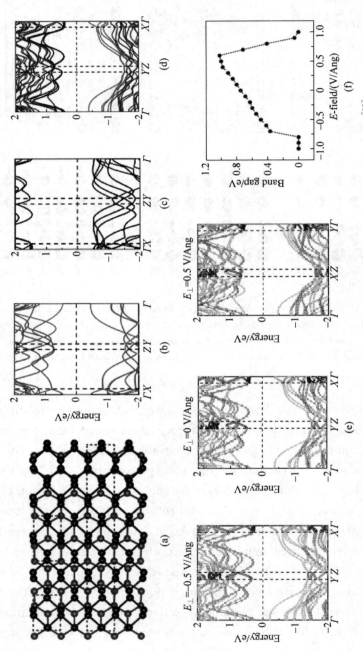

Figure 9-5 Heterostructure and electronic structure diagram of BP and blue phosphorus[18]

(a) Schematic diagram of a heterogeneous crystal structure composed of BP and blue phosphorus; (b) pure blue phosphorus (c), and their heterostructures (d); (e) The electronic energy band of a heterostructure with different applied electric fields; (f) The curve of band-gap value of the heterogeneous structure varies with the intensity of the different disturbing electric fields

method of 2D BP electronic structure still relies on first-principles calculation. The more popular method now is density-functional theory (DFT), a method based on single-electron approximation that may not be well described for interactions between multiple particles, so other methods are needed for correction.

9.2 Preparation for BP

The preparation process for BP is complicated, mainly because BP is easily oxidized and deteriorated in the air. These are the reasons why BP is expensive. There are many preparation methods for BP, most of which require high temperature and pressure. In the early stage, white phosphorus (1.2 GPa, 200℃) or red phosphorus (8 GPa) are used to prepare BP under high temperature and pressure[47]. However, the size of the BP prepared was very small. The BP bar was prepared at room temperature using the single-crystal-growth method of bismuth solution, and the length reached 5 mm[48]. Other researchers slowly cooled crystals of BP to 600℃ after melting them at 900℃[49]. In recent years, high-energy mechanical ball milling has been used to prepare BP particles with a diameter of 3.3 μm at room temperature[50]. However, some of these methods for preparing BP take too long, while others have more impurities. Researchers are presently attempting to develop a preparation method for 2D BP.

9.2.1 Mechanical exfoliation

The van der Waals force between layers of 2D materials is weak, so the mechanical stripping method is used to prepare 2D materials. Zhang et al. repeatedly peeled BP onto a thin layer with adhesive tape and then placed it on an SiO_2 substrate for rapid stripping to obtain BP nanosheets with a thickness of 5 nm[13]. Experimental results show that the mobility of carriers is related to the thickness of 2D BP. The researchers then repeatedly peeled the BP off with tape and pressed it onto a polydimethylsiloxane (PDMS) substrate, resulting in two-layer BP nanosheets. Figure 9-6 shows an optical image of BP obtained via mechanical exfoliation. Lu et al. obtained double-layer BP nanosheets using the mechanical exfoliation method[51] and then used Ar^+ plasmon to peel double-layer BP nanosheets to obtain single-layer BP[52].

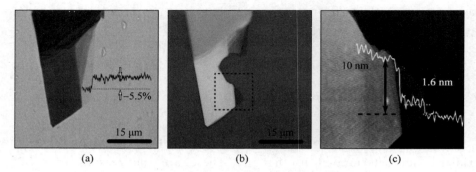

Figure 9-6 Isolation of few-layer BP flakes

(a) A transmission-mode optical microscopy image of a few-layer BP flake exfoliated onto a polydimethylsiloxane (PDMS) substrate. An optical transmittance line profile is included to highlight the reduction of approximately 5.5 % in the optical transmittance in the thinner part of the flake. (b) A bright field optical microscopy image of the same flake after transferring it onto a SiO_2/Si substrate. Note that part of the flake was broken during the transfer. (c) An atomic force microscopy topography image of the region highlighted with a dashed square in (b). A topographic line profile, acquired along the horizontal dashed black line, is included[52]

The mechanical exfoliation method requires relatively simple experimental conditions, but is time-consuming and low yield. The crystal size prepared using this method is not easy to control, and it is impossible to produce black phosphene of appropriate length under control. This method is inadequate for industrial production and is suitable only for the preparation of black phosphene in the laboratory to meet the requirements of the test. In addition, BP nanosheets cannot exist stably in the atmosphere and must be prepared without water or oxygen; otherwise, they will immediately react and disappear.

9.2.2 Liquid phase exfoliation (LPE)

The LPE method can achieve 2D material stripping using specific solvents, increasing the material layer spacing, and then providing energy using ultrasound. Compared to mechanical stripping, liquid phase stripping can be prepared in large quantities, and the concentration and size of the obtained sheet layer can be better regulated by centrifugation and other means. Since the 2D materials obtained are dispersed in the solution, further modification and coating in the later stage are more convenient. For BP, liquid phase stripping using organic solvents can effectively isolate the oxygen and facilitate storage.

Brent et al. used LPE to prepare 2D BP for the first time[53]. First, an n-methylpyrrolidone (NMP) solution of BP was subjected to water bath ultrasound for 24 hours with the temperature lower than 30℃. The large BP slices were removed via centrifuge, and a light-yellow stable dispersion was obtained. The sample prepared is shown in Figure 9-7. Kang et al. obtained 2D BP using ultrasonic probe stripping in NMP solution[54]. Yasaei et al. compared the stripping effects of alcohol, ketone, chlorinated organic solvents, and other organic solvents on BP. They used dimethylformamide (DMF) and dimethyl sulfoxide (DMSO) as stripping solvents to prepare BP and obtained a BP nanosheet with a thickness of 11.8 nm[55]. Guo et al. compared the stripping effects of NMP solution and NaOH + NMP standard solution on BP, and found that the latter made BP nanosheets more stable due to the presence of OH^-. After centrifugation, they obtained 2~4 layers of BP nanosheets[56]. Hanlon et al. used a centrifuge to separate the CHP solution probe of BP via ultrasound for 5 hours at different centrifugal speeds and obtained BP nanosheets with fewer than 10 layers[57].

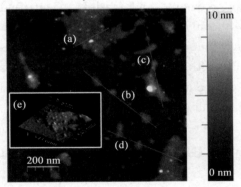

Figure 9-7　An atomic force microscopy height profile image of few-layer phosphorene sheets produced via the ultrasonic exfoliation of bulk phosphorus in NMP for 24 hours and spin-coated onto an SiO_2/Si substrate. (a), (b), (c), and (d) show z profiles of phosphorene flakes ($N = 11$) along the lines marked in the AFM relief image. Inset: (e) a 3D representation of the large three-layer phosphorene flake observed at point (a)[53]

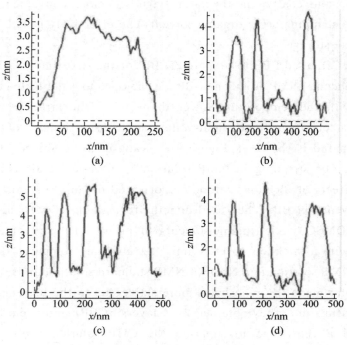

Figure 9-7(Continued)

The LPE method is simple and feasible. Centrifuging can filter high-quality nanosheets with only a few layers of thickness according to different sizes, which have high controllability. This technique enables the mass production of BP nanosheets. In addition, the aforementioned studies have demonstrated that CHP and NMP can be used as solvents in a liquid environment to steadily peel the BP nanosheets and prevent their oxidation.

9.3 Anisotropy of BP's properties and application

The lattice structure of BP determines the unique characteristics of anisotropy in the BP plane. BP belongs to the D_{2h} point group structure of an orthogonal crystal system. The effective carriers along the longitudinal zigzag shape are 10 times greater than those along the transverse structure, which directly leads to the strong anisotropy of BP electrons in its plane[44,58].

9.3.1 Anisotropic characteristics of band structures

The unique crystal structure of BP determines that the dispersion curves of the electron structure of BP's BZ are very different, and the curvature of the energy band near the Fermi energy level varies greatly in all directions. Under physical and chemical modulation, the band structure of BP can be accurately regulated[45,59-60].

Researchers doped potassium atoms into BP using a surface-doping technique[61]. They used an electric field perpendicular to the BP plane to regulate the band structure of BP. The results show that, through the quantitative doping of potassium atoms, the energy-band structure of BP changes from a typical semiconductor energy-band structure to an anisotropic Dirac semi-metal energy-band structure, and the energy-band structure of the second Dirac cones (DCs) appears along the zigzag boundary. Then, along the zigzag lattice's direction, the effective mass of the carriers is far less than the effective mass along the armchair lattice's direction. ARPES is a very sophisticated means of measuring the material-band structure and is widely used in the field of 2D materials and topological materials. Figure 9-8 shows the ARPES results. Figures 9-8(a) and (b) show experimental results and finite-width spectral simulation results for high-resolution ARPES, respectively. It can be seen that the results are very similar. In Figure 9-8(c), red represents the valence band and blue line represents the conduction band. Figures 9-8(d)~(g) shows the energy-band section of the upper and lower areas of the Fermi level. It can be concluded in conjunction with Figure 9-8(a) that a small amount of BP has a Dirac semimetal property.

9.3.2 Anisotropic mechanical properties

2D BP is significantly brittle. Theoretical studies show that the intensity of 2D BP along the zigzag direction of a single layer is as high as 18 GPa and the strain variable is as high as 27%. The intensity along the niche direction is as high as 8 GPa and the dependent variable is as high as 30%[62-63].

Figure 9-9 shows the results of the calculation. The fold-style layered structure of 2D BP decreases its Young's modulus and shows obvious stress-strain anisotropy. BP has a negative Poisson effect (NPE) in the direction

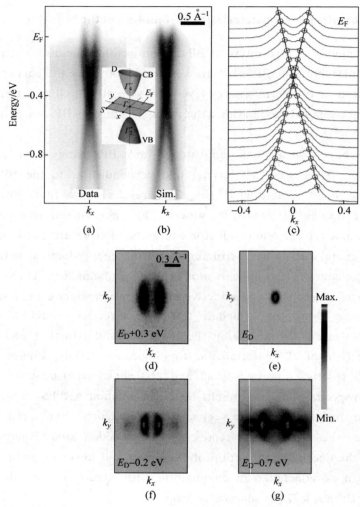

Figure 9-8　An anisotropic Dirac semimetal state at a critical density[61]
(a) High-resolution ARPES data obtained at 15 K along the armchair direction k_x. (b) Corresponding spectral simulation with two linearly crossing bands and finite broadening. (c) Normalized momentum-distribution curves (0.05 eV steps from E_F) with their peak positions marked by open circles. The red and blue lines are linear fits to the VB and CB, respectively. (d)~(g) A series of ARPES intensity maps at constant energies (marked at the bottom of each panel) shown over a whole surface BZ (inset picture)

perpendicular to its plane structure. The Stillinger-Weber potential was used to describe the NPE, and the hyperbolic rule of bending strain energy was closely combined with the strain energy density of continuum mechanics to systematically study the mechanical behavior of single-layer BP. The results demonstrate that the bending deformation of 2D BP also shows obvious anisotropy, and the excessive bending stiffness of the zigzag type is the reason for the premature occurrence of an instability zone in zigzag type directional bending[64]. The repulsion of the electron pairs in BP is the main physical mechanism of anisotropy in bending deformation.

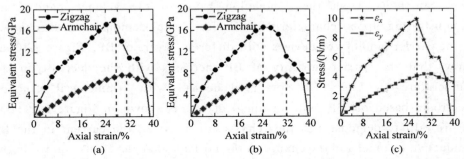

Figure 9-9 The strain-stress relation for (a) monolayer and (b) two-layer phosphorene structures. The monolayer can sustain stress up to 18 GPa and 8 GPa in the zigzag and armchair directions, respectively. The corresponding critical strains are 27 % (zigzag) and 30 % (armchair). The ideal strengths for the multilayer phosphorene are 16 GPa and 7.5 GPa in the zigzag and armchair directions, respectively, and their critical strains are 24 % (zigzag) and 32 % (armchair)[63]. (c) The stress as a function of the tensile load for phosphorene. Phosphorene can withstand a critical tensile strain up to 30 % in the armchair direction[62]

9.3.3 Anisotropic electrical properties

The anisotropy of electrical properties of BP is revealed by theoretical calculations and experimental studies based on the first principles. Theoretical studies show that the average effective mass of the electrons in black phosphorus is 0.22 m_0 and the effective mass of the holes is 0.24 m_0, significantly less than the effective mass of the silicon carriers[65].

The experimental results show that the resistivity of BP decreases with the increase in the temperature. The conductivity of BP is dominated by holes, not electrons. At room temperature, the carrier mobility of BP holes is 350 cm^2/V·S, significantly higher than that of the electrons, 220 cm^2/V·S[66]. At standard atmospheric pressure, the band gap of BP decreases with the increase in the external pressure, and BP changes from a semiconductor to a metal and finally to a superconductor. Among all materials, only BP can undergo phase transition under changes in pressure, which again demonstrates the anisotropy of BP' electrical properties[67].

Figure 9-10(a) demonstrates the intrinsic p-type doping characteristics of low-layer BP. Figure 9-10(b) shows that the FET based on BP regulates its Fermi energy level by regulating the electric field so that the Fermi energy level can move from the conduction band to the valence band, thus demonstrating bipolar characteristics. Figure 9-10(c) demonstrates the changes in the carrier mobility of devices with different layers of BP. Figure 9-10(d) shows that the carrier mobility of BP increases with the decrease in the temperature[13]. Liu et al. prepared devices with symmetrical structure electrodes based on BP with an angle of 45° as shown in Figure 9-10(e). Figure 9-10(f) demonstrates the relationship between the current in the BP conductive channel and the change in the horizontal angle[68]. As shown in the figure, the BP current changes periodically with the angle of the electrode. The anisotropy of the electrical properties of BP has been demonstrated.

The energy-band characteristics and electrical properties of BP are affected by the number of layers, temperature, pressure, doping, external stress, and other factors. Researchers have studied the changes in the energy band structure of BP under different stress levels. The results demonstrate the anisotropy of BP' electrical and band structure. Figures 9-11(a)～(d) show that when stress is applied in the xy direction[69], the band structure of BP changes from a semiconductor to a metal. As the stress parallel to the surface of BP increases, the valence band gradually increases. When $h/h_0 = 0.20$, the top of the valence band in the T direction intersects the bottom of the conduction band, causing BP to become a metal without a band gap. The effect of stress is also anisotropic for the effective mass of BP. Figure 9-11(e) shows the relationship between BP' effective mass and the crystal orientation, and Figure 9-11(f) demonstrates the distribution of BP' effective mass under biaxial stretching. As shown in the figure, the effective mass of BP in the y direction (armchair) increased from 0.146 me to 1.26 me. In the x direction (zigzag), the effective mass of BP decreased from 1.246 me to 0.158 me[70]. With the decrease in the effective mass, the mobility of BP carriers in different directions also changes.

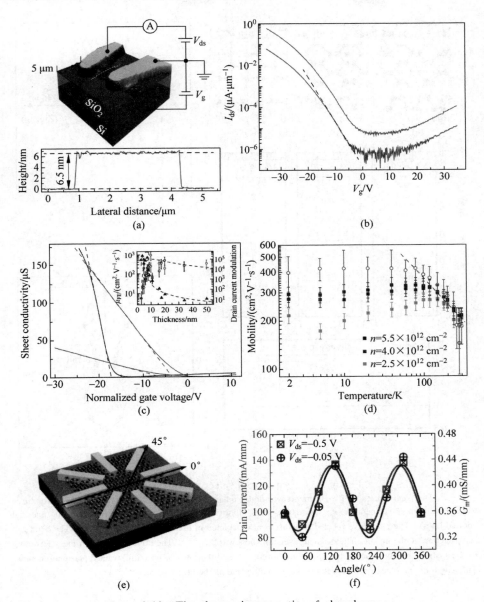

Figure 9-10 The electronic properties of phosphorene
(a) A schematic diagram of a BP FET device; (b) The transfer characteristic curve of FET at different bias voltages; (c) The relationship between the carrier mobility and the thickness of the BP device; (d) The relationship between the carrier and the temperature of the BP device[13]; (e) Device structure used to determine the angle-dependent transport behavior; (f) Angular dependence of the drain current and the transconductance G_m of a device with a film thickness about 10 nm[68].

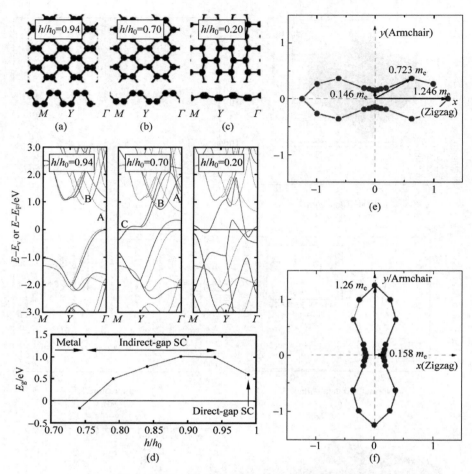

Figure 9-11 Anisotropic electrical properties

(a)~(c) The band structure of a BP monolayer under uniaxial compression along the z direction for three values of the imposed height h/h_0 (continuous line) along with the band structure of the pristine material (dotted lines). The respective relaxed structure is also depicted in the top and side views. (d) The band gap as a function of the height. The original layer thickness was $2\,h_0$ [69]. The effective electron mass according to the spatial direction, that of intrinsic phosphorene (e) and that of 5 % biaxially strained phosphorene (f). The length of the blue arrow represents the absolute value of the effective mass[70]

9.3.4 Anisotropic thermal and thermoelectric properties

The performance of thermoelectric devices mainly depends on the Seebeck effect of thermoelectric materials. Based on its anisotropic transport characteristics, BP is considered an excellent material for improving the thermoelectric properties of traditional nanomaterials.

The first-principles method was used to calculate the anisotropic electron heat conductivity of a single layer of BP: the direction with the best thermal conductivity of 2D BP was orthogonal to the direction with the best electrical conductivity[71-72]. When T = 500 K, the ZT value along the direction of the armchair configuration of BP is as high as 2.5; however, when T = 300 K, BP can still complete thermal energy conversion with high efficiency as indicated in Figure 9-12(a)[71]. Another group of researchers calculated the thermal conductivity of BP[36], which varies with its thickness. The results of the study once again show that, due to the anisotropy of BP' crystal structure, the thermal conductivity of the zigzag orientation (red points) is significantly higher than that of the armchair orientation (blue points), no matter if it is single-layer BP or block BP, see Figures 9-12(b) and (c). This is mainly due to the anisotropy of the phonon dispersion relation that the thermal conductivity of BP depends on direction. As demonstrated by ZT = $S^2 T\sigma/\kappa$ (where S is the Seebeck coefficient, T is the temperature, σ is the conductivity, and κ is the thermal conductivity), BP has higher conductivity and thermal conductivity in the niche direction than in the zigzag direction, so it has better thermoelectric properties. In addition, the thermoelectric effect of BP can be adjusted by doping other atoms in the BP film, applying an external electric field, external force, and other regulatory means[35,73].

9.3.5 Anisotropic optical properties

2D BP has many peculiar optical properties, including linear optical absorption and nonlinear saturation absorption. There are also different excitons involved in these peculiar optical properties, exhibiting anisotropic exciton properties. Of course, there is many theoretical and applied research results on these peculiar optical properties. For example, the study of surface plasmons is a hot spot in modern optical research. BP has a unique tunable wide direct band gap, which gives it broad applications in optics (especially in the infrared-wave region and the middle infrared region).

1. Linear optical absorption characteristics

The linear optical absorption characteristics of BP depend on its electron

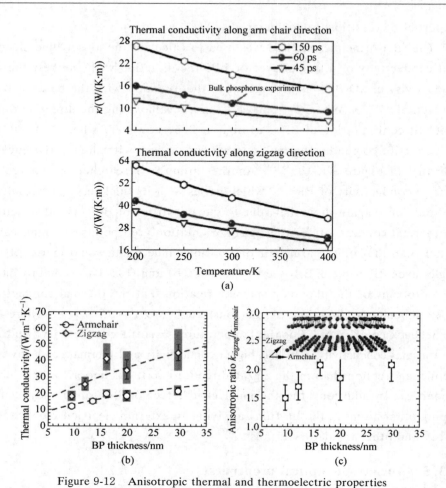

Figure 9-12 Anisotropic thermal and thermoelectric properties
(a) The thermal conductance of monolayer phosphorene along the armchair and zigzag directions, respectively[71]. (b) The extracted armchair and zigzag in-plane thermal conductivities ($k_{armchair}$ and k_{zigzag}) of multiple BP films. The dashed lines are the results of theoretical modeling. (c) The anisotropic ratio of armchair $k_{armchair}/k_{zigzag}$ at different BP thicknesses. The ratio at a thickness of 12 nm is calculated using linearly interpolated armchair thermal conductivity from adjacent thicknesses[36]

band structure. The interaction between the electrons and the light at long wavelengths can be characterized by absorption measurements. The linear optical absorption characteristics of BP were obtained by analyzing its absorption spectra in the visible, near-infrared, and mid-infrared regions.

Figure 9-13(a) shows the output spectrum of incident light polarized in different directions. For incident light, there was a significant peak value in the extinction spectrum at approximately 2,400 cm^{-1}, which was consistent with the band gap of multilayer BP of 0.3 eV. Figure 9-13(b) shows the angular resolution dc conductance of BP[58]. The researchers confirmed the black phosphorescence absorption spectra in the range of visible to near-infrared wavelengths using a high-magnification objective lens and a high-

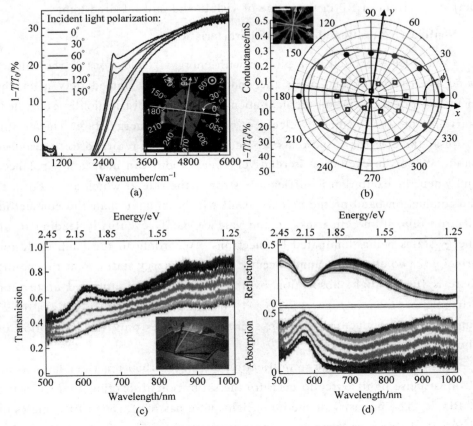

Figure 9-13　Optical anisotropy of BP

(a) Polarization-resolved infrared relative extinction spectra when light is polarized along the six directions as shown in the inset. Inset: an optical micrograph of a BP flake with a thickness of around 30 nm. (b) DC conductivity and IR relative extinction measured along the same six directions on this BP flake and plotted in polar coordinates[58]. (c) The polarization-dependent transmission spectra of a 70 nm thick BP flake. The polarization direction of the incident light for each curve is shown in the inset by arrows of the same color. The light interacts more strongly with the material along the atomic buckling (a axis) direction. (d) The reflection (top) and absorption (bottom) spectra of the 70 nm BP flake with the same set of input polarization angles[74]

sensitivity spectrometer micro-optical experimental device. By changing the incident angle of the incident light to control its linear polarization, the output, reflection, and absorption spectra are measured under different polarization states. Here, as the polarization of the incident light travels in the x direction, the light's absorption increases. Figures 9-13(c) and (d) show the output spectrum of BP in the range of visible and mid-infrared light[74].

2. Nonlinear saturation absorption characteristics

When BP is excited by a photon with energy E, electrons in the valence band absorb energy and enter the conduction band. Immediately after the electron optical excitation thermalization, Fermi Dirac distribution is established, starting from the low-energy electrons occupying an energy state. During this process, the electronic and relaxation conduction band returned to the valence band. When the excitation is relatively weak, both can maintain a balance, and when the excitation is sufficiently strong, the rate at which an electron to the conduction band of the valence band will be greater than the conducting electron relaxation rate backs to the valence band, eventually leading to all energy states being dominated by electrons. According to the Pauli exclusion principle, two electrons cannot occupy the same energy state, so at this point, there is further light absorption by the block and the saturated absorption principle as shown in Figures 9-14(a), (b), and (c)[75]. Since BP is a direct band-gap semiconductor, electron excitation requires only the absorption of photons[8,75-76].

Li et al. used a power amplification ultrafast 1,550 nm fiber laser as an ultrafast pulse light source to measure the saturation absorption characteristics of BP[76]. The intensity of the input light pulse passing through BP samples of different thicknesses changes. The transmittance of a light pulse passing through BP is shown in Figures 9-14(d) and (e). When the light pulse intensity is greater than 100 $\mu J/cm^2$, the light transmittance increases nonlinearly with the increase in the light pulse intensity. This is because BP has an anisotropic energy-band structure, strong polarization, and thickness independent of its nonlinear absorption characteristics.

3. Anisotropic optical properties

BP belongs to the orthorhombic crystal system, and has quadratic symmetry

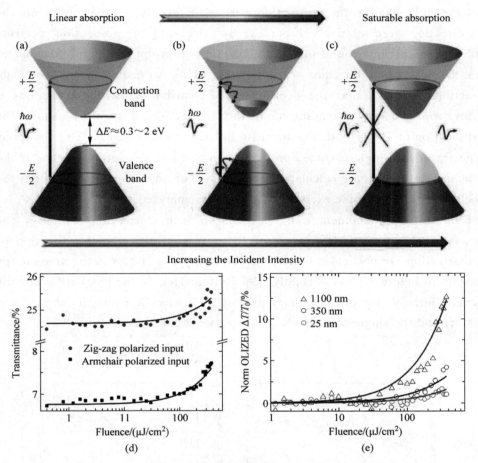

Figure 9-14 Nonlinear saturation absorption characteristics
(a), (b), and (c) The schematic diagrams of the saturable absorption in multilayer BP NPs[75]. (d) Fluence-dependent transmittance of an 1,100 nm thick BP film measured with ultrafast pulses in two orthogonal polarized light directions. (e) The relative transmittance change measured from 25 nm, 350 nm, and 1,100 nm thick BP films as a function of the input pulse fluence. The input polarization direction is along the armchair direction of the BP films[76]

only in the 2D plane. Based on the unique anisotropic crystal structure with low symmetry, 2D BP has the optical characteristics of anisotropy.

The anisotropy of 2D BP was studied theoretically using first-principle calculations. After calculation, researchers found that the energy bands near the top of the valence band and the bottom of the guide band were highly anisotropic as shown in Figure 9-3(c)[45]. The energy bands from the bottom of the guide band to the y direction were relatively flat, while those from the x direction were relatively dispersed. By calculating the absorption spectrum of BP, researchers found that the absorption spectrum was highly dependent on the polarization direction. An illustration of the wave function of the bright excitons indicates that the overall spatial distribution of these excitons is anisotropic, mainly extending along the x direction. Figure 9-15 shows the absorption of BP in both the armchair and zigzag directions[45]. The absorption spectra of the single-particle approximation and the absorption spectra of the electron-hole pairs are calculated separately. The advantage of this is that the behavior of anisotropic excitons in BP can be analyzed very intuitively. A lot of interesting phenomena can be found in such calculations. Firstly, the exciton behavior in a few layers of BP is very sensitive to the direction of polarization. In particular, exciton absorption peaks appear in the high-energy region in Figure 9-3(a). Secondly, the exciton effect in the block BP is already small. Finally, for the imaginary part of the dielectric constant, the exciton effect and the single-particle effect are very similar.

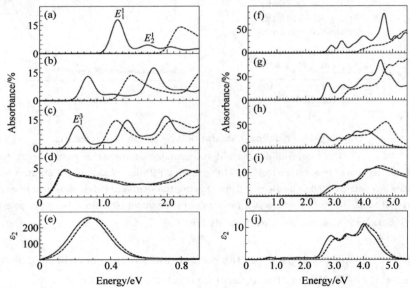

Figure 9-15 The optical absorption spectra of monolayer (a), bilayer (b), trilayer (c), and bulk phosphorene (d) and (e) for incident light polarized along the x (armchair) direction. The optical absorption spectra of monolayer (f), bilayer (g), trilayer (h), and bulk phosphorene (i) and (j) for incident light polarized along the y (zigzag) direction[45]

The Raman spectra of BP have been measured. The Raman scattering intensity is the strongest among the modes corresponding to the three peaks of the Raman scattering spectra and the modes related to atoms moving in the armchair direction[59]. The results also provide an effective method to determine the crystal direction of single-layer BP. The photoluminescence spectrum shows that when the detection and excitation directions are aligned with the x direction, the photoluminescence intensity is the highest. However, no matter what the excitation direction is, the emission intensity along the y axis is always less than 3 % of that along the x axis. In addition, the optical conductivity of BP on the y axis is theoretically zero, while the highly anisotropic excitons also cause the carriers along the x axis (armchair) and y axis (zigzag) to differ.

Tang et al. vertically stacked molybdenum disulfide and BP to form a van der Waals heterostructure and theoretically studied its electronic structure and light absorption properties using the density functional theory method[22]. As shown in Figure 9-16(a), molybdenum disulfide and BP are vertically stacked to form a van der Waals heterostructure. The crystal structure of BP is an orthorhombic system, and molybdenum disulfide is a hexagonal system. When the two constitute a heterojunction, it is necessary to unify their symmetry groups. Therefore, an orthorhombic system is uniformly used. The Brillouin zone of the heterostructure is shown in Figure 9-16(c). The high-symmetry-point path in the Brillouin zone is also consistent with an orthorhombic system. Tang et al. calculated the projected energy band of a heterostructure. The energy band indicates that the valence band of the heterostructure is mainly dominated by BP, and the conduction band is mainly dominated by the contribution of molybdenum disulfide as demonstrated in Figure 9-16(b). The entire system is smaller due to the insertion of BP. This may be the main cause of the absorption of heterostructures in the near-infrared region. Tang et al. also calculated the absorption spectrum of the heterostructure based on a calculation of its electronic structure. In general, pure 2D materials are uniaxial crystals. That is to say, its absorption spectrum is equal in the x and y directions. However, the heterostructure of BP and molybdenum disulfide loses symmetry in the x and y directions. Therefore, the absorption spectra in the x and y directions are not equal. As shown in Figures 9-16(d) and (e), in

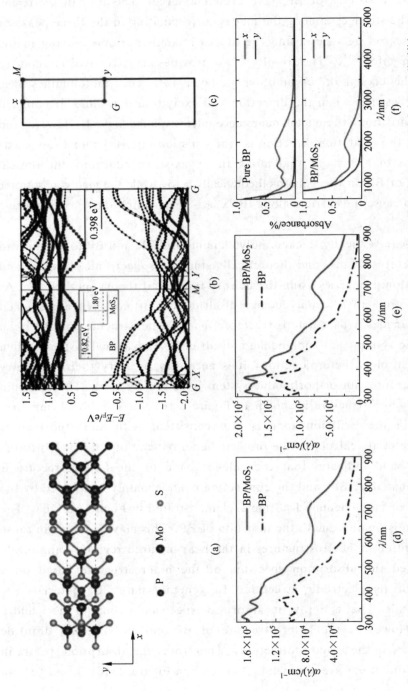

Figure 9-16 The electronic structure and optical properties of a heterojunction composed of BP and molybdenum disulfide (a) A schematic diagram of the heterojunction's crystal structure; (b) An-energy band structure diagram of the heterostructure; (c) The new cubic Brillouin zone and its high-symmetry-point path after the BP and molybdenum disulfide form a heterostructure; Comparisons of the visible-light-absorption spectrum of the BP-MoS$_2$ heterostructure in the x direction (c) and the y direction (d) and the absorption spectrum in the near-infrared region (e)[22]

the visible region, the light absorption of the heterostructure is significantly greater than that of pure BP. In the near-infrared region, the absorption of the heterostructure is only approximately 2,000 nm larger than the absorption of pure BP.

4. Dynamic light response characteristics

Polarization-resolved transient absorption measurement based on ultrafine light pulse excitation is a powerful technique to study the transient light response characteristics of 2D materials[77-79].

Figure 9-17(a) is a schematic diagram of an experimental device for the time resolution of optical pumping detection based on ultrafast light pulses[77]. Linearly polarized pulse light with a wavelength of 800 nm was used as the pump light source, and linearly polarized pulse light with a wavelength of 1,940 nm was used as the detection light source. The delay time of detection light was controlled by a delay line, and the recovery time of carrier was studied using the delay time as a variable. The time resolution of experimental device was 200 fs. At this point, the typical transient reflection spectrum of the device was obtained as shown in Figure 9-17(b). The experimental results showed the dynamic anisotropic response of BP. The polarization of detected light not only results in anisotropic absorption but also influences the carrier motion. Figures 9-17(c) and (d) show the relationship between R and the delay time when the incident light is at different angles. The polarization of detected light has no effect on the carrier relaxation rate, indicating that the carrier relaxation process is isotropic. Figure 9-17(e) shows the relationship between the delay time and the $\Delta R/R$ under different pump-light sources. Figure 9-17(f) demonstrates the relationship between σ_{xx}/σ_{yy} and the delay time. Figure 9-17(g) shows the evolution of the polarization correlation of the detected light. The figure demonstrates that the size of the differential reflection is related to the polarization of the detected light.

5. Resonance characteristics of anisotropic plasmons

Since all of the electrons in the BP layer exhibit strong in-plane anisotropy, this also depends on the unique anisotropy of the resonance vector of the plasmon device based on BP[80-83]. Low et al. adjusted the resonance frequency of the

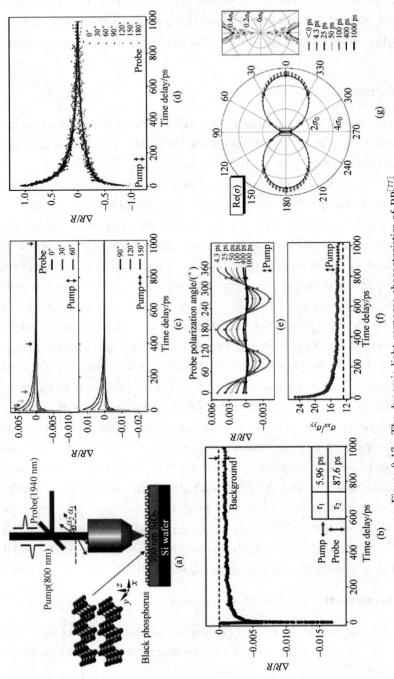

Figure 9-17 The dynamic light response characteristics of BP[77]

(a) A schematic diagram of the polarization-resolved transient reflection experiment; (b) A representative transient reflection spectrum with pump polarization along 0° ($\sim x$ direction in panel c) and 90° ($\sim y$ direction in panel c). The data is fitted by the biexponential decay function with two decay time constants τ_1 and τ_2; (c) The probe polarization dependence of the transient reflection spectrum with pump polarization fixed along 0° and 90°; (d) The normalized transient reflection spectrum (normalized with $\Delta R/R|_{t=4.5ps}$) with different probe polarization angles; (e) The transient reflection spectra with different probe polarizations at fixed delays; (f) The dynamic evolution of $\mathrm{Re}(\sigma_{xx})/\mathrm{Re}(\sigma_{yy})$ at different delays with the pump polarized along 90°; (g) The dynamic evolution of a $\mathrm{Re}(\sigma)$ ellipse at different delays. $\Delta \mathrm{Re}(\sigma)$ is doubled for clarity. Inset: an enlarged plot of the area marked by a red rectangle

plasma by changing the polarization direction of the incident light. The adjustment range depends on the anisotropic electron conductivity in the x direction and y direction of the electrons in the layer[84]. The results confirm the resonance properties of the anisotropic plasmon of phosphorus atoms in BP. Conventional metal plasma exciters and graphene plasma devices cannot achieve the polarization of external photoelectrons, which is the unique characteristic of plasma devices based on BP that distinguishes them from their performance. Figure 9-18(a) shows the energy loss and plasmon dispersion, and Figure 9-18(b) demonstrates plasmon scaling with carrier concentration[84].

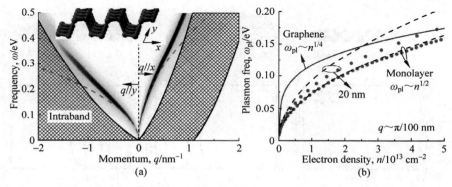

Figure 9-18 Resonance characteristics of anisotropic plasmons
(a) Energy loss and plasmon dispersion. The loss function calculated for monolayer BP for electron doping of $1 \times 1,013/cm^2$ for q along the two crystal x (right) and y (left) axes. (b) Plasmon scaling with carrier concentration. The plasmon energies ω_{pl} as a function of the density n calculated for the monolayer and for a 20 nm thick BP film at a specified q along x (Graphene plasmons are shown for comparison)[84]

Hong et al. designed a periodic metamaterial with a heterostructure of a two dimensional material mixed with a substrate[85]. The periodic repeating unit is shown in Figure 9-19(a). The light blue disc in the figure is a 2D material and its heterostructure. Hong et al. used a numerical simulation approach[85]. That is to say, Maxwell's equations were solved in 3D space using finite element technology (COMSOL multiphysics software). The electromagnetic parameters of the material used the Drude-Lorentz model. The benefit of this model is that the refractive index of the material is resolved and smooth in the frequency domain. However, since the model is based on classical electromagnetic field theory, the interlayer electron interaction between heterojunctions cannot be fully described. Figures 9-19(e)~(g) show

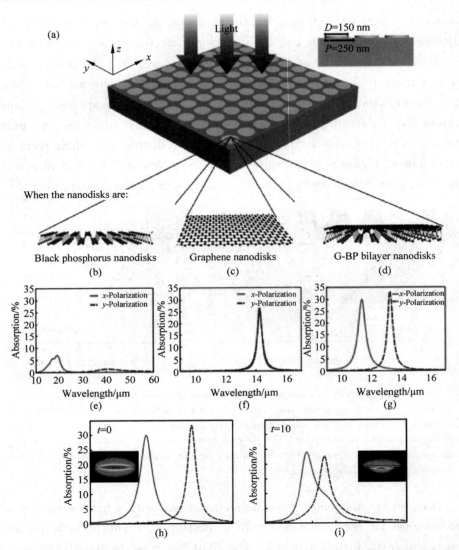

Figure 9-19 A schematic diagram of a high-performance 2D hybrid material surface plasmon component and a performance simulation diagram[85]

(a) A device schematic and standard orientation. (b) A schematic diagram of the periodic structure of 2D BP. (c) A schematic diagram of the periodic structure of graphene. (d) A schematic diagram of the periodic structure of a van der Waals heterostructure composed of graphene and BP. (e), (f) and (g) Three graphs representing the absorption spectra of the three 2D structures (BP alone, graphene alone, and the van der Waals heterostructure) under an electric field in the direction of polarization x (blue curve) and y (red dotted line). (h)~(k) A color-filling diagram of the absorption spectrum of the heterostructure in (a) at different separate distances and its electric field strength mode (inserted map)

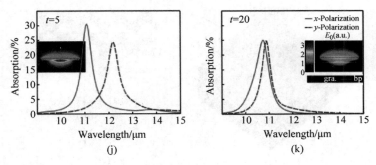

Figure 9-19(Continued)

the absorption spectrum of the device when the disk in the metamaterial is a different material. From left to right, there is BP, graphene alone, and the van der Waals heterojunction composed of the two. The absorption of the device exhibits a certain anisotropy when the disk is along BP as shown in Figure 9-19(e). But the absorption is weak compared to the other two cases. When the disk material is along graphene, the absorption of the material is enhanced and exhibits isotropy as shown in Figure 9-19(f). If the disc material is replaced with a van der Waals heterostructure of graphene and BP, the device exhibits two absorption peaks with the strongest absorption coefficient shown in Figure 9-19(g). Hong et al. also analyzed the relationship of the distance between graphene and BP and the light absorption coefficient of the device as shown in Figures 9-19(h) ~ (k). As the spacing increases, the absorption peaks of different polarization directions tend to approach each other. This close proximity is mainly due to the absorption peak of the y polarization peak polarized near the x direction. The absorption peak in the x direction is almost unchanged. However, the closer the distance between the two, the stronger the absorption. The change in the absorption intensity of the two different polarization directions is similar.

Zhou et al. used the finite-difference time-domain (FDTD) method and coupled mode theory (CMT) to study the electromagnetic field behavior of BP on the surface of SiO_2[86]. As shown in Figure 9-20(a), the black phosphor quantum dots are placed on the SiO_2 substrate. Electromagnetic waves are perpendicularly incident to the device surface along the z axis. When using CMT to analyze the properties of BP nanostructures as demonstrated in Figure 9-20(b), a single-point response model as shown in Figure 9-20(c) is

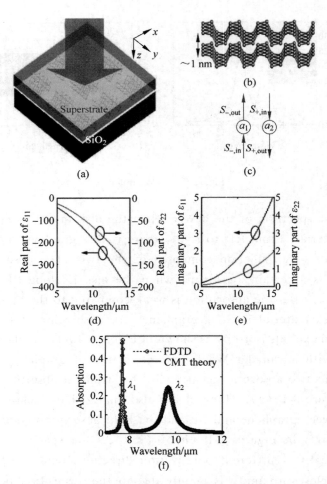

Figure 9-20 The electromagnetic field behavior of black phosphorus on the surface of SiO_2[86] (a) A schematic diagram of the model used in the simulation; (b) A schematic representation of the crystal structure of a few layers of BP; (c) The CMT model with two resonant modes of the nanostructured BP system; (d) The real part of the relative dielectric constant in different directions; (e) The imaginary part of the relative dielectric constant in different directions; (f) The device resonance frequency calculated using the CMT and FDTD methods (the red curve is the result of CMT and the black circle is the result of FDTD)[86]

used. Due to the surface plasmon resonance (SPR) effect, the light wave energy can be coupled into the BP nanostructure when the electromagnetic wave $S_{+,in}$ passes through a BP-like grating structure. The resonance mode of BP SP can be excited by the energy amplitude a_m (m = 1, 2) in the entire region of the black phosphor grating. a_m is the mth mode of the resonance mode with a resonance frequency ω_m. λ_m represents the coupling between the cavity modes. The input (output) light is represented by $S_{\pm, in(out)}$. The material optical parameters in the paper were derived using the Drude-Lorentz model. The dielectric constants in two different directions are shown in Figures 9-20(d) and (e).

Coupled mode theory is an analytical tool for analyzing the electromagnetic field mode of a waveguide. The FDTD method is a universal finite element method for solving partial differential equations. When using modern electromagnetic field theory to analyze the optical behavior of materials or devices, the finite element scheme based on the FDTD method is universal, but it has some shortcomings. First, since the finite element method is based on meshing, the accuracy of the mesh and its properties directly affect the accuracy of the calculation or simulation. Second, if the mesh density increases to such an extent that it does not affect the results, the amount of calculation is very large. However, the coupled mode theory is superior. This method can be used to analyze and calculate the local behavior of electromagnetic field propagation through a certain physical approximation. In this paper, the surface plasmon polarization (SPP) resonance wavelength of BP is analyzed using FDTD and coupled mode theory. As shown in Figure 9-20(f), the red curve is the result of the coupled mode theory, and the black circle is the result of the FDTD method. The two sets of theories agree very well. The two-day curve almost completely coincides. The results of both methods show that BP has two SPP resonance wavelengths on an SiO_2 substrate.

9.3.6 Optoelectronic properties

The PN junction is the most basic component of electrical and optoelectronic devices. BP is a P-type semiconductor material, and molybdenum disulfide is an N-type semiconductor material[87]. Buscema et al. designed a gate based on BP-defined PN junction and through experimental research observed zero-bias photocurrents and open-circuit voltages caused by the photovoltaic effect[88]. The small band gap of the material allows power generation for illumination wavelengths up to 940 nm, which is attractive for energy harvesting in the near-infrared region. Figures 9-21(a) ~ (c) show the results of both the device, and the photovoltaic effect of this device[88]. Chen et al. transferred

BP to MoS_2 to form a van der Waals heterostructure. After testing, they found that the rectifying characteristics of the heterostructure were completely regulated by the grid, demonstrating good PN junction characteristics. Especially under laser irradiation of 633 nm, the short circuit current and open circuit voltage of the device increase with the increase in the laser power. Figures 9-21(d)~(e) show the results of the experiment[89]. The anisotropic photoelectric properties of BP are completely determined by its low symmetry crystal structure and electrical and optical properties.

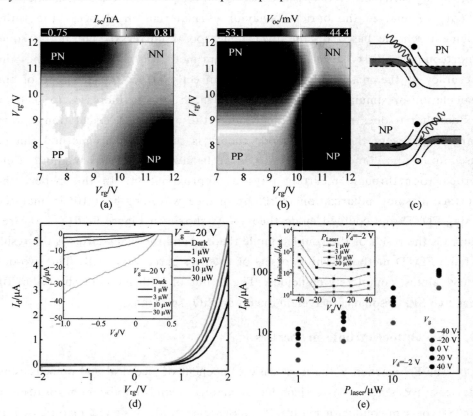

Figure 9-21　Optoelectronic properties

(a) A false-color map of the short-circuit current (I_{sc}, V_{ds} = 0) as the voltages of the two local gates are changed independently. (b) A false-color map of the open-circuit voltage (V_{oc}, I_{ds} = 0) as the voltages on the two local gates are changed independently. (c) Band diagrams illustrating the photovoltaic mechanism in PN and NP configurations: an impinging photon (red arrow) generates an electron-hole pair that is swept away from the junction region by the built-in electric field[88]. (d) I-V characteristics of the P-N diode under various incident laser powers. The inset shows the details in the reverse bias region[89]

1. Optoelectronic devices of BP and graphene heterostructures

Graphene is the most common and popular 2D material. Heterostructures based on graphene have been the focus of many studies[90-92]. Graphene is a zero-band-gap material. Due to its unusual electronic structure at the Dirac point, the extraordinary electrical, magnetic, optical, and mechanical properties of graphene and its heterostructures are extensive and diverse. Heterojunctions composed of graphene and BP have good applications in the field of optoelectronics[19].

Liu et al. used BP and graphene to form a van der Waals heterostructure. This structure formed the gate of the MOS device and was used to receive electromagnetic waves as demonstrated in Figures 9-22(a) and (c)[19]. When graphene and BP form an ohmic contact, the band can bend upward due to the lower BP Fermi level, see Figure 9-22(b). When BP is excited by light, the valence-band electrons transition to the conduction band. Because of band bending, the holes in the BP valence band are transferred to the valence band of graphene. This process enhances the photoelectric conversion of graphene, see Figure 9-22(c). Figure 9-22(d) demonstrates the aforementioned assumptions with different device configurations. Significant photocurrents can be observed only when graphene and BP form a van der Waals heterostructure and are exposed to light. This heterogeneous structure is wavelength selective for light responsiveness. In the near-infrared region, the absorption of graphene is almost negligible. However, the heterostructure of BP and graphene has strong absorption in the near-infrared region, and absorption peaks at 1,350 nm and 1,550 nm as shown in Figure 9-22(e). The switching effects of this optoelectronic device were tested in two heterostructure devices utilizing wavelengths of 1,550 nm and 980 nm, which are widely used in the communication field. The device was more responsive to light signals at 1,550 nm and weaker at 980 nm as shown in Figure 9-22(f). This device can therefore be used as a wavelength-dependent photodetector.

9.3.7 Magnetic properties

Intrinsic BP is not magnetic. There are no isolated electrons within the monatomic thickness range of BP. All of the electrons participate in sp^3 hybridization, so there is no residual magnetic moment. However, it is inevitable that some impure atoms or defects will be introduced during the preparation of BP.

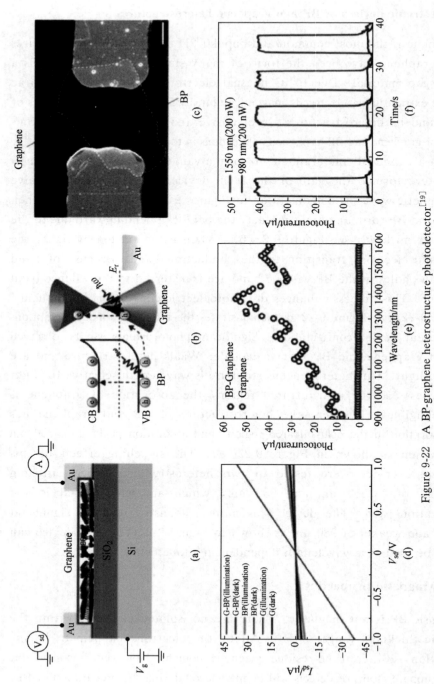

Figure 9-22 A BP-graphene heterostructure photodetector[19]

(a) A schematic diagram of a graphene-BP heterostructure photodetector device; (b) A schematic diagram of the band change when the graphene-BP heterojunctions are in contact; (c) A scanning electron micrograph of the device; (d) Graphene alone, BP alone, and their composition of heterostructure devices yield a comparison of the photocurrent and dark current; (e) The graphene and graphene-BP heterostructure devices alone respond to photocurrents at different wavelengths; (f) The optical switching effect curve of the optical signals of the communication wavelengths (1,550 nm and 980 nm) at 200 mW

These can be adsorbed or doped or cause vacancy defects that may cause residual magnetic moment in the system and show magnetism[93,94].

Chen et al. observed the Shubnikov-de Haas (SdH) oscillation effect of electrons and holes in BP for the first time by testing FETs based on BP. The specific measurement of the SdH oscillation effect of BP was realized in a later work[95]. Figures 9-23 (a) and (b) show the relevant experimental results. When the magnetic field intensity is 8 T, the electron displays obvious SdH oscillation behaviors in the x axis direction, but this oscillation does not occur in the y axis direction. From the dotted line in Figure 9-23(c), it can be seen intuitively that the oscillation period of SdH also exhibits anisotropy.

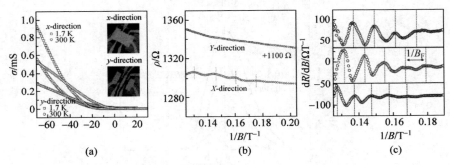

Figure 9-23 SdH oscillations in BP

(a) The conductivity of BP along the x and y directions at 1.7 K and 300 K. The insets are optical images of sample C before and after reshaping; (b) The SdH oscillations are measured at $V_g = 60$ V from the samples along the x and y directions; (c) dR/dB plotted as a function of 1/B. The dashed lines indicate an oscillation period of $1/B_F$[95]

Cyclotron resonance is a type of magneto-optical effect wherein electromagnetic waves and static magnetic fields (B) interact. When the carriers in the material system move under the action of an electromagnetic wave power plant, due to the magnetic field, the carriers engage in cyclotron motion. When the type of carrier is the electron, in a system with isotropic effective mass, the angular frequency of the electron cyclotron motion is $\omega_c = eB/m^*$, where m^* is the effective electron mass. The effective mass of the carrier is determined by the cyclotron resonance effect. Since the frequency of the electromagnetic wave can be adjusted, when the frequency of the electromagnetic wave is the same as the frequency of the electron cyclotron motion, $\omega = \omega_c$, the electrons can absorb energy from the electromagnetic field, thus generating the cyclotron resonance effect[96]. Figure 9-24 shows the cyclotron resonance light absorption spectrum of N-type BP. When the magnetic field (**B**) is perpendicular to different planes, the magnetic field intensity and optical circular frequency corresponding to the resonance absorption peak differ. p-BP has a similar cyclotron resonance absorption spectrum. Three cyclotron resonance masses were used to obtain the effective mass of the electron in three directions (ab-crystal plane, bc-crystal plane, and ac-crystal plane).

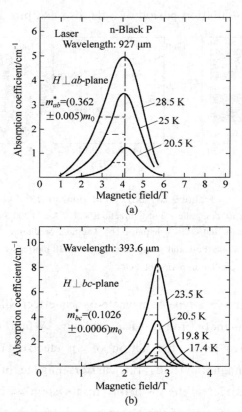

Figure 9-24 The cyclotron resonance absorption spectra of N-type BP in three samples with the largest surface area parallel to the ab-crystal plane (a), the bc-plane (b), and the ac-plane (c) at various temperatures[96]

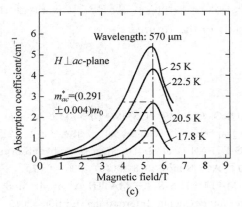

Figure 9-24(Continued)

9.4 Summary and outlook

BP differs from the traditional 2D material graphene, and its applications are less extensive. However, BP has an incomparable advantage over graphene in the fields of electricity, optics, optoelectronics, and thermoelectricity, namely in-plane anisotropy. Secondly, BP has good properties in the field of optics. Wide-spectrum optical absorption, especially in the infrared region, can be used for photoelectric detection. The ultrafast optical response and saturated absorption can be used as a mode-locked laser. BP has a superior method of regulating energy bands than other 2D materials. Chemical vapor deposition and other variants are major trends in the preparation of 2D materials. However, the BP CVD method has not been reported. Finally, the instability of BP severely limits its development and application. Methods of improving the stability of BP remain to be developed. As a low-cost and high-stability preparation method, BP is applicable to various fields. Although experimentally it may be difficult to prepare and process BP, various theoretical methods are in full bloom. In recent years, researchers have used various methods under the framework of the first principles to theoretically study the various properties of BP. The physical mechanism of various properties exhibited by BP has been deeply revealed. It can be said that the use of first-principles methods, quantum Monte Carlo methods and linear response theory will make a great deal of research on 2D BP. It is also a trend in the research of small-layer materials in the future.

References

[1] ABANIN D A, MOROZOV S V, PONOMARENKO L A, et al. Giant nonlocality near the Dirac point in graphene[J]. Science, 2011, 332(6027): 328-330.
[2] WANG K, WANG J, FAN J, et al. Ultrafast saturable absorption of two-dimensional MoS_2 nanosheets[J]. ACS Nano, 2013, 7(10): 9260-9267.
[3] RAMAKRISHNA MATTE H S S, GOMATHI A, et al. MoS_2 and WS_2 analogues of graphene[J]. Angewandte Chemie International Edition, 2010, 49(24): 4059-4062.
[4] ZHANG X, ZHANG S, CHANG C, et al. Facile fabrication of wafer-scale MoS_2 neat films with enhanced third-order nonlinear optical performance[J]. Nanoscale, 2015, 7(7): 2978-2986.
[5] CASTELLANOS-GOMEZ A, BARKELID M, GOOSSENS A M, et al. Laser-thinning of MoS_2: on demand generation of a single-layer semiconductor[J]. Nano Letters, 2012, 12(6): 3187-3192.
[6] LEE Y H, ZHANG X Q, ZHANG W, et al. Synthesis of large-area MoS_2 atomic layers with chemical vapor deposition[J]. Advanced Materials, 2012, 24(17): 2320-2325.
[7] LI L, KIM J, JIN C, et al. Direct observation of the layer-dependent electronic structure in phosphorene[J]. Nature Nanotechnology, 2017, 12(1): 21-25.
[8] ZHANG Y, SUN W, LUO Z Z, et al. Functionalized few-layer black phosphorus with super-wettability towards enhanced reaction kinetics for rechargeable batteries[J]. Nano Energy, 2017, 40: 576-586.
[9] WANG Y, HUANG G, MU H, et al. Ultrafast recovery time and broadband saturable absorption properties of black phosphorus suspension[J]. Applied Physics Letters, 2015, 107(9): 091905.
[10] BUSCEMA M, GROENENDIJK D J, BLANTER S I, et al. Fast and broadband photoresponse of few-layer black phosphorus field-effect transistors[J]. Nano Letters, 2014, 14(6): 3347-3352.
[11] LONG G, XU S, SHEN J, et al. Type-controlled nanodevices based on encapsulated few-layer black phosphorus for quantum transport[J]. 2D Materials, 2016, 3(3): 031001.
[12] AVSAR A, VERA-MARUN I J, TAN J Y, et al. Air-stable transport in graphene-contacted, fully encapsulated ultrathin black phosphorus-based field-effect transistors [J]. ACS Nano, 2015, 9(4): 4138-4145.
[13] LI L, YU Y, YE G J, et al. Black phosphorus field-effect transistors[J]. Nature Nanotechnology, 2014, 9(5): 372-377.
[14] AVSAR A, TAN J Y, LUO X, et al. van der Waals bonded Co/h-BN contacts to ultrathin black phosphorus devices[J]. Nano Letters, 2017, 17(9): 5361-5367.

[15] CHEN Y, SHI T, LIU P, et al. Insights into the mechanism of the enhanced visible-light photocatalytic activity of black phosphorus/BiVO$_4$ heterostructure: a first-principles study[J]. Journal of Materials Chemistry A, 2018, 6(39): 19167-19175.

[16] DING Y, SHI J, XIA C, et al. Enhancement of hole mobility in InSe monolayer via an InSe and black phosphorus heterostructure[J]. Nanoscale, 2017, 9(38): 14682-14689.

[17] DING Y, SHI J, ZHANG M, et al. Electric field modulation of electronic structures in InSe and black phosphorus heterostructure[J]. Solid State Communications, 2018, 269: 112-117.

[18] HUANG L, LI J. Tunable electronic structure of black phosphorus/blue phosphorus van der Waals pn heterostructure[J]. Applied Physics Letters, 2016, 108(8): 083101.

[19] LIU Y, SHIVANANJU B N, WANG Y, et al. Highly efficient and air-stable infrared photodetector based on 2D layered graphene-black phosphorus heterostructure[J]. ACS applied materials & interfaces, 2017, 9(41): 36137-36145.

[20] LUO Z Z, ZHANG Y, ZHANG C, et al. Multifunctional 0D-2D Ni$_2$P nanocrystals-black phosphorus heterostructure[J]. Advanced Energy Materials, 2017, 7(2): 1601285.

[21] ROBBINS M C, NAMGUNG S, OH S H, et al. Cyclical thinning of black phosphorus with high spatial resolution for heterostructure devices[J]. ACS Applied Materials & Interfaces, 2017, 9(14): 12654-12662.

[22] TANG K, QI W, LI Y, et al. Electronic properties of van der Waals heterostructure of black phosphorus and MoS$_2$ [J]. The Journal of Physical Chemistry C, 2018, 122(12): 7027-7032.

[23] ZHENG Y, YU Z, OU H, et al. Black phosphorus and polymeric carbon nitride heterostructure for photoinduced molecular oxygen activation [J]. Advanced Functional Materials, 2018, 28(10): 1705407.

[24] WANG J, MA F, LIANG W, et al. Optical, photonic and optoelectronic properties of graphene, h-BN and their hybrid materials[J]. Nanophotonics, 2017, 6(5): 943-976.

[25] WANG J, MA F, LIANG W, et al. Electrical properties and applications of graphene, hexagonal boron nitride (h-BN), and graphene/h-BN heterostructures[J]. Materials Today Physics, 2017, 2: 6-34.

[26] WANG J, XU X, MU X, et al. Magnetics and spintronics on two-dimensional composite materials of graphene/hexagonal boron nitride [J]. Materials Today Physics, 2017, 3: 93-117.

[27] WANG J, MU X, WANG X, et al. The thermal and thermoelectric properties of in-plane C-BN hybrid structures and graphene/h-BN van der Waals heterostructures[J]. Materials Today Physics, 2018, 5: 29-57.

[28] WANG J, MA F, SUN M. Graphene, hexagonal boron nitride, and their heterostructures: properties and applications[J]. RSC Advances, 2017, 7(27): 16801-16822.

[29] WANG J, LIN W, XU X, et al. Plasmon-exciton coupling interaction for surface catalytic reactions[J]. The Chemical Record, 2018, 18(5): 481-490.

[30] CHEN X L, XIA F N. Enabling novel device functions with black phosphorus/MoS$_2$ van der Waals heterostructures[J]. Sci Bull, 2017, 62(23): 1557-1558.

[31] HUANG L, HUO N, LI Y, et al. Electric-field tunable band offsets in black phosphorus and MoS$_2$ van der Waals pn heterostructure[J]. The Journal of Physical Chemistry Letters, 2015, 6(13): 2483-2488.

[32] HUANG L, LI Y, WEI Z, et al. Strain induced piezoelectric effect in black phosphorus and MoS$_2$ van der Waals heterostructure[J]. Scientific Reports, 2015, 5(1): 1-7.

[33] WU L, GUO J, WANG Q, et al. Sensitivity enhancement by using few-layer black phosphorus-graphene/TMDCs heterostructure in surface plasmon resonance biochemical sensor[J]. Sensors and Actuators B: Chemical, 2017, 249: 542-548.

[34] QIAO J, KONG X, HU Z X, et al. Few-layer black phosphorus: emerging 2D semiconductor with high anisotropic carrier mobility and linear dichroism[J]. Nature Communications, 2014, 5: 4475.

[35] TAO J, SHEN W, WU S, et al. Mechanical and electrical anisotropy of few-layer black phosphorus[J]. ACS Nano, 2015, 9(11): 11362-11370.

[36] LUO Z, MAASSEN J, DENG Y, et al. Anisotropic in-plane thermal conductivity observed in few-layer black phosphorus[J]. Nature Communications, 2015, 6(1): 1-8.

[37] SMITH B, VERMEERSCH B, CARRETE J, et al. Temperature and thickness dependences of the anisotropic in-plane thermal conductivity of black phosphorus[J]. Advanced Materials, 2017, 29(5): 1603756.

[38] ZHANG C D, LIAN J C, YI W, et al. Surface structures of black phosphorus investigated with scanning tunneling microscopy[J]. The Journal of Physical Chemistry C, 2009, 113(43): 18823-18826.

[39] LING X, WANG H, HUANG S, et al. The renaissance of black phosphorus[J]. Proceedings of the National Academy of Sciences, 2015, 112(15): 4523-4530.

[40] DAI J, ZENG X C. Bilayer phosphorene: effect of stacking order on bandgap and its potential applications in thin-film solar cells[J]. The Journal of Physical Chemistry Letters, 2014, 5(7): 1289-1293.

[41] POPOVIĆ Z S, KURDESTANY J M, SATPATHY S. Electronic structure and anisotropic Rashba spin-orbit coupling in monolayer black phosphorus[J]. Physical Review B, 2015, 92(3): 035135.

[42] TAKAO Y, ASAHINA H, MORITA A. Electronic structure of black phosphorus in tight binding approach[J]. Journal of the Physical Society of Japan, 1981, 50(10): 3362-3369.

[43] QIAO J, KONG X, HU Z X, et al. High-mobility transport anisotropy and linear dichroism in few-layer black phosphorus[J]. Nature Communications, 2014, 5(1): 4475.

[44] KUMAR P, BHADORIA B S, KUMAR S, et al. Thickness and electric-field-dependent polarizability and dielectric constant in phosphorene[J]. Physical Review B, 2016, 93(19): 195428.

[45] TRAN V, SOKLASKI R, LIANG Y, et al. Layer-controlled band gap and anisotropic excitons in few-layer black phosphorus[J]. Physical Review B, 2014, 89(23): 235319.

[46] LIANG L, WANG J, LIN W, et al. Electronic bandgap and edge reconstruction in phosphorene materials[J]. Nano Letters, 2014, 14(11): 6400-6406.

[47] JAMIESON J C. Crystal structures adopted by black phosphorus at high pressures[J]. Science, 1963, 139(3561): 1291-1292.

[48] MARUYAMA Y, SUZUKI S, KOBAYASHI K, et al. Synthesis and some properties of black phosphorus single crystals[J]. Physica B, 1981, 105 (1): 99-102.

[49] SHIROTANI I. Growth of large single crystals of black phosphorus at high pressures and temperatures, and its electrical properties[J]. Molecular Crystals and Liquid Crystals, 1982, 86(1): 203-211.

[50] PARK C M, SOHN H J. Black phosphorus and its composite for lithium rechargeable batteries[J]. Advanced Materials, 2007, 19(18): 2465-2468.

[51] LU W, NAN H, HONG J, et al. Plasma-assisted fabrication of monolayer phosphorene and its Raman characterization[J]. Nano Research, 2014, 7(6): 853-859.

[52] CASTELLANOS-GOMEZ A, VICARELLI L, PRADA E, et al. Isolation and characterization of few-layer black phosphorus[J]. 2D Materials, 2014, 1(2): 025001.

[53] BRENT J R, SAVJANI N, LEWIS E A, et al. Production of few-layer phosphorene by liquid exfoliation of black phosphorus[J]. Chemical Communications, 2014, 50(87): 13338-13341.

[54] KANG J, WOOD J D, WELLS S A, et al. Solvent exfoliation of electronic-grade, two-dimensional black phosphorus[J]. ACS Nano, 2015, 9: 3596-3604.

[55] YASAEI P, KUMAR B, FOROOZAN T, et al. High-quality black phosphorus atomic layers by liquid-phase exfoliation[J]. Advanced Materials, 2015, 27(11): 1887-1892.

[56] GUO Z, ZHANG H, LU S, et al. From black phosphorus to phosphorene: basic solvent exfoliation, evolution of Raman scattering, and applications to ultrafast photonics[J]. Advanced Functional Materials, 2015, 25(45): 6996-7002.

[57] HANLON D, BACKES C, DOHERTY E, et al. Liquid exfoliation of solvent-stabilized few-layer black phosphorus for applications beyond electronics[J]. Nature Communications, 2015, 6(1): 1-11.

[58] XIA F, WANG H, JIA Y. Rediscovering black phosphorus as an anisotropic layered material for optoelectronics and electronics[J]. Nature Communications, 2014, 5(1): 4458.

[59] WANG X, JONES A M, SEYLER K L, et al. Highly anisotropic and robust excitons in monolayer black phosphorus[J]. Nature Nanotechnology, 2015, 10(6): 517-521.

[60] BAIK S S, KIM K S, YI Y, et al. Emergence of two-dimensional massless dirac fermions, chiral pseudospins, and berry's phase in potassium doped few-layer black phosphorus[J]. Nano Letters, 2015, 15(12): 7788-7793.

[61] KIM J, BAIK S S, RYU S H, et al. Observation of tunable band gap and anisotropic Dirac semimetal state in black phosphorus[J]. Science, 2015, 349(6249): 723-726.

[62] PENG X, WEI Q, COPPLE A. Strain-engineered direct-indirect band gap transition and its mechanism in two-dimensional phosphorene[J]. Physical Review B, 2014, 90(8): 085402.

[63] WEI Q, PENG X. Superior mechanical flexibility of phosphorene and few-layer black phosphorus[J]. Applied Physics Letters, 2014, 104(25): 251915.

[64] JIANG J W, RABCZUK T, PARK H S. A Stillinger-Weber potential for single-layered black phosphorus, and the importance of cross-pucker interactions for a negative Poisson's ratio and edge stress-induced bending[J]. Nanoscale, 2015, 7(14): 6059-6068.

[65] TAKAO Y, ASAHINA H, MORITA A. Electronic structure of black phosphorus in tight binding approach[J]. Journal of the Physical Society of Japan, 1981, 50(10): 3362-3369.

[66] BABA M, IZUMIDA F, TAKEDA Y, et al. Two-dimensional Anderson localization in black phosphoruscrystals prepared by bismuth-flux method[J]. Journal of the Physical Society of Japan, 1991, 60(11): 3777-3783.

[67] BABA M, NAKAMURA Y, TAKEDA Y, et al. Hall effect and two-dimensional electron gas in black phosphorus[J]. Journal of Physics: Condensed Matter, 1992, 4(6): 1535-1544.

[68] LIU H, NEAL A T, ZHU Z, et al. Phosphorene: an unexplored 2D semiconductor with a high hole mobility[J]. ACS Nano, 2014, 8(4): 4033-4041.

[69] XIA F, WANG H, JIA Y. Rediscovering black phosphorus as an anisotropic layered material for optoelectronics and electronics[J]. Nature Communications, 2014, 5(1): 4458.

[70] FEI R, YANG L. Strain-engineering the anisotropic electrical conductance of few-layer black phosphorus[J]. Nano Letters, 2014, 14(5): 2884-2889.

[71] FEI R, FAGHANINIA A, SOKLASKI R, et al. Enhanced thermoelectric efficiency via orthogonal electrical and thermal conductances in phosphorene[J]. Nano Letters, 2014, 14(11): 6393-6399.

[72] KONABE S, YAMAMOTO T. Significant enhancement of the thermoelectric performance of phosphorene through the application of tensile strain[J]. Applied Physics Express, 2014, 8(1): 015202.

[73] ZHANG J, LIU H J, CHENG L, et al. Phosphorene nanoribbon as a promising candidate for thermoelectric applications[J]. Scientific Reports, 2014, 4(1): 6452.

[74] LAN S, RODRIGUES S, KANG L, et al. Visualizing optical phase anisotropy in black phosphorus[J]. ACS Photonics, 2016, 3(7): 1176-1181.

[75] LU S B, MIAO L L, GUO Z N, et al. Broadband nonlinear optical response in multi-layer black phosphorus: an emerging infrared and mid-infrared optical material [J]. Optics Express, 2015, 23(9): 11183-11194.

[76] LI D, JUSSILA H, KARVONEN L, et al. Polarization and thickness dependent absorption properties of black phosphorus: new saturable absorber for ultrafast pulse

generation[J]. Scientific Reports, 2015, 5(1): 15899.

[77] GE S, LI C, ZHANG Z, et al. Dynamical evolution of anisotropic response in black phosphorus under ultrafast photoexcitation[J]. Nano Letters, 2015, 15(7): 4650-4656.

[78] HE J, HE D, WANG Y, et al. Exceptional and anisotropic transport properties of photocarriers in black phosphorus[J]. ACS Nano, 2015, 9(6): 6436-6442.

[79] SUESS R J, JADIDI M M, MURPHY T E, et al. Carrier dynamics and transient photobleaching in thin layers of black phosphorus[J]. Applied Physics Letters, 2015, 107(8): 081103.

[80] SABERI-POUYA S, VAZIFEHSHENAS T, SALEH M, et al. Plasmon modes in monolayer and double-layer black phosphorus under applied uniaxial strain[J]. Journal of Applied Physics, 2018, 123(17): 174301.

[81] VENUTHURUMILLI P K, YE P D, XU X. Plasmonic resonance enhanced polarization-sensitive photodetection by black phosphorus in near infrared[J]. ACS Nano, 2018, 12(5): 4861-4867.

[82] LAM K T, GUO J. Plasmonics in strained monolayer black phosphorus[J]. Journal of Applied Physics, 2015, 117(11): 113105.

[83] FO Q, PAN L, CHEN X, et al. Anisotropic plasmonic response of black phosphorus nanostrips in terahertz metamaterials[J]. IEEE Photonics Journal, 2018, 10(3): 1-9.

[84] LOW T, ROLDÁN R, WANG H, et al. Plasmons and screening in monolayer and multilayer black phosphorus[J]. Physical Review Letters, 2014, 113(10): 106802.

[85] HONG Q, XIONG F, XU W, et al. Towards high performance hybrid two-dimensional material plasmonic devices: strong and highly anisotropic plasmonic resonances in nanostructured graphene-black phosphorus bilayer[J]. Optics Express, 2018, 26(17): 22528-22535.

[86] ZHOU R, PENG J, YANG S, et al. Lifetime and nonlinearity of modulated surface plasmon for black phosphorus sensing application[J]. Nanoscale, 2018, 10(39): 18878-18891.

[87] GEIM A K, NOVOSELOV K S. The rise of graphene[M]//Nanoscience and Technology: a collection of reviews from nature journals, 2010: 11-19.

[88] BUSCEMA M, GROENENDIJK D J, STEELE G A, et al. Photovoltaic effect in few-layer black phosphorus PN junctions defined by local electrostatic gating[J]. Nature Communications, 2014, 5(1): 4561.

[89] CHEN P, XIANG J, YU H, et al. Gate tunable MoS_2-black phosphorus heterojunction devices[J]. 2D Materials, 2015, 2(3): 034009.

[90] LHERBIER A, LIANG L, CHARLIER J C, et al. Charge carrier transport and separation in pristine and nitrogen-doped graphene nanowiggle heterostructures[J]. Carbon, 2015, 95: 833-842.

[91] DI LECCE V, GRASSI R, GNUDI A, et al. Graphene base heterojunction transistor: an explorative study on device potential, optimization, and base parasitics[J]. Solid-State Electronics, 2015, 114: 23-29.

[92] NISANCı B, GANJEHYAN K, METIN Ö, et al. Graphene-supported NiPd alloy nanoparticles: a novel and highly efficient heterogeneous catalyst system for the reductive amination of aldehydes[J]. Journal of Molecular Catalysis A: Chemical, 2015, 409: 191-197.

[93] WANG G, PANDEY R, KARNA S P. Effects of extrinsic point defects in phosphorene: B, C, N, O, and F adatoms[J]. Applied Physics Letters, 2015, 106(17): 173104.

[94] KHAN I, HONG J. Manipulation of magnetic state in phosphorene layer by non-magnetic impurity doping[J]. New Journal of Physics, 2015, 17(2): 023056.

[95] CHEN X, WU Y, WU Z, et al. High-quality sandwiched black phosphorus heterostructure and its quantum oscillations[J]. Nature Communications, 2015, 6(1): 7315.

[96] NARITA S, TERADA S, MORI S, et al. Far-infrared cyclotron resonance absorptions in black phosphorus single crystals[J]. Journal of the Physical Society of Japan, 1983, 52(10): 3544-3553.

Chapter 10

Graphitic Carbon Nitride Nanostructures

10.1 Introduction

Over the past few decades, the graphitic carbon nitride ($g-C_3N_4$) as an organic semiconductor has attracted extensive attention in research fields due to its similarity to graphene[1-2]. Graphene, as a typical two dimensional (2D) material, has huge surface area, high electron mobility and other outstanding advantages. But lacking of band gap inhibits its applications in catalytic reactions. In contrast to graphene, $g-C_3N_4$ not only has the graphite-like structure, but more importantly is a medium-band-gap semiconductor which contributes to catalyze chemical reactions. However, $g-C_3N_4$ has the low electronic conductivity due to its band-gap energy (2.7 eV) and specific surface area which are greatly limited in general electrochemical hydrolysis reactions (1.4~2.8 eV). Therefore, the modification of $g-C_3N_4$ is one of the most important solutions in the catalytic field.

In fact, the explorations of $g-C_3N_4$-based nanocomposites have undergone a long-term process. The carbon nitrides were synthesized and obtained from a series of ammonocarbonic acids by Franklin in 1922. In 1937, Pauling and her co-workers investigated triazine and tri-s-triazine which are the structural units of $g-C_3N_4$. Until 1990s, researchers studied the result that $g-C_3N_4$ is the most stable structure of C_3N_4. Here, both tri-s-triazine and triazine were considered as the potential tectonic units of $g-C_3N_4$ (Figure 10-1). The tri-s-triazine has more nitrogen atoms than triazine, which not only provides more anchoring sites but also has more active sites for catalytic reactions. More importantly, the tri-s-triazine based connection patterns of $g-C_3N_4$ shows the excellent stability compared with triazine. Therefore, tri-s-triazine has been widely studied and considered as a promising candidate for modification of g-C_3N_4. Note that Liebig in 1844 named it Melon, and recently as polymeric carbon nitride[3-4], where the NH/NH_2 signals around 3200 cm^{-1}. Algara-Sille achieved the successful synthesis of a triazine-based 2D material in 2014[5].

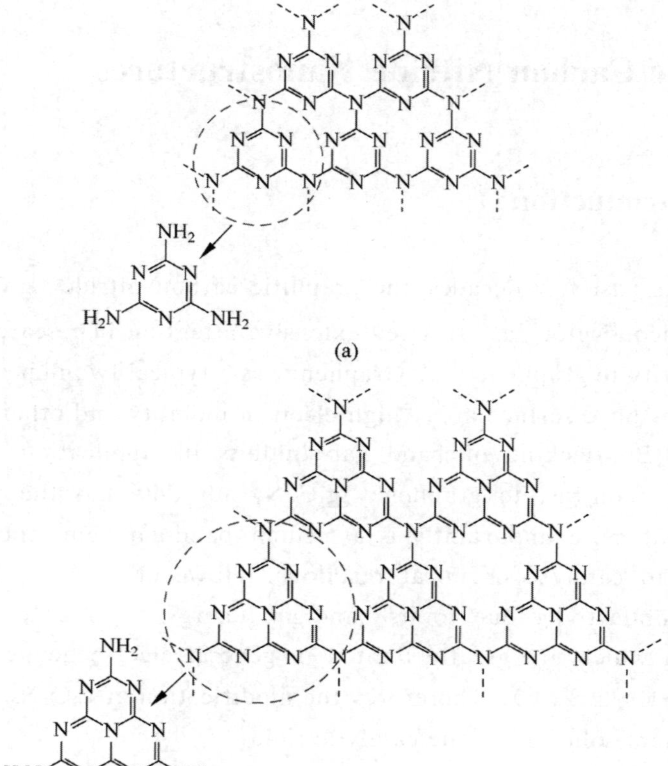

Figure 10-1 Structures of g-C_3N_4

(a) Triazine and (b) tri-s-triazine (heptazine) structures of g-C_3N_4

To overcome the shortcomings of g-C_3N_4 in catalytic reactions, researchers have spared no effort to explore several studies, such as the optimization of precursor, preparation process, and even nano-modification. And the main approaches involve either introducing an isolated impurity state into the forbidden band or narrowing the band gap by doping a foreign element into nanostructure of g-C_3N_4. The synthesized samples are called based on g-C_3N_4

nanostructures, which exhibit the surprising catalytic performance in oxygen reduction reactions (ORRs) and hydrogen evolution reactions (HERs), especially[6-10]. However, the systematic introduction and specific description of g-C_3N_4-based nanostructures are extremely rare, from materials, synthesis methods and characterization to applications, especially in ORR HER nitrogen photofixation reactions, degradation of CO_2, NO_x, 2-propanol, degradation of organic pollutants and photoreduction removal of Cr(Ⅵ).

Therefore, combined with previous studies, we concentrate mainly on them in this paper. With regard to materials and synthesis methods, they are introduced in detail by classification. The characterization techniques are also thoroughly summarized and described, including structural characterization and performance characterization. Moreover, the applications of g-C_3N_4-based nanostructures are explained in detail, which are revealed specifically by the mechanism and measurement of catalytic performance.

10.2 Materials and synthesis methods

10.2.1 Materials

Currently, there are many kinds of materials based on g-C_3N_4 nanostructures with a certain characteristic, such as the element species, morphology structures and band-gap energy. Generally, these categories would bring us great challenges to studying their structures, especially when we want to optimize their performance. To avoid the occurrence of similar questions, here, two major categories based on g-C_3N_4 nanostructures according to the synthesis principle are introduced, which are compounding and doping. Specifically, there are many challenges in the application process of g-C_3N_4 materials, including the fast recombination electron-hole, the low quantum efficiency, etc. Composite modification is the most convenient method currently, which introduces an isolated impurity state into the forbidden band of g-C_3N_4. And due to the difference in the position of the conduction band and the valence band, g-C_3N_4 fully contacts the certain material to form the heterojunction which enormously promotes electron-hole separation and leads to the decrease in the recombination rate, so making the more active particles catalyze chemical reactions. The selected composites are mainly metal materials (such as common metals, noble metal and bimetallic materials), semiconductor materials (such as metal oxides, metals hydroxides, metal sulfides, metal complexes, and metal organic frameworks). On the other hand, doping modification can excellently change the electronic structure of g-C_3N_4, which can be divided into metal doping and nonmetal doping. Usually, doping the metal ions into g-C_3N_4 structural unit can generate the traps of electron-hole pairs as to prolong the time for electron and hole recombination, leading to the catalytic performance improving greatly. Non-metal doping mainly includes O, N, P, S, B, F, etc. It is generally believed that the C, N, and H elements in the 3-s-triazine structural units are replaced by these non-metal elements, thereby forming lattice defects of g-C_3N_4 to achieve efficient separation of electron-hole pairs, resulting in catalytic properties promoted enormously.

Besides, the surface area based on g-C_3N_4 nanostructures can be also greatly developed compared to g-C_3N_4. And the nanocomposites can carry some unique properties. For example, combining g-C_3N_4 with Fe_3O_4 has magnetic properties, which facilitates the recycling of catalysts. As mentioned above, g-C_3N_4 basically has all the prerequisites for heterogeneous photocatalysts, due to its unique electronic structure with an appropriate band gap of 2.7 eV[11]. This band gap not only overcome the endothermic character of catalytic reactions, but interestingly, is more conducive to combine with other substances for emerging and manipulating better photoelectrocatalytic properties[12-13]. Generally, carbon nitride nanocompounds are broadly divided into two major categories, optimization design with metal-free elements and metal elements and in all aspects of the application, as shown in Figure 10-2.

Specifically, the metal-free carbon nitride nanocompounds contain mainly adulterating with the elements including boron, fluorine, phosphorus and sulfur, especially with carbon[14-18]. As we all know, the nonmetal materials are easy to prepare and have lower cost compared with noble metals, which are beneficial to widely used in actual industrial production[19-21]. Besides, the nanocomposites synthesized by g-C_3N_4 nonmetal can diminish or even eliminate the abundance and inertness of C-H bonds in organic structures which can seriously affect the molecular selectivity, and avoid excessive oxidation of molecular oxygen[22-23]. More importantly, doping nonmetal elements into

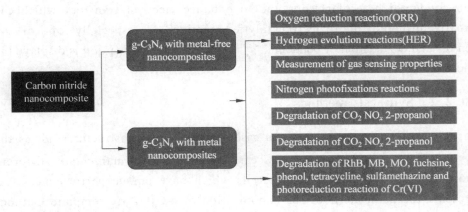

Figure 10-2 The types and applications of Carbon Nitride Nanocompounds

g-C_3N_4 regarded as the framework, can slightly develop the molecular orbital shape and position, and lead to the adjustment of electronic properties and huge enhancement ionic conductivity[24], indicating the enormous promotion of photocatalytic performances. At present, due to the outstanding biocompatibility, stability and amazing electronic conductivity, g-C_3N_4 nanocompounds, especially mesoporous g-C_3N_4/carbon, have been extensively used for highly efficient photoelectrochemical catalytic reactions, such as water splitting, energy conversation and the ORR.

Meanwhile, the hybrids of g-C_3N_4 and inorganic semiconductors not merely reflect on metal-free g-C_3N_4 nanocomposites, but more reveal in g-C_3N_4 nanocomposites with metal elements that couple g-C_3N_4 with metal-oxide semiconductors (ZnO, $BiPO_4$, Ag_3PO_4, TiO_2, Fe_2O_3, SnO_2, $ZnWO_4$, $BiWO_6$)[25-33]. The g-C_3N_4/metal nanocomposites exhibit the more measurement of the gas sensing properties and cathodic electrochemiluminescent activity compared with pure g-C_3N_4[6, 8], because the metal elements play an important role in capturing and storing the electrons from the conduction band of g-C_3N_4, and can prevent electrode passivation caused by high-energy electrons injected into the conduction band of g-C_3N_4. Besides, the metal nanoparticles also have a high surface area, and can be easily separated and recycled with catalytic activity as much as possible in catalytic reaction[34], such as deposition with Cu, Pd, Pt and Au[35-36]. And it was found that combining the g-C_3N_4 substrate with supported metal nanoparticles can lead to controlled access to metal-semiconductor heterojunctions and to enhance electron transfer, which can greatly promote photoelectrocatalytic reaction[37], especially optical gas sensors and degradation of pollutants[38-39]. The detailed description is displayed in the later application.

10.2.2 Synthesis methods

Generally, synthesis methods based on g-C_3N_4 nanostructures act as the pivotal characters, because they can synthesize and manipulate different morphological and nano-size materials which have unique performances. At present, the main methods for nanonization of g-C_3N_4 are template methods or non-template methods. However, there are many synthetic methods for the formation based on g-C_3N_4 nanostructures.

Currently, carbon nitride nanocompounds mainly derived from the composite of several materials as precursors via a specific technology[40-41]. There are some means to be applied in the process of synthesizing carbon nitride nanostructures, as shown in Table 10-1. For instance, the materials including carbon nitride would be inserted into highly ordered mesoporous carbon by using facile nanocasting techniques[42], or into graphene-based mesoporous silica nanosheets[43] for conductivity improvement. And the chemical vapor deposition method[44-47], a highly popular approach to obtaining graphene, makes the inoculate and growth of graphene by rationally using the atomic structure of the metal substrate. Furthermore, with regard to other synthetic methods (Table 10-1), there are more detailed explanations in References[39],[48-51].

Table 10-1 The various methods of synthesizing carbon nitride nanocompounds[7, 45, 52-62]

Methods of synthesis	Advantages	Disadvantages	References
The silica microspheres as hard template method	Controlling easily product appearance	Dangerous for work and unfriendly for environment	[7]
CVD	Maintaining the high purity of the material	Difficult to maintain altitude and sustained pressure	[45]
Solvent thermal polymerization method	Presenting an exceptional high surface area and overall porosity	Suppressing transmission and diffusion performance	[47]

Continued

Methods of synthesis	Advantages	Disadvantages	References
The microcontact printing technique	Easy to operate and low cost of equipment	Limiting to sub-micron scale	[52]
The facile calcination method	Can greatly improve the mechanical strength of the sample	Thermal dissociation of compounds or crystal transition process is not easy to observe and control	[44-55]
The facile solid-state method	Prepared samples exhibit high purity and stability	High equipment requirements	[56-58]
The ultrasound-assised liquid-phase exfoliation method	The sample can be provided with large size, multi-sized pores in order to rapidly transport the reactants to the electroactive sites	Sample morphology is not easy to specific regulation	[59]
The sample vacuum filtration method	The samples are of high purity and strong flexibility	Equipment requirements are high and sample preparation conditions are limited	[60]
The melt blending method	The synergetic distribution effect between different hybrids is easier to visualize	Sample morphology is not easy to control	[61]
The facile hydrothermal method	Synthetic sample particles are high purity, good dispersion	It is difficult to observe the growth process of sample	[62]

In short, these methods not only ensure progress of the study but also contribute to improving the dispersion and the conductivity of the composite materials which appear better catalytic performances.

10.2.3 Characterization methods

To investigate the morphology and properties, the synthetic carbon nitride nano composites have been characterized by various means of measurement. In this review, the characterization methods are introduced mainly focusing on two aspects: structure characterization and performance characterization. The former is more inclined to use various spectrograms or morphology data to display the structural properties, including X-ray diffraction (XRD) patterns, scanning electron microscopy (SEM), atomic force microscopy (AFM), energy-dispersive X-ray spectroscopy (EDS) pattern, Fourier transform infrared (FT-IR) spectrometer techniques and so on. Some specific introductions of characterization methods are shown in Table 10-2.

Table 10-2 The various methods of characterization and their functions

Methods of characterization	Function
Scanning electron microscopy (SEM)	Showing specific morphology and nanostructure, as well as the surface dispersion of nanoparticles, stacking situation of sample
Transmission electron microscopy (TEM)	Further revealing the morphology and nanostructure of the sample, especially the size of the nanopore
Atomic force microscopy (AFM)	Testing the thickness and nanolayer of sample
Energy dispersive X-ray spectroscopy (EDS)	Investigating the element composition and distribution of sample
X-ray photoelectron spectroscopy (XPS)	Studying the chemical composition formation, element binding energies and heterojunction of sample
X-ray diffraction (XRD)	Demonstrating the characteristic peak and lattice constant (crystal surface peak)
Thermogravimetry-differential thermal analysis (TG-DTA)	Presenting the change of sample' weight with increasing temperature to obtain the optimum temperature for the best properties
Fourier transform infrared spectrometer (FT-IR) Techniques	Exhibiting the heterocyclic stretches of typical structure and the chemical composition formation
Brunauer-Emmett-Teller (BET) and Barrett-Joyner-Halenda (BJH) models	The specific surface areas and pore size distribution plots were calculated according to the BET and BJH models, respectively

There are more specific introductions in References [63] ~ [65]. In addition, the above characterization method is just the part of the structure characterization. There are many optical spectroscopy methods widely used to show the structural properties of the sample, including the UV-Vis diffuse reflection spectra, Raman and photoluminescence spectra[66]. As for performance characterization, it more tends to explain the internal performance in practical applications, which is the focus of characterization in this review. Common characterization methods contain cyclic voltammetry, chronoamperometry, constant potential intermittent titration, tafel curves and so on. The detailed introduction would be revealed in the next application.

10.3 Applications

Metal-free g-C_3N_4[67], has received increasing attention owing to its potential application in ORR[68], photocatalysis[69-70], bioimaging[71-72], etc. The g-C_3N_4 nanocompounds have been developed for diversified catalysis in many fields, such as photoelectrocatalytic reactions, solar energy conversion and gas sensors[2,73-76]. Currently, the efficient use of renewable energy and the purification of the environment are one of the most important factors in modern sustainable development, therefore, the gas reduction reaction in the field of photoelectrocatalysis and gas sensors are emphatically considered by the scientific communities.

Besides, there are several reports to elaborate minutely many other applications revealed by carbon nitride nanomaterials as the photoelectrocatalysts are used to other applications, and energy storage/conversion, including the water splitting, photodegradation of pollutants. There are several reports to reveal and introduce in[15],[77]~[80].

10.3.1 Based on g-C_3N_4 nanostructures nanocatalysts driven highly the ORR

The ORR acts as a critical role in renewable energy technologies, but there has been a big limitation that the kinetics of the ORR are too sluggish to achieve the better conversion on efficiency and the performance of electrochemical energy, especially[81-82]. Conventionally, with the development of precious metals, the Pt has been assigned to be an efficacious electrocatalyst, due to the high-electrocatalytic activity[83].

One of the key steps of fuel cells is ORR. ORR is limited by the catalysts[11,84-87]. Pt/C is the most traditional catalyst which the precious metal Pt is doomed to not be applied in long-term and large-scale commercial production, this result attributes completely to its the high cost, limited supply and poor durability[87]. Other kinds of catalysts are mostly environmentally unfriendly intermetallic compounds and alloyed metals[88-90]. At present, research studies have focused on nonmetallic materials, like fullerene[91], nonmetallic-element-doped graphene[92], and organic compounds[93]. The g-C_3N_4 is a plane conjugate structure. This structure is conducive to the movement of electrons on the surface of g-C_3N_4[94-100]. Hence, g-C_3N_4 can be used as the catalyst for ORR, and some researchers have reported that it possesses ORR activities[101]. However, its performance is not good enough because of the small specific surface area[102]. Usually, g-C_3N_4 has a bulk structure with lamellar stacking[103], unsatisfactory electrical conductivity between the layers, and limited ORR[104]. In previous works, the ORR mainly occurs on the conduction band (CB) of g-C_3N_4. To improve the electrical conductivity and the band structure, elements doping is a popular method that is easier to control and saves energy. Nonmetal elements doping is an effective method to change redox potentials, absorbance, and photogenerated transport efficiency of g-C_3N_4[105-111]. Elements belonging to the VIA family (O, S, and Se) are commonly used to dope g-C_3N_4. Li et al. reported that oxygen doping g-C_3N_4 can restrain the recombination of electron-hole pairs and improve light absorption[112-113]. Lee et al. suggested that S/Se/graphene co-doping g-C_3N_4 can prolong the lifetime of the charge carrier and improve the band structure of g-C_3N_4[114]. Thus, doping-using elements of the VIA family (O, S, and Se) can change the structure of the CB, which can promote the ORR performance of g-C_3N_4. Besides, it shows significantly superior catalytic performance than the Pt/C catalyst owing to the larger surface area and the high nitrogen content which can provide more reactive reaction sites.

In fact, the doping of heteroatoms of nitrogen-doped carbon materials can induce the redistribution of charges, reduce the ORR potential, change effectively the chemisorption mode of O_2 and greatly promote the ORR process[82]. To further understand clearly the effect of doping electrons absorbed on g-C_3N_4 in the ORR reaction, we present the standard process of electronic participation in ORR reaction, which are based on the first-principles calculations in alkaline solutions by a two-step $2e^-$ pathway or a direct $4e^-$ pathway[115].

As follows: the two step $2e^-$ pathways,

$$O_2 + 2H_2O + 2e^- \longrightarrow HO_2^- + OH^-, \qquad (10\text{-}1)$$

$$HO_2^- + H_2O + 2e^- \longrightarrow 3OH^-. \qquad (10\text{-}2)$$

Or a direct 4e⁻ pathway,
$$O_2 + 2H_2O + 4e^- \longrightarrow 4OH^-. \tag{10-3}$$

Generally, there are mainly three free-energy states of ORR and can generate three products which are absorbed by g-C_3N_4 in the solution, including the initial product O_2, the intermediate reagent OOH^- and the final reactant OH^- (Figure 10-3). With the participation of zero, two, and four electrons in the reaction system, the paths Ⅰ, Ⅱ, and Ⅲ exhibit the specific ORR processes, respectively. We can clearly observe that oxygen can not be spontaneously reduced on the surface of pristine g-C_3N_4 in the absence of electronic participation (Figures 10-3 (a) and (b)). This phenomenon is caused by the barriers that are hard to overcome at the states of intermediate and final products in the free energy. However, in the case of two electrons participations (part Ⅱ), the free energy of OOH^- and OH^- show an obvious reduction compared with part Ⅰ, especially at the free energy of OOH^-. This result indicates that the first 2e⁻ reaction (Equation 10-1) can smoothly and spontaneously carry out, but the second 2e⁻ reaction (Equation 10-2) proceed hardly, due to the existence of the barrier at the final state of OH^-/g-C_3N_4. As a result, lots of intermediate products OOH^- accumulate on g-C_3N_4 leading to an enormous ORR resistance (Figure 10-3 (c)). Note that, there is a very fast and direct formation on the surface of g-C_3N_4 from the initial reagent O_2 to the final product OOH^- (Figure 10-3 (d)), due to the introduction of four electrons which can completely eliminate the barriers of ORR reaction (part Ⅲ in Figure 10-3(a)). Combined with Equation 10-3, the OOH^- accumulation that limits the rate of electron participation on g-C_3N_4 can be perfectly eliminated via an efficient 4e⁻ pathway[42].

In short, the poor electron transfer efficiency which limits the active sites, results in an unsatisfactory performance of ORR. With the addition of doping electrons and accumulation from the composites on the surface of the g-C_3N_4, the active sites can across the entire surface of the catalyst and extend the concentration, as shown the red areas which represent the active sites in Figure 10-3 (d), which lead to the increase of the electron transfer efficiency and facilitate greatly the catalytic properties of ORR. In order to verify the prediction above, the hybrid which incorporates g-C_3N_4 into the mesoporous carbon (CMK-3) that are highly ordered and synthesized facile nanocasting

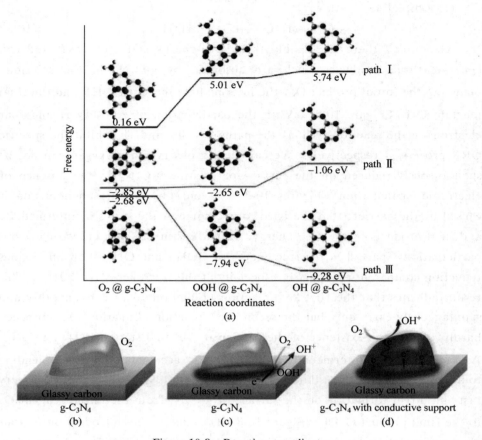

Figure 10-3 Reaction coordinates
(a) Free energy of ORR and optimized configurations of adsorbed species on the g-C_3N_4 surface with the participation of zero, two, and four electrons showed as paths I, II, and III, respectively. Schemes of ORR's pathway on pristine g-C_3N_4. (b) without electron participation. (c) with 2e$^-$ participation. (d) conductive support composite with 4e$^-$ participation[100]

method, recognized as g-C_3N_4 at CMK-3, are measured some electrocatalytic performances compared with g-C_3N_4 and mixed g-C_3N_4 + CMK-3 that just is a physical mixture of CMK-3 and g-C_3N_4 with the same catalyst content as g-C_3N_4 at CMK-3 (Figure 10-4). It is noteworthy that the reason for choosing porous carbon is that it has more excellent adsorption properties than normal carbon.

As we all see, the electrocatalytic activity is reflected by cyclic voltammograms (CVs) in O_2-saturated 0.1 M KOH solution (Figure 10-4(a)). The pristine mesoporous g-C_3N_4 has two obvious peaks of ORR at the relatively low

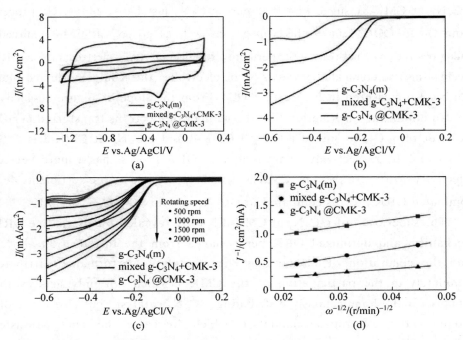

Figure 10-4 Cyclic voltammograms of ORR and LSV of various electrocatalysts
(a) Cyclic voltammograms of ORR on various electrocatalysts in O_2-saturated 0.1 M KOH solution;
(b) LSV of various electrocatalysts on RDE at 1,500 r/min in O_2-saturated 0.1 M KOH solution;
(c) and (d) LSV of various electrocatalysts on RDE at different rotating rates (500 r/min to 2000 r/min) and corresponding Koutecky-Levich plots at 0.6 V[100]

cathode voltage in CV curve, but the CV curve of g-C_3N_4 at CMK-3 just with a single ORR peak at -0.25 V[43, 115-118]. That is to say, the composites undergo 4e⁻ pathway, which are faster than the 2e⁻ pathways that pure g-C_3N_4 experiences. More importantly, compared with the CV curve of g-C_3N_4 and mixed g-C_3N_4 + CMK-3, g-C_3N_4 at CMK-3 reveals a stronger ORR peak and much larger capacitance for ORR. And the superiority of catalytic performances is also further demonstrated in linear sweep voltammograms (LSVs) on a rotating disk electrode (RDE) (Figure 10-4 (b)). The g-C_3N_4 at CMK-3 electrodes has the lower onset potential and higher ORR current density than g-C_3N_4 and mixed g-C_3N_4 + CMK-3 and this result is more adequately reflected in Figure 10-4 (c) that is investigated at different rotating speeds. Besides, the LSV of g-C_3N_4 exhibits a reduction peak at -0.49 V without any current plateau, however, there are the wide current plateau on g-C_3N_4 at CMK-3, suggesting that pure g-C_3N_4 just has a 2e⁻ ORR process from O_2 to OOH^- under this voltage, as well as g-C_3N_4 at CMK-3 directly undergoes the 4e⁻ pathway. Furthermore, for deeper study electrocatalytic ORR mechanisms and dominated processes of the electrode, the Koutecky-Levich plots can be also measured in Figure 10-4 (d). From the slope of Figure 10-4 (d), we can obtain the information that the number of electrons transferred per O_2 molecule on g-C_3N_4, mixed g-C_3N_4 + CMK-3 and g-C_3N_4 at CMK-3 is 2.6, 1.7, and 4.0, respectively. That is to say, ORR process has a more perfect selectivity with g-C_3N_4 at CMK-3 which possess a more efficient 4e⁻ dominated ORR process compared to g-C_3N_4 and mixed g-C_3N_4 + CMK-3[42].

The superiority of g-C_3N_4 at CMK-3 is fully revealed by alternative ORR mechanism and optimized ORR performance from the introduction above. And the generation of outstanding electrocatalytic performance attributes completely to the participation of the CMK-3 which not only acts as the support for the homogenously distributing g-C_3N_4 catalyst, but also markedly promotes the electron accumulation which increases the energy-transfer efficiency on the surface of g-C_3N_4 catalyst. Besides, the nitrogen-doped carbon nanostructures also show excellent electrochemical stability, as well as strong tolerance ability of the methanol compared with other materials, especially Pt/C that is the combination of Pt nanoparticles and carbon support (Figure 10-5).

Here, the highly ordered mesoporous carbon nitride (OMCN) is different from above-mentioned g-C_3N_4 at CMK-3, and is constructed with g-C_3N_4 frameworks and ordered arrays of uniform mesopores. Importantly, it possesses the high-surface area and great catalytic activity density of nitrogen groups. Usually, the durability of OMCN and Pt/C (20wt% Pt) is revealed by comparing changes before and after an accelerated durability test (ADT) in the ORR activity[119]. From Figures 10-5 (a) and (b) which was continuously cycled from 0.05 V to 1.4 V in the ADT in an acidic environment, the Pt/C catalyst manifests remarkable degradation behavior from 2 cycles to 200 cycles of ADT (Figure 10-5 (a)), but there is hardly any significant change for CVs of OMCN (Figure 10-5 (b)). In other words, the Pt/C catalyst has a substantial loss of electrochemical surface area from 94 m^2/g to 69 m^2/g which is attributed to the Ostwald ripening phenomenon[120], so it shows very

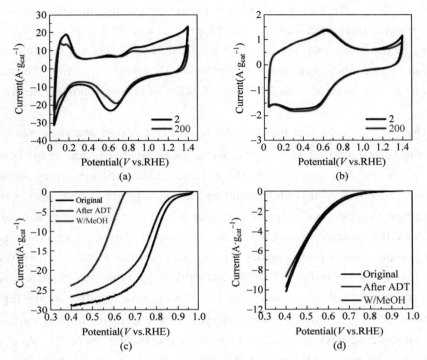

Figure 10-5 CV curves of (a) Pt/C and (b) OMCN during the ADT (numbers in the legend indicate cycle numbers). ORR activity of (c) Pt/C and (d) OMCN before and after ADT and in the presence of MeOH[100]

unstable electrochemical properties. On the contrary, the OMCN catalyst does not undergo a similar dissolution/redeposition process to Pt, suggesting the distinguished durability of OMCN with respect to the Pt/C catalyst in an acidic medium.

Meanwhile, compared with the ORR activities of OMCN and Pt/C after the ADT and methanol additions at 900 r/min in the oxygen-saturated electrolyte, the amazing stability and tolerance in the methanol can be reflected entirely by the OMCN (Figures 10-5 (c) and (d)). The Pt/C demonstrates an obvious reduction in ORR activity after the ADT, as well as extremely poor tolerance in the methanol (Figure 10-5 (c)). By contrast, the OMCN brings into correspondence with ORR activity after the ADT with original condition, similarly, keeping with in the methanol (Figure 10-5 (d)). Due to the existence of simultaneously reaction of methanol and oxygen on the surfaces of

Pt particles[121], the mixed potential is generated during the ORR process of Pt/C, formed in an actual cathode in direct methanol fuel cells (DMFCs)[122], leading to the loss of cell voltage. But the OMCN has excellent resistance with methanol and cannot occur other side effects. Therefore, selecting the suitable catalysts for the methanol in ORR, the OMCN is priority choice compared to Pt/C[123].

The g-C_3N_4, with rich pyridine-like nitrogen, can trap abundant transitional metals to form potential active sites, while graphene, with similar sp^2 bonding structure to g-C_3N_4, can promote the electron transfer through strong electronic coupling. Inspired by this understanding, an assumption is proposed that the assembly of Co-doped g-C_3N_4 and graphene would give an easy access to highly efficient ORR process. So, Liu et. al. developed Co-g-C_3N_4 at graphene, a novel nonprecious metal-heterocyclic polymer as ORR catalyst with high activity and good stability. This is much different from the reported 3D g-C_3N_4 structures, all of which become almost invalid after removing 3D skeletons[111, 124-127]. For such a 2D composite, the significant enhancement of ORR catalytic activity is mainly ascribed to the formation of Co-N_x unites, which is a new but highly active ORR species in g-C_3N_4-based materials. The comparable ORR activity as well as better durability and methanol tolerance ability in comparison to commercial Pt/C, indicate that Co-g-C_3N_4 at graphene would be a promising non-noble cathode catalyst in alkaline fuel cells.

g-C_3N_4 polymer was doped with cobalt species and supported on a similar sp^2 structure graphene, to form a novel nitrogen-metal macrocyclic catalyst for the ORR in alkaline fuel cells. The structural characterizations confirmed the formation of Co-N bonds and the close electron coupling between Co-g-C_3N_4 and graphene sheets. The electrocatalytic measurements demonstrated Co-g-C_3N_4-catalyzed reduction of oxygen mainly in a four-electron pathway. The improvement of ORR activity is closely related to the abundant accessible Co-N_x active sites and fast charge transfer at the interfaces of Co-g-C_3N_4/graphene. Also, Co-g-C_3N_4 at graphene exhibited comparable ORR activity, better durability, and methanol tolerance ability in comparison to Pt/C, and bodes well for a promising non-noble cathode catalyst for the application of direct methanol fuel cells. The chemical doping strategy in Liu et al.[127] would be helpful in improving other present catalysts for fuel cell applications.

First, the electrocatalytic activity of Co-g-C_3N_4 at rGO was estimated through CV curves in O_2^- or Ar-saturated electrolyte. The current response shows an intense reduction peak compared with the case in the absence of O_2, which is the characteristic ORR signal of Co-g-C_3N_4 at rGO initiated at -0.054 V vs Ag/AgCl. Then systematic polarization measurements were carried out by the rotating disk electrode (RDE) method to reveal the ORR kinetics of Co-g-C_3N_4 at rGO hybrid in O_2-saturated 0.1 M KOH (Figure 10-6(a)). The corresponding Koutecky-Levich (K-L) plots (j^{-1} vs $\omega^{-1/2}$) show good linearity at various potentials and near parallelism to the case of 20 % Pt/C (Figure 10-6(b)). This suggests similar electron transfer numbers (n) in the ORR process, and most importantly, the dominance of the efficient $4e^-$ transfer process over the range of tested potentials. This indicates high efficiency for oxygen reduction to OH^- with negligible formation of HO_2^-, thus suggesting excellent intrinsic ORR catalytic ability. It is also an indication of first-order reaction kinetics in regard to the concentration of resolved oxygen. For better understanding the electrocatalytic performance of Co-g-C_3N_4 at rGO, the chemical doping effect toward ORR was appraised. Because of the stiff texture, bulk g-C_3N_4 and Co-gC_3N_4 are difficult to disperse well in water-nafion solution and are not suitable for further electrocatalytic testing. Therefore, rGO and g-C_3N_4 at rGO electrodes were prepared in comparison to revealing the influence of Co species and graphene on ORR catalytic activity (Figure 10-6(c) and (d)). Co-g-C_3N_4 at rGO displays obviously better activity than the undoped g-C_3N_4 at rGO, considering both the current density and the onset potential. This indicates that a significant enhancement of ORR catalytic activity is achieved by the introduction of Co-N species. rGO electrode shows negligible ORR current, suggesting that the excellent activity originates from Co-g-C_3N_4, and rGO functions as the synergist and conductive support. This confirms a good chemical interaction between the catalyst and the support, which has been considered essential not only for improvement of catalyst efficiency and decrease of catalyst loss but also an inspiration for charge transfer. Herein, the decisive role of cobalt supports the assumption of Co-N species being the active sites in the supported Co-g-C_3N_4 composite; the synergistic function of graphene points to the existence of electron-hole puddles on the electron coupling interfaces of Co-g-C_3N_4 and graphene sheets.

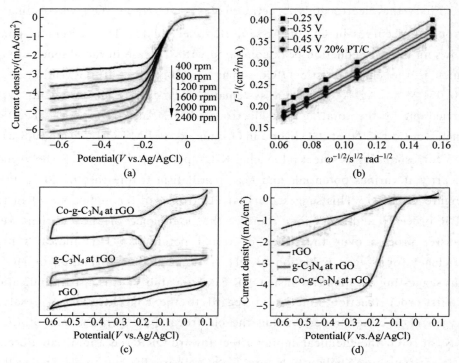

Figure 10-6　The electrocatalytic activity of Co-g-C_3N_4 at rGO
(a) RDE curves of Co-g-C_3N_4 at rGO in O_2-saturated 0.1 M KOH with a sweep rate of 5 mV/s at different rotation rates; (b) Corresponding K-L plots at different potentials for RDE curves in (a); (c) CV curves of rGO, g-C_3N_4 at rGO, and Co-g-C_3N_4 at rGO with a sweep rate of 50 mV/s; (d) RDE curves of these samples with a sweep rate of 5 mV/s at 1,600 rpm in O_2-saturated 0.1 M KOH[127]

Liu et al. developed a graphene-supported cobalt-doped heterocyclic polymer (Co-g-C_3N_4) cathode catalyst via a direct polycondensation process. In comparison with 20% Pt/C catalyst, as-synthesized Co-g-C_3N_4 at rGO delivers great ORR activity and kinetics, as well as better stability in the presence of methanol or for long testing time (10 h) in alkaline condition. Such a 2D Co-g-C_3N_4 composite would give a promising Pt-substitute for fuel cell applications after further optimizations. The apparent enhancement of ORR activity is accompanied with the formation of abundant Co-N_x moieties, which is considered as the ORR active sites on the Co-g-C_3N_4 catalyst. The electron transfer is effectively promoted by the electron-hole puddle formed on the sheet interfaces between g-C_3N_4 and graphene. Therefore, in this work, the chemical doping strategy has been demonstrated to be highly effective in improving the electron structure and thus the ORR catalytic performance, in terms of the introduction of metal ions and similarly structured carbon support. This work not only offers an initial insight into g-C_3N_4-based metal-macrocyclic catalyst on ORR performance, but also contributes to the design of highly efficient catalysts based on current materials in energy conversion devices.

Doping heteroatoms such as N into carbon-based materials significantly contributed to the better selectivity for ORR since the N atom with five valence electrons could easily shift the neutrally charged adjacent carbon atoms and create positively charged sites, favoring O_2 adsorption and reduction[128-132]. The catalytic activity of N-doped carbon is affected by the pH value of the electrolytes, four-electron pathway is more favorable in alkaline electrolytes as compared to acidic electroytes[133-134]. In general, N-doped carbon-based materials are prepared mainly by two methods: post-doping[128-129, 134-136], and in situ direct-doping[137-139]. However, these pathways have obvious disadvantages including low efficiency, residual metallic impurity, and requirement of multiple organic precursors[134]. The post-doping method usually involves a chemical pretreatment with HNO_3, NH_3, or HCN to introduce the N atom, which may lead to a poor control over the chemical homogeneity, reproducibility, and crystal structure[140]. Therefore, it is urgent to develop a green, highly efficient and environmentally benign route to synthesize N-doped carbon-based materials with high ORR catalytic activity. Notably, even with great effort toward the ORR origin, the nature of catalytic sites has been heavily debated because of the following two technological barriers: ① N-doping level and type are varied significantly depending on the synthesis conditions; and ② in most cases, an inevitable operation to remove the template or activate the samples using potassium hydroxide deteriorates the crystallinity of the samples, thwarting the attempts to get an unambiguous understanding of the catalytic nature[141-142]. Thus, the development of a synthesis route for metal-free N-doped carbon materials that can both exhibit superior activity toward ORR and reveal the nature of catalytic sites is highly desired. g-C_3N_4 was often used as the N source to synthesize N-doped materials or incorporate into mesoporous or conductive carbon as the potential candidate for ORR[143]. The unique crystal structure with no other metal impurity element makes it an ideal self-sacrificing template to synthesize N-doped graphene-like carbon materials. However, unsuccessful control over the heat treatment process of g-C_3N_4 usually leads to gaseous components ($C_2N_2^+$, $C_3N_2^+$, N_2 etc.) as the final products[144-145].

Li et. al demonstrated a one-pot direct transformation from graphitic-phase C_3N_4(g-C_3N_4) to nitrogen-doped graphene. g-C_3N_4, containing only C and N elements, acts as a self-sacrificing template to construct the framework of nitrogen-doped graphene. The relative contents of graphitic and pyridinic N can be well-tuned by the controlled annealing process. The resulting nitrogen-doped graphene materials show excellent electrocatalytic activity toward ORR, and much-enhanced durability and tolerance to methanol in contrast to the conventional Pt/C electrocatalyst in alkaline medium. It is determined that a higher content of N does not necessarily lead to enhanced electrocatalytic activity; rather, at a relatively low N content and a high ratio of graphitic-N/pyridinic-N, the nitrogen-doped graphene obtained by annealing at 900℃ (NGA900) provides the most promising activity for ORR.

The electrocatalytic activities were evaluated by CV measurements in 0.1M KOH solution saturated with O_2 or N_2 at a scanning rate of 100 mV/s. As shown in Figure 10-7(a), NGA900 shows no peak in N_2 saturated solution but an obvious reduction peak in O_2-saturated solution. Here, it is believed that enhanced-graphitization degree, tuned relative ratio of different N species, and a proper total content of N in NGA900 have mainly contributed to its superior electrocatalytic activity. To further gain insight into the ORR kinetics, the kinetic current density (J_k) was analyzed based on the RDE tests with various rotating speeds (ω) (Figure 10-7(c)). By using the Koutecky-Levich (K-L) equation, the number of electron transfers (n) per O_2 molecule involved in the ORR can be determined (inset of Figure 10-7(b)). The straight-and parallel-fitted curves suggest a first-order reaction with the dissolved oxygen on NGA900 from 0.57 V to 0.72 V vs. RHE. From the slopes of the K-L plots in Figure 10-7(c), the n per O_2 molecule in ORR was calculated to be 3.9 for NGA900 (at 0.365 V vs. RHE), indicating a one-step four-electron reduction pathway to produce water as the main product, which will benefit the construction of fuel cells with high efficiency. As shown in

Figure 10-7(c), an increase in n from 3.2 (NG730) to 3.9 (NGA900) was observed. With the increase in the annealing temperature from 800℃ to 1,000℃, the kinetic current density (J_k, at 0.365 V vs. RHE) keeps at a relatively low value except for NGA900 (Figure 10-7(e)). The highest current density about 29.3 mA/cm^2 was observed for NGA900, which is even superior to that of the commercial Pt/C (23.7 mA/cm^2) under the same testing conditions, further indicating the unique properties of NGA900. Considering the potential applications of NGA900 as effective ORR catalysts by replacing the commercial Pt/C, the electrochemical stability was also investigated. The NGA900 showed a better long-term durability than the commercial Pt/C. As shown in Figure 10-8(d), after 35,000s of reaction at 0.715 V vs. RHE, 86%

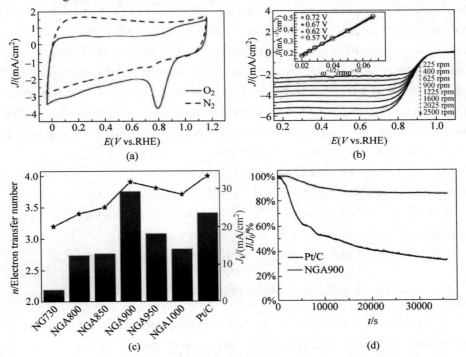

Figure 10-7 Electrocatalytic performance of the N-doped graphene samples

(a) CV curves of NGA900 in a N_2-and O_2-saturated 0.1 M KOH aqueous solution at a scan rate of 50 mV/s. (b) LSV curves of NGA900 at different rotating rates. The inset shows the corresponding Koutecky-Levich plots at different potentials. (c) Summary of the kinetic limiting current density (J_k) and electron-transfer number (n) based on the RDE data on various samples (at 0.365 V vs. RHE). (d) Electrochemical stability of the catalysts[148]

of the current density toward ORR can be maintained for NGA900, which is much higher than that of the commercial Pt/C catalyst (33 %). Owing to the involvement of metal-free direct transformation, different electrocatalytic activities of the as-prepared N-doped graphene for ORR can be attributed exclusively to the total contents of N and relative ratio of various N species. However, with respect to the nature of catalytic sites in the metal-free catalysts for ORR, inconsistencies are still present.

This work demonstrated a simple and cost-effective approach to the synthesis of N-doped graphene via a direct transformation process. The synthesis strategy offers a chance to investigate the nature of catalytic sites for ORR in a *clean* binary system, with only carbon and nitrogen elements. The electrochemical tests in combination with XPS analysis proved that the high-electrocatalytic activity could be attributed to the improved graphitization degree, high ratio of graphitic-N/pyridinic-N, and proper total content of N. This study may help to elucidate the nature of catalytic sites in metal-free N-doped carbon materials, which is very important for catalyst design and property modulation. Superior ORR performance with good durability rendered the as-prepared N-doped graphene a promising candidate to replace the commercial Pt/C for practical applications.

In addition, the mechanism and outstanding superior electrocatalytic performances of heteroatom-doped graphitic carbon catalysts have been verified by numerous studies[43, 82, 149-151]. And the same principle has also been applied to the development of various other efficient catalysts for using not only in ORRs of fuel cells but also HER in water-splitting systems[53, 59, 152-154].

10.3.2 Based on g-C_3N_4 nanostructures driven for HER

Currently, the world is facing the common issues of energy crisis and environmental pollution caused by large energy consumption. Although researchers and other workers have studied many measures to avoid and reduce energy consumption, the energy crisis is also growing. To eliminate these problems from the source, hydrogen is recognized as clear energy to deal with them, which comes from HER. However, there are still some problems in the actual application process, such as the low efficiency of visible light utilization and rapid recombination of electrons and holes. Therefore, nanocomposites based g-C_3N_4 have been developed as the new promising metal-free catalysts for hydrogen evolution reduction, due to the tunable band gap and the large surface area, importantly, the excellent performance which inhibits recombination of electrons and holes. Based on these advantages, it can greatly drive the catalytic reactions and promote energy efficiency. In the following, we would introduce them in detail.

Photocatalytic water splitting on semiconductors is a green and friendly route to directly convert solar energy into renewable and clean hydrogen energy[155]. Fully photocatalytic water splitting requires semiconductors with good stability and appropriate band gap and edges. The band gap must be larger than 1.23 eV but not too large, because too large band gap would limit the utilization of solar energy[156-158]. On the other hand, the conduction (valence) band edge must be more negative (positive) than the redox potential of H^+/H_2 (O_2/H_2O). Although many semiconductors have been explored for photocatalytic water splitting[159-164], most of them either are instable or have improper band structure[160, 165-166]. In the past three decades, photocatalytic water splitting to produce hydrogen has attracted extensive concerns. Notably, as a typical semiconductor photocatalyst, TiO_2 photocatalyst is still a major aspect since its low cost and high chemical stability. Unfortunately, due to its large band gap, TiO_2 can only respond to UV light that accounts for about 4% of solar energy, which limits its practical applications[167-168].

Due to its non-toxicity, chemical and thermal stability, and good photocatalytic hydrogen production ability, g-C_3N_4 has received increasing attention both in the fields of photocatalysis and photoelectrochemistry. However, the rapid recombination rate of photo-generated electron-hole pairs of pure g-C_3N_4 leads to its low quantum efficiency. From the point of view of the enhancement of electrical conductivity, chemical doping is an effective strategy[169]. Coupling with other semiconductors (such as $BiOX$[170], CdS[171], Al_2O_3[172], and TiO_2[173]) is an effective way for lowering the recombination rate. It is worth noting that the band edge position of TiO_2 is matched well with that of g-C_3N_4 during the process of photocatalytic water splitting to produce hydrogen. Thus, many studies have been reported on the fabrication and photocatalytic applications of TiO_2/g-C_3N_4 composites[174-176]. Wang et. al synthesized the TiO_2/g-C_3N_4 heterojunction via microwave assistance, in which the highly dispersed interface between TiO_2 and $g\text{-}C_3N_4$ facilitated the migration of photo-generated electrons. Nevertheless, most of the synthetic methods of $TiO_2/g\text{-}C_3N_4$ composites need organic solvents, titanium alkoxides and high temperature. Moreover, the physical properties of $g\text{-}C_3N_4$ in the composites are unchanged for the most part. This in-situ hydrothermal synthetic route results in coordination between Ti species and $g\text{-}C_3N_4$, which improves the electronic conductivity of the composites, and the hybridization of TiO_2 and $g\text{-}C_3N_4$ facilitates the separation of charge carriers. The photocatalytic performances are systematically evaluated by photocatalytic hydrogen production under both visible-light and simulated sunlight. According to the results of the experiments, the photocatalytic mechanism of Ti species modified $g\text{-}C_3N_4$ is proposed. This work could provide deeper insight into the enhanced mechanisms of π-conjugated molecules hybridized semiconductors.

A Ti specie modified $g\text{-}C_3N_4$ photocatalysts was synthesized via an in-situ hydrothermal route and the following low temperature calcination. The hydrothermal process results in not only the fabrication of $TiO_2/g\text{-}C_3N_4$ heterojunctions, but also the coordination between Ti species and $g\text{-}C_3N_4$. The PL and photocurrent measurements exhibit that the hybridization enhances the separation efficiency of photo-induced electrons and holes. The Ti species modified $g\text{-}C_3N_4$ photocatalysts exhibit much higher photocatalytic H_2 evolution than the simplex heterojunction of $TiO_2/g\text{-}C_3N_4$ obtained via microwave method and the mechanical mixture of TiO_2 and $g\text{-}C_3N_4$ under visible-light irradiation[177-180].

In electrochemical spectra, the high-frequency arc corresponds to the charge transfer limiting process and can be attributed to the double-layer capacitance in parallel with the charge transfer resistance in the sample tablet. As seen from Figure 10-8(a), the arc radius from the electrochemical impedance spectroscopy (EIS) Nyquist plots of 0TC-300 is slightly smaller than bare $g\text{-}C_3N_4$, suggesting that the process of hydro-thermal treatment in the $H_2O_2\text{-}NH_4OH$ solvent can improve the conductivity of $g\text{-}C_3N_4$ inconspicuously. Besides, the introduction of TiO_2 can also enhance the electric conductivity. By comparison, 20TC-300 possesses the highest electric conductivity. Thus, it is speculated that the highly electric conductivity of 20TC-300 not only comes from the introduction of TiO_2, but also from the coordination between Ti species with the lone-pair electrons on N atom in g-C_3N_4. In a word, the micro-structural change of g-C_3N_4 via coordination with Ti species during the hydrothermal treatment improves the electric conductivity of g-C_3N_4 and makes charge transfer easier[181].

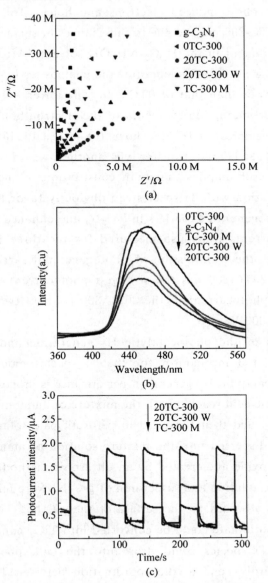

Figure 10-8　Photocatalysts characterizations
(a) EIS Nyquist plots of the photocatalysts; (b) Photoluminescence (PL) spectra of the photocatalysts; (c) Photocurrent responses of the photocatalyst under visible-light irradiation[180]

To further investigate the charge carrier transfer behavior, the photoluminescence measurement was employed (Figure 10-8(b)). It is interesting that the main emission wavelength of bare g-C_3N_4 at 460nm is shifted to 450 nm after the hydrothermal treatment. The PL signal is often considered as the radioactive recombination of photo-induced electrons and holes. The shift of emission wavelength for 0TC-300 may be due to the changes of surface states during the hydrothermal treatment in the H_2O_2-NH_4OH solvent. After the introduction of the Ti specie, the main wavelength of the composites is still at 450 nm and the intensity is lower than that of 0TC-300. The lower PL emission intensity indicates lower recombination efficiency of the electron-hole pair[182]. Among the three hybrid samples, 20TC-300 shows the lowest PL intensity, indicating more excitons of 20TC-300 transfer via another way instead of radiative paths[183]. This result implies that both constitutions of heterojunction with TiO_2 and coordination with Ti species can obviously favor the separation and transition of photogenerated carriers in g-C_3N_4 and enhance the photocatalytic activity. Photocurrents were also measured for the three hybrid samples to further investigate the transmission of photogenerated carriers. As shown in Figure 10-8(c), 20TC-300 has the strongest photocurrent under visible-light irradiation. The photocurrent of the 20TC-300 is about two times as high as that of the 20TC-300W.

On the basis of the photocatalytic H_2 generation and test results, the proposed process and mechanism for the Ti species modification and the promotion photocatalytic H_2 generation performance is proposed in Figure 10-9. Firstly, metatitanic acid reacted with the mixture of ammonium hydroxide and hydrogen peroxide, and then converted to peroxotitanate solution. Meanwhile, bare g-C_3N_4 was dispersed into the mixture solution of ammonium hydroxide and hydrogen peroxide and treated by an ultrasonic method. Secondly, when the peroxotitanate solution and suspension of g-C_3N_4 were mixed together, the peroxotitanate was adsorbed on the surface of the g-C_3N_4. Under hydrothermal conditions, peroxotitanate could be converted into TiO_2 nanoparticles, and in the meanwhile, Ti species could drop into the bulk phase of g-C_3N_4 by coordination. Thirdly, due to the coordination between Ti species with g-C_3N_4, the photo-induced electrons in g-C_3N_4 are becoming easier to transfer to the heterointerface of the composite. Then, because the lowest unoccupied molecular orbital (LUMO) level of g-C_3N_4 (-1.20 eV) is more negative than the conduction band (CB) edge of TiO_2 (-0.50 eV), the excited electrons on LUMO of g-C_3N_4 can inject into the CB of TiO_2. Finally, H^+ is reduced to hydrogen molecule after accepting the electrons on the Pt particles or CB of TiO_2.

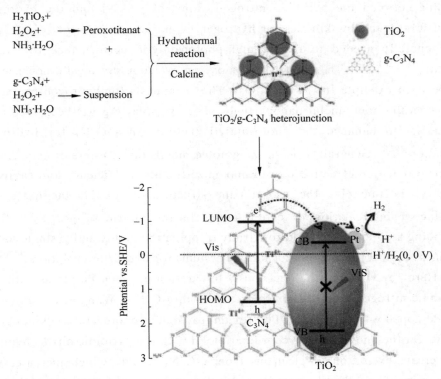

Figure 10-9　The formation and photocatalytic mechanism scheme of the Ti species modified g-C_3N_4[180]

A Ti specie modified g-C_3N_4 photocatalyst was synthesized via an in-situ hydrothermal route and calcinations. This synthetic method not only results in the fabrication of TiO_2/g-C_3N_4 heterojunctions but also leads to a microstructure modification of g-C_3N_4 by Ti species. The coordination between Ti species and g-C_3N_4 improves the electric conductivity of the composite and makes charge transfer easier. Meanwhile, the heterojunction between TiO_2 and g-C_3N_4 can also improve the separation efficiency of photo-generated carriers in the process of photocatalysis. This work could provide deeper insight into the enhanced mechanisms of π-conjugated molecules hybridized semiconductors.

The kinetics of g-C_3N_4 for photocatalytic water splitting is very slow because of the fast electron-hole recombination rate and slow hydrogen and oxygen evolution rate. The catalytic active site for hydrogen evolution on g-C_3N_4 is believed to be the pyridinic N atoms[184-185]. The relatively strong binding between the pyridinic nitrogen atoms of g-C_3N_4 and the hydrogen atoms results in slow kinetics for hydrogen evolution. As a result, cocatalysts are generally needed to realize high-photocatalytic performance for water splitting[96, 186-191]. The cocatalysts for oxygen evolution are usually noble metal oxides, for example IrO_x and RhO_x. The cocatalysts for hydrogen evolution are generally noble metal nanoparticles. For example, Ag has been used as a cocatalyst to enhance the photocatalytic activity of g-C_3N_4 for hydrogen evolution[192]. Generally, downsizing noble metals into clusters or even single atoms provide an effective way to maximize the atom efficiency and catalytic activity. Fortunately, the tri-s-triazine structure of g-C_3N_4 facilitates the binding or intercalation of exotic atoms in the matric of g-C_3N_4[193-195], providing a potential scaffold for firmly trapping the highly active single-metal atoms as high-efficiency photocatalytic systems for hydrogen evolution[196-198].

Here, by virtue of the tenaciously interaction between Pd atoms with the pyridinic nitrogen atoms of six-fold cavities in g-C_3N_4, Wang et al. designed the Pd-doped g-C_3N_4 (g-C_3N_4-Pd) for improving the photocatalytic kinetics of HER. On the one hand, the coordination Pd with the pyridinic nitrogen atoms may create excitation of electrons from g-C_3N_4 to Pd, which increases the separation of electron-hole pairs. On the other hand, the doped Pd can accelerate the hydrogen evolution kinetics. They found that the Pd doping does not induce obvious change of the main structure of g-C_3N_4, while enhances the photocatalytic performance for hydrogen evolution by 15 times. Further experimental and theoretical studies confirm their speculation that the Pd doping alters the excitation manner of electrons and accelerates the kinetics of hydrogen evolution, which enhances the photocatalytic activity for hydrogen evolution.

Figure 10-10(a) shows the absorbance of the g-C_3N_4 samples with different Pd doping contents. Clearly, the Pd doping extends the absorption of g-C_3N_4 toward longer wavelength. The increase of absorption at longer wavelength is directly reflected by the color of the samples. As Pd doping content is increased, the color of the samples changes from yellow to light brown and then to dark brown. The absorption at longer wavelength indicates that the Pd doping forms midgap energy levels. Under light excitation, electrons transfer from g-C_3N_4 to the doped-Pd ions, which is similar to the ligand-to-metal transfer in complex compound. The extended and enhanced absorption of g-C_3N_4 in the longer wavelength is highly desired for photocatalytic water splitting. Figure 10-10(b) gives the amount of hydrogen produced by the g-C_3N_4 doped with different Pd contents within three hours. The Pd-doping clearly improved the photocatalytic hydrogen evolution performance of g-C_3N_4. The performance of hydrogen evolution first increases with the increase of Pd-doping amount and then decreases. During three cycles, the catalyst does not show any degradation in photocatalytic performance (Figure 10-10(c)), indicating that the Pd-doped g-C_3N_4 has very good stability during photocatalytic hydrogen evolution. In contrast, the photocatalytic activity increases in the three cycles. The increase of photocatalytic activity can be attributed to the fact that the dispersion of the catalyst is improved as the photocatalytic reaction takes place. The photoelectrochemical performance of the Pd-doped g-C_3N_4 was also investigated (Figure 10-10(d)). When light is turned on, the photoelectrochemical currents of all samples have very sharp edges and then gradually decrease to steady values. This phenomenon may arise from the fast photo-oxidation of defects or edges of g-C_3N_4. Unlike hydrogen evolution, the photocurrent is reduced with the increase of Pd-doping amount. The different performance of Pd-doped g-C_3N_4 in hydrogen evolution and photocurrent can be understood by the reaction difference between hydrogen evolution and photocurrent.

In HER, the reduction and oxidation reactions simultaneously occur on the Pd-doped g-C_3N_4. The Pd-doping accelerates the electron-hole separation and the kinetics of hydrogen evolution, leading to the improvement of the photocatalytic performance. In contrast, in the photocurrent measurements, g-C_3N_4 acts as photoanode because it is a N-type semiconductor. Only oxidation reaction takes place on the electrode. The advantage of Pd-doped g-C_3N_4, improving kinetics of hydrogen evolution, cannot be reflected. Contrarily, Pd ions act as trap centers for electrons in the photoelectrode and therefore inhibit the transfer of electrons to F-doped SnO_2-coated (FTO) glass and increase the electron-hole recombination rate. As a result, the Pd-doping

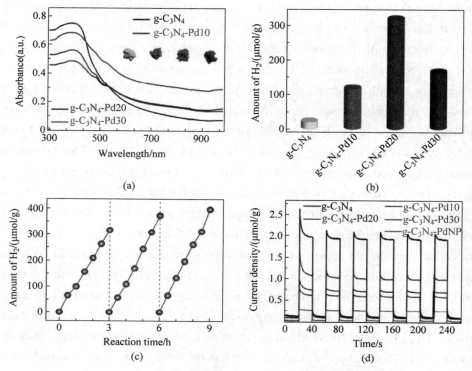

Figure 10-10 Photocatalytic performance by Pd-doped g-C_3N_4

(a) Absorption spectra of g-C_3N_4 and differently Pd-doped g-C_3N_4 samples. Inset is the photographs of the g-C_3N_4, g-C_3N_4-Pd10, g-C_3N_4-Pd20, and g-C_3N_4-Pd30 samples from left to right; (b) Amount of hydrogen produced within three hours for the four samples; (c) Cycling photocatalytic H_2 production activity of g-C_3N_4-Pd20; (d) Photoelectrochemical current of g-C_3N_4, g-C_3N_4-Pd10, g-C_3N_4-Pd20, g-C_3N_4-Pd30, and g-C_3N_4-PdNP samples[199]

reduces the photoelectrochemical current of g-C_3N_4. Because of the same reason, the g-C_3N_4-PdNP also exhibits smaller photoelectrochemical current than g-C_3N_4.

At the same time the DFT calculations on g-C_3N_4 with different Pd-doping contents are carried out. The doping sites of Pd were first studied. It is found that only the Pd doped at the cavity site is stable. When Pd is set at other sites, Pd will move to the cavity site after the geometry optimization.

The hydrogen atom adsorption on g-C_3N_4 and Pd-doped g-C_3N_4 was further studied to unravel the effect of Pd doping on the hydrogen evolution kinetics. On g-C_3N_4, hydrogen atom can adsorb stably at the nitrogen atoms of the six-fold cavities with very large adsorption energy of 4.00 eV (Figure 10-11 (a)). Hydrogen atoms can adsorb stably on Pd atoms of Pd-doped g-C_3N_4 (Figure 10-11 (b)). Compared with g-C_3N_4, the adsorption energy of hydrogen on Pd atoms is greatly reduced to 2.11 eV. As shown in Figure 10-11(c), the Gibbs free-energy of hydrogen adsorbed on Pd is greatly reduced to -1.82 eV compared with hydrogen adsorbed on g-C_3N_4 (-3.71 eV)[120-122]. The huge reduction of Gibbs free-energy indicates that the Pd doping improves the hydrogen evolution kinetics on g-C_3N_4. On the basis of above studies, we believe that the improvement of hydrogen evolution performance by Pd doping arises from the alternation of electron excitation manner and the improvement of hydrogen evolution kinetics. The Pd doping results in electrons excited directly from g-C_3N_4 to the doped Pd atoms which are just the active sites for hydrogen evolution (Figure 10-11(d)). This mechanism is different from the improvement mechanism of metal nanoparticles loaded on g-C_3N_4. In g-C_3N_4 loaded with metal nanoparticles, electrons are first excited into conduction band of g-C_3N_4 and then tranferred to metal nanoparticles. The transfer of excited electrons to metal nanoparticles needs to diffuse a certain distance, which would increase the electron-hole recombination.

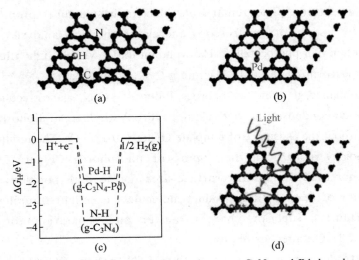

Figure 10-11 The hydrogen atom adsorption on g-C_3N_4 and Pd-doped g-C_3N_4
(a) Adsorption configuration of H on g-C_3N_4; (b) Adsorption configuration of H on Pd-doped g-C_3N_4; (c) Free-energy diagram of H adsorbed on g-C_3N_4 and differently Pd-doped g-C_3N_4 samples; (d) Mechanism of photocatalytic hydrogen evolution in Pd-doped g-C_3N_4 [199]

The Pd doping extends the absorption of g-C_3N_4 toward longer wavelength through forming excitation from g-C_3N_4 to Pd atoms. The performance of photocatalytic hydrogen evolution of g-C_3N_4 is improved by 15 times by the Pd doping. The improvement of photocatalytic hydrogen evolution is found to arise from the alternation of electron excitation manner and the improvement of hydrogen evolution kinetics.

Polymeric nature of g-C_3N_4 makes the ample choice of chemical protocols possible to introduce foreign atoms into the graphitized backbone so as to alter the optical and electronic properties[203-204]. For instance, doping g-C_3N_4 with heteroatoms, such as B, S, P, and I[104, 110, 205, 206] is expected to present a homogeneous functionalization through the bulk matrixes and extended electronic possibilities. Nevertheless, it remains a challenge to simultaneously achieve the nanostructuring and doping of g-C_3N_4 through a facile and cost-effective strategy.

To exhibit the above considerations, flower-like P-doped g-C_3N_4 (P-CN) nanostructure is synthesized by a template-free phosphonic-mediated route, where in P can affect the polycondensation process and conformation of the resulting g-C_3N_4, thereby offering a direction role to modify the texture and electronic structures. The remarkable electronic structural properties were evidenced by a series of spectroscopy and electrochemical analysis, suggesting the great potency in photocatalytic H_2 evolution under visible-light illumination, largely outperforming the bulk pristine g-C_3N_4 counterpart.

They obtained in-plane mesoporous P-doped g-C_3N_4 nanostructured flowers following a co-condensation route using (hydroxyethylidene) diphosphonic acid (HEDP) without the assistance of template (Figure 10-12(a)). The resulting flower-like morphology along with the doping of phosphorus (Figure 10-12(b)) markedly increased the specific surface area due to high porosity, improved mass transfer of reactant and product molecules, excellent trapping of light; and importantly, superior charge transfer and separation for excellent reduction of H_2O under visible light.

The introduction of P atoms into the g-C_3N_4 polymeric skeleton alters the structural and electronic aspects and thus the resultant optical/electronic properties. This influence can be reflected by the color change from pale yellow for undoped g-C_3N_4 to dark brown after P doping (Figure 10-13(a)

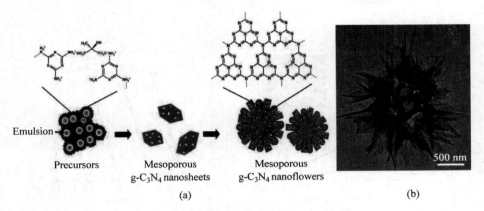

Figure 10-12　Mesoporous P-doped g-C_3N_4

(a) Formation mechanism of mesoporous P-doped g-C_3N_4 with a flower-like morphology; (b) TEM image of P-doped g-C_3N_4[210].

inset), which can further be quantified by the optical absorption spectra (Figure 10-13(a)). Pure g-C_3N_4 displays typical semiconductor absorption of g-C_3N_4 in the blue region, together with a band gap of about 2.7 eV. In contrast, it is found that the optical absorption band edge for P-CN is shifted to lower band gap energies, implying a band gap narrowing and the electronic integration of P atoms in the lattice of g-C_3N_4 as well. This modification accompanied by the unique nanostructures, in principle, can improve the capability to harvest visible light and favor the mass transfer and charge transport, creating a platform in efficient heterogeneous photocatalysis.

　　The synthesized g-C_3N_4 semiconductors are therefore evaluated in an assay of H_2 production from water under visible-light irradiation using triethanolamine as sacrificial agent. As shown in Figure 10-13(b), a remarkably enhanced H_2 evolution rate of 104.1 μmol/h is achieved on P-CN, tremendously exceeding that of pure g-C_3N_4 reference (11.2 μmol/h). The corresponding activity virtually outperforms a number of inorganic semiconductor photocatalysts. In order to present the superiority of P-CN, two additional g-C_3N_4 samples, mesoporous g-C_3N_4 and g-C_3N_4 nanosheets, were investigated under the identical conditions (Figure 10-13(b)). Although these two g-C_3N_4 samples are of higher specific surface areas (216 m^2/g for mesoporous g-C_3N_4 and 273 m^2/g for g-C_3N_4 nanosheet), the catalytic activities are inferior compared with P-CN, which indicates that the synergic effect of nanostructures and P doping is

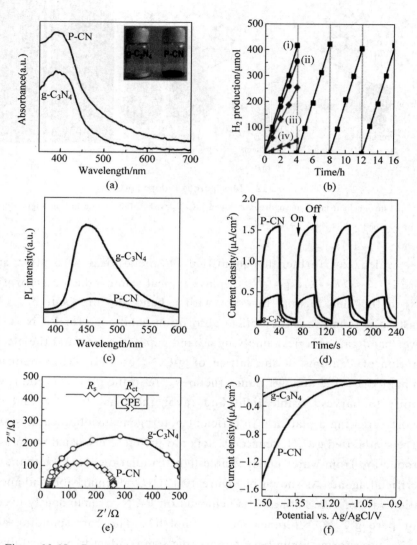

Figure 10-13　P atoms into the g-C_3N_4 polymeric skeleton for photocatalysis

(a) UV-vis diffuse reflectance spectra of g-C_3N_4 and P-CN and (inset) photographs of (left) g-C_3N_4 and (right) P-CN; (b) Photocatalytic water reduction with the use of the synthesized g-C_3N_4 materials under visible-light illumination: (i) P-CN, (ii) g-C_3N_4 nanosheet, (iii) mesoporous g-C_3N_4, and (iv) g-C_3N_4; (c) Photoluminescence spectra; (d) photocurrent response; (e) EIS Nyquist plots; (f) Polarization curves of P-CN and g-C_3N_4[210]

able to attain increased surface reactivity for the photocatalytic reaction. Durability is another crucial criterion for practical application. The effective generation and instant separation of photoexicited charge carriers are the prerequisite for photocatalytic reactions, which can be analyzed by PL emission spectroscopy (Figure 10-13(c)). Relative to g-C_3N_4, the broad PL band at approximately 450 nm quenches quickly for P-CN, indicating that the recombination of electron-hole pairs occurring in the polymeric substrates is largely impeded. This is ascribed to the promoted electron relocalization on surface terminal sites after doping and structural advantages of P-CN, as the interconnected nanosheets can greatly curtail the diffusion distance of charge migration[211-212]. The low recombination rate is proposed to contribute to increase photoactivities and quantum yield. The transient photocurrent measurement was carried out to qualitatively investigate the separation efficiency of photoinduced charges during the photoreactions. It can be obviously seen that rapid and stable photocurrent responses are detected in both electrodes, and the photoresponsive phenomenon is entirely reversible upon each light excitation (Figure 10-13(d)). benefiting from more efficient light harvesting, the photocurrent for P-CN is much higher than that of pure bulky g-C_3N_4, revealing that the mobility of the charge carriers is efficaciously elevated EIS is a valid electrochemical approach to illustrate the electron-transfer efficiency at the electrodes[213]. Figure 10-13(e) discloses similar semicircular Nyquist plots for both g-C_3N_4 and P-CN. However, P-CN features significantly smaller diameter, corresponding to a smaller contact and charge-transfer impedance[214]. The larger semicircle radius for g-C_3N_4 can be due to the poorer electrical conductivity impeding the electron migration from carbon nitride to the back-contact electrode. In addition, the polarization curve of P-CN in the potential region of -1.5 to -0.9 V vs. Ag/AgCl attributed to H_2 evolution was measured with that of g-C_3N_4 as a control. As for P-CN, the much more pronounced cathodic current density and low over potential indicate the boosted catalytic activity and reaction kinetics over g-C_3N_4 (Figure 10-13(f)), which is in accordance with the apparently improved photocatalytic hydrogen production rate detected on P-CN. Both the PL spectra and electrochemical tests present the identical trend as the H_2 evolution performance, reflecting that generation, separation, and transfer efficiency of electrons in the g-C_3N_4

matrixes play a decisive role to influencing the photocatalytic activity. Here, the P species in the carbon nitride framework can chemically bond with the C and N neighbors and force planar coordination, and the lone electron pair can delocalize to the π-conjugated tri-s-triazine of P-doped g-C_3N_4 that can serve as reinforcing active sites to some extent, accordingly enhancing the conductivity and electron transfer capability. The thin nanosheets constructed P-CN are quite propitious for photochemical reactions, because the reduced thickness can shorten charge transfer length from the bulk to the interface[68, 215-216], where the photoredox reaction takes place. Furthermore, the well-developed in-plane mesopores can also facilitate the charge and mass transport in the catalytic process.

A template-free methodology is presented to prepare mesoporous P-doped g-C_3N_4 nanostructured flowers through the condensation and the rmolysis of melamine and organophosphosnic acid. The low P concentration well retains the g-C_3N_4 polymeric framework, but the electronic properties have been seriously changed, causing not only red-shifted light absorption but also improved electronic conductivity. Additionally, the textual and structural features of high-surface area associated with well-structured mesoporosity and nanosheet construction units ensure the excellent photocatalytic water reduction performance under visible-light irradiation. These findings provide a promising way to improve the photocatalytic performance for hydrogen evolution and pave a new avenue for the development of highly efficient and cost-effective photocatalysts for water splitting.

10.3.3　g-C_3N_4 measurement of the gas sensing properties

As is well known, the global environment is facing a deteriorating crisis, due to the massive leakage of toxic and harmful gases from chemical industries. Meanwhile, there are many organic poisonous gases inevitably wandering in our lives, such as methylbenzene and formaldehyde, attributed to the new painted furniture and newly renovated houses. Those poisonous gases not only pollute seriously the environment but also act as a killer that threatens human health[217-218]. Therefore, we need to find urgently out a highly effective and extremely sensitive sensor to detect and degrade the evaporated substances.

In the past few decades, the metal-oxide semiconductor (MOS) material-based gas sensors have received sizeable concern and widespread availability, such as SnO_2, CuO, ZnO and MnO_2[219-224], putting down to their unique sensing performances that include the low cost, fast response and recovery time. Especially, SnO_2 is extensively applied to detect diverse gases[225-227]. It has unique chemical properties and crystal structure, a typical N-type MOS with a rutile crystalline structure. Nonetheless, there are still defects which confine its applications in gas sensors, such as high resistance, high temperature of working and easy agglomeration. To overcome these defects, researchers have tried to fabricate the coupling SnO_2 with other metal semiconductors to create heterostructures, including SnO_2/ZnO, SnO_2/Fe_2O_3 and SnO_2/NiO[225, 228-229], etc. Although most of the drawbacks have been eliminated, their catalytic activity and stable cycling are not satisfactory. Until $SnO_2/g-C_3N_4$ nanocomposites, designing $g-C_3N_4$ as the support intermediate to SnO_2 nanoparticles, exhibits a higher response value, outstanding stable property and other remarkable gas-sensing characteristics.

In order to understand clearly $SnO_2/g-C_3N_4$ nanostructures how to influence the gas-sensing properties of the sensor, we must primarily investigate its gas-sensing mechanism. In fact, there are several different types of gas-sensing mechanisms. And as mentioned above, SnO_2, a typical N-type metal-oxide semiconductor, exists abundant electronics in energy band which can more obviously embody gas-sensing mechanisms. Therefore, the $SnO_2/g-C_3N_4$ compounds usually are considered as an example to further introduce specific gas-sensing mechanism. When the sensor is exposed to air, oxygen molecules would adsorb numerous electrons on the surface of SnO_2, as well as capture lots of electrons from the conduction band of SnO_2. And then, the oxygen molecules are ionized into O_2^-, the O^- and O_2^- (Equation (10-4)), would promptly generate the depletion layer which brings about an increase of the electrical resistance in the composite sensor (Figure 10-14 (a))[230-231]. Differently, when the sensor is exposed to ethanol gas, there are some chemical reactions between ethanol molecules and oxygen ions under high temperature. Consequently, acetaldehyde is oxidized by ethanol molecules (Equation (10-5)) and then creates carbon dioxide and water by oxidation reaction (Equation (10-6))[228]. Finally, the captured electrons were released back to the depletion layer of the sensing film, leading to an obvious decrease of electrical resistance in the composite-based sensor

(Figure 10-14 (b)). According to the analysis above, $SnO_2/g\text{-}C_3N_4$ would occur prodigious change attributing to the large amount of electrons transfer[232].

$$O_2 + e^- \longrightarrow O_2^-, \quad (10\text{-}4)$$

$$2CH_3CH_2OH + O_2^- \longrightarrow 2CH_3CHO + 2H_2O + e^-, \quad (10\text{-}5)$$

$$2CH_3CHO + 5O_2^- \longrightarrow 4CO_2 + 4H_2O + 5e^-. \quad (10\text{-}6)$$

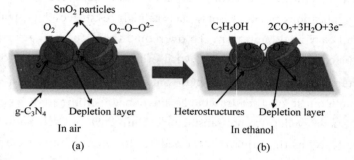

Figure 10-14　The gas-sensing mechanism of $SnO_2/g\text{-}C_3N_4$ nanocomposite[230]

Moreover, it is worth mentioning that the $SnO_2/g\text{-}C_3N_4$ nanocomposites are more beneficial to adsorb and diffuse ethanol molecules compared with other coupling systems, due to the unique material structure which can prevent the aggregation of SnO_2 nanoparticles and make the reaction between ethanol and cations more intense. At the same time, SnO_2 and $g\text{-}C_3N_4$ are N-type semiconductors and their band gaps are 3.71 eV and 2.7 eV, respectively. Based on these facts, $SnO_2/g\text{-}C_3N_4$ nanocompounds would form heterojunction structure which prompts more electrons to inflow into the conduction band of SnO_2 from the conduction band of $g\text{-}C_3N_4$[233]. That is to say, a higher potential barrier would be produced, thus conducive to the separation of electrons and holes which pressurize electrons transfer rapidly from the ethanol vapor to the surface of $SnO_2/g\text{-}C_3N_4$[234]. As a result, these phenomena would come into being a higher gas-sensing response. Besides, the repeatability, stability and the high selectivity under different gases of $SnO_2/g\text{-}C_3N_4$ hybrids are revealed by Figure 10-15 and Figure 10-16.

Here, $SnO_2/g\text{-}C_3N_4$ nanosheet synthesized by a facile calcination method under the condition of $SnCl_4 \cdot 5H_2O$ and urea as the precursor, makes the SnO_2 nanoparticles disperse highly on the surface of the $g\text{-}C_3N_4$ nanosheet. And the dispersibility contributes to avoiding the excessive agglomeration of

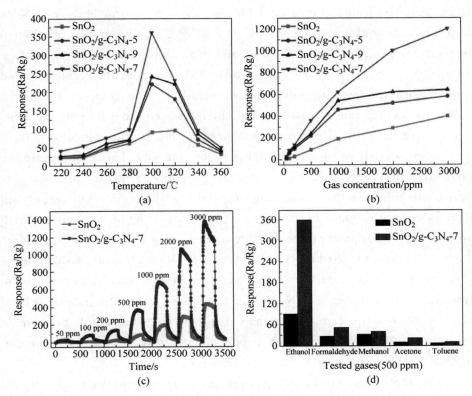

Figure 10-15 Response values of the sensors based on SnO_2, $SnO_2/g\text{-}C_3N_4\text{-}5$, $SnO_2/g\text{-}C_3N_4\text{-}7$, and $SnO_2/g\text{-}C_3N_4\text{-}9$

(a) In 500 ppm ethanol with the increase of the operating temperature and (b) in different concentrations of ethanol at 300°C; (c) Real-time response curves of the pure SnO_2 and $SnO_2/g\text{-}C_3N_4\text{-}7$ to ethanol in the range of 50~3,000 ppm; (d) Responses of SnO_2 and $SnO_2/g\text{-}C_3N_4\text{-}7$-based sensors to 500 ppm different reducing gases at 300°C [230]

SnO_2 particles and greatly expands surface area which helps to demonstrate better gas sensitivity. As we can observe that $SnO_2/g\text{-}C_3N_4$ sensor has a much higher response values in the ethanol compared with pure SnO_2 (Figure 10-14). Specifically, Figure 10-15 (a) exhibits that the response values of pure SnO_2 and $SnO_2/g\text{-}C_3N_4$ sensors at different working temperature in 500 ppm ethanol. We can visibly see the response values of $SnO_2/g\text{-}C_3N_4$ sensors distinctly and integrally greater than SnO_2 in the same conditions, which shows the maximum response especially at 300°C. And among the $SnO_2/g\text{-}C_3N_4$ sensors, the $SnO_2/g\text{-}C_3N_4\text{-}7$ that the mass percentage of $g\text{-}C_3N_4$ in the composites is 7%, has the highest response value compared with $SnO_2/g-C_3N_4$-5 and $SnO_2/g-C_3N_4$-9. Meanwhile, the gas-sensing superiority of $SnO_2/g-C_3N_4$-7 exhibited at different temperatures is uniformly revealed at different concentrations of ethanol gas (Figure 10-15 (b)). As shown in Figure 10-15(b), under the response values increase with the growing ethanol concentrations, the slope of the curves quickly rises at the range of concentration from 50 ppm to 500 ppm. But, they slowly increase with an increasing concentration in the range of 500 ~ 3,000 ppm. Based on these phenomena, on the one hand, we definitely understand that $SnO_2/g-C_3N_4$ nanosheet generates a great influence and enormously enhances the gas-sensing properties at different temperatures and different concentrations of ethanol gas. On the other hand, we can distinctly conclude the importance of the suitable amount of $g-C_3N_4$ in the compound, which not only can attribute to the dispersity of SnO_2 nanoparticles on $g-C_3N_4$, but also accelerate the formation of heterojunctional structure on the interface region between $g-C_3N_4$ and SnO_2. Equally, these also indicate that the $SnO_2/g-C_3N_4$-7 sensor can achieve the balance between the speeds of chemical adsorption and desorption as much as possible at a lower temperature to create a higher response than other measured sensors[235].

To further understand other gas sensing properties of $SnO_2/g-C_3N_4$-18, the recovery response time and selectivity were surveyed. Figure 10-15 (c) reveals the real-time response recovery curves of the pure SnO_2 and $SnO_2/g-C_3N_4$-7 to 500 ppm ethanol at 300℃ in the range of 50~3,000 ppm. There is a significant increasing trend of the response values with the growing concentrations and the response value of the $SnO_2/g-C_3N_4$-7-based sensor is much higher than that of the pure SnO_2-based sensor in the uniform conditions. Namely, the composites can lead to much greater enhancement on the gas-sensing properties. Besides, the pure SnO_2 and $SnO_2/g-C_3N_4$-7 sensors have been measured at the same concentration and in different gases including methanol, ethanol, toluene, formaldehyde and acetone. It is no doubt that the $SnO_2/g-C_3N_4$-7 sensor shows a higher response than pure SnO_2 sensor at different gases and the ethanol is assigned to the best gas, as displayed in Figure 10-15 (d). This result is caused by the reason that the ethanol is more likely to lose electrons in the process of the redox reaction with the absorbed oxygen and hydroxyl group, as well as much easier to be oxidized at the suitable working temperature[233].

This also indicates that the $SnO_2/g\text{-}C_3N_4\text{-}7$ nanohybrid has a higher selectivity for different gases, and more importantly, exhibits higher tolerance in methanol.

Furthermore, the repeatability and stability are revealed in Figure 10-16, which are the indispensable features of any normal material. Figure 10-16 (a) show the repeatability of the $SnO_2/g\text{-}C_3N_4\text{-}7$ sensor to 500 ppm ethanol at 300℃. The response values and recovery response cycles at every stage are brought into correspondence with each other, indicating the carbon nitride nanocomposite sensors have an admirable repeatability for ethanol gas sensing. Meanwhile, the measured results once every five days are displayed Figure 10-16 (b) within 30 consecutive days, and are maintained at same value around 155. This is also an intuitive suggestion that the $SnO_2/g\text{-}C_3N_4\text{-}7$-based sensor has an unexceptionable stability for ethanol gas sensing[236].

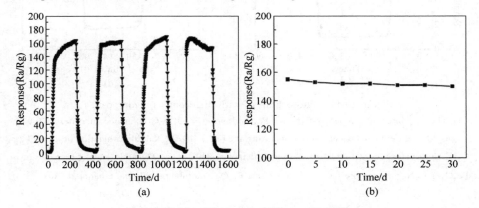

Figure 10-16　The repeatability and stability
(a) Repeatability and (b) stability measurements of the $SnO_2/g\text{-}C_3N_4\text{-}7$-based sensor to 500 ppm ethanol at 340℃ [230]

Additionally, $\alpha\text{-}Fe_2O_3$, an n-type semiconductor with a band gap of 2.2 eV, is also paid more attention to gas sensing applications due to its low cost, low toxicity and high chemical stability[237]. But pure $\alpha\text{-}Fe_2O_3$ is far from satisfaction due to its low sensitivity and slow responses. Therefore, $\alpha\text{-}Fe_2O_3/g\text{-}C_3N_4$ nanocomposites have been widely considered, which exhibit much higher sensitivity, faster response and better stability than those of commonly being consisted of pure semiconductor materials (Figure 10-17).

Here, $\alpha\text{-}Fe_2O_3/g\text{-}C_3N_4$ nanocomposites are synthesized via porous $\alpha\text{-}Fe_2O_3$ nanotubes wrapped by lamellar $g\text{-}C_3N_4$ nanostructures, which have the

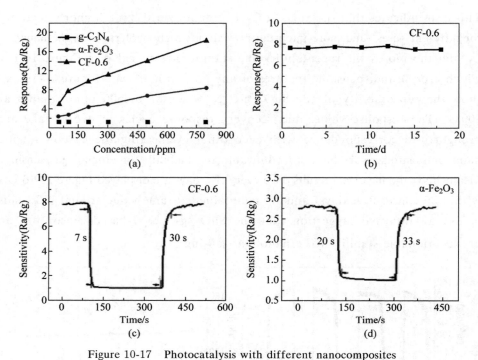

Figure 10-17 Photocatalysis with different nanocomposites

(a) The responses of the pure g-C_3N_4, α-Fe_2O_3, and α-Fe_2O_3/g-C_3N_4 nanocomposites (CF-0.6) versus different ethanol concentrations operating at 340℃; (b) the stability evaluation of α-Fe_2O_3/g-C_3N_4 nanocomposites (CF-0.6) to 100 ppm ethanol at 340℃; (c) Dynamic sensing response of α-Fe_2O_3/g-C_3N_4 nanocomposites (CF-0.6); (d) α-Fe_2O_3 nanotube to 100 ppm ethanol at 340℃ [236]

large surface areas and can form the heterojunction structures, providing efficient diffusion paths and adsorption sites for gas molecules[233]. The curves of Figure 10-17(a) obviously show the responses of the pure g-C_3N_4, α-Fe_2O_3, and α-Fe_2O_3/g-C_3N_4 nanocomposites (CF-0.6) versus different ethanol concentrations operating at 340℃. It was found that the gas sensor based on α-Fe_2O_3/g-C_3N_4 heterostructural nanocompounds show a higher response than pure α-Fe_2O_3 and g-C_3N_4 in ethanol at 340℃. Meanwhile, Figure 10-17 (b) exhibits the extreme stability evaluation of α-Fe_2O_3/g-C_3N_4 (CF-0.6) in 100 ppm ethanol at 340℃. (Note that, the CF-0.6 means that the mass fraction of α-Fe_2O_3 is 0.6 in composites.) Besides, there are other factors required to be considered in practical application, such as fast response time and recovery time. Figures 10-17(c) and (d) describe the dynamic sensing response of α-Fe_2O_3/g-C_3N_4

nanocomposites (CF-0.6) and α-Fe$_2$O$_3$ in 100 ppm ethanol at 340℃, respectively. According to the comparison of Figures 10-17 (c) and (d), the response time and recovery time of the α-Fe$_2$O$_3$/g-C$_3$N$_4$ are 7 s and 30 s, however, those of α-Fe$_2$O$_3$ are 7 s and 30 s, respectively. Obviously, the response time and recovery time of the α-Fe$_2$O$_3$/g-C$_3$N$_4$ are faster and shorter than those of α-Fe$_2$O$_3$, especially for response time.

From the discussion above, we can obviously conclude that SnO$_2$/g-C$_3$N$_4$-7 and α-Fe$_2$O$_3$/g-C$_3$N$_4$ exhibit powerful enhancement of gas-sensing sensitivity and outstanding gas-sensing properties, including the speed response, the recovery response time, selectivity stability, repeatability and response stability, etc. Of course, these excellent gas sensing properties are also revealed by other carbon nitride nanocomposites in different gases[238-242], such as g-C$_3$N$_4$/MnO$_2$ sandwich nanocompounds, and CeO$_2$/g-C$_3$N$_4$ nanohybrids.

In another similar work, Aziz et al. studied transition metals (abbreviated as TM, including Ni, Pd, and Pt) embedded g-C$_3$N$_4$ for adsorption of HCN gas using density functional theory calculations. The results indicated that upon adsorption of HCN gas, the initial planar structures of the pure g-C$_3$N$_4$ and TM-embedded g-C$_3$N$_4$ systems (except Ni-embedded g-C$_3$N$_4$) are changing and the structures become corrugated. Furthermore, it was found that the d-orbitals of TM atoms hybridize with the p-orbitals of g-C$_3$N$_4$, leading to the decrease of the band gap energy for the embedded systems when compared with the pure g-C$_3$N$_4$. In addition, the results of polarized band structure plots revealed that the asymmetry between the majority spin and the minority spin branches of the Pt-embedded g-C$_3$N$_4$ gives rise to the spontaneous magnetization of this system with a net magnetic moment of about 1.35μB. Among the applied systems, the Pt-embedded g-C$_3$N$_4$ displayed substantially enhanced adsorption energy relative to the pure g-C3N4 (-0.297 eV), which is -1.98 eV. To the best of our knowledge the Pt-embedded g-C$_3$N$_4$ displayed the highest affinity for adsorption of HCN gas among the proposed adsorbents. Thus, the Pt-embedded g-C$_3$N$_4$ could be a low cost and an excellent candidate for sensing HCN gas and its removal from the atmosphere[243].

1. g-C$_3$N$_4$ nanostructure used to photocatalytic CO$_2$ reduction

CO$_2$ reduction is a much more complex and difficult reaction compared to the HER because CO_2 reduction requires a large overpotential to drive the reaction at a sufficient rate. In addition, the reaction pathway is variable to produce a variety of intermediate species, thus making CO_2 photoreduction complicated. Furthermore, the selectivity for CO_2 reduction is also difficult due to the relatively complicated reaction pathways.

The CB edge of g-C_3N_4 is sufficiently negative to reduce CO_2 in principle. However, there are few concerns on CO_2 reduction by using pristine g-C_3N_4 due to the fast recombination of photogenerated holes and electrons. To overcome this issue, the g-C_3N_4/SnS_2 was formed to support SnS_2 quantum dots on-site on g-C_3N_4 by a simple one-step hydrothermal treatment, especially about photocatalytic CO_2 reduction[244].

Here, FTIR spectra were measured to explore the structure of g-C_3N_4/SnS_2 in the Figure 10-18 (a). As we can clearly observe that the FTIR spectrum of g-C_3N_4/SnS_2 has no obvious difference from that of pure g-C_3N_4[245-246]. But for SnS_2, the adsorption peak mainly depends on the Sn-S vibration at 545 cm^{-1}[247]. The characteristic band of SnS_2 is not significant in the spectrum of the hybrid because of the strong absorption peak of g-C_3N_4 and the low content of SnS_2. Meanwhile, the optical absorption properties were also investigated by UV-vis diffuse reflectance measurement, as shown in Figure 10-18 (b). The g-C_3N_4/SnS_2 shows the remarkable absorption abilities combining with g-C_3N_4 and SnS_2. We can obviously observe that there is the markedly long-wavelength enhancement compared to individual g-C_3N_4. Besides, the color of g-C_3N_4 (the light yellow) was deepened after the deposition of SnS_2 (the brown color) in the inset of Figure 10-18 (b), further suggesting that SnS_2 was successfully deposited onto the g-C_3N_4. Furthermore, the photocatalytic reduction of CO_2 was carried out under visible-light irradiation. As we all know that CH_4 and CH_3OH are the main products of photocatalytic CO_2 reduction in corresponding reduction potentials, whose conversion amount directly revealing the activity of photocatalytic reduction of CO_2. As displayed in Figure 10-18 (c), an unambiguous comparison of the photocatalytic generation yield of CH_4 and CH_3OH, and that of g-C_3N_4/SnS_2 is much higher than g-C_3N_4 and SnS_2. This phenomenon can also reveal that g-C_3N_4/SnS_2 has the surprising properties of photocatalytic reduction of CO_2. And g-C_3N_4/SnS_2 also shows the astonishing photoelectrochemical performance. Transient photocurrent responses for SnS_2, g-C_3N_4 and g-C_3N_4/SnS_2 were measured by switching the irradiation light on and off under several cycles in Figure 10-18 (d). The g-C_3N_4/SnS_2 displays the higher current density compared with individual SnS_2 and g-C_3N_4. This result indicates that g-C_3N_4/SnS_2 has the outstanding photocurrent catalytic activity which matches perfectly with the consequence of Figure 10-18 (c).

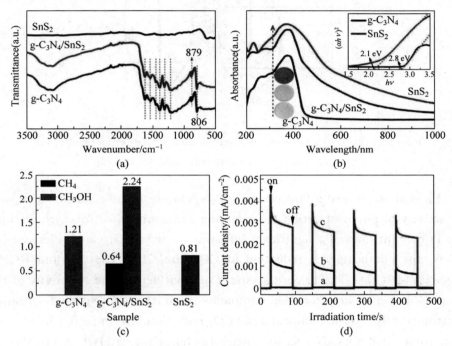

Figure 10-18 g-C_3N_4 nanostructures used to photocatalytic CO_2 reduction
(a) FTIR spectra of SnS_2, g-C_3N_4/SnS_2 and g-C_3N_4; (b) UV-vis diffuse reflectance spectra and the inset is the plots of SnS_2 and g-C_3N_4 against photon energy; (c) Hydrocarbon generation rate after one-hour illumination; (d) Transient photocurrent responses under LED light irradiation (λ = 420 nm) for (a) SnS_2, (b) g-C_3N_4, and (c) g-C_3N_4/ SnS_2 in 0.5 M Na_2SO_4 aqueous solution[244]

To understand the charge transfer processes in the g-C_3N_4/SnS_2 binary photocatalytic system, the band-edge positions of g-C_3N_4 and SnS_2 are presented in Figure 10-19. As described above, the Eg of gC_3N_4 and SnS_2 is 2.8 eV and 2.1 eV, respectively. The E_{CB} of g-C_3N_4 and SnS_2 is 1.52 V and 0.56 V vs. NHE (pH 7), respectively, which is more negative than the reduction potentials of CO_2 to CH_4 and CO_2 to CH_3OH. Thus, from a thermodynamic viewpoint, both semiconductors should be able to be photoexcited and thus reduce CO_2 to

CH_3OH or CH_4.

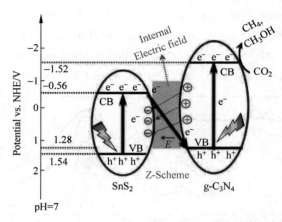

Figure 10-19　IEF-induced direct Z-scheme charge transfer mechanism for $g-C_3N_4/SnS_2$ in photocatalytic reduction of CO_2 [244]

Yu et al. fabricated a Pt-deposited $g-C_3N_4$ (Figure 10-20(a)) and investigated the activity of photocatalytic CO_2 reduction under simulated solar irradiation. The Pt content showed a significant influence on the activity and selectivity of $g-C_3N_4$ for photocatalytic reduction of CO_2 into CH_4, CH_3OH and HCHO (Figure 10-20(b)). The Pt cocatalyst not only influenced the selectivity of the product, but also affected the photoactivity of the $g-C_3N_4$ by facilitating electron transfer to the photocatalytic CO_2 reduction. On the other hand, Zou et al. constructed a $g-C_3N_4/NaNbO_3$ heterojunction photocatalyst by introducing polymeric $g-C_3N_4$ on $NaNbO_3$ nanowires. It was found that the activity of $g-C_3N_4/NaNbO_3$ composite photocatalyst for CO_2 photoreduction was higher than that of $g-C_3N_4$ and $NaNbO_3$ alone. This was mainly attributed to the improved separation and transfer of photogenerated electron-hole pairs at the intimate interface of $g-C_3N_4/NaNbO_3$ heterojunctions. Very recently, Fan et al. fabricated a solid-state Z-scheme hybrid of $Ag_3PO_4/g-C_3N_4$ for CO_2 photoconversion into fuels by using visible light. In comparison with the conventional heterojunctions which would decrease the reducing ability of the photoexcited electrons, this solid-state Z-scheme can keep the strong reducing ability of the photogenerated electrons in $g-C_3N_4$ and thereby result in an enhanced CO_2 reduction efficiency. The optimal $Ag_3PO_4/g-C_3N_4$ photocatalyst shows a CO_2 conversion rate of 57.5 mol·h^{-1}·$gcat^{-1}$, 6.1 and 10.4 times higher

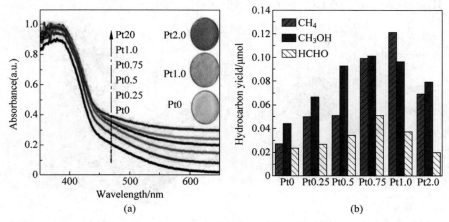

Figure 10-20 Pt-deposited g-C_3N_4 and the activity of photocatalytic CO_2 reduction
(a) UV-vis diffuse reflectance spectra of Ptx-loaded g-C_3N_4, x refers to the Pt loading amount;
(b) photoconversion of CO_2 into CH_4, CH_3O and HCHO over Ptx-loaded g-C_3N_4 [248]

than those of g-C_3N_4 and P25, respectively. More recently, Maeda and co-workers reported a heterogeneous photocatalyst system consisting of a ruthenium complex and g-C_3N_4. In this system, the ruthenium complex and g-C_3N_4 serve as the catalytic and light-harvesting units, respectively (Figure 10-21(a)). This hybrid system can photoreduce CO_2 into formic acid with a turnover number over 1,000 and apparent quantum yield of 5.7 % at 400 nm. This result is the highest conversion efficiency in heterogeneous photocatalysts for CO_2 reduction under visible-light irradiation to date. The high CO_2 conversion efficiency is attributed to synergistic effect of the electron injection from excited g-C_3N_4 to ruthenium unit as well as the strengthened electronic interactions of g-C_3N_4 and ruthenium complex. Recently, Wang developed an integrated g-C_3N_4 with a cobalt-containing zeolitic imidazolate framework (Co-ZIF-9). Here, g-C_3N_4 is considered as a semiconductor photocatalyst, while Co-ZIF-9 is a cocatalyst that both can facilitate the capture/concentration of CO_2 and promote light-induced charge separation (Figure 10-21(b)). This hybrid system efficiently enhanced CO_2-to-CO conversion under visible-light illumination with an apparent quantum yield of 0.9 %. By coupling g-C_3N_4 with elemental photocatalyst red phosphor, we also found that an enhanced CH_4 production by CO_2 photoreduction under a 500 W xenon arc lamp irradiation. This work is the first case of coupling elemental photocatalyst with g-C_3N_4 to achieve the improved activity for H_2 generation and CO_2 photoconversion.

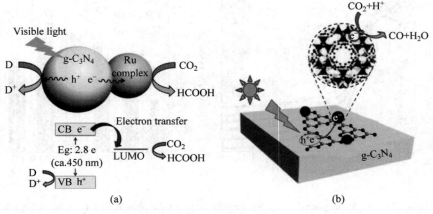

Figure 10-21 Ru complex/g-C_3N_4 hybrid photocatalyst
(a) Scheme of CO_2 reduction using a Ru complex/g-C_3N_4 hybrid photocatalyst (CB, conduction band; VB, valence band.)[249]; (b) Scheme of the cooperation of Co-ZIF-9 and g-C_3N_4 for the CO_2 reduction under visible-light irradiation[250]

2. g-C_3N_4 nanostructure used to degrade 2-propanol

In another similar work, Kong et al. reported the highly efficient experimental photocatalysis by the N vacancy in graphic carbon nitride with tris-s-triazine structures (g-C_3N_4), where 2-propanol is photocatalyzed to acetone. Experimental analysis reveals that the N vacancy is localized on the side N atom of tris-s-triazine units, which results in the highly efficient photocatalysis. The N vacancy is an essential factor for determining the defecting element and vacancy site for rational design of the highly efficient photocatalyst, using 2D semiconductor materials on the nanoscale. These results can promote a deeper understanding of the contribution of the defecting element and the vacant site to photocatalysis.

A facile method was developed to synthesize a g-C_3N_4 nanosheet by acid treatment of pristine g-C_3N_4. Meanwhile, g-C_3N_4 was exfoliated into nanosheets with abundant in-plane nitrogen vacancies due to strong chemical exfoliation and an etching effect. The photocatalytic activities of the exfoliated g-C_3N_4 nanosheet is evaluated by oxidative dehydrogenation of 2-propanol to acetone under visible-light irradiation ($\lambda > 420$ nm). 2-propanol photodegradation to acetone can serve as a good reaction model to evaluate the photo-catalytic activities of semiconductors[251]. The synthesized N defecting g-C_3N_4 nanosheets possess large specific surface areas, and improved electron transport ability, exhibit high photocatalytic activities as much as 9.6 times more than bulk g-C_3N_4 for the photodegradation of gas pollute under visible light. The results demonstrate a significant progress for low-cost fabrication of high-quality porous 2D g-C_3N_4 with excellence photocatalytic activity.

The XRD patterns of the prepared sample are displayed in Figure 10-22 (a). From Figure 10-22 (a), the (002) diffraction peak evidently shifts from 27.7° to 27.9°, implying the decreasing interlayer distance in the g-C_3N_4, which can be explained by the undulated single layer in g-C_3N_4 being planarized by treating in the HNO_3, leading to a denser stacking and decreasing the interlayer distance. Figure 10-22 (b) shows the nitrogen adsorption-desorption isotherms of g-C_3N_4 and ECNC-X catalysts. Obviously, the nitrogen adsorption-desorption isotherms for g-C_3N_4 and ECNC-X catalysts belonged to type Ⅳ, indicating the presence of mesopores. The BET specific surface areas (S_{BET}) of g-C_3N_4, ECNC-1.25, ECNC-2.5 and ECNC-3.75 were calculated to be 5.4 m^2/g, 41 m^2/g, 43 m^2/g, and 52 m^2/g, respectively. This result demonstrating that nitric acid mediation could be used to construct nanosheet structures with significant increased specific surface area to promote photocatalytic activity. As shown in Figure 10-22(c), for g-C_3N_4, a strong symmetrical resonance signal at g = 2.003 is obtained, which confirms the presence of unpaired electrons on the carbon atoms of the aromatic units and nitrogen vacancies. Generally, when nitrogen-containing species of g-C_3N_4 were lost, extra electrons will be rapidly redistributed to their closest carbon atoms by the delocalized p-conjugated networks of g-C_3N_4. A weak analogous signal was also detected for CN, since the thermal polymerization of CN may contribute to the existence of partial disorders, i.e., nitrogen defects. Figure 10-22 (d) presents the UV-vis spectra of the as-prepared catalysts. Compared with g-C_3N_4, ECNV-X exhibits a slightly blue shift in the absorption edge, which could be attributed to the quantum confinement induced by the ultrathin g-C_3N_4 nanosheets. Though the quantum size effect limits the visible-light absorption, a necessary result of increasing the specific surface area of the sample, the wall of the large specific surface area structure is usually constructed of small nanocrystals, and it is

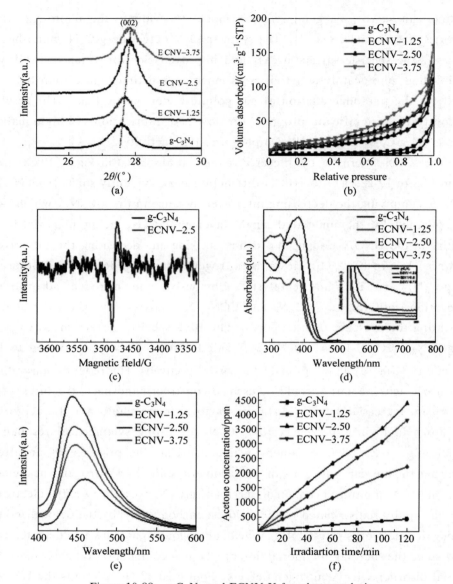

Figure 10-22　g-C_3N_4 and ECNV-X for photocatalysis

(a) The transformed amplification of g-C_3N_4 and ECNV-X; (b) Nitrogen adsorption/desorption isotherm of g-C_3N_4 and ECNV-X; (c) EPR spectra of g-C_3N_4 and ECNV-2.5; (d) UV-vis absorption spectra of g-C_3N_4 and ECNV-X; (e) Photoluminescence spectra of them excited at 330 nm of g-C_3N_4 and ECNV-X; (f) The concentration of evolved acetone over different photocatalysts under visible-light irradiation[252]

significant to develop large specific surface area photocatalysts, as increasing the surface area can greatly enhance the photocatalytic activity even absorbing less light. In addition, samples from ECNV-2.5 show a slight absorption in the range from 450 nm to 800 nm (see the inset in Figure 10-22 (d)), which could be related to the increased defects. Figure 10-22 (e) shows the PL spectra of the as-synthesized catalysts excited by light at 330 nm at room temperature. The emission peak of the pristine g-C_3N_4 appears at 445 nm, attributed to the band-band recombination of the charge carriers with emission photon energy equal to the band-gap energy. Compared with g-C_3N_4, ECNV-X possesses a much weaker emission peak, implying that the separation efficiency of photogenerated carriers is improved, and the probability of electron-hole recombination may be decreased by nitrogen vacancies. Figure 10-22 (f) shows the concentration of evolved acetone over different photocatalysts under visible-light irradiation. Acetone can be evolved over all the tested catalysts under light irradiation, but obviously the photocatalytic activities for all of the ECNV-X samples obtained from nitric acid intermediate g-C_3N_4 are higher than the photocatalytic activity of g-C_3N_4 obtained from melamine. The photocatalytic results show that porous samples exhibit much higher activities than samples prepared by the conventional approach. Enhancement in photocatalytic activity of the porous g-C_3N_4 can be attributed to the large specific surface area and small particle size that provides more active sites for the catalysis reaction and short bulk diffusion length for reducing the recombination probability of photoexcited charge carriers.

Sun et al. proposed a new template-free approach for the synthesis of porous g-C_3N_4 nanosheets as efficient photocatalysts. This new template-free method includes a one-step hydrothermal treatment and a sequential two-step thermal treatment. This method can provide massively porous g-C_3N_4 nanosheets without external physical defects (such as atom doping and/or atom vacancies) in the g-C_3N_4 nanosheets. The BET surface areas and pore volumes of the porous g-C_3N_4 nanosheets can reach 125 m^2/g and 1.2 cm^3/g, respectively, thereby showing significant increases of 22 and 15 times, respectively, compared with those of the pristine g-C_3N_4 nanosheets. The synthesized porous g-C_3N_4 nanosheets have large specific surface areas and exhibit high-photocatalytic activities that are enhanced by as much as 8 times compared to those of the pristine g-C_3N_4 for the photodegradation of gas pollutants (using the photocatalysis of 2-propanol to acetone as an example) under visible light ($\lambda > 420$ nm). Experimental and theoretical results reveal the suitable porous size of porous g-C_3N_4 plays an essential role for the efficient photocatalysis. This multistep template-free approach can provide a new method for the preparation of porous g-C_3N_4 nanosheets as efficient photocatalysts that are better than those obtained by destroying the physical structures of the pristine g-C_3N_4 nanosheets.

A possible mechanism for obtaining the larger specific surface area of photocatalyst is suggested as follows. As illustrated in Figure 10-23, in the hydrothermal treatment process, citric acid reacts with some of the amino groups of melamine through an intermolecular dehydrolysis reaction, and an amide bond (—NH—CO—) is formed by the reaction of —NH_2 and —COOH. Citric-acid-treated melamine used as the precursor inhibits crystal growth, while the designed *two-step* thermal treatment method can avoid the decomposition

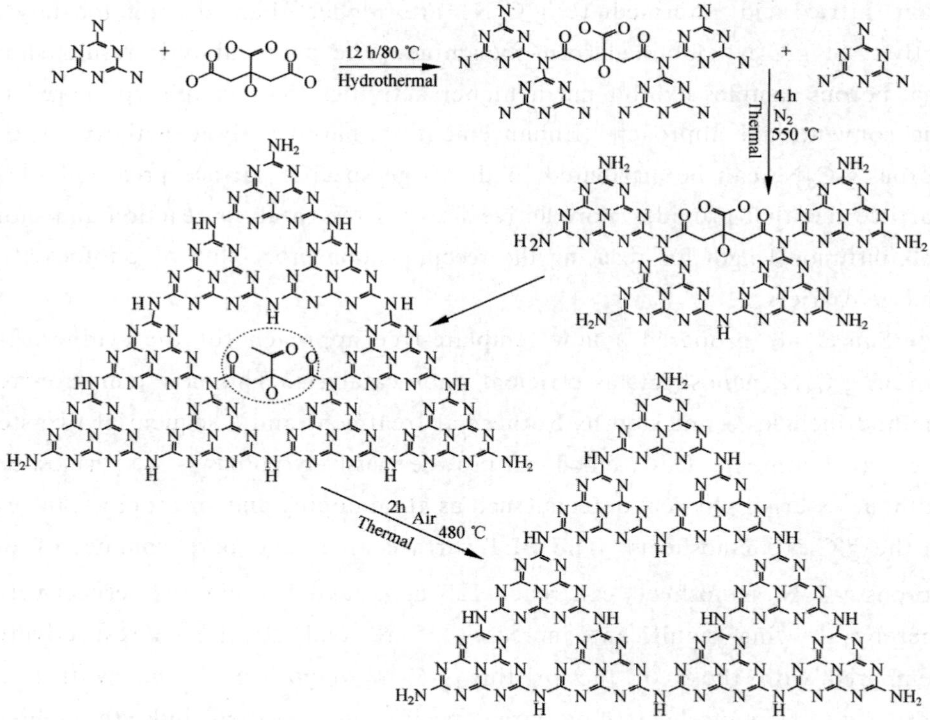

Figure 10-23 The complete synthesis process of porous g-C_3N_4 [253]

of the pore-maker material at a too early stage. The synergy between these two procedures leads to the formation of porous g-C_3N_4 with negligible residual carbon content. The resultant porous nanosheets have not only a large surface area and low sheet thickness but also a decreased band gap. The high porosity and large surface area of g-C_3N_4 are favorable for inhibiting the recombination of the photogenerated electrons and holes and improving the electron transport ability along the in-plane direction due to the quantum confinement effect. As a result, when used as photocatalysts, the porous nanosheets show photocatalytic activities superior to those of the bulk g-C_3N_4. The results reveal that the suitable porous size of porous g-C_3N_4 plays an essential role for the efficient photocatalysis.

3. g-C_3N_4 nanostructure used to degrade NO

NO_x liberated into atmosphere from automobile exhausts and fossil fuel combustion, comprise the major air pollutants. They are responsible for serious environmental problems such as acid rain, ozone accumulation, haze and photochemical smog. Besides they contribute to the deterioration of human health by causing decrease of the lung function and respiratory problems. The application of photocatalytic methods in order to mitigate the presence of NO_x in the atmosphere is preferable as they are environmentally friendly, mild and low cost. Therefore, in this portion, the photocatalytic activity of g-C_3N_4 based composites toward NO_x removal was discussed. The effect of g-C_3N_4 doping and copolymerization with metals/semiconductors on its photocatalytic activity toward NO_x oxidation was thoroughly discussed.

As far as the fundamental mechanism for photocatalytic NO_x removal is concerned, when the photocatalyst absorbs a photon, with energy equal or higher than the band-gap energy, an electron is excited from the VB to CB leaving a positive hole (h^+) in the VB. As oxygen is the second most abundant gas present in the atmosphere of the Earth, at a percentage equal to 21%, the reduction of the oxygen adsorbed on the surface of the photocatalyst, is the most important reaction of the photocatalytic mechanism. Namely, superoxide ($O_2^{\cdot -}$) radicals are produced via the reaction of photogenerated electrons (e^-) with adsorbed O_2 molecules as shown below:

$$O_2 + e^- \longrightarrow O_2^{\cdot -}, \tag{10-7}$$

whereas hydroxyl (OH.) radicals are generated, usually in a lower extent, by the reaction of h+ with H_2O/OH^- and/or by redox reactions of superoxide radicals via the formation of H_2O_2, as depicted in Equation (10-8)~(10-10):

$$h^+ + H_2O/OH^- \longrightarrow OH^·, \qquad (10\text{-}8)$$

$$O_2^{·-} + H + e^- \longrightarrow H_2O_2, \qquad (10\text{-}9)$$

$$H_2O_2 + e^- \longrightarrow OH^·. \qquad (10\text{-}10)$$

Afterwards, NO may be oxidized by superoxide radicals and/or hydroxyl radicals and/or photogenerated positive holes mainly to nitrate ion through the formation of NO_2 and HNO_2 intermediates, as shown in Equation (10-11)~(10-13)[254]:

$$NO + O_2^{·-} \longrightarrow NO_3^-, \qquad (10\text{-}11)$$

$$NO + OH^· \longrightarrow NO_3^-, \qquad (10\text{-}12)$$

$$NO + h^+ \longrightarrow NO_3^-. \qquad (10\text{-}13)$$

The NO removal ratio of different reactions is summarized in Table 10-3. Dong et al. presenteted a facile chemical approach to form internal van der Walls heterostructures (IVDWHs) within $g\text{-}C_3N_4$, as Figure 10-24(a) *cake model* and structure of OCN-K-CN, which enhance the interlayer coulomb interaction and facilitate the spatially oriented charge separation. Such a structure, generated through simultanous $g\text{-}C_3N_4$ intralayer modification by O and interlayer intercalation by K, enables the oriented charge flow between the layers, enhancing the accumulation of the localized electrons and promoting the production of active radicals for the activation of reactants as suggested by density functional theory calculations. The resultant O, K-functionalized $g\text{-}C_3N_4$ with IVDWHs shows an enhanced photocatalytic activity, with nearly 100 % enhancement of NO purification efficiency compared to pristine $g\text{-}C_3N_4$. The reaction mechanism for NO purification is also provided, in which the functionality of IVDWHs could effectively restrain the production of toxic intermediates and promote the selectivity for final products. This work provides a strategy to engineer heterostructures within 2D materials for tunable charge carrier separations and migrations for advanced energy and environmental catalysis.

Table 10-3 g-C_3N_4-based structures for removal of NO

Photocatalysts	Application	Pollutant concentration	Removal ratio	Reference(year)
g-C_3N_4/O/La co-functionalization	Removal of NO	NO: 500 ppm	NO: 50.4% in 30 min	[256](2019)
Ca-intercalated g-C_3N_4	Removal of NO	NO: 100 ppm	NO: 54.78% in 30 min	[257](2018)
g-C_3N_4/O/Ba co-functionalized	Removal of NO and suppressing toxic NO_2 generation	NO: 500 ppm	NO: 57% in 30 min	[258](2018)
Porous g-C_3N_4 nanosheets	Removal of NO and NO_2	NO: 600 ppm NO_2: 400 ppm	NO: 65 % in 30 min NO_2: 33.8 % in 30 min	[259](2018)
KCl-doped g-C_3N_4	Removal of NO	NO: 50 ppm	NO: 38.4 % in 30 min	[260](2018)
Sr-intercalated g-C_3N_4	Removal of NO	NO: 600 ppb	NO: 53.1 % in 30 min	[261](2018)
MnO_x/ g-C_3N_4	Removal of NO	NO: 500 ppb	NO: 44 % in 30 min	[262](2018)
alkalis intercalated g-C_3N_4	Removal of NO		NO: 51.11 % in 30 min	[263](2017)
SrO clusters at g-C_3N_4	Removal of NO	NO: 100 ppm	NO: 50 % in 30 min	[264](2017)
Bi/ g-C_3N_4	Removal of NO	NO: 600 ppb	NO: 59.7 % in 30 min	[265](2015)

DFT calculations are used to understand the electronic structure details and property advantages of OCN-K-CN as depicted in Figure 10-24. Upon the addition of O on CN, OCN band structure is evidently adjusted by the O adjustor atoms (Figures 10-24(b)), leading to a favorable band offset between the OCN layer and the CN sub-layer and therefore a spatial charge carrier separation (light-generated hole migration to OCN and electron transfer to CN, Figure 10-24(c)).

By comparing the electronic structures of pristine CN and OCN-K-CN, we find, as shown in Figures 10-25(a) and (b), that the potential energies of the OCN layer and CN sublayer in OCN-K-CN are significantly increased after the incorporation of O and K. The potential difference between layers provides the driving force for electron transfer from the OCN layer to the CN sublayer through the interlayer K channel, which effectively realizes the spatial charge separation. We further considered the Bader effective charge35 (Figures 10-25(c) and (d)) to understand the role of O adjuster atoms. It is found that the local charge distribution in the OCN layer is altered with the incorporation of O,

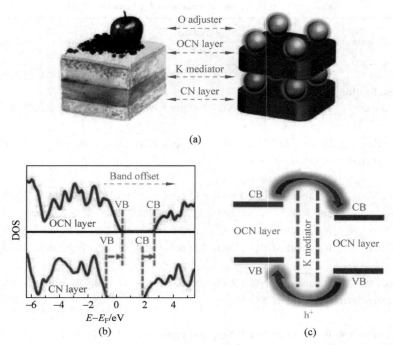

Figure 10-24 Schematic illustration of the internal van der Waals heterostructure (IVDWH)
(a) *Cake model* and structure of OCN-K-CN; (b) Calculated total density of states (TDOS) of CN and OCN layers; (c) Band sketch of the OCN-K-CN IVDWH[255]

resulting in the increased electron depletion from the adjacent C and N atoms to the O adjuster atom. Moreover, the interlayer electron transfer channel (Figures 10-25(e) and (f)) is strengthened with the introduction of O adjuster atoms. These accumulated electrons around the O adjuster atoms, plus the mediation of intercalated K, can reinforce the internal VdW force to effectively facilitate the interlayer charge flow and accelerate the spatial separation of electron-hole pairs.

Due to the O attachment on CN, the OCN layer and the adjacent CN layer form a heterostructure to direct the charge carrier transfer. Assisted by the K mediated interlayered electrons channel, the directional electrons can expeditiously transfer crossing multiple layers. Under visible-light irradiation, the accumulated localized electrons facilitate the product generation and reactants activation, hence leading to more efficient NO conversion with the suppressed production of toxic intermediates. This work provides a new design of IVDWHs within

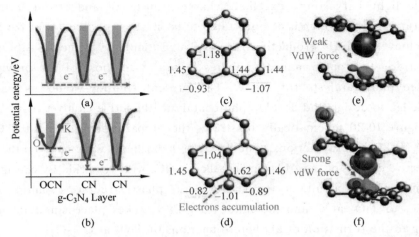

Figure 10-25　The potential energies of the OCN layer and CN sublayer in OCN-K-CN after the incorporation of O and K

(a) and (b) The layered electrostatic potential energy for pristine CN (a) and OCN-K-CN (b); (c) and (d) Calculated Bader effective charge for pristine CN (c) and OCN-K-CN (d); (e) and (f) Charge density difference of K-CN (e) and OCN-K-CN (f) Blue, green, red and gold spheres depict N, C, K and O atoms. Charge accumulation is labeled in blue and depletion in yellow, and the isosurfaces were both set to 0.005 eV/Å³ for (e) and (f)[255]

2D material and may advance the utilization of functionalized 2D materials in the vast fields associated with catalysis, energy and environmental sciences.

4. Based on g-C_3N_4 nanostructures driven for nitrogen photofixations reactions

Developing robust catalysis for nitrogen reduction reaction (NRR) is an ongoing scientific challenge, which needs activation and cleavage of the extremely strong N-N triple bond (bond energy: 940.95 kJ/mol). The famous Harber-Bosch process requires enormous energy and extreme conditions in the presence of an iron catalyst to achieve this process. Environment-friendly solar nitrogen fixation at room temperature and atmospheric pressure is an appealing way, but its efficiency is far from satisfied for industrialization. To solve these problems, Liu et al. reported for the first time on the design and preparation of MXene-derived TiO_2 at C/g-C_3N_4 heterojunctions where 2D planar carbon nanosheet-supported TiO_2 is thermally derived from MXene $Ti_3C_2T_x$ and effectively coupled with in situ formed g-C_3N_4 nanosheets. Such structure features abundant surface defects, high electron-donating ability, suitable light harvesting, excellent charge transport, and strong nitrogen activation ability, which is demonstrated to be a robust photocatalyst for NRR and achieves an NH_3 production rate of 250.6 μmol/g·h under visible-light irradiation. This performance compares favorably to those of the previously reported photocatalysts for NRR. The present finding would enrich our knowledge to design and develop photocatalytically active catalysts for NRR.

Figure 10-26 schematically illustrates the preparation process of TiO_2 at $C/g\text{-}C_3N_4$ 2D heterojunctions. The $Ti_3C_2T_x$ nanosheets were first synthesized by selective etching of the Al layers in bulk Ti_3AlC_2 with HF acid. By co-sintering with melamine at a suitable temperature, the phase structure evolution from $Ti_3C_2T_x$ to TiO_2 at C and melamine to $g\text{-}C_3N_4$ takes place simultaneously, leading to the formation of 2D heterojunctions of TiO_2 at $C/g\text{-}C_3N_4$.

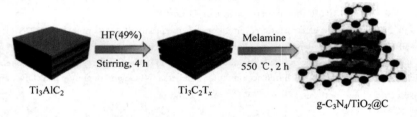

Figure 10-26 A schematic illustration of the synthesis process of TiO_2 at $C/g\text{-}C_3N_4$[266]

Figure 10-27 illustrates the schematic diagram of the energy band structure according to above estimation and the principle of charge carrier separation in TiO_2 at $C/g\text{-}C_3N_4$. As depicted in literature, the reduction potential of N_2 to NH_4^+ (N_2/NH_4^+: −0.276 V vs. NHE) is actually higher than that of CB bottom of TiO_2 and $g\text{-}C_3N_4$. Therefore, TiO_2 at $C/g\text{-}C_3N_4$ is capable of activating and reducing of nitrogen. Under visible-light irradiation, the electrons are able to jump from the VB to the CB of $g\text{-}C_3N_4$ and TiO_2, leaving behind the holes on the VB. Due to the more positive potentials of the CB and VB for TiO_2 in comparison with $g\text{-}C_3N_4$, the photoinduced electrons thermodynamically transfer from the CB of $g\text{-}C_3N_4$ to that of TiO_2, while holes transfer from the VB of TiO_2 to that of $g\text{-}C_3N_4$. Such behavior of interfacial electron transition is helpful to reduce recombination of photoinduced electrons and holes. Meanwhile, Ti^{3+} species in TiO_2 at $C/g\text{-}C_3N_4$ serve as adsorption sites for activating N_2 and trapping sites for the photoexcited electrons to promote efficient nitrogen reduction.

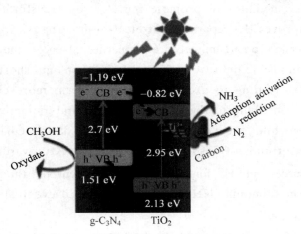

Figure 10-27　A schematic illustration of the energy-band structure and electron-hole separation of TiO_2 at $C/g-C_3N_4$[266]

In Aziz et al. research, novel $g-C_3N_4$ nanosheets/carbon dots/CdS (denoted as CN-NS/CDs/CdS) nanocomposites by prominent nitrogen photofixation capability under simulated solar irradiation were fabricated with decoration of CDs and CdS nanoparticles over CN-NS thorough a microwave-assisted method. Interestingly, the CN-NS/CDs/CdS nanocomposites showed superior photocatalytic performance both in nitrogen photofixation and in degradation of MB in comparison with the bulk CN.

The photocatalysts are offered, as observed in Figure 10-28. Theoretically, the CB positions for CN-NS and CdS are negative than the redox potential of N_2/NH_3 (-0.0922 V vs. NHE). Hence, both semiconductors can reduce N_2 to NH_3. It is noteworthy that upon the simulated solar irradiation, both the CN-NS and CdS are excited to generate electrons and holes. Because of the suitable overlapping band structures and strictly contacted interfaces of the semiconductors, the photogenerated electrons desire to migrate from CN-NS to CdS driven by the CB offset of 0.75 eV, while the holes spontaneously move from CdS to CN-NS driven by the VB offset of 0.42 eV. Additionally, when CDs are deposited onto the surface of CN-NS, the integrated CDs play the following roles in the photocatalytic process. Firstly, transfer of the photoinduced electrons from CN-NS to CDs is thermodynamically favorable, which makes electrons to simply transfer from CN-NS to CDs. Secondly, CDs can act as electron transfer mediator to promote the electron transfer from CN-NS to CdS, which improves the separation of electron-hole pairs. In addition, CDs possess upconversion photoluminescence properties. Hence, they absorb longer wavelengths from visible light ($\lambda > 550$ nm) and then emit shorter wavelengths ($\lambda < 460$ nm), leading to exciting CN-NS to form more charge carriers, which can also contribute to the enhanced photocatalytic activity of CN-NS/CDs/CdS. Meanwhile, the electrons on the CB of CdS immediately transfer to the antibonding orbitals of adsorbed N_2 molecules. Afterwards, the nitrogen molecules gain successively H^+ ions from the environment and the photogenerated electrons to form ammonia (NH_3), and there-upon it eventually forms NH_4^+.

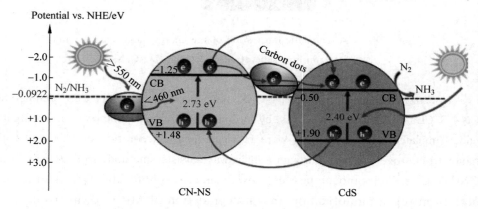

Figure 10-28 Diagram of the separation and transfer of photogenerated electron-hole pairs over the CN-NS/CDs/CdS nanocomposites[267]

To further investigate the latest issue, the effect of solution pH on the ammonia concentration over the CN-NS/CDs/CdS (30%) nanocomposite was studied. By decrease of solution pH, the NH_4^+ concentration was increased from 3.72×10^3 μmol/L·g (at pH 8) to 5.81×10^3 μmol/L·g (at pH 3). This is due to availability of excess protons to increase the reaction rate in acidic media. The reactions are shown in Equations (10-14) and (10-15)

$$N_2 + 6H^+ + 6e^- \longrightarrow 2NH_3, \qquad (10\text{-}14)$$

$$NH_3 + H_2O \longrightarrow NH_3 \cdot H_2O \rightleftharpoons NH_4^+ + OH^-. \qquad (10\text{-}15)$$

However, with further decrease of the solution pH to 2, the generation of NH_4^+ was decreased, which can be attributed to the decrease of photocatalyst stability in more acidic media. The synergistic influences of CN-NS, CDs, and

CdS collaborated in impressively enhancing solar-light absorption, reducing recombination of the charge carriers, and increasing mobility of electrons, which are beneficial in improving the photocatalytic activity. Moreover, the mechanistic investigations indicated that the generation rate of NH_4^+ by the ternary nanocomposite is almost related to the presence of electrons and protons, as the main reactive species in the nitrogen reduction reactions. As a consequence, the CN-NS/CDs/CdS (30 %) nanocomposite has remarkable potential to be used in photofixation reaction.

10.3.4 g-C_3N_4 nanostructure used to wastewater treatments

Water is one of the most important substances on the earth, which is critical for life-sustaining of all living creatures. Although about 71 % of the earth surface is covered by water, only about 2.5 % is fresh water. On the other hand, water consumption in various sectors has been dramatically overdone in the past decades, which caused diminution in fresh water required for wildlife and human life. In addition, billions of people access to limited clean water resources, that makes them vulnerable to water-borne infections. It is estimated that prior to 2015, approximately 70 % of the globally generated wastewater was not properly treated so that their discharge caused the pollution of receiving natural water-bodies. Correspondingly, by 2025, it is predicted that 50 % of people will face clean water crisis. In light of this, there is an urgent need to develop effective, dependable and economically viable methods to deal with the emerging contaminants and address the safety problems caused by them. In the last decades, several technologies have been established for wastewater treatments. The well-known methods are coagulation, sedimentation-flocculation, ion exchange, molecular sieves, reverse osmosis, membrane filtration, adsorption processes, ozonation, chlorination, chemical precipitation, and other chemical and electrochemical techniques. The conventional treatment technologies are reportedly not cost and energy-efficient and inadequate for complete degradation of recalcitrant contaminants in wastewater. Accordingly, novel advanced treatment methods are needed to destroy the organic portion of wastewaters. Among the various contaminants, persistent organic pollutants, such as polycyclic aromatic hydrocarbons, polychlorinated biphenyls, and pentachlorophenol, have led to severe environmental pollution. With persistence, bioaccumulation, mobility, and high toxicity, typical persistent organic pollutants can exist in the environment for a long time and migrate everywhere. Therefore, persistent organic pollutants have become the research hot spot in the fields of environmental, chemical, and environmental toxicology. For photoinduced charge-transfer mechanisms of g-C_3N_4-based heterojunction. Unlike from metals, the energy band of semiconductors is discontinuous and a band gap exists between the VB and the CB. When a semiconductor is located in the ground state, its chemical stability is outstanding. As shown in Figure 10-29(a), when g-C_3N_4 is irradiated with photons, whose energy is no less than the band-gap energy of g-C_3N_4, the electrons get excited and then jump to the CB, producing corresponding holes on the VB. Then, the photoinduced electron-hole pairs can be segregated and moved to the semiconductor surface under the control of an electric field, resulting in highly active electrons and holes on the surface of the semiconductor. The electrons in the CB with a chemical potential of $+0.5$ V to -1.5 V (vs. normal hydrogen electrode (NHE)) and holes in the VB with a chemical potential of $+1.0$ V to $+3.5$ V (vs. NHE) reveal excellent reduction and oxidation abilities, respectively. Therefore, electrons and holes can be used as a reductant and oxidant to react with electron acceptors and donors attached to the surface of the semiconductor, respectively. However, the photogenerated electrons and holes may also be directly combined in the interior or surface of photocatalyst and disperse the input energy in terms of heat or emitted light, resulting in a decrease of the utilization rate of photoinduced carriers for desired photoreactions. For the sake of improving the photocatalytic ability, the separation process of electron-hole pairs should be effective and photoinduced carriers should be rapidly transferred across the surface/interface to inhibit the recombination. The universally applicable method is the formation of heterojunctions by coupling with other substances. When g-C_3N_4 makes contact with other semiconductors, band bending occurs at the interface and leads to the formation of in-built electric field. Therefore, the separation rate of photogenerated electron-hole pairs is greatly increased in well-designed g-C_3N_4-based heterojunction in the presence of an in-built electric field.

According to the charge-transfer mechanisms, the frequently reported g-C_3N_4-based heterojunctions in the existing research can be mainly divided into

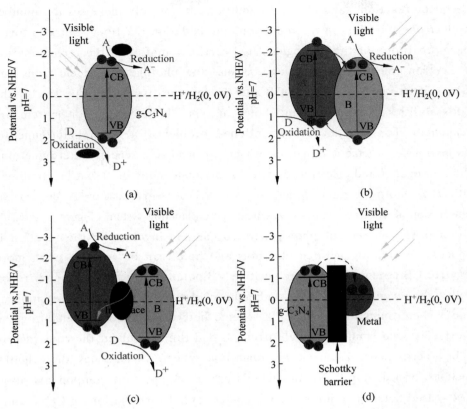

Figure 10-29　Photocatalytic mechanism of pure g-C_3N_4; schematic illustration of the transfer of photoinduced carriers for various types of heterojunction

(a) Photocatalytic mechanism of pure g-C_3N_4; schematic illustration of the transfer of photoinduced carriers for various types of heterojunction nanocomposites; (b) Type Ⅱ heterojunction; (c) Z-scheme heterojunction; (d) Schottky junction[273]

two categories: type Ⅱ heterojunction system and Z-scheme heterojunction system. The type Ⅱ heterojunction system is a widely used type at present. As shown in Figure 10-29(b), the band-edge potentials of type Ⅱ heterojunctions are staggered between semiconductor A and semiconductor B. Under sunlight irradiation, photogenerated electrons move from the CB of semiconductor A to the CB of semiconductor B and holes transfer to the VB of semiconductor A from the VB of semiconductor B, which make the photogenerated electrons and holes enriched in two semiconductors, respectively. Moreover, the rates of oxidation and reduction reactions increased due to the space separation where the reactions take place. In addition, it not only increases the number of electrons and holes in the reduction and oxidation reactions, but also forms space separation of the photogenerated carriers, inhibiting the recombination of electron-hole pairs. However, because the photogenerated electrons and holes shift to the less negative CB and the less positive VB, respectively, the redox abilities of electrons and holes in type II heterojunction system are weakened. To solve the aforementioned problems, a novel heterojunction charge-transfer scheme is proposed, which is named as Z-scheme heterojunction. The photogenerated electrons on the CB of semiconductor B can be combined with the photogenerated holes on the VB of semiconductor A due to the transmission of interface phase in Z-scheme heterojunction system (Figure 10-29(c)). Although the number of photogenerated carriers involved in the reaction is reduced, the photogenerated electrons and holes can be enriched with more negative CB potential and more positive VB potential, indicating higher redox ability. In recent years, some studies have been reported about switching of type II heterojunction to direct Z-scheme heterojunction via modulating the interfacial band bending, which offers a good thinking about the construction of high-performance Z-scheme heterojunction system. In addition, the Schottky junctions formed by $g-C_3N_4$ and metals are also important components of $g-C_3N_4$-based heterojunction systems. Under sunlight irradiation, $g-C_3N_4$ could be excited and resulting electrons migrate from the higher Fermi levels of $g-C_3N_4$ to metal until the two levels match (Figure 10-29(d)). Metal particles can improve charge separation at the interface of the metal—semiconductor heterojunction, which in turn contributes to an enhancement in the catalytic properties. Moreover, $g-C_3N_4$ not only acts as a support, but can also confine the aggregation of the metal particles, further promoting the selectivity and activity. The Schottky barrier and space charge region were formed at the interface between metal and $g-C_3N_4$ because of the different Fermi levels and work functions of metal and $g-C_3N_4$ when they were in close contact. The Schottky junctions with rectifying characteristics and lower interface voltage can modulate the generation and flow of photogenerated electrons and separate electron-hole pairs more effectively. Meanwhile, the surface plasmon resonance of metal nanoparticles is a key factor to increasing the photocatalytic performance of $g-C_3N_4$-based heterojunction. Obviously, there will be a good prospect in the future of g-C_3N_4-based heterojunction in redox reactions according to these mechanisms[268-272].

Hence, photodegradation of persistent organ pollutants over semiconductor solids has attracted extensive research attention recently. A more comprehensive list of persistent organ pollutants treatment by g-C_3N_4-based heterojunction is summarized in Table 10-4.

Table 10-4 Degradation of persistent organ policlutants by g-C_3N_4-Based Nanostructures

Photocatalysts	Application	Pollutant concentration	Degradation efciency	Reference (year)
g-C_3N_4/CDs/ BiOCl	Degradation of RhB, MB, MO, fuchsine dyes phenol, and photoreduction reaction of Cr(VI)	RhB: 1×10^{-5} M MB: 1×10^{-5} M MO: 1×10^{-5} M fuchsine: 0.77×10^{-5} M phenol: 5×10^{-5} M Cr(VI): 100 mg/L	RhB: 100 % in 45min	[274](2019)
g-C_3N_4/Fe_3O_4/ $CoWO_4$	Degradation of RhB, MB, Mo, and fuchsine dyes	RhB: 1×10^{-5} M MB: 1×10^{-5} M MO: 1×10^{-5} M fuchsine: 0.77×10^{-5} M	RhB: 100 % in 240 min	[275](2018)
g-C_3N_4/Fe_3O_4/ $MnWO_4$	Degradation of RhB, MB, Mo, and fuchsine dyes	RhB: 1×10^{-5} M MB: 1×10^{-5} M MO: 1×10^{-5} M fuchsine: 0.77×10^{-5} M	RhB: 100 % in 360 min	[276](2018)
g-C_3N_4/AgBr/ Fe_3O_4	Degradation of RhB	RhB: 2.5×10^{-5} M	RhB: 98.3 % in 420 min	[277](2018)
g-C_3N_4/Fe_3O_4/ $CuWO_4$	Degradation of RhB, MB, Mo, and fuchsine dyes	RhB: 1×10^{-5} M MB: 1×10^{-5} M MO: 1×10^{-5} M fuchsine: 0.77×10^{-5} M	RhB: 100 % in 270 min	[278](2018)
g-C_3N_4/CDs/ BiOI	Degradation of RhB, MB, Mo, and fuchsine dyes	RhB: 2.5×10^{-5} M MB: 2.5×10^{-5} M MO: 2.5×10^{-5} M fuchsine: 9.20×10^{-6} M	RhB: 100 % in 120 min	[279](2018)
g-C_3N_4/Fe_3O_4/ $CoMoO_4$	Degradation of RhB, MB, MO, fuchsine dyes, and photoreduction reaction of Cr(VI)	RhB: 1×10^{-5} M MB: 1×10^{-5} M MO: 1×10^{-5} M fuchsine: 0.77×10^{-5} M Cr(VI): 100 mg/L	RhB: 100 % in 300 min	[280](2019)

Continued

Photocatalysts	Application	Pollutant concentration	Degradation efciency	Reference (year)
g-C_3N_4/CDs	Degradation of RhB, MB, fuchsine dyes, phenol, and photoreduction reaction of Cr(Ⅵ)	RhB: 1×10^{-5} M MB: 1×10^{-5} M fuchsine: 0.77×10^{-5} M phenol: 5×10^{-5} M Cr(Ⅵ): 100 mg/L	RhB: 100 % in 60 min	[281](2019)
g-C_3N_4/$CuCr_2O_4$	Degradation of RhB, MB dyes and phenol	RhB: 2.5×10^{-5} M MB: 2.5×10^{-5} M phenol: 5×10^{-5} M	RhB: 98.8 % in 210 min	[282](2017)
g-C_3N_4/Fe_3O_4/Ag/Ag_2SO_3	Degradation of RhB, and fuchsine dyes	RhB: 2.5×10^{-5} M fuchsine: 9.2×10^{-6} M	RhB: 99 % in 270 min	[283](2017)
g-C_3N_4/Fe_3O_4/Ag_3PO_4/Co_3O_4	Degradation of RhB, MB, MO and phenol dyes	RhB: 1×10^{-5} M MB: 1.3×10^{-5} M MO: 1.05×10^{-5} M phenol: 5×10^{-5} M	RhB: 100 % in 150 min	[284](2017)
g-C_3N_4/Fe_3O_4/AgI/Ag_2CrO_4	Degradation of RhB and fuchsine dyes	RhB: 2.5×10^{-5} M fuchsine: 9.2×10^{-6} M	RhB: 99.4 % in 150 min	[285](2016)
g-C_3N_4/Fe_3O_4/Ag_3PO_4/AgCl	Degradation of RhB, MO, and fuchsine dyes	RhB: 1×10^{-5} M MO: 1.05×10^{-5} M fuchsine: 0.77×10^{-5} M	RhB: 100 % in 150 min	[286](2016)
g-C_3N_4/Fe_3O_4/AgI/Bi_2S_3	Degradation of RhB, MB and fuchsine dyes	RhB: 2.5×10^{-5} M MB: 2.5×10^{-5} M fuchsine: 9.2×10^{-6} M	RhB: 99 % in 60 min	[287](2016)
g-C_3N_4/Fe_3O_4/BiOI	Degradation of RhB, MB and MO	RhB: 1×10^{-5} M MB: 1.3×10^{-5} M MO: 1.05×10^{-5} M	RhB: 100 % in 120 min	[288](2016)
g-C_3N_4/CDs/AgCl	Degradation of RhB, MB, MO dyes and phenol	RhB: 2.5×10^{-5} M MB: 2.5×10^{-5} M MO: 2.5×10^{-5} M Phenol: 5×10^{-5} M	RhB: 100 % in 120 min	[289](2018)
g-C_3N_4/Ag_2SO_4	Degradation of RhB, MB and fuchsine dyes	RhB: 2.5×10^{-5} M MB: 1.94×10^{-5} M fuchsine: 9.2×10^{-6} M	RhB: 99 % in 390 min	[290](2016)
Salicylic acid/g-C_3N_4	Degradation of tetracycline (TC) sulfamethazine (SMZ)	TC: 5 mg/L	TC: 92 % in 60 min	[291](2019)
C-doped g-C_3N_4/$Bi_2O_{17}Cl_2$	Degradation of tetracycline (TC)	TC: 5 mg/L	TC: 94 % in 60 min	[292](2018)
K-doped g-C_3N_4	Degradation of tetracycline (TC)	TC: 10 mg/L	TC: 92.06 % in 60 min	[293](2018)

For example, Aziz et al. reported the novel hybrid g-C_3N_4/Fe_3O_4/$NiWO_4$ (gCN/M/$NiWO_4$) photocatalysts were synthesized by the integration of Fe_3O_4 and $NiWO_4$ with gCN using refluxing calcination method (Figure 10-30). The activity of the hybrid photocatalysts was explored by degradations of rhodamine B (RhB), methylene blue (MB), methyl orange (MO), fuchsine, and phenol under visible-light illumination. After the investigation about the effect of various operational factors on the photocatalytic activity, a possible degradation mechanism for the excellent photocatalytic performance of the gCN/M/$NiWO_4$ nanocomposites was proposed.

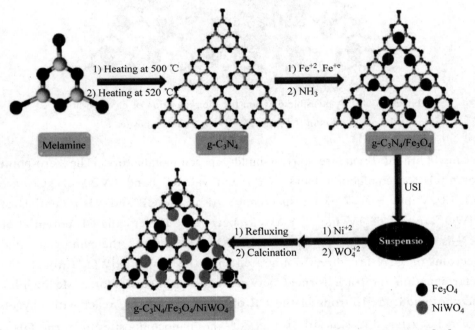

Figure 10-30 Preparation procedure for the gCN/M/$NiWO_4$ nanocomposites[294]

Using the structural characterizations and photocatalytic performance, a possible mechanism for the superior photocatalytic activity of gCN/M/$NiWO_4$ nanocomposites was proposed and is illustrated in Figure 10-31. As known, driving force for charge transfers and consequently increase of the photocatalytic activity originate in the matched band potentials of the hybridized semiconductors. The energy-gap (Eg) values of gCN and $NiWO_4$ are 2.7 eV and 2.2 eV, respectively. Under the visible light, gCN and $NiWO_4$ produce e^--h^+ pairs,

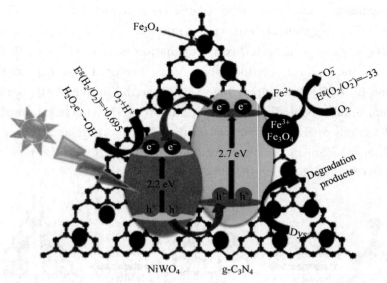

Figure 10-31　A plausible diagram for the separation of electron-hole pairs in the gCN/M/NiWO$_4$ nanocomposites[294]

because both of them are narrow-band-gap semiconductors. The zero-point potentials for conduction band (CB) and valence band (VB) of gCN are −1.13 eV and +1.57 eV, respectively. Also, the CB and VB potentials of NiWO$_4$ are +0.68 eV and 2.88 eV, respectively. Hence, the CB potential of gCN is more negative than that of NiWO$_4$. Therefore, the photogenerated electrons of gCN are directly migrated into the CB of NiWO$_4$ through the heterojunction interface formed between these semiconductors. Meanwhile, the photoinduced electrons in the CB of gCN and NiWO$_4$ react with oxygen molecules after diffusion to surface of the nanocomposite to form OH· radicals in successive reduction reactions because the CB potentials of gCN and NiWO$_4$ are more negative than that of the reduction potential of oxygen to hydrogen peroxide. The produced hydroxyl radicals, with the reduction potential of +2.80 eV/NHE, have great oxidation power to decompose pollutants. Meanwhile, some of the photogenerated electrons on the CB of gCN can react with O$_2$ to form O$_2^-$, because the CB potential of gCN is more negative than the potential of O$_2$/O$_2^-$ (E^0(O$_2$/O$_2^-$) = −0.33 eV/NHE). As depicted in Figure 10-32, the excited holes produced over NiWO$_4$ are injected into the VB of gCN. The photoinduced holes over the VB of gCN can oxidize the adsorbed pollutants. Moreover, nanoparticles of Fe_3O_4 can contribute to the separation of the e^-/h^+ pairs. The CB potential of Fe_3O_4 is positive than that of gCN. Thus, these electrons can easily migrate to the CB of Fe_3O_4. The injected electrons can react with Fe^{3+} ions of Fe_3O_4 to produce Fe^{2+} ions. Afterward, the formed Fe^{2+} ions react with O_2 to produce Fe^{3+} ions and O_2^- species.

The project confirmed that heterojunction formation between gCN, Fe_3O_4, and $NiWO_4$ increased the interfacial charge transfer and inhibited the recombination of the e^-/h^+ pairs, leading to increase of the produced reactive species to participate in the degradation reactions. In addition, the gCN/M/$NiWO_4$ (30 %) photocatalyst also had remarkable stability. This study demonstrated that the enhanced visible-light absorption, largely reduced recombination of the e^-/h^+ pairs, increased surface area, ability to degrade different pollutants, and excellent stability ensure that the novel gCN/M/$NiWO_4$ photocatalysts are promising photocatalysts for environmental remediation.

As the main source of environmental pollution, the water pollution caused by organic compounds is considered as one of the most challenging issues in the environment. For instance, the total amount of tetracyclines and sulfonamides used in China in 2013 was estimated to be 12,000 t and 7920 t, respectively. In addition, they have not only been found in water and sediment, but also in epitopes, especially for tetracycline. It is necessary to remove these contaminants from the polluted environment due to the bioaccumulation and persistence.

Zhou et al. synthesized a metal-free carbon doping-carbon nitride (BCM-C_3N_4) nanocomposite by introducing barbituric acid and cyanuric acid during the polymerization of melamine. The BCM-C_3N_4 sample exhibited higher-surface area, lower fluorescence intensity, better photocurrent signals and more efficient charge transfer in comparison to pure g-C_3N_4. The BCM-C_3N_4 exhibits the excellent photocatalytic degradation ability of sulfamethazine (SMZ) under visible-light irradiation. Much superior photocatalytic activity and high pollutant mineralization rate was achieved by BCM-C_3N_4, where it was 5 times more than that of pristine C_3N_4. The effect of initial SMZ concentrations on photocatalyst was also investigated. Additionally, the trapping experiments and electron spin resonance tests de monstrated that the main active species, such as $\cdot O_2^-$ and h^+, could be produced under light irradiation.

The morphological g-C_3N_4 and BCM-C_3N_4 have been characterized by the TEM images. As shown in Figures 10-32(a) and (b), the pristine g-C_3N_4 presented stacked sheets structure and smooth surface. After the addition to the barbituric acid and cyanuric acid, the BCM-C_3N_4 obviously displayed thin nanosheet structures and randomly distributed mesopores, which significantly differed from the mainly dense and stacked sheets of the pristine g-C_3N_4 (Figure 10-32(c)). As shown in Figure 10-32(d), many pores, approximately 5~10 nm, were embedded in a porous BCM-C_3N_4. These surface pores existed on BCM-C_3N_4 can provide a lot of active sites and improve the mass transportation, which can be further utilized for photocatalytic degradation.

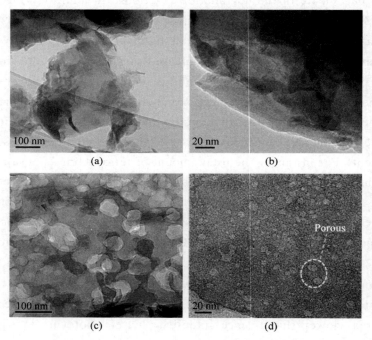

Figure 10-32 TEM images of as-prepared photocatalysts
(a) low and (b) high resolution of g-C_3N_4, (c) low and (d) high resolution of BCM-C_3N_4[295]

A possible mechanism is proposed to explain the high-photocatalytic activity of the as-prepared BCM-C_3N_4 in Figure 10-33. BCM-C_3N_4 has higher specific surface area and numerous active sites for the reaction process. Under visible-light irradiation, the BCM-C_3N_4 could generate electrons and holes. The electrons on the CB of BCM-C_3N_4(-0.79 eV), which are more negative than E (O_2/·O_2^-) (−0.33 eV), are a good reductant that could efficiently reduce oxygen molecules adsorbed on the surface of the photoanode to ·O_2^-. Moreover, super-oxide radicals ·O_2^- are the most important oxidizing species and then induce the SMZ degradation. The standard redox potential of ·OH/OH^- is +2.38 eV, which is more positive than the VB position of BCM-C_3N_4 (+1.23 eV). Thus, the photogenerated hole (h^+) cannot react with H_2O to produce ·OH over BCM-C_3N_4 nanosheet. The generated ·OH may be attributed to the conversion of the ·O_2^-. In conclusion, the predominant active species (·O_2^- and h^+) as well as the converted products (·OH) could effectively degrade SMZ into CO_2 and H_2O, etc. According to the above analysis, the reaction process can be listed as following equations:

$$BCM\text{-}C_3N_4 + h\nu \longrightarrow BCM\text{-}C_3N_4(h^+ + e^-), \tag{10-16}$$

$$O_2 + e^- \longrightarrow \cdot O_2^-, \tag{10-17}$$

$$\cdot O_2^- + e^- - + 2H^+ \longrightarrow H_2O_2, \tag{10-18}$$

$$H_2O_2 + e^- \longrightarrow OH - + \cdot OH, \tag{10-19}$$

$$(\cdot O_2^-, h^+ \text{ and } \cdot OH) + SMZ \longrightarrow \text{degradation products } (CO_2 + H_2O). \tag{10-20}$$

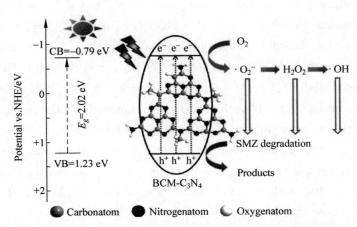

Figure 10-33　A schematic illustration of degradation of SMZ on BCM-C_3N_4 under visible-light irradiation[295]

The similar work is summarized in Table 10-4. The study may provide new insight into the strategies for the design and utilization of highly efficient and stable nonmetal photocatalyst for visible-light-driven degradation of pollutants.

Cr is a common pollutant in industrial wastewater produced by tanning, printing and dyeing, and the fabrication of medicine and preservatives. As Cr-polluted wastewater is highly toxic and complex in chemical composition, its pollution control has become an important part of water treatment research. In water, Cr exists in the two main states of trivalent Cr(III) and hexavalent Cr(VI). Their morphological distributions are not only determines their behavioral characteristics and toxicities in different media, but also affect their treatment methods. Cr(VI) is both mobile and highly toxic, while Cr(III) is a necessary trace element of the human body 100 times less toxic than Cr(VI); it can be removed from water by the simple methods of precipitation and adsorption. Therefore, the typical process for treating Cr(VI) is reducing Cr(VI) to low-toxicity Cr(III) and then removing Cr(III) by precipitation. Photocatalysis is an environmentally friendly technology that effectively reduces Cr(VI) to Cr(III). In this process, the photocatalyst determines the catalytic efficiency, so it is important to find efficient photocatalysts for removing Cr(VI) from wastewater.

The C and N atoms in the g-C_3N_4 structure are sp^2-hybridized, forming a highly delimited p-conjugated system. The N p_z and C p_z orbits form the highest occupied molecular orbital (HOMO) and the lowest unoccupied molecular orbital (LUMO), with the band positions of +1.4 eV and -1.3 eV relative to a normal hydrogen electrode (NHE), respectively. The standard reduction potential of Cr(VI)/Cr(III) is +1.38 eV, below the g-C_3N_4 conduction band (CB). Therefore, from a thermodynamic perspective, the use of g-C_3N_4 for reducing Cr(VI) to Cr(III) is feasible.

However, the application of g-C_3N_4 in the field of photocatalytic treatment retains the following two problems: First, the quantum efficiency is low because of the high recombination rate of photogenerated electron-hole pairs; to improve the photocatalytic activity of g-C_3N_4, many new materials have been prepared, such as Ag-$Sr_{0.25}H_{1.5}Ta_2O_6 \cdot H_2O$/g-$C_3N_4$, g-$C_3N_4$-$TiO_2$, g-$C_3N_4$/ZnO, F-g-$C_3N_4$ and F-doping ultrathin g-C_3N_4. Second, g-C_3N_4 is difficult to separate and recycle from solutions. Usually, the specific surface area is inversely proportional to the particle size. Thus, the particle size of g-C_3N_4 is minimized to increase its specific surface area and thereby improve its catalytic efficiency. The separation of catalysts and the purification processes for catalytic systems are complex and costly. For these reasons, some magnetic materials, such as Fe_3O_4 and a-Fe_2O_3, have been composited with g-C_3N_4 to improve the solid-liquid separation ability of the photocatalyst.

There are different views on the reduction mechanism of Cr(Ⅵ) by g-C_3N_4-based materials. Most studies suggested that Cr(Ⅵ) is converted directly into Cr(Ⅲ) by photogenerated electrons. Hu et al. attributed the reduction of Cr(Ⅵ) to the direct transfer of photogenerated electrons and the indirect transfer of electrons by ·O_2^-. In addition, the study by Dong et al. showed that F-g-C_3N_4 formed by fixing the formate anion on the surface of g-C_3N_4 could change the surface potential and increase the adsorption of Cr(Ⅵ) ions. The F-g-C_3N_4 could also change ·O_2^--mediated indirect reduction to direct photogenerated electron reduction, thereby improving the Cr(Ⅵ) photoreduction. Likewise, Wei et al. also propos that hydrothermally treated g-C_3N_4 in HNO_3 aqueous solution could induce one-step Cr(Ⅵ) reduction directly by electrons owing to the change of its surface chemistry. In summary, although many studies on the photocatalytic treatment of Cr(Ⅵ) have been conducted, the photocatalytic mechanism of Cr(Ⅵ) on the surface of g-C_3N_4-based materials requires further study. Zhao et al. also proposed that magnetically responsive $BiFeO_3$ was coupled with monolayer GO and g-C_3N_4 to form a g-C_3N_4/GO/$BiFeO_3$ ternary coupling material (CNGB), which has the characteristics of good stability, a reasonable band gap, high-catalytic activity, easy solid-liquid separation, and good visible-light response. The effects of environmental conditions on the photoreduction of Cr(Ⅵ) by the CNGB were studied, and the microscopic mechanism of Cr(Ⅵ) photoreduction was deeply investigated.

Figure 10-34 shows the possible mechanism of the generation, transfer, and reaction of the photoelectron-hole pairs. In the photocatalytic reaction system, both g-C_3N_4 and $BiFeO_3$ can be excited by visible light to generate photoelectrons and holes. The photogenerated electrons can transfer from the CB of g-C_3N_4 to that of $BiFeO_3$, because of the difference between the CB edge potentials of g-C_3N_4 and $BiFeO_3$ at -1.33 eV and $+0.58$ eV, respectively. GO can act as an electron sink in collecting and delivering electrons, thereby facilitating the charge transfer and separation from holes. Hence, the Cr(Ⅵ) ions adsorbed on CNGB-2 are reduced to Cr(Ⅲ) by the electrons dispersed on the surfaces of GO nanosheets. Meanwhile, the photoinduced holes can be migrated from VB of $BiFeO_3$ to that of $g-C_3N_4$ as a result of the matching of VBs edge potentials of $g-C_3N_4$ and $BiFeO_3$ at 1.57 eV and 2.51 eV, respectively. As photogenerated holes are retained on the VB of $g-C_3N_4$ and photoinduced electrons accumulate on the CB of $BiFeO_3$, the recombination of photoelectron-hole pairs is largely inhibited. Photogenerated electrons on CBs of $g-C_3N_4$ partly react with O_2 to form $·O_2^-$ which is believed to be a mediator to reduce Cr(Ⅵ). And the photoexcited holes are consumed by H_2O or OH^- to produce $·OH$ which is proved to play a minor role in oxidizing Cr(Ⅲ) back into Cr(Ⅵ), further strengthening the electron-hole separation and improving Cr(Ⅵ) decontamination efficiency. The combined effects of effective visible-light utilization, high separation rate of photoelectron-hole pairs and accelerated charge transfer contribute to high-efficiency photocatalytic reduction capacity[297-298]. This study has important practical significance for improving the efficiency of Cr-contaminated wastewater treatment.

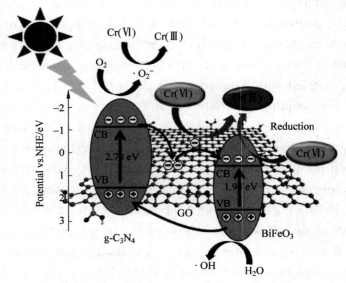

Figure 10-34　Illustration of possible mechanism for Cr(Ⅵ) photoreduction over CNGB-2 heterojunction under visible-light irradiation[296]

10.4 Summary and outlook

Over the past ten years, the effort dedicated to the $g-C_3N_4$-based hybrid nanocomposites has led to a rich knowledge and database for their smart engineering, characterizations, and versatile energy and environmental-related applications. The boundless breakthroughs in the arena of $g-C_3N_4$-based nanomaterials have undoubtedly witnessed novel appealing properties with remarkably ameliorated photocatalytic activity. In this work, on the one hand, we specifically describe the exploration process of graphite carbon nitride nanomaterials; on the other hand, the detailed study route of graphite carbon nitride nanomaterials is introduced systematically, including the types of materials, the methods and techniques of synthesis, characterization, as well as the application. As a result, the doping graphite carbon nitride nanostructures not only show excellent material properties in morphology and optical characterization, but more importantly, also show remarkable catalytic performance in applications, especially in the ORR, HER, measurement of the gas sensing properties, nitrogen photofixations reactions, degradation of CO_2, NO_x, 2-propanol, degradation of organic pollutants and photoreduction removal of Cr(Ⅵ). To this end, the enhancement of the photocatalytic performance of $g-C_3N_4$-based nanostructures was by and large recognized by three main far-reaching criteria, including ①extending the absorption toward the visible-light region (visible-light harvesting), ②hindering the recombination rate of photoinduced charge carriers (charge migration and separation), and ③increasing the amount of adsorbed reactant species on the photocatalyst surface (surface adsorption and reaction). Since the pioneering work of Wang et al. in 2009, ample studies have put emphasis on the construction of $g-C_3N_4$ with a large surface area, a high chemical tunability, prolonged visible-light absorption, reduced band gap, extended π-conjugated structure, highly porous architecture, and enhanced charge transfer and separation. Seeing that there is an abundance of literature related to $g-C_3N_4$-based photoactive materials, this critical review covers a large timeline of scientific literature, as this is essential to elucidate a whole framework and agenda of the progress in this emerging field. Typically, the article presents the state-of-the-art advancements in the development of diverse synthesis strategies to render g-C_3N_4 with improved crystal structure, increased optical absorption, excellent structural design, superior electronic properties, and optimized band arrangements to promote the photocatalytic applications. Although g-C_3N_4 can be facilely synthesized by thermal polymerization of nitrogen-rich precursors, the bulk counterparts are disadvantageous to the photochemical efficiency, as a result of low surface area, limited surface reactive sites, and inadequate utilization of a wide spectrum of solar light. The aforementioned shortcomings can be alleviated by several means. To name a few, these include selection of appropriate g-C_3N_4 precursors, optimized reaction temperatures and durations for the condensation process, an exfoliation-assisted strategy, hard and soft templating approaches, a supramolecular preorganization route, and template-free methods to account for highly porous g-C_3N_4 nanostructures with controllable morphologies. Equally important, g-C_3N_4 can also be modulated at an atomic level via elemental nonmetal and metal doping and also at a molecular level through copolymerization to modify the band alignments, broaden the visible-light response, improve redox ability, increase electron-hole mobility, as well as create surface dyadic heterostructures at the interfaces. Following that, the development of heterojunction nanocomposites by hybridizing g-C_3N_4 with another well-matched energy levels of semiconductors or metals as cocatalysts to form different types of heterojunctions is regarded as a new paradigm toward targeting pronounced enhancement of artificial photoredox applications. In this review, we have divided the heterojunction nanohybrids into several major classes, namely metal-g-C_3N_4 heterojunctions, inorganic semiconductor-g-C_3N_4 heterojunctions, graphitic carbon-g-C_3N_4 heterojunctions, conducting polymer-g-C_3N_4 heterojunctions and g-C_3N_4-based multicomponent nanocomposites. Among all these types, most of the g-C_3N_4-based nanomaterials possess a Type II heterojunction or a Z-scheme heterojunction system, which have been proven to be outstanding for a cornucopia of potential applications such as water splitting to H_2 and O_2, conversion of CO_2 to energy-bearing fuels, pollutant decontamination and bacteria disinfection. Hence, engineering g-C_3N_4-based heterostructured composites at different nanoscales will undeniably enrich the family of visible-light-driven photocatalysts in a more rational manner. In spite of some promising results reported thus far, the studies in this field are still in preliminary stages and further developments are prominently required. Notably, the works still suffer from the relatively low efficiency and low stability of the hybrid nanocomposites, which are far from the requirements of industrial needs. To date, there are a number of issues, which need to be resolved for promising solar-to-fuel conversion before practical applications at commercial scale are possible in the future. Until now, the design of g-C_3N_4 based nanoarchitectures is of extreme significance for superior photoredox efficiency. These include: ①to exploit new strategies to increase the light harvesting ability of g-C_3N_4 to utilize higher wavelengths of light ($\lambda > 500$ nm) to imitate natural photosynthesis in plants, ② to increase the stability of the hybrid photocatalysts, and ③to develop novel and scalable exfoliation approaches for obtaining uniform single or few-layered g-C_3N_4 nanosheets for efficient solar-to-chemical applications. This can give rise to unexpected and captivating results by modifying the existing molecular configuration of g-C_3N_4. Apart from that, g-C_3N_4 nanosheets act as anchoring sites for the decoration of nanoparticles to form a hybrid system. However, it is challenging to control the particle size on the g-C_3N_4 substrate at the molecular level, resulting in severe aggregation. Consequently, the well-contacted heterojunction interface between g-C_3N_4 and the nanoparticles is difficult to accomplish. This markedly affects the intimate interfacial interaction between g-C_3N_4 and the other composites for effective charge transport and separation. To address this shortfall, surface functionalization of g-C_3N_4 to tune the surface charge with specific surface groups will be a positive point to strengthen the anchoring ability of g-C_3N_4. As such, the design of visible-light-active g-C_3N_4-based semiconductor photocatalysts with rational design of nanostructures should be given more emphasis to attain a high quantum efficiency and superior turnover number. Not only that, the development of an effective hybrid photocatalyst system without employing noble metal cocatalysts is necessary to achieve a viable and cost-effective photocatalyst for practical benefits. Therefore, more works that explore an efficient non-noble-metal cocatalyst have become an urgent necessity in order to match well or even outperform the widely used noble metals-modified g-C_3N_4. There have been plenty of recent findings devoted to applications in H_2 evolution and degradation of pollutants using g-C_3N_4-based photocatalysts.

The work on the platforms of O_2 evolution from the other half-reaction of H_2O splitting and CO_2 photoconversion has conceivably gained less attention than being warranted. Furthermore, some key issues that account for the high photocatalytic activity, including optical absorption, electronic band structure, and interfacial charge transfer across the g-C_3N_4-based heterojunction nanocomposites, should be exhaustively investigated to gain theoretical insights by means of first-principles DFT calculations. In terms of experimental work, reactant adsorption sites, charge transfer dynamics, and molecular orbitals should also be further researched. Therefore, the synergy cooperation of both experimental and computational simulations is indispensable to be analyzed together to provide a logical framework for further advancing the current state of knowledge on the photocatalysis. As a whole, the relationship of the key factors in nanoarchitecture design, functionalization, and heterostructuring design can be realisticaly depicted in Figure 10-35 as a mind map to all the readers for the state-of-the-art research in g-C_3N_4-based heterostructure composites. Looking at the future, there is a limitless scope of opportunities and challenges presenting for this booming research hotspot. We hope that this review article will be a good guiding star for the next decade of research in the field of g-C_3N_4-based nanomaterials. We are confident that expanding the knowledge on fundamental aspects via more physical chemistry research accompanied with the experimental results by the joint collaboration between experiment and theory will positively bring new findings in materials science and technology. Without doubt, this will further advance the development road of light-harvesting g-C_3N_4-based nanohybrids in the posthype era. Therefore, with more advances in materials science and engineering, the bottleneck of energy and environmental-related subjects can substantially be addressed. Following that, solar energy conversion applications will be enriched by more practical means. As mentioned earlier, joint efforts from both parties, including chemists (academia) and industrial engineers, are mandatory to deliver the prospects of g-C_3N_4-based nanostructures for practical use to open a revolution of renewable energy. Last but not least, with the synergy from all disciplines of researchers worldwide, including materials scientists, physicists, and chemists, it is highly anticipated that the objectives of accomplishing a clean environment and overcoming the crisis of fossil fuel depletion through the production of chemical fuels from green photocatalysis will no longer be a dream and, more importantly, will turn dreams into reality for heading toward a sustainable future. With that successful accomplishment in years to come, it will unquestionably be very far beyond what has been comprehensively described in this review article.

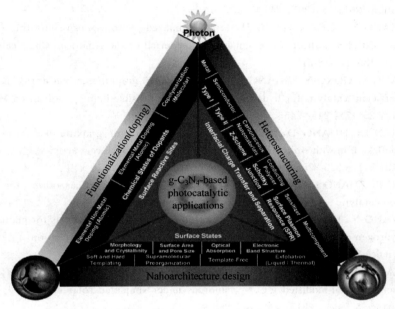

Figure 10-35　Interrelationship of predominant factors in nanoarchitecture design, functionalization, and heterostructuring of g-C_3N_4-based hybrid nanocomposites for advanced solar energy conversion and environmental remediation toward achieving sustainability[299]

References

[1] ZHANG Y, MORI T, NIU L, et al. Non-covalent doping of graphitic carbon nitride polymer with graphene: controlled electronic structure and enhanced optoelectronic conversion[J]. Energy & Environmental Science, 2011, 4(11): 4517-4521.

[2] THOMAS A, FISCHER A, GOETTMANN F, et al. Graphitic carbon nitride materials: variation of structure and morphology and their use as metal-free catalysts[J]. Journal of Materials Chemistry, 2008, 18(41): 4893-4908.

[3] KESSLER F K, ZHENG Y, SCHWARZ D, et al. Functional carbon nitride materials—design strategies for electrochemical devices[J]. Nature Reviews Materials, 2017,

2(6): 1-17.
[4] WANG X, MAEDA K, THOMAS A, et al. A metal-free polymeric photocatalyst for hydrogen production from water under visible light[J]. Nature materials, 2009, 8(1): 76-80.
[5] ALGARA-SILLER G, SEVERIN N, CHONG S Y, et al. Triazine-based graphitic carbon nitride: a two-dimensional semiconductor[J]. Angewandte Chemie International Edition, 2014, 53(29): 7450-7455.
[6] ZHANG X, XIE X, WANG H, et al. Enhanced photoresponsive ultrathin graphitic-phase C3N4 nanosheets for bioimaging[J]. Journal of the American Chemical Society, 2013, 135(1): 18-21.
[7] YANG S, GONG Y, ZHANG J, et al. Exfoliated graphitic carbon nitride nanosheets as efficient catalysts for hydrogen evolution under visible light[J]. Advanced Materials, 2013, 25(17): 2452-2456.
[8] CHEN L, HUANG D, REN S, et al. Preparation of graphite-like carbon nitride nanoflake film with strong fluorescent and electrochemiluminescent activity[J]. Nanoscale, 2013, 5(1): 225-230.
[9] NIU P, ZHANG L, LIU G, et al. Graphene-like carbon nitride nanosheets for improved photocatalytic activities[J]. Advanced Functional Materials, 2012, 22(22): 4763-4770.
[10] WANG Y, HONG J, ZHANG W, et al. Carbon nitride nanosheets for photocatalytic hydrogen evolution: remarkably enhanced activity by dye sensitization[J]. Catalysis Science & Technology, 2013, 3(7): 1703-1711.
[11] TAHIR B, TAHIR M, AMIN N A S. Photo-induced CO2 reduction by CH4/H2O to fuels over Cu-modified g-C_3N_4 nanorods under simulated solar energy[J]. Applied Surface Science, 2017, 419: 875-885.
[12] LIAO G, CHEN S, QUAN X, et al. Graphene oxide modified gC_3N_4 hybrid with enhanced photocatalytic capability under visible light irradiation[J]. Journal of Materials Chemistry, 2012, 22(6): 2721-2726.
[13] YE L, LIU J, JIANG Z, et al. Facets coupling of BiOBr-g-C_3N_4 composite photocatalyst for enhanced visible-light-driven photocatalytic activity [J]. Applied Catalysis B: Environmental, 2013, 142: 1-7.
[14] WANG Y, ZHANG J, WANG X, et al. Boron-and fluorine-containing mesoporous carbon nitride polymers: metal-free catalysts for cyclohexane oxidation[J]. Angewandte Chemie International Edition, 2010, 49(19): 3356-3359.
[15] WANG Y, LI H, YAO J, et al. Synthesis of boron doped polymeric carbon nitride solids and their use as metal-free catalysts for aliphatic C-H bond oxidation [J]. Chemical Science, 2011, 2(3): 446-450.
[16] WANG Y, DI Y, ANTONIETTI M, et al. Excellent visible-light photocatalysis of fluorinated polymeric carbon nitride solids[J]. Chemistry of Materials, 2010, 22(18): 5119-5121.
[17] LIU G, NIU P, SUN C, et al. Unique electronic structure induced high photoreactivity of

sulfur-doped graphitic C_3N_4[J]. Journal of the American Chemical Society, 2010, 132(33): 11642-11648.

[18] ZHANG J, SUN J, MAEDA K, et al. Sulfur-mediated synthesis of carbon nitride: band-gap engineering and improved functions for photocatalysis[J]. Energy & Environmental Science, 2011, 4(3): 675-678.

[19] SHILOV A E, SHUL'PIN G B. Activation of C-H bonds by metal complexes[J]. Chemical Reviews, 1997, 97(8): 2879-2932.

[20] DAS S, INCARVITO C D, CRABTREE R H, et al. Molecular recognition in the selective oxygenation of saturated CH bonds by a dimanganese catalyst[J]. Science, 2006, 312(5782): 1941-1943.

[21] LEE S J, CHO S H, MULFORT K L, et al. Cavity-tailored, self-sorting supramolecular catalytic boxes for selective oxidation[J]. Journal of the American Chemical Society, 2008, 130(50): 16828-16829.

[22] GUENGERICH F P. Common and uncommon cytochrome P450 reactions related to metabolism and chemical toxicity[J]. Chemical research in toxicology, 2001, 14(6): 611-650.

[23] NAM W. Dioxygen activation by metalloenzymes and models[J]. 2007, 40(7): 465.

[24] ZHANG Y, ANTONIETTI M. Photocurrent generation by polymeric carbon nitride solids: an initial step towards a novel photovoltaic system[J]. Chemistry-An Asian Journal, 2010, 5(6): 1307-1311.

[25] PAN C, XU J, WANG Y, et al. Dramatic activity of $C_3N_4/BiPO_4$ photocatalyst with core/shell structure formed by self-assembly[J]. Advanced Functional Materials, 2012, 22(7): 1518-1524.

[26] KUMAR S, SURENDAR T, BARUAH A, et al. Synthesis of a novel and stable g C_3N_4-Ag_3PO_4 hybrid nanocomposite photocatalyst and study of the photocatalytic activity under visible light irradiation[J]. Journal of Materials Chemistry A, 2013, 1(17): 5333-5340.

[27] ZHOU X, JIN B, LI L, et al. A carbon nitride/TiO_2 nanotube array heterojunction visible-light photocatalyst: synthesis, characterization, and photoelectrochemical properties [J]. Journal of Materials Chemistry, 2012, 22(34): 17900-17905.

[28] WANG Y, SHI R, LIN J, et al. Enhancement of photocurrent and photocatalytic activity of ZnO hybridized with graphite-like C_3N_4[J]. Energy & Environmental Science, 2011, 4(8): 2922-2929.

[29] SUN J X, YUAN Y P, QIU L G, et al. Fabrication of composite photocatalyst g C_3N_4-ZnO and enhancement of photocatalytic activity under visible light[J]. Dalton Transactions, 2012, 41(22): 6756-6763.

[30] LI M, HUANG X, WU C, et al. Fabrication of two-dimensional hybrid sheets by decorating insulating PANI on reduced graphene oxide for polymer nanocomposites with low dielectric loss and high dielectric constant[J]. Journal of Materials Chemistry, 2012, 22(44): 23477-23484.

[31] WANG Y, WANG Z, MUHAMMAD S, et al. Graphite-like C_3N_4 hybridized $ZnWO_4$ nanorods: synthesis and its enhanced photocatalysis in visible light [J]. CrystEngComm, 2012, 14(15): 5065-5070.

[32] GE L, HAN C, LIU J. Novel visible light-induced g-C_3N_4/Bi_2WO_6 composite photocatalysts for efficient degradation of methyl orange[J]. Applied Catalysis B: Environmental, 2011, 108: 100-107.

[33] MI X, WANG Y, LI R, et al. Multiple surface plasmon resonances enhanced nonlinear optical microscopy[J]. Nanophotonics, 2019, 8(3): 487-493.

[34] FUKUI T, MURATA K, OHARA S, et al. Morphology control of Ni-YSZ cermet anode for lower temperature operation of SOFCs[J]. Journal of Power Sources, 2004, 125(1): 17-21.

[35] WANG Y, YAO J, LI H, et al. Highly selective hydrogenation of phenol and derivatives over a Pd at carbon nitride catalyst in aqueous media[J]. Journal of the American Chemical Society, 2011, 133(8): 2362-2365.

[36] DATTA S J, YOON K B. Synthesis of Ideal AM-6 and Elucidation of V4+-to-O Charge Transfer in Vanadate Quantum Wires [J]. Angewandte Chemie, 2010, 122(29): 5091-5095.

[37] LI X H, WANG X, ANTONIETTI M. Mesoporous g C_3N_4 nanorods as multifunctional supports of ultrafine metal nanoparticles: hydrogen generation from water and reduction of nitrophenol with tandem catalysis in one step[J]. Chemical Science, 2012, 3(6): 2170-2174.

[38] CHEN L, ZENG X, SI P, et al. Gold nanoparticle-graphite-like C_3N_4 nanosheet nanohybrids used for electrochemiluminescent immunosensor[J]. Analytical chemistry, 2014, 86(9): 4188-4195.

[39] CAO J, GONG Y, WANG Y, et al. Cocoon-like ZnO decorated graphitic carbon nitride nanocomposite: hydrothermal synthesis and ethanol gas sensing application[J]. Materials Letters, 2017, 198: 76-80.

[40] SHI Y, WAN Y, ZHAO D. Ordered mesoporous non-oxide materials[J]. Chemical Society Reviews, 2011, 40(7): 3854-3878.

[41] MEILIKHOV M, YUSENKO K, ESKEN D, et al. Metals at MOFs-loading MOFs with metal nanoparticles for hybrid functions[J]. European Journal of Inorganic Chemistry, 2010, 2010(24): 3701-3714.

[42] ZHENG Y, JIAO Y, CHEN J, et al. Nanoporous graphitic-C_3N_4 at carbon metal-free electrocatalysts for highly efficient oxygen reduction[J]. Journal of the American Chemical Society, 2011, 133(50): 20116-20119.

[43] YANG S, FENG X, WANG X, et al. Graphene-based carbon nitride nanosheets as efficient metal-free electrocatalysts for oxygen reduction reactions[J]. Angewandte Chemie, 2011, 123(23): 5451-5455.

[44] BERGER C, SONG Z, LI X, et al. Electronic confinement and coherence in patterned epitaxial graphene[J]. Science, 2006, 312(5777): 1191-1196.

[45] SRIVASTAVA A, GALANDE C, CI L, et al. Novel liquid precursor-based facile synthesis of large-area continuous, single, and few-layer graphene films[J]. Chemistry of Materials, 2010, 22(11): 3457-3461.

[46] LI X, CAI W, AN J, et al. Large-area synthesis of high-quality and uniform graphene films on copper foils[J]. Science, 2009, 324(5932): 1312-1314.

[47] LIN W, CAO Y, WANG P, et al. Unified treatment for plasmon-exciton co-driven reduction and oxidation reactions[J]. Langmuir, 2017, 33(43): 12102-12107.

[48] ZHANG S, HANG N T, ZHANG Z, et al. Preparation of g-C_3N_4/graphene composite for detecting NO_2 at room temperature[J]. Nanomaterials, 2017, 7(1): 12.

[49] ZHAO Y, ZHAO F, WANG X, et al. Graphitic carbon nitride nanoribbons: graphene-assisted formation and synergic function for highly efficient hydrogen evolution [J]. Angewandte Chemie International Edition, 2014, 53(50): 13934-13939.

[50] KUHN P, FORGET A, SU D, et al. From microporous regular frameworks to mesoporous materials with ultrahigh surface area: dynamic reorganization of porous polymer networks[J]. Journal of the American Chemical Society, 2008, 130(40): 13333-13337.

[51] KUHN P, ANTONIETTI M, THOMAS A. Porous, covalent triazine-based frameworks prepared by ionothermal synthesis[J]. Angewandte Chemie International Edition, 2008, 47(18): 3450-3453.

[52] LIU J, WANG H, CHEN Z P, et al. Microcontact-printing-assisted access of graphitic carbon nitride films with favorable textures toward photoelectrochemical application[J]. Advanced Materials, 2015, 27(4): 712-718.

[53] LIU Y, CHEN M, HU M, et al. In-situ anchoring sulfiphilic silica nanoparticles onto macro-mesoporous carbon framework for cost-effective Li-S cathodes[J]. Chemical Engineering Journal, 406: 126781.

[54] BERGER C, SONG Z, LI X, et al. Electronic confinement and coherence in patterned epitaxial graphene[J]. Science, 2006, 312(5777): 1191-1196.

[55] WU X, YIN S, DONG Q, et al. Synthesis of high visible light active carbon doped TiO_2 photocatalyst by a facile calcination assisted solvothermal method[J]. Applied Catalysis B: Environmental, 2013, 142: 450-457.

[56] ZHANG Z, XIAO F, QIAN L, et al. Facile Synthesis of 3D MnO_2-graphene and carbon nanotube-graphene composite networks for high-performance, flexible, all-solid-state asymmetric supercapacitors [J]. Advanced Energy Materials, 2014, 4(10): 1400064.

[57] WU Z S, WINTER A, CHEN L, et al. Three-dimensional nitrogen and boron Co-doped graphene for high-performance all-solid-state supercapacitors[J]. Advanced Materials, 2012, 24(37): 5130-5135.

[58] DING Y L, KOPOLD P, HAHN K, et al. Facile solid-state growth of 3D well-interconnected nitrogen-rich carbon nanotube-graphene hybrid architectures for lithium-sulfur batteries[J]. Advanced Functional Materials, 2016, 26(7): 1112-1119.

[59] HUANG H, YANG S, VAJTAI R, et al. Pt-decorated 3D architectures built from graphene and graphitic carbon nitride nanosheets as efficient methanol oxidation catalysts[J]. Advanced Materials, 2014, 26(30): 5160-5165.

[60] DUAN J, CHEN S, JARONIEC M, et al. Porous C_3N_4 nanolayers at N-graphene films as catalyst electrodes for highly efficient hydrogen evolution[J]. ACS Nano, 2015, 9(1): 931-940.

[61] SHI Y, FU L, CHEN X, et al. Hypophosphite/graphitic carbon nitride hybrids: preparation and flame-retardant application in thermoplastic polyurethane [J]. Nanomaterials, 2017, 7(9): 259.

[62] XIAO D, DAI K, QU Y, et al. Hydrothermal synthesis of α-Fe_2O_3/g-C_3N_4 composite and its efficient photocatalytic reduction of Cr (Ⅵ) under visible light[J]. Applied Surface Science, 2015, 358: 181-187.

[63] HAN X, TIAN L, JIANG H, et al. Facile transformation of low cost melamine-oxalic acid into porous graphitic carbon nitride nanosheets with high visible-light photocatalytic performance[J]. RSC Advances, 2017, 7(24): 14372-14381.

[64] HAN X, WANG Y, LV J, et al. An artful and simple synthetic strategy for fabricating low carbon residual porous g C_3N_4 with enhanced visible-light photocatalytic properties[J]. RSC Advances, 2016, 6(87): 83730-83737.

[65] HAN D, LIU J, CAI H, et al. High-yield and low-cost method to synthesize large-area porous g-C_3N_4 nanosheets with improved photocatalytic activity for gaseous nitric oxide and 2-propanol photodegradation[J]. Applied Surface Science, 2019, 464: 577-585.

[66] DING Q, LI R, CHEN M, et al. Ag nanoparticles-TiO_2 film hybrid for plasmon-exciton co-driven surface catalytic reactions[J]. Applied Materials Today, 2017, 9: 251-258.

[67] ALGARA-SILLER G, SEVERIN N, CHONG S Y, et al. Triazine-based graphitic carbon nitride: a two-dimensional semiconductor[J]. Angewandte Chemie International Edition, 2014, 53(29): 7450-7455.

[68] MA T Y, DAI S, JARONIEC M, et al. Graphitic carbon nitride nanosheet-carbon nanotube three-dimensional porous composites as high-performance oxygen evolution electrocatalysts[J]. Angewandte Chemie, 2014, 126(28): 7409-7413.

[69] ZHANG Q, UCHAKER E, CANDELARIA S L, et al. Nanomaterials for energy conversion and storage[J]. Chemical Society Reviews, 2013, 42(7): 3127-3171.

[70] LI Y, WANG J, YANG Y, et al. Seed-induced growing various TiO_2 nanostructures on g-C_3N_4 nanosheets with much enhanced photocatalytic activity under visible light [J]. Journal of hazardous materials, 2015, 292: 79-89.

[71] ZHANG X, XIE X, WANG H, et al. Enhanced photoresponsive ultrathin graphitic-phaseC_3N_4 nanosheets for bioimaging[J]. Journal of the American Chemical Society, 2013, 135(1): 18-21.

[72] ZHANG X, WANG H, WANG H, et al. Single-layered graphitic-C_3N_4 quantum dots

for two-photon fluorescence imaging of cellular nucleus[J]. Advanced Materials, 2014, 26(26): 4438-4443.

[73] KROKE E, SCHWARZ M. Novel group 14 nitrides[J]. Coordination Chemistry Reviews, 2004, 248(5-6): 493-532.

[74] LI J, CAO C, ZHU H. Synthesis and characterization of graphite-like carbon nitride nanobelts and nanotubes[J]. Nanotechnology, 2007, 18(11): 115605.

[75] LEE E Z, JUN Y S, HONG W H, et al. Cubic mesoporous graphitic carbon (Ⅳ) nitride: an all-in-one chemosensor for selective optical sensing of metal ions[J]. Angewandte Chemie, 2010, 122(50): 9900-9904.

[76] ZHANG J, DAI L. Heteroatom-doped graphitic carbon catalysts for efficient electrocatalysis of oxygen reduction reaction[J]. ACS Catalysis, 2015, 5(12): 7244-7253.

[77] KOHL S W, WEINER L, SCHWARTSBURD L, et al. Consecutive thermal H_2 and light-induced O_2 evolution from water promoted by a metal complex[J]. Science, 2009, 324(5923): 74-77.

[78] MAEDA K, WANG X, NISHIHARA Y, et al. Photocatalytic activities of graphitic carbon nitride powder for water reduction and oxidation under visible light[J]. The Journal of Physical Chemistry C, 2009, 113(12): 4940-4947.

[79] YAN S C, LI Z S, ZOU Z G. Photodegradation of rhodamine B and methyl orange over boron-doped g-C_3N_4 under visible light irradiation[J]. Langmuir, 2010, 26(6): 3894-3901.

[80] ZHANG J, ZHU Z, TANG Y, et al. Titania nanosheet-mediated construction of a two-dimensional titania/cadmium sulfide heterostructure for high hydrogen evolution activity[J]. Advanced Materials, 2014, 26(5): 734-738.

[81] WU G, MORE K L, JOHNSTON C M, et al. High-performance electrocatalysts for oxygen reduction derived from polyaniline, iron, and cobalt[J]. Science, 2011, 332(6028): 443-447.

[82] GONG K, DU F, XIA Z, et al. Nitrogen-doped carbon nanotube arrays with high electrocatalytic activity for oxygen reduction[J]. Science, 2009, 323(5915): 760-764.

[83] GASTEIGER H A, MARKOVIĆ N M. Just a dream—or future reality[J]. Science, 2009, 324(5923): 48-49.

[84] SUN C, LI F, MA C, et al. Graphene-Co_3O_4 nanocomposite as an efficient bifunctional catalyst for lithium-air batteries[J]. Journal of Materials Chemistry A, 2014, 2(20): 7188-7196.

[85] YU D, XU C, SU Y, et al. Nitrogen-doped graphene aerogels-supported cobaltosic oxide nanocrystals as high-performance bi-functional electrocatalysts for oxygen reduction and evolution reactions[J]. Journal of Electroanalytical Chemistry, 2017, 787: 46-54.

[86] ARMAND M, TARASCON J M. Building better batteries[J]. Nature, 2008, 451(7179): 652-657.

[87] DEBE M K. Electrocatalyst approaches and challenges for automotive fuel cells[J]. Nature, 2012, 486(7401): 43-51.

[88] CHEN T W, KANG J X, ZHANG D F, et al. Ultralong PtNi alloy nanowires enabled by the coordination effect with superior ORR durability[J]. RSC Advances, 2016, 6(75): 71501-71506.

[89] GAN L, RUDI S, CUI C, et al. Size-controlled synthesis of sub-10 nm PtNi$_3$ alloy nanoparticles and their unusual volcano-shaped size effect on ORR electrocatalysis[J]. Small, 2016, 12(23): 3189-3196.

[90] PICKRAHN K L, PARK S W, GORLIN Y, et al. Active MnO$_x$ electrocatalysts prepared by atomic layer deposition for oxygen evolution and oxygen reduction reactions[J]. Advanced Energy Materials, 2012, 2(10): 1269-1277.

[91] DAI L, XUE Y, QU L, et al. Metal-free catalysts for oxygen reduction reaction[J]. Chemical Reviews, 2015, 115(11): 4823-4892.

[92] WU R, CHEN S, ZHANG Y, et al. Template-free synthesis of hollow nitrogen-doped carbon as efficient electrocatalysts for oxygen reduction reaction[J]. Journal of Power Sources, 2015, 274: 645-650.

[93] WANG R, ZHOU T, WANG H, et al. Lysine-derived mesoporous carbon nanotubes as a proficient non-precious catalyst for oxygen reduction reaction[J]. Journal of Power Sources, 2014, 269: 54-60.

[94] ONG W J, TAN L L, NG Y H, et al. Graphitic carbon nitride (g-C$_3$N$_4$)-based photocatalysts for artificial photosynthesis and environmental remediation: are we a step closer to achieving sustainability[J]. Chemical Reviews, 2016, 116(12): 7159-7329.

[95] WANG X, MAEDA K, THOMAS A, et al. A metal-free polymeric photocatalyst for hydrogen production from water under visible light[J]. Nature Materials, 2009, 8(1): 76-80.

[96] MAEDA K, WANG X, NISHIHARA Y, et al. Photocatalytic activities of graphitic carbon nitride powder for water reduction and oxidation under visible light[J]. The Journal of Physical Chemistry C, 2009, 113(12): 4940-4947.

[97] XU Q, CHENG B, YU J, et al. Making co-condensed amorphous carbon/g-C$_3$N$_4$ composites with improved visible-light photocatalytic H$_2$-production performance using Pt as cocatalyst[J]. Carbon, 2017, 118: 241-249.

[98] WANG K, LI Q, LIU B, et al. Sulfur-doped g-C$_3$N$_4$ with enhanced photocatalytic CO$_2$-reduction performance[J]. Applied Catalysis B: Environmental, 2015, 176: 44-52.

[99] LI K, SU F Y, ZHANG W D. Modification of g-C$_3$N$_4$ nanosheets by carbon quantum dots for highly efficient photocatalytic generation of hydrogen[J]. Applied Surface Science, 2016, 375: 110-117.

[100] JIAN X, LIU X, YANG H, et al. Construction of carbon quantum dots/proton-functionalized graphitic carbon nitride nanocomposite via electrostatic self-assembly strategy and its application[J]. Applied Surface Science, 2016, 370: 514-521.

[101] LIANG J, ZHENG Y, CHEN J, et al. Facile oxygen reduction on a three-dimensionally ordered macroporous graphitic C_3N_4/carbon composite electrocatalyst [J]. Angewandte Chemie International Edition, 2012, 51(16): 3892-3896.

[102] SANO T, TSUTSUI S, KOIKE K, et al. Activation of graphitic carbon nitride (g-C_3N_4) by alkaline hydrothermal treatment for photocatalytic NO oxidation in gas phase[J]. Journal of Materials Chemistry A, 2013, 1(21): 6489-6496.

[103] WANG Y, HONG J, ZHANG W, et al. Carbon nitride nanosheets for photocatalytic hydrogen evolution: remarkably enhanced activity by dye sensitization[J]. Catalysis Science & Technology, 2013, 3(7): 1703-1711.

[104] ZHANG Y, MORI T, YE J, et al. Phosphorus-doped carbon nitride solid: enhanced electrical conductivity and photocurrent generation[J]. Journal of the American Chemical Society, 2010, 132(18): 6294-6295.

[105] CHEN F, YANG Q, WANG Y, et al. Novel ternary heterojunction photcocatalyst of Ag nanoparticles and g-C_3N_4 nanosheets co-modified $BiVO_4$ for wider spectrum visible-light photocatalytic degradation of refractory pollutant[J]. Applied Catalysis B: Environmental, 2017, 205: 133-147.

[106] SHANG Y, CHEN X, LIU W, et al. Photocorrosion inhibition and high-efficiency photoactivity of porous g-C_3N_4/Ag_2CrO_4 composites by simple microemulsion-assisted co-precipitation method[J]. Applied Catalysis B: Environmental, 2017, 204: 78-88.

[107] ZHANG W, XIAO X, LI Y, et al. Liquid-exfoliation of layered MoS_2 for enhancing photocatalytic activity of TiO_2/g-C_3N_4 photocatalyst and DFT study[J]. Applied Surface Science, 2016, 389: 496-506.

[108] ZHANG L, CHEN X, GUAN J, et al. Facile synthesis of phosphorus doped graphitic carbon nitride polymers with enhanced visible-light photocatalytic activity[J]. Materials Research Bulletin, 2013, 48(9): 3485-3491.

[109] LU C, CHEN R, WU X, et al. Boron doped g-C_3N_4 with enhanced photocatalytic UO_2^{2+} reduction performance[J]. Applied Surface Science, 2016, 360: 1016-1022.

[110] HAN Q, HU C, ZHAO F, et al. One-step preparation of iodine-doped graphitic carbon nitride nanosheets as efficient photocatalysts for visible light water splitting [J]. Journal of Materials Chemistry A, 2015, 3(8): 4612-4619.

[111] MA X, LV Y, XU J, et al. A strategy of enhancing the photoactivity of g-C_3N_4 via doping of nonmetal elements: a first-principles study[J]. The Journal of Physical Chemistry C, 2012, 116(44): 23485-23493.

[112] LIU S, LI D, SUN H, et al. Oxygen functional groups in graphitic carbon nitride for enhanced photocatalysis[J]. Journal of Colloid and Interface Science, 2016, 468: 176-182.

[113] WEN J, XIE J, CHEN X, et al. A review on g-C_3N_4-based photocatalysts[J]. Applied Surface Science, 2017, 391: 72-123.

[114] AI B, DUAN X, SUN H, et al. Metal-free graphene-carbon nitride hybrids for photodegradation of organic pollutants in water[J]. Catalysis Today, 2015, 258: 668-675.

[115] ZHANG J, GUO C, ZHANG L, et al. Direct growth of flower-like manganese oxide on reduced graphene oxide towards efficient oxygen reduction reaction[J]. Chemical Communications, 2013, 49(56): 6334-6336.

[116] LIU Z W, PENG F, WANG H J, et al. Phosphorus-doped graphite layers with high electrocatalytic activity for the O_2 reduction in an alkaline medium[J]. Angewandte Chemie International Edition, 2011, 50(14): 3257-3261.

[117] SUN Y, LI C, XU Y, et al. Chemically converted graphene as substrate for immobilizing and enhancing the activity of a polymeric catalyst[J]. Chemical Communications, 2010, 46(26): 4740-4742.

[118] LIU R, WU D, FENG X, et al. Nitrogen-doped ordered mesoporous graphitic arrays with high electrocatalytic activity for oxygen reduction[J]. Angewandte Chemie, 2010, 122(14): 2619-2623.

[119] KWON K, JIN S, PAK C, et al. Enhancement of electrochemical stability and catalytic activity of Pt nanoparticles via strong metal-support interaction with sulfur-containing ordered mesoporous carbon[J]. Catalysis Today, 2011, 164(1): 186-189.

[120] SMITH M C, GILBERT J A, MAWDSLEY J R, et al. In situ small-angle X-ray scattering observation of Pt catalyst particle growth during potential cycling[J]. Journal of the American Chemical Society, 2008, 130(26): 8112-8113.

[121] LEE K, SAVADOGO O, ISHIHARA A, et al. Methanol-tolerant oxygen reduction electrocatalysts based on Pd-3D transition metal alloys for direct methanol fuel cells [J]. Journal of the Electrochemical Society, 2005, 153(1): A20.

[122] CARRETTE L, FRIEDRICH K A, STIMMING U. Fuel cells-fundamentals and applications[J]. FuelCells, 2001, 1(1): 5-39.

[123] KWON K, SA Y J, CHEON J Y, et al. Ordered mesoporous carbon nitrides with graphitic frameworks as metal-free, highly durable, methanol-tolerant oxygen reduction catalysts in an acidic medium[J]. Langmuir, 2012, 28(1): 991-996.

[124] CHEN X, ZHANG J, FU X, et al. Fe-g-C_3N_4-catalyzed oxidation of benzene to phenol using hydrogen peroxide and visible light[J]. Journal of the American Chemical Society, 2009, 131(33): 11658-11659.

[125] KWON K, SA Y J, CHEON J Y, et al. Ordered mesoporous carbon nitrides with graphitic frameworks as metal-free, highly durable, methanol-tolerant oxygen reduction catalysts in an acidic medium[J]. Langmuir, 2012, 28(1): 991-996.

[126] ZHENG Y, JIAO Y, CHEN J, et al. Nanoporous graphitic-C_3N_4 at carbon metal-free electrocatalysts for highly efficient oxygen reduction[J]. Journal of the American Chemical Society, 2011, 133(50): 20116-20119.

[127] LIU Q, ZHANG J. Graphene supported Co-g-C_3N_4 as a novel metal-macrocyclic electrocatalyst for the oxygen reduction reaction in fuel cells[J]. Langmuir, 2013, 29(11): 3821-3828.

[128] WANG S, IYYAMPERUMAL E, ROY A, et al. Vertically aligned BCN nanotubes as efficient metal-free electrocatalysts for the oxygen reduction reaction: a synergetic

effect by co-doping with boron and nitrogen[J]. Angewandte Chemie International Edition, 2011, 50(49): 11756-11760.

[129] LI Y, ZHAO Y, CHENG H, et al. Nitrogen-doped graphene quantum dots with oxygen-rich functional groups[J]. Journal of the American Chemical Society, 2012, 134(1): 15-18.

[130] PARK H W, LEE D U, ZAMANI P, et al. Electrospun porous nanorod perovskite oxide/nitrogen-doped graphene composite as a bi-functional catalyst for metal air batteries[J]. Nano Energy, 2014, 10: 192-200.

[131] UNNI S M, ILLATHVALAPPIL R, GANGADHARAN P K, et al. Layer-separated distribution of nitrogen doped graphene by wrapping on carbon nitride tetrapods for enhanced oxygen reduction reactions in acidic medium [J]. Chemical Communications, 2014, 50(89): 13769-13772.

[132] HIGGINS D C, HOQUE M A, HASSAN F, et al. Oxygen reduction on graphene-carbon nanotube composites doped sequentially with nitrogen and sulfur[J]. ACS Catalysis, 2014, 4(8): 2734-2740.

[133] ZHANG Y, FUGANE K, MORI T, et al. Wet chemical synthesis of nitrogen-doped graphene towards oxygen reduction electrocatalysts without high-temperature pyrolysis[J]. Journal of Materials Chemistry, 2012, 22(14): 6575-6580.

[134] WANG D W, SU D. Heterogeneous nanocarbon materials for oxygen reduction reaction[J]. Energy & Environmental Science, 2014, 7(2): 576-591.

[135] XING T, ZHENG Y, LI L H, et al. Observation of active sites for oxygen reduction reaction on nitrogen-doped multilayer graphene [J]. ACS Nano, 2014, 8 (7): 6856-6862.

[136] GENG D, CHEN Y, CHEN Y, et al. High oxygen-reduction activity and durability of nitrogen-doped graphene[J]. Energy & Environmental Science, 2011, 4 (3): 760-764.

[137] YE T N, LV L B, LI X H, et al. Strongly veined carbon nanoleaves as a highly efficient metal-free electrocatalyst[J]. Angewandte Chemie, 2014, 126(27): 7025-7029.

[138] HE W, JIANG C, WANG J, et al. High-rate oxygen electroreduction over graphitic-n species exposed on 3D hierarchically porous nitrogen-doped carbons [J]. Angewandte Chemie, 2014, 126(36): 9657-9661.

[139] MENG Y, VOIRY D, GOSWAMI A, et al. N-, O-, and S-tridoped nanoporous carbons as selective catalysts for oxygen reduction and alcohol oxidation reactions[J]. Journal of the American Chemical Society, 2014, 136(39): 13554-13557.

[140] LIN Z, WALLER G, LIU Y, et al. Facile synthesis of nitrogen-doped graphene via pyrolysis of graphene oxide and urea, and its electrocatalytic activity toward the oxygen-reduction reaction[J]. Advanced Energy Materials, 2012, 2(7): 884-888.

[141] ZHANG P, SUN F, XIANG Z, et al. ZIF-derived in situ nitrogen-doped porous carbons as efficient metal-free electrocatalysts for oxygen reduction reaction[J]. Energy & Environmental Science, 2014, 7(1): 442-450.

[142] CHEN P, WANG L K, WANG G, et al. Nitrogen-doped nanoporous carbon nanosheets derived from plant biomass: an efficient catalyst for oxygen reduction reaction[J]. Energy & Environmental Science, 2014, 7(12): 4095-4103.

[143] LI X H, KURASCH S, KAISER U, et al. Synthesis of monolayer-patched graphene from glucose[J]. Angewandte Chemie International Edition, 2012, 51(38): 9689-9692.

[144] ZHAO H, LEI M, YANG X, et al. Route to GaN and VN assisted by carbothermal reduction process[J]. Journal of the American Chemical Society, 2005, 127(45): 15722-15723.

[145] FISCHER A, ANTONIETTI M, THOMAS A. Growth confined by the nitrogen source: synthesis of pure metal nitride nanoparticles in mesoporous graphitic carbon nitride[J]. Advanced Materials, 2007, 19(2): 264-267.

[146] QU L, LIU Y, BAEK J B, et al. Nitrogen-doped graphene as efficient metal-free electrocatalyst for oxygen reduction in fuel cells[J]. ACS Nano, 2010, 4(3): 1321-1326.

[147] LI X, WANG H, ROBINSON J T, et al. Simultaneous nitrogen doping and reduction of graphene oxide[J]. Journal of the American Chemical Society, 2009, 131(43): 15939-15944.

[148] LI J, ZHANG Y, ZHANG X, et al. Direct transformation from graphitic C_3N_4 to nitrogen-doped graphene: an efficient metal-free electrocatalyst for oxygen reduction reaction[J]. ACS Applied Materials & Interfaces, 2015, 7(35): 19626-19634.

[149] SILVA R, VOIRY D, CHHOWALLA M, et al. Efficient metal-free electrocatalysts for oxygen reduction: polyaniline-derived N-and O-doped mesoporous carbons[J]. Journal of the American Chemical Society, 2013, 135(21): 7823-7826.

[150] CHEN H, SUN F, WANG J, et al. Nitrogen doping effects on the physical and chemical properties of mesoporous carbons[J]. The Journal of Physical Chemistry C, 2013, 117(16): 8318-8328.

[151] DAEMS N, SHENG X, VANKELECOM I F J, et al. Metal-free doped carbon materials as electrocatalysts for the oxygen reduction reaction[J]. Journal of Materials Chemistry A, 2014, 2(12): 4085-4110.

[152] ZHENG D, WANG X. Integrating CdS quantum dots on hollow graphitic carbon nitride nanospheres for hydrogen evolution photocatalysis[J]. Applied Catalysis B: Environmental, 2015, 179: 479-488.

[153] XIE Y P, YU Z B, LIU G, et al. CdS-mesoporous ZnS core-shell particles for efficient and stable photocatalytic hydrogen evolution under visible light[J]. Energy & Environmental Science, 2014, 7(6): 1895-1901.

[154] MA T Y, DAI S, JARONIEC M, et al. Graphitic carbon nitride nanosheet-carbon nanotube three-dimensional porous composites as high-performance oxygen evolution electrocatalysts[J]. Angewandte Chemie, 2014, 126(28): 7409-7413.

[155] FUJISHIMA A, HONDA K. Electrochemical photolysis of water at a semiconductor electrode[J]. Nature, 1972, 238(5358): 37-38.

[156] CHEN X, SHEN S, GUO L, et al. Semiconductor-based photocatalytic hydrogen

generation[J]. Chemical Reviews, 2010, 110(11): 6503-6570.
[157] WALTER M G, WARREN E L, MCKONE J R, et al. Solar water splitting cells[J]. Chemical Reviews, 2010, 110(11): 6446-6473.
[158] TONG H, OUYANG S, BI Y, et al. Nano-photocatalytic materials: possibilities and challenges[J]. Advanced Materials, 2012, 24(2): 229-251.
[159] SCHNEIDER J, MATSUOKA M, TAKEUCHI M, et al. Understanding TiO_2 photocatalysis: mechanisms and materials[J]. Chemical Reviews, 2014, 114(19): 9919-9986.
[160] NOLAN M, IWASZUK A, LUCID A K, et al. Design of novel visible light active photocatalyst materials: surface modified TiO_2 [J]. Advanced Materials, 2016, 28(27): 5425-5446.
[161] LI Q, LI X, WAGEH S, et al. CdS/Graphene nanocomposite photocatalysts[J]. Adv Energy Mater, 2015, 5: 1500010.
[162] TRAN P D, WONG L H, BARBER J, et al. Recent advances in hybrid photocatalysts for solar fuel production[J]. Energy & Environmental Science, 2012, 5(3): 5902-5918.
[163] CHEN S, WANG L W. Thermodynamic oxidation and reduction potentials of photocatalytic semiconductors in aqueous solution[J]. Chemistry of Materials, 2012, 24(18): 3659-3666.
[164] SIVULA K, VAN DE KROL R. Semiconducting materials for photoelectrochemical energy conversion[J]. Nature Reviews Materials, 2016, 1(2): 15010.
[165] GUO S, LI X, ZHU J, et al. Au NPs at MoS_2 sub-micrometer sphere-ZnO nanorod hybrid structures for efficient photocatalytic hydrogen evolution with excellent stability[J]. Small, 2016, 12(41): 5692-5701.
[166] LI X, GUO S, KAN C, et al. Au Multimer at MoS_2 hybrid structures for efficient photocatalytical hydrogen production via strongly plasmonic coupling effect[J]. Nano Energy, 2016, 30: 549-558.
[167] OBREGÓN S, COLÓN G. Improved H_2 production of Pt-TiO_2/g-C_3N_4-MnO_x composites by an efficient handling of photogenerated charge pairs[J]. Applied Catalysis B: Environmental, 2014, 144: 775-782.
[168] CHENG X, YU X, LI B, et al. Enhanced visible light activity and mechanism of TiO_2 codoped with molybdenum and nitrogen[J]. Materials Science and Engineering: B, 2013, 178(7): 425-430.
[169] DONG G, ZHAO K, ZHANG L. Carbon self-doping induced high electronic conductivity and photoreactivity of g C_3N_4[J]. Chemical Communications, 2012, 48(49): 6178-6180.
[170] WANG X, WANG Q, LI F, et al. Novel BiOCl-C_3N_4 heterojunction photocatalysts: in situ preparation via an ionic-liquid-assisted solvent-thermal route and their visible-light photocatalytic activities[J]. Chemical Engineering Journal, 2013, 234: 361-371.
[171] PAWAR R C, KHARE V, LEE C S. Hybrid photocatalysts using graphitic carbon nitride/cadmium sulfide/reduced graphene oxide (g-C_3N_4/CdS/RGO) for superior photodegradation of organic pollutants under UV and visible light [J]. Dalton

Transactions, 2014, 43(33): 12514-12527.
[172] LI F, ZHAO Y, WANG Q, et al. Enhanced visible-light photocatalytic activity of active Al_2O_3/g-C_3N_4 heterojunctions synthesized via surface hydroxyl modification [J]. Journal of Hazardous Materials, 2015, 283: 371-381.
[173] GU L, WANG J, ZOU Z, et al. Graphitic-C_3N_4-hybridized TiO_2 nanosheets with reactive {0 0 1} facets to enhance the UV-and visible-light photocatalytic activity[J]. Journal of Hazardous Materials, 2014, 268: 216-223.
[174] SRIDHARAN K, JANG E, PARK T J. Novel visible light active graphitic C_3N_4-TiO_2 composite photocatalyst: synergistic synthesis, growth and photocatalytic treatment of hazardous pollutants [J]. Applied Catalysis B: Environmental, 2013, 142: 718-728.
[175] ZHOU S, LIU Y, LI J, et al. Facile in situ synthesis of graphitic carbon nitride (g-C_3N_4)-N-TiO_2 heterojunction as an efficient photocatalyst for the selective photoreduction of CO_2 to CO[J]. Applied Catalysis B: Environmental, 2014, 158: 20-29.
[176] ZOU X X, LI G D, WANG Y N, et al. Direct conversion of urea into graphitic carbon nitride over mesoporous TiO_2 spheres under mild condition[J]. Chemical Communications, 2011, 47(3): 1066-1068.
[177] TAUC J. Absorption edge and internal electric fields in amorphous semiconductors [J]. Materials Research Bulletin, 1970, 5(8): 721-729.
[178] HOU Y D, WANG X C, WU L, et al. N-doped SiO_2/TiO_2 mesoporous nanoparticles with enhanced photocatalytic activity under visible-light irradiation[J]. Chemosphere, 2008, 72(3): 414-421.
[179] YE L, LIU J, JIANG Z, et al. Facets coupling of BiOBr-g-C_3N_4 composite photocatalyst for enhanced visible-light-driven photocatalytic activity[J]. Applied Catalysis B: Environmental, 2013, 142: 1-7.
[180] WANG X, TIAN X, LI F, et al. The synergy between Ti species and g-C_3N_4 by doping and hybridization for the enhancement of photocatalytic H_2 evolution[J]. Dalton Transactions, 2015, 44(40): 17859-17866.
[181] ZHANG M, BAI X, LIU D, et al. Enhanced catalytic activity of potassium-doped graphitic carbon nitride induced by lower valence position[J]. Applied Catalysis B: Environmental, 2015,164: 77-81.
[182] CHEN L, ZHANG W, FENG C, et al. Replacement/etching route to ZnSe nanotube arrays and their enhanced photocatalytic activities [J]. Industrial & Engineering Chemistry Research, 2012, 51(11): 4208-4214.
[183] SHALOM M, INAL S, FETTKENHAUER C, et al. Improving carbon nitride photocatalysis by supramolecular preorganization of monomers[J]. Journal of the American Chemical Society, 2013, 135(19): 7118-7121.
[184] ZHENG Y, JIAO Y, ZHU Y, et al. Hydrogen evolution by a metal-free electrocatalyst [J]. Nature Communications, 2014, 5(1): 1-8.
[185] GAO G, JIAO Y, MA F, et al. Metal-free graphitic carbon nitride as mechano-

catalyst for hydrogen evolution reaction[J]. Journal of Catalysis, 2015, 332: 149-155.

[186] XU J, ZHANG L, SHI R, et al. Chemical exfoliation of graphitic carbon nitride for efficient heterogeneous photocatalysis[J]. Journal of Materials Chemistry A, 2013, 1(46): 14766-14772.

[187] ZHANG G, ZHANG M, YE X, et al. Iodine modified carbon nitride semiconductors as visible light photocatalysts for hydrogen evolution[J]. Advanced Materials, 2014, 26(5): 805-809.

[188] ZHANG J, ZHANG M, LIN L, et al. Sol processing of conjugated carbon nitride powders for thin-film fabrication[J]. Angewandte Chemie, 2015, 127(21): 6395-6399.

[189] CHEN F, YANG H, LUO W, et al. Selective adsorption of thiocyanate anions on Ag-modified g-C_3N_4 for enhanced photocatalytic hydrogen evolution[J]. Chinese Journal of Catalysis, 2017, 38(12): 1990-1998.

[190] YANG J, WANG D, HAN H, et al. Roles of cocatalysts in photocatalysis and photoelectrocatalysis[J]. Accounts of Chemical Research, 2013, 46(8): 1900-1909.

[191] FENG J, LV F, ZHANG W, et al. Iridium-based multimetallic porous hollow nanocrystals for efficient overall-water-splitting catalysis[J]. Advanced Materials, 2017, 29(47): 1703798.

[192] NI Z, DONG F, HUANG H, et al. New insights into how Pd nanoparticles influence the photocatalytic oxidation and reduction ability of g-C_3N_4 nanosheets[J]. Catalysis Science & Technology, 2016, 6(16): 6448-6458.

[193] DING Z, CHEN X, ANTONIETTI M, et al. Synthesis of transition metal-modified carbon nitride polymers for selective hydrocarbon oxidation[J]. Chem Sus Chem, 2011, 4(2): 274-281.

[194] VILÉ G, ALBANI D, NACHTEGAAL M, et al. A stable single-site palladium catalyst for hydrogenations[J]. Angewandte Chemie International Edition, 2015, 54(38): 11265-11269.

[195] GAO G, JIAO Y, WACLAWIK E R, et al. Single atom (Pd/Pt) supported on graphitic carbon nitride as an efficient photocatalyst for visible-light reduction of carbon dioxide[J]. Journal of the American Chemical Society, 2016, 138(19): 6292-6297.

[196] LI Y, WANG Z, XIA T, et al. Implementing metal-to-ligand charge transfer in organic semiconductor for improved visible-near-infrared photocatalysis[J]. Advanced Materials, 2016, 28(32): 6959-6965.

[197] DONG F, ZHAO Z, XIONG T, et al. In situ construction of g-C_3N_4/g-C_3N_4 metal-free heterojunction for enhanced visible-light photocatalysis[J]. ACS Applied Materials & Interfaces, 2013, 5(21): 11392-11401.

[198] VANYSEK P. CRC handbook of chemistry and physics[M]. Boca Raton: CRC Press, 2008.

[199] WANG N, WANG J, HU J, et al. Design of palladium-doped g-C_3N_4 for enhanced photocatalytic activity toward hydrogen evolution reaction[J]. ACS Applied Energy

Materials, 2018, 1(6): 2866-2873.

[200] HINNEMANN B, MOSES P G, BONDE J, et al. Biomimetic hydrogen evolution: MoS_2 nanoparticles as catalyst for hydrogen evolution[J]. Journal of the American Chemical Society, 2005, 127(15): 5308-5309.

[201] CHOU S S, SAI N, LU P, et al. Understanding catalysis in a multiphasic two-dimensional transition metal dichalcogenide [J]. Nature Communications, 2015, 6(1): 1-8.

[202] Deng J, Li H, Xiao J, et al. Triggering the electrocatalytic hydrogen evolution activity of the inert two-dimensional MoS_2 surface via single-atom metal doping[J]. Energy & Environmental Science, 2015, 8(5): 1594-1601.

[203] Gong Y, Li M, Li H, et al. Graphitic carbon nitride polymers: promising catalysts or catalyst supports for heterogeneous oxidation and hydrogenation [J]. Green Chemistry, 2015, 17(2): 715-736.

[204] ZHAO Z, SUN Y, DONG F. Graphitic carbon nitride based nanocomposites: a review[J]. Nanoscale, 2015, 7(1): 15-37.

[205] YAN S C, LI Z S, ZOU Z G. Photodegradation of rhodamine B and methyl orange over boron-doped g-C_3N_4 under visible light irradiation[J]. Langmuir, 2010, 26(6): 3894-3901.

[206] LIU G, NIU P, SUN C, et al. Unique electronic structure induced high photoreactivity of sulfur-doped graphitic C_3N_4 [J]. Journal of the American Chemical Society, 2010, 132(33): 11642-11648.

[207] GU H, GU Y, LI Z, et al. Low-temperature route to nanoscale P_3N_5 hollow spheres [J]. Journal of Materials Research, 2003, 18(10): 2359-2363.

[208] MITORAJ D, KISCH H. On the mechanism of urea-induced titania modification[J]. Chemistry-A European Journal, 2010, 16(1): 261-269.

[209] CHOI C H, PARK S H, WOO S I. Phosphorus-nitrogen dual doped carbon as an effective catalyst for oxygen reduction reaction in acidic media: effects of the amount of P-doping on the physical and electrochemical properties of carbon[J]. Journal of Materials Chemistry, 2012, 22(24): 12107-12115.

[210] ZHU Y P, REN T Z, YUAN Z Y. Mesoporous phosphorus-doped g-C_3N_4 nanostructured flowers with superior photocatalytic hydrogen evolution performance[J]. ACS Applied Materials & Interfaces, 2015, 7(30): 16850-16856.

[211] LIN Z, WANG X. Nanostructure engineering and doping of conjugated carbon nitride semiconductors for hydrogen photosynthesis[J]. Angewandte Chemie, 2013, 125(6): 1779-1782.

[212] WANG X, MAEDA K, CHEN X, et al. Polymer semiconductors for artificial photosynthesis: hydrogen evolution by mesoporous graphitic carbon nitride with visible light[J]. Journal of the American Chemical Society, 2009, 131(5): 1680-1681.

[213] ZHU Y P, REN T Z, LIU Y P, et al. In situ simultaneous reduction-doping route to synthesize hematite/N-doped graphene nanohybrids with excellent photoactivity[J].

RSC Advances, 2014, 4(60): 31754-31758.

[214] RAN J, ZHANG J, YU J, et al. Enhanced visible-light photocatalytic H_2 production by $Zn_xCd_{1-x}S$ modified with earth-abundant nickel-based cocatalysts[J]. ChemSusChem, 2014, 7(12): 3426-3434.

[215] XIE G, ZHANG K, GUO B, et al. Graphene-based materials for hydrogen generation from light-driven water splitting[J]. Advanced Materials, 2013, 25(28): 3820-3839.

[216] XIANG Q, YU J, JARONIEC M. Synergetic effect of MoS_2 and graphene as cocatalysts for enhanced photocatalytic H_2 production activity of TiO_2 nanoparticles [J]. Journal of the American Chemical Society, 2012, 134(15): 6575-6578.

[217] HOU C, LI J, HUO D, et al. A portable embedded toxic gas detection device based on a cross-responsive sensor array[J]. Sensors and Actuators B: Chemical, 2012, 161(1): 244-250.

[218] ARIYAGEADSAKUL P, VCHIRAWONGKWIN V, KRITAYAKORNUPONG C. Determination of toxic carbonyl species including acetone, formaldehyde, and phosgene by polyaniline emeraldine gas sensor using DFT calculation[J]. Sensors and Actuators B: Chemical, 2016, 232: 165-174.

[219] KANETI Y V, YUE J, MORICEAU J, et al. Experimental and theoretical studies on noble metal decorated tin oxide flower-like nanorods with high ethanol sensing performance[J]. Sensors and Actuators B: Chemical, 2015, 219: 83-93.

[220] BHUVANESHWARI S, GOPALAKRISHNAN N. Hydrothermally synthesizedcopper oxide (CuO) superstructures for ammonia sensing[J]. Journal of Colloid and Interface Science, 2016, 480: 76-84.

[221] MIRZAEI A, JANGHORBAN K, HASHEMI B, et al. Synthesis, characterization and gas sensing properties of Ag at α-Fe_2O_3 core-shell nanocomposites [J]. Nanomaterials, 2015, 5(2): 737-749.

[222] DENG S, LIU X, CHEN N, et al. A highly sensitive VOC gas sensor using p-type mesoporous Co_3O_4 nanosheets prepared by a facile chemical coprecipitation method [J]. Sensors and Actuators B: Chemical, 2016, 233: 615-623.

[223] YANG C, DENG W, LIU H, et al. Turn-on fluorescence sensor for glutathione in aqueous solutions using carbon dots-MnO_2 nanocomposites[J]. Sensors and Actuators B: Chemical, 2015, 216: 286-292.

[224] AN X, JIMMY C Y, WANG Y, et al. WO_3 nanorods/graphene nanocomposites for high-efficiency visible-light-driven photocatalysis and NO_2 gas sensing[J]. Journal of Materials Chemistry, 2012, 22(17): 8525-8531.

[225] CHOI K S, PARK S, CHANG S P. Enhanced ethanol sensing properties based on SnO_2 nanowires coated with Fe_2O_3 nanoparticles[J]. Sensors and Actuators B: Chemical, 2017, 238: 871-879.

[226] LI Y, CHEN N, DENG D, et al. Formaldehyde detection: SnO_2 microspheres for formaldehyde gas sensor with high sensitivity, fast response/recovery and good

[227] LI P, CUI Y, BEHAN G, et al. Room temperature synthesis and one-dimensional self-assembly of interlaced Ni nanodiscs under magnetic field[J]. Journal of Physics D: Applied Physics, 2010, 43(27): 275002.

[228] ZHANG D, LIU J, CHANG H, et al. Characterization of a hybrid composite of SnO_2 nanocrystal-decorated reduced graphene oxide for ppm-level ethanol gas sensing application[J]. 2015.

[229] LIU J, WANG T, WANG B, et al. Highly sensitive and low detection limit of ethanol gas sensor based on hollow ZnO/SnO_2 spheres composite material[J]. Sensors and Actuators B: Chemical, 2017, 245: 551-559.

[230] CHEN Y, ZHANG W, WU Q. A highly sensitive room-temperature sensing material for NH3: SnO_2-nanorods coupled by rGO[J]. Sensors and Actuators B: Chemical, 2017, 242: 1216-1226.

[231] ZHANG Z, ZHU L, WEN Z, et al. Controllable synthesis of Co_3O_4 crossed nanosheet arrays toward an acetone gas sensor [J]. Sensors and Actuators B: Chemical, 2017, 238: 1052-1059.

[232] CAO J, QIN C, WANG Y, et al. Solid-state method synthesis of SnO_2-decorated g-C_3N_4 nanocomposites with enhanced gas-sensing property to ethanol[J]. Materials, 2017, 10(6): 604.

[233] ZHOU Y, ZHANG L, HUANG W, et al. N-doped graphitic carbon-incorporated g-C_3N_4 for remarkably enhanced photocatalytic H_2 evolution under visible light[J]. Carbon, 2016, 99: 111-117.

[234] ZANG Y, LI L, LI X, et al. Synergistic collaboration of g-C_3N_4/SnO_2 composites for enhanced visible-light photocatalytic activity[J]. Chemical Engineering Journal, 2014, 246: 277-286.

[235] CAO J, QIN C, WANG Y, et al. Calcination method synthesis of SnO_2/g-C_3N_4 composites for a high-performance ethanol gas sensing application[J]. Nanomaterials, 2017, 7(5): 98.

[236] WANG Y, CAO J, QIN C, et al. Synthesis and enhanced ethanol gas sensing properties of the g-C_3N_4 nanosheets-decorated tin oxide flower-like nanorods composite [J]. Nanomaterials, 2017, 7(10): 285.

[237] SONG H J, JIA X H, QI H, et al. Flexible morphology-controlled synthesis of monodisperse α-Fe_2O_3 hierarchical hollow microspheres and their gas-sensing properties[J]. Journal of Materials Chemistry, 2012, 22(8): 3508-3516.

[238] ZHOU L, SHEN F, TIAN X, et al. Stable Cu_2O nanocrystals grown on functionalized graphene sheets and room temperature H_2S gas sensing with ultrahigh sensitivity[J]. Nanoscale, 2013, 5(4): 1564-1569.

[239] CAO S W, LIU X F, YUAN Y P, et al. Solar-to-fuels conversion over In_2O_3/g-C_3N_4 hybrid photocatalysts[J]. Applied Catalysis B: Environmental, 2014, 147: 940-946.

[240] ZENG B, ZHANG L, WAN X, et al. Fabrication of α-Fe_2O_3/g-C_3N_4 composites

for cataluminescence sensing of H_2S[J]. Sensors and Actuators B: Chemical, 2015, 211: 370-376.

[241] HU Y, LI L, ZHANG L, et al. Dielectric barrier discharge plasma-assisted fabrication of g-C_3N_4-Mn_3O_4 composite for high-performance cataluminescence H_2S gas sensor[J]. Sensors and Actuators B: Chemical, 2017, 239: 1177-1184.

[242] ZHANG X L, ZHENG C, GUO S S, et al. Turn-on fluorescence sensor for intracellular imaging of glutathione using g-C_3N_4 nanosheet-MnO_2 sandwich nanocomposite[J]. Analytical Chemistry, 2014, 86(7): 3426-3434.

[243] XU H, ZHONG H, TANG Q, et al. A novel collector 2-ethyl-2-hexenoic hydroxamic acid: flotation performance and adsorption mechanism to ilmenite [J]. Applied Surface Science, 2015, 353: 882-889.

[244] DI T, ZHU B, CHENG B, et al. A direct Z-scheme g-C_3N_4/SnS_2 photocatalyst with superior visible-light CO_2 reduction performance[J]. Journal of Catalysis, 2017, 352: 532-541.

[245] HONG J, XIA X, WANG Y, et al. Mesoporous carbon nitride with in situ sulfur doping for enhanced photocatalytic hydrogen evolution from water under visible light [J]. Journal of Materials Chemistry, 2012, 22(30): 15006-15012.

[246] YU J, WANG K, XIAO W, et al. Photocatalytic reduction of CO_2 into hydrocarbon solar fuels over g-C_3N_4-Pt nanocomposite photocatalysts[J]. Physical Chemistry Chemical Physics, 2014, 16(23): 11492-11501.

[247] ZHANG Z, SHAO C, LI X, et al. Hierarchical assembly of ultrathin hexagonal SnS_2 nanosheets onto electrospun TiO_2 nanofibers: enhanced photocatalytic activity based on photoinduced interfacial charge transfer[J]. Nanoscale, 2013, 5(2): 606-618.

[248] YU J, WANG K, XIAO W, et al. Photocatalytic reduction of CO_2 into hydrocarbon solar fuels over g C_3N_4-Pt nanocomposite photocatalysts[J]. Physical Chemistry Chemical Physics, 2014, 16(23): 11492-11501.

[249] KURIKI R, MATSUNAGA H, NAKASHIMA T, et al. Nature-inspired, highly durable CO_2 reduction system consisting of a binuclear ruthenium (II) complex and an organic semiconductor using visible light[J]. Journal of the American Chemical Society, 2016, 138(15): 5159-5170.

[250] WANG S, LIN J, WANG X. Semiconductor-redox catalysis promoted by metal-organic frameworks for CO_2 reduction[J]. Physical Chemistry Chemical Physics, 2014, 16(28): 14656-14660.

[251] FAN X, GAO J, WANG Y, et al. Effect of crystal growth on mesoporous $Pb_3Nb_4O_{13}$ formation, and their photocatalytic activity under visible-light irradiation [J]. Journal of Materials Chemistry, 2010, 20(14): 2865-2869.

[252] KONG L, MU X, FAN X, et al. Site-selected N vacancy of g-C_3N_4 for photocatalysis and physical mechanism[J]. Applied Materials Today, 2018, 13: 329-338.

[253] KONG L, WANG J, MU X, et al. Porous size dependent g-C_3N_4 for efficient photocatalysts: regulation synthesizes and physical mechanism[J]. Materials Today Energy, 2019, 13: 11-21.

[254] WANG H, HE W, WANG H, et al. In situ FT-IR investigation on the reaction mechanism of visible light photocatalytic NO oxidation with defective g-C_3N_4 [J]. Science Bulletin, 2018, 63(2): 117-125.

[255] LI J, ZHANG Z, CUI W, et al. The spatially oriented charge flow and photocatalysis mechanism on internal van der Waals heterostructures enhanced g-C_3N_4[J]. ACS Catalysis, 2018, 8(9): 8376-8385.

[256] CHEN P, WANG H, LIU H, et al. Directional electron delivery and enhanced reactants activation enable efficient photocatalytic air purification on amorphous carbon nitride co-functionalized with O/La[J]. Applied Catalysis B: Environmental, 2019, 242: 19-30.

[257] LI J, SUN Y, JIANG G, et al. Tailoring the rate-determining step in photocatalysis via localized excess electrons for efficient and safe air cleaning[J]. Applied Catalysis B: Environmental, 2018, 239: 187-195.

[258] CUI W, LI J, SUN Y, et al. Enhancing ROS generation and suppressing toxic intermediate production in photocatalytic NO oxidation on O/Ba co-functionalized amorphous carbon nitride[J]. Applied Catalysis B: Environmental, 2018, 237: 938-946.

[259] LI Y, SUN Y, HO W, et al. Highly enhanced visible-light photocatalytic NO_x purification and conversion pathway on self-structurally modified g-C_3N_4 nanosheets [J]. Science Bulletin, 2018, 63(10): 609-620.

[260] XIONG T, WANG H, ZHOU Y, et al. KCl-mediated dual electronic channels in layered g-C_3N_4 for enhanced visible light photocatalytic NO removal[J]. Nanoscale, 2018, 10(17): 8066-8074.

[261] LI J, XING Q, ZHOU Y, et al. The activation of reactants and intermediates promotes the selective photocatalytic NO conversion on electron-localized Sr-intercalated g-C_3N_4[J]. Applied Catalysis B: Environmental, 2018, 232: 69-76.

[262] CHEN P, DONG F, RAN M, et al. Synergistic photo-thermal catalytic NO purification of MnO_x/g-C_3N_4: enhanced performance and reaction mechanism[J]. Chinese Journal of Catalysis, 2018, 39(4): 619-629.

[263] LI J, CUI W, SUN Y, et al. Directional electron delivery via a vertical channel between g-C_3N_4 layers promotes photocatalytic efficiency[J]. Journal of Materials Chemistry A, 2017, 5(19): 9358-9364.

[264] CUI W, LI J, DONG F, et al. Highly efficient performance and conversion pathway of photocatalytic NO oxidation on SrO-clusters at amorphous carbon nitride[J]. Environmental Science & Technology, 2017, 51(18): 10682-10690.

[265] DONG F, ZHAO Z, SUN Y, et al. An advanced semimetal-organic Bi spheres-g-C_3N_4 nanohybrid with SPR-enhanced visible-light photocatalytic performance for NO

purification[J]. Environmental Science & Technology, 2015, 49(20): 12432-12440.

[266] LIU Q, AI L, JIANG J. MXene-derived TiO_2 at $C/g-C_3N_4$ heterojunctions for highly efficient nitrogen photofixation[J]. Journal of Materials Chemistry A, 2018, 6 (9): 4102-4110.

[267] DIARMAND-KHALILABAD H, HABIBI-YANGJEH A, SEIFZADEH D, et al. g-C_3N_4 nanosheets decorated with carbon dots and CdS nanoparticles: novel nanocomposites with excellent nitrogen photofixation ability under simulated solar irradiation[J]. Ceramics International, 2019, 45(2): 2542-2555.

[268] MOUSAVI M, HABIBI-YANGJEH A, POURAN S R. Review on magnetically separable graphitic carbon nitride-based nanocomposites as promising visible-light-driven photocatalysts[J]. Journal of Materials Science: Materials in Electronics, 2018, 29(3): 1719-1747.

[269] PIRHASHEMI M, HABIBI-YANGJEH A, POURAN S R. Review on the criteria anticipated for the fabrication of highly efficient ZnO-based visible-light-driven photocatalysts[J]. Journal of Industrial and Engineering Chemistry, 2018, 62: 1-25.

[270] SHEKOFTEH-GOHARI M, HABIBI-YANGJEH A, ABITORABI M, et al. Magnetically separable nanocomposites based on ZnO and their applications in photocatalytic processes: a review[J]. Critical Reviews in Environmental Science and Technology, 2018, 48(10-12): 806-857.

[271] HUANG D, LI Z, ZENG G, et al. Megamerger in photocatalytic field: 2D g-C_3N_4 nanosheets serve as support of 0D nanomaterials for improving photocatalytic performance[J]. Applied Catalysis B: Environmental, 2019, 240: 153-173.

[272] LIU J, WANG H, ANTONIETTI M. Graphitic carbon nitride "reloaded": emerging applications beyond (photo) catalysis[J]. Chemical Society Reviews, 2016, 45(8): 2308-2326.

[273] HUANG D, YAN X, YAN M, et al. Graphitic carbon nitride-based heterojunction photoactive nanocomposites: applications and mechanism insight[J]. ACS Applied Materials & Interfaces, 2018, 10(25): 21035-21055.

[274] ASADZADEH-KHANEGHAH S, HABIBI-YANGJEH A, YUBUTA K. Novel g-C_3N_4 nanosheets/CDs/BiOCl photocatalysts with exceptional activity under visible light[J]. Journal of the American Ceramic Society, 2019, 102(3): 1435-1453.

[275] MOUSAVI M, HABIBI-YANGJEH A. Decoration of Fe_3O_4 and $CoWO_4$ nanoparticles over graphitic carbon nitride: novel visible-light-responsive photocatalysts with exceptional photocatalytic performances[J]. Materials Research Bulletin, 2018, 105: 159-171.

[276] MOUSAVI M, HABIBI-YANGJEH A, SEIFZADEH D. Novel ternary g-C_3N_4/Fe_3O_4/$MnWO_4$ nanocomposites: synthesis, characterization, and visible-light photocatalytic performance for environmental purposes[J]. Journal of Materials Science & Technology, 2018, 34(9): 1638-1651.

[277] AKHUNDI A, HABIBI-YANGJEH A. Novel magnetically separable g-C_3N_4/AgBr/Fe_3O_4 nanocomposites as visible-light-driven photocatalysts with highly enhanced

activities[J]. Ceramics International, 2015, 41(4): 5634-5643.
[278] HABIBI-YANGJEH A, MOUSAVI M. Deposition of $CuWO_4$ nanoparticles over g-C_3N_4/Fe_3O_4 nanocomposite: novel magnetic photocatalysts with drastically enhanced performance under visible-light[J]. Advanced Powder Technology, 2018, 29(6): 1379-1392.
[279] ASADZADEH-KHANEGHAH S, HABIBI-YANGJEH A, SEIFZADEH D. Graphitic carbon nitride nanosheets coupled with carbon dots and BiOI nanoparticles: boosting visible-light-driven photocatalytic activity[J]. Journal of the Taiwan Institute of Chemical Engineers, 2018, 87: 98-111.
[280] HABIBI-YANGJEH A, MOUSAVI M, NAKATA K. Boosting visible-light photocatalytic performance of g-C_3N_4/Fe_3O_4 anchored with $CoMoO_4$ nanoparticles: novel magnetically recoverable photocatalysts[J]. Journal of Photochemistry and Photobiology A: Chemistry, 2019, 368: 120-136.
[281] ASADZADEH-KHANEGHAH S, HABIBI-YANGJEH A, NAKATA K. Decoration of carbon dots over hydrogen peroxide treated graphitic carbon nitride: exceptional photocatalytic performance in removal of different contaminants under visible light [J]. Journal of Photochemistry and Photobiology A: Chemistry, 2019, 374: 161-172.
[282] AKHUNDI A, HABIBI-YANGJEH A. Graphitic carbon nitride nanosheets decorated with $CuCr_2O_4$ nanoparticles: novel photocatalysts with high performances in visible light degradation of water pollutants[J]. Journal of Colloid and Interface Science, 2017, 504: 697-710.
[283] AKHUNDI A, HABIBI-YANGJEH A. High performance magnetically recoverable g-$C_3N_4/Fe_3O_4/Ag/Ag_2SO_3$ plasmonic photocatalyst for enhanced photocatalytic degradation of water pollutants[J]. Advanced Powder Technology, 2017, 28(2): 565-574.
[284] MOUSAVI M, HABIBI-YANGJEH A. Novel magnetically separable g-$C_3N_4/Fe_3O_4/Ag_3PO_4/Co_3O_4$ nanocomposites: visible-light-driven photocatalysts with highly enhanced activity[J]. Advanced Powder Technology, 2017, 28(6): 1540-1553.
[285] AKHUNDI A, HABIBI-YANGJEH A. Codeposition of AgI and Ag2CrO4 on g-C_3N_4/Fe_3O_4 nanocomposite: novel magnetically separable visible-light-driven photocatalysts with enhanced activity[J]. Advanced Powder Technology, 2016, 27(6): 2496-2506.
[286] MOUSAVI M, HABIBI-YANGJEH A, ABITORABI M. Fabrication of novel magnetically separable nanocomposites using graphitic carbon nitride, silver phosphate and silver chloride and their applications in photocatalytic removal of different pollutants using visible-light irradiation[J]. Journal of Colloid and Interface Science, 2016, 480: 218-231.
[287] HABIBI-YANGJEH A, AKHUNDI A. Novel ternary g-$C_3N_4/Fe_3O_4/Ag_2CrO_4$ nanocomposites: magnetically separable and visible-light-driven photocatalysts for degradation of water pollutants[J]. Journal of Molecular Catalysis A: Chemical, 2016, 415: 122-130.

[288] MOUSAVI M, HABIBI-YANGJEH A. Magnetically separable ternary g-C_3N_4/Fe_3O_4/BiOI nanocomposites: novel visible-light-driven photocatalysts based on graphitic carbon nitride[J]. Journal of Colloid and Interface Science, 2016, 465: 83-92.

[289] ASADZADEH-KHANEGHAH S, HABIBI-YANGJEH A, ABEDI M. Decoration of carbon dots and AgCl over g-C_3N_4 nanosheets: novel photocatalysts with substantially improved activity under visible light[J]. Separation and Purification Technology, 2018, 199: 64-77.

[290] AKHUNDI A, HABIBI-YANGJEH A. Novel g-C_3N_4/Ag_2SO_4 nanocomposites: fast microwave-assisted preparation and enhanced photocatalytic performance towards degradation of organic pollutants under visible light[J]. Journal of Colloid and Interface Science, 2016, 482: 165-174.

[291] ZHOU C, HUANG D, XU P, et al. Efficient visible light driven degradation of sulfamethazine and tetracycline by salicylic acid modified polymeric carbon nitride via charge transfer[J]. Chemical Engineering Journal, 2019, 370: 1077-1086.

[292] ZHOU C, LAI C, XU P, et al. Rational design of carbon-doped carbon nitride/$Bi_{12}O_{17}C_{12}$ composites: a promising candidate photocatalyst for boosting visible-light-driven photocatalytic degradation of tetracycline[J]. ACS Sustainable Chemistry & Engineering, 2018, 6(5): 6941-6949.

[293] WANG W, XU P, CHEN M, et al. Alkali metal-assisted synthesis of graphite carbon nitride with tunable band-gap for enhanced visible-light-driven photocatalytic performance[J]. ACS Sustainable Chemistry & Engineering, 2018, 6(11): 15503-15516.

[294] MOUSAVI M, HABIBI-YANGJEH A. Integration of $NiWO_4$ and Fe_3O_4 with graphitic carbon nitride to fabricate novel magnetically recoverable visible-light-driven photocatalysts[J]. Journal of Materials Science, 2018, 53(12): 9046-9063.

[295] ZHOU C, LAI C, HUANG D, et al. Highly porous carbon nitride by supramolecular preassembly of monomers for photocatalytic removal of sulfamethazine under visible light driven[J]. Applied Catalysis B: Environmental, 2018, 220: 202-210.

[296] HU X, WANG W, XIE G, et al. Ternary assembly of g-C_3N_4/graphene oxide sheets/$BiFeO_3$ heterojunction with enhanced photoreduction of Cr(Ⅵ) under visible-light irradiation[J]. Chemosphere, 2019, 216: 733-741.

[297] XU H, ZHONG H, TANG Q, et al. A novel collector 2-ethyl-2-hexenoic hydroxamic acid: flotation performance and adsorption mechanism to ilmenite[J]. Applied Surface Science, 2015, 353: 882-889.

[298] AMIRI M, SALEHNIYA H, HABIBI-YANGJEH A. Graphitic carbon nitride/chitosan composite for adsorption and electrochemical determination of mercury in real samples[J]. Industrial & Engineering Chemistry Research, 2016, 55(29): 8114-8122.

[299] ONG W J, TAN L L, NG Y H, et al. Graphitic carbon nitride (g-C_3N_4)-based photocatalysts for artificial photosynthesis and environmental remediation: are we a step closer to achieving sustainability?[J]. Chemical Reviews, 2016, 116(12): 7159-7329.

Acknowledgements

This book is supported by the Natural Science Foundation of Guangdong Province in China (Grant No: 2017A030313022, 2019A1515011132), the Talent Scientific Research Fund of LSHU (No. 2018XJJ-007), Key Scientific Research Platforms and Projects in Guangdong Universities (Grant No: 2018KZDXM046 and 2019KTSCX090).